- If you think radio listening is limited to AM and FM, you're missing a whole world that's waiting for you out there.
- If you tune in the same humdrum local stations every day, you need this

PASSPORT TO WORLD BAND RADIO

Courtesy China Pictorial

to open these exciting new vistas

Stations from more than 160 different countries are on the air right now . . . beaming programs on music, nature, tourism, culture, and politics from every part of our planet.

Passport to World Band Radio gives you everything you need to explore and enjoy this exceptional realm. The first part, **How to Tune in the World,** looks at just a few of the countries and subjects available on world band radio. Part two, **1988 Buyer's Guide to World Band Radio,** supplies authoritative "hands-on" and laboratory tests of the latest models and comparative ratings of just about every major portable or tabletop receiver. Finally, **Worldscan** runs through all the frequencies of the world band spectrum and shows who's broadcasting what, from where, at what times, and in which languages.

With **Passport** as your guide, you have everything you need to tune in the world.

SATELLIT 650 INTERNATIONAL – THE EAR TO THE WORLD

Exclusive Licensee & Distributor of Grundig AG in U.S.A.
Lextronix, INC.
P.O. BOX 2307
Menlo Park, CA 94026
Tel. (415) 361-16 11

Top Class World Receivers with German Technology

Microcomputer controlled PLL-frequency-synthesizer tuning for AM and FM. Direct frequency key-in for all wave bands with automatic follow-up preselector and three-stage fading control for AM. Short Wave covering 1.6–30.0 MHz. Double superhet for exceptional selectivity and image rejection, switchable SSB/BFO section (adaptable to USB/LSB). AC 110–240 V, 50/60 Hz, DC-operation 9 V with batteries or external power supply.

GRUNDIG

PASSPORT TO

HOW TO TUNE IN THE WORLD

Listening In: How Sweet It Is! 11
The Great One was a world band radio fanatic

Riding the Wave of the Future 17
It's as easy as using a pushbutton phone

Canada by Radio 19
One nation that has made good use of world band to reach its vast territory

Canada Airs Popular Program for World Band Listeners 22
SWL Digest is devoted to the understanding and enjoyment of world band radio

Parcourir le Canada par les bandes 23
L'une des nations qui ait le plus profité de la radiodiffusion à ondes courtes

News ... and "News" ... over World Band Radio 36
Sorting the facts from the propaganda

Vladimir Posner: Not Your Average Russian 41
Is he a peace missionary or a propagandist?

Ghana: Radio Gold Worth Digging For 45
The rhythms of "Ghana Is Free" mirror African pride throughout the world

Tuning in the Middle East Powder Keg 49
Eleven countries broadcast their versions of the news in English

China's Potpourri of World Band Stations 51
Reaching a billion people over 3.6 million square miles

Latin America, Land of Traditional Music 54
The eerie, haunting sounds of true Latin American folk music

How to Listen to Worldwide Weather Information 59
VOLMET and NOAA broadcast weather over land and sea

1988 BUYER'S GUIDE TO WORLD BAND RADIO

How to Buy a World Band Radio 68

Comparative Ratings of Portable World Band Radios 69

Why You Should—or Should Not—Buy a Tabletop Receiver 78

Comparative Ratings of Tabletop World Band Receivers 79

The Perfect Starter Radio 85
Sangean ATS-803 / EEB 2020 / Realistic DX-440 / Eskab RX33

High-Tech Radio for the Technically Timid 87
Sony ICF-7700 / Sony ICF-7600DA

Best of the Easy-to-Operate Portables 89
Panasonic RF-B60

Best Value in a Portable 90
Philips D2935 / Magnavox D2935

Travel Favorite in a New Package 91
Sony ICF-2003 / Sony 7600DS

Superior Audio in a Portable 92
Grundig Satellit 400

Handheld Scanner Doubles as World Band Radio 93
Sony ICF-PRO80 / Sony ICF-PRO70

Europe's Finest Portable 95
Grundig Satellit 650

The Prince of Portables 96
Sony ICF 2010 / Sony ICF 2001D

Superior Audio from a Top Performer 98
Kenwood R-5000 / Trio R-5000

WORLD BAND RADIO

1988 EDITION

One of the Best Is Now Even Better **100**
 Japan Radio NRD-525

A Favorite for Hearing the Tough Ones **102**
 ICOM IC-R71

An Affordable,
Well-Balanced Performer **104**
 Yaesu FRG-8800

Britain's New Bulldog **105**
 Lowe HF-125

A Tough American-Made Receiver **106**
 Ten-Tec RX-325

Antennas to Hear the World By **107**

WORLDSCAN

Worldwide Broadcasts 115
 Quick-Access Guide to World Band
 Schedules by Country

**Summary of Worldwide
Broadcasting Activity 127**

Worldscan—The Blue Pages 129
 Hours and Languages of Broadcast by
 Frequency

LEXICONS AND GUIDES

Lexicon of Terms **387**

Lexique **390**

Guide de Worldscan **390**

Léxico de Términos **393**

Como utilizar el Worldscan **393**

Glossar und Abkürzungen **396**

Worldscan: Gebrauchsanweisung **398**

Directory of Advertisers **399**

PASSPORT TO WORLD BAND RADIO

1988

Editor-in-Chief	Lawrence Magne
Editor	Tony Jones
Features Editor	Larry Miller
Contributing Editors	Tim Akester, Geoff Cosier, Noel Green
Consulting Editors	John Campbell, Don Jensen
Special Contributors	Rogildo Fontenelle Aragão, J. M. Brinker, Stephen J. Bohac, James A. Conrad, Gordon Darling, Antonio Ribeiro da Motta, DXFL/Isao Ugusa, Ruth M. Hesch, Robert J. Hill, Edward J. Insinger, David Klopfenstein, Konrad Kroszner, Ibrahim Mansour, Toshimichi Ohtake, Al Quaglieri, RNM/Tetsuya Hirahara, Don Swampoe, URSC/Shigenori Aoki, David L. Walcutt
Database Software	Richard Mayell
Laboratory	Sherwood Engineering Inc.
Marketing & Distribution	Mary W. Kroszner
Publicity	Consultech Communications, Inc.
Graphic Preparation	The Bookmakers, Incorporated Wilkes-Barre, Pennsylvania
Designer	John Beck
Composition Supervisor	Annette Thomas
Production Supervisor	Kevin McMullen
Layout Artist	Ken Sampson
Typesetter	Valerie Bucan
Airbrush Artist	Tom Price, Jr.

Library of Congress Cataloging-in-Publication Data

Passport to world band radio.

1. Radio, Short wave—Amateurs' manuals. 2. Radio stations, Short wave—Directories. I. Magne, Lawrence, 1941–
TK9956.P27   1987   384.54'5   87-22739
ISBN 0-914941-15-1

Copyright © 1987 International Broadcasting Services, Ltd.
P.O. Box 300, Penn's Park, PA 18943 USA

Manufactured in the United States of America

No part of this publication may be reproduced, stored in a retrieval system, or transmitted in any form or by any means without the prior written consent of the publisher.

Title page and page 9 show part of Canada's scenic riches: Prince Edward Island and Toronto

ICOM RECEIVERS
The World at Your Fingertips

Only ICOM brings the world into your living room...HF, VHF, UHF, and low band receptions. ICOM is the professional's choice to receive international broadcasts, aircraft, marine, business, emergency services, television, and government bands. Tune in with ICOM's IC-R7000 25-2000MHz* and IC-R71A 0.1-30MHz commercial quality scanning receivers for full spectrum coverage.

Incomparable Frequency Control. Both the IC-R71A and IC-R7000 feature **direct frequency access** via their front keypad, main tuning dial, optional infrared remote control and/or computer interface adapter. **Flexibility of this nature can only be accomplished with an ICOM!**

Full Coverage, Maximum Performance. The superb **IC-R71A** is your front row seat to worldwide SSB, CW, RTTY, AM, and FM (optional) communications and foreign broadcasts in the 100kHz to 30MHz range. It features passband, IF Notch, low noise mixer circuits, and 100dB dynamic range. The pacesetting **IC-R7000** receives today's hot areas of interest, including aircraft, marine, public services, amateur, and satellite transmissions in the 25MHz to 2000MHz* range. It includes **all mode operation** low noise circuits plus outstanding sensitivity and selectivity. The combined IC-R71A/IC-R7000 pair creates a full radio window to the world!

The IC-R71A is a shortwave listener's delight. Its **32 tunable memories** store frequency and mode information, and they are single-button reprogrammable **independent of VFO A or VFO B's operations!** This HF reception is further enhanced by a dual width and level adjustable noise blanker, panel selectable RF preamp, selectable AGC, **four scan modes,** and all-mode squelch.

The IC-R7000 is a high band monitor's masterpiece. Its **99 tunable memories** are complemented by **six scanning modes.** It even scans a band and loads memories 80 to 99 with active frequencies without operator assistance! Additional features include selectable scan speed and pause delays, wide/narrow FM reception, and high frequency stability. Many professional services use IC-R7000's as calibration references.

Options. IC-R7000: RC-12 remote control, EX-310 voice synthesizer, CK-70 DC adapter, MB-12 mobile bracket. IC-R71A: RC-11 remote control, EX-310 voice synthesizer, FM module, CK-70 DC adapter, MB-12 mobile bracket, FL-32A 500Hz, FL-63A 250Hz, and FL-44A filters.

See the IC-R7000 and IC-R71A at your local authorized ICOM dealer.

* Specifications of IC-R7000 guaranteed from 25-100MHz and 1260-1300MHz. No coverage from 1000-1025MHz

First in Communications

ICOM America, Inc., 2380-116th Ave. N.E., Bellevue, WA 98004 **Customer Service Hotline (206) 454-7619**
3150 Premier Drive, Suite 126, Irving, TX 75063 / 1777 Phoenix Parkway, Suite 201, Atlanta, GA 30349
ICOM CANADA, A Division of ICOM America, Inc., 3071 - #5 Road Unit 9, Richmond, B.C. V6X 2T4 Canada
All stated specifications are approximate and subject to change without notice or obligation. All ICOM radios significantly exceed FCC regulations limiting spurious emissions. RCVRS587.

McKay Dymek, division of:
stoner communications, inc.
9119 Milliken Avenue
Rancho Cucamonga, California 91730 USA
Telephone: (714) 987-4624   Telex: 676-468

DYMEK DA 100D

The World's Most Respected All-Wave Receiving Antenna

- Frequency Coverage: 50 KHz to 30 MHz, continuous
- Output Impedance: Switch selectable 50, 100, 500 Ohms
- Attenuation: Switch selectable 0, 10, 20 dB
- Power Required: Switch selectable 110 or 240VAC, 50–60 Hz, 4 Watts, or 12VDC

DA100DM Marine version also available with fiberglass whip antenna for use on or near salt water.

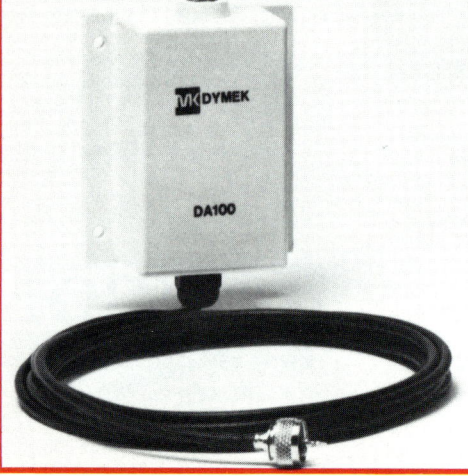

THE BEST GETS BETTER AT EEB

The Most Exotic VHF-UHF Receiver Ever Offered To General Public

EEB introduces the new ICOM R7000 VHF/UHF receiver, which also is available as the R7000 HP (high performance) with EEB-designed modifications that remarkably improve the sensitivity, audio frequency response, selectivity, and more.

Service Manual, Order: SMR7000

R7000 CALL FOR PRICE

In conformance with their record as a responsible manufacturer, ICOM has done everything possible to hold down production costs while leaving open the potential for post-production modification (e.g., R71A) that can substantially improve performance and give the discriminating operator a more cost effective unit with enhanced specifications.

You have read about EEB's R71A modification in *World Radio TV Handbook*. Now we have done it again with the ICOM R7000 VHF/UHF receiver.

When you are making an investment in this kind of equipment, if only makes sense to maximize your options by selecting the ICOM R7000 and then to go that last yard toward ultimate performance by choosing the R7000 HP from EEB.

ICOM R-7000 Specs

- 25-2000 MHz coverage
- Precise frequency entry via keyboard
- 99 programmable memories
- Scan-memory-mode-select memory-frequency
- 5 tuning speeds: 1, 1.0, 5, 10, 12.5, 25 KHz
- Narrow/wide filter selection
- Memory back-up • Noise blanker
- "S" meter or center meter for FM
- AM & FM wide, FM narrow, SSB, CW

EEB HP OPTIONS

1. Front end upgrade improves sensitivity.
2. Audio mod for better volume, less distortion.
3. Spike protection on AC line.
4. 24 hour bench test, Final alignment & overall checkout. Numbers 1 through 4 above for $200.00
5. Power Supply mod: A completely new power supply reduces the heat buildup & lowers the noise floor for longer component life and increased sensitivity. Price is $150.00.
6. Multiplex output mod for SCA and Subcarrier Analysis. Price: $50.00

EEB'S FAMOUS R71A HP

EEB is ICOM's #1 R71A dealer and there is good reason. We offer more modification to enhance your listening pleasure and take better care of you. This is our 17th year. Buy with confidence.

Service Manual ORDER: SMR71A

R71A CALL FOR PRICE

ICOM R71A Specs

- 100 KMz to 30 MHz • 32 memories
- SSB-CW-AM-RTTY (FM optional)
- Computer control • Scanning
- DC Kit • FM • Filters
- Voice Synthesizer
- Computer interface
- Remote control

- 24 hour bench test extended warranty
- 4 filter options
- Front end upgrade
- Audio output mode
- AC time constant
- Spike protection
- Final alignment, checkout

EEB HP Options

- Mech filt HP Add $200
- 8 pole xtal HP Add $250
- 8 pole super HP Add $300

TELEX: 62915985
FAX: 703-938-6911
• We ship world-wide
• Shipping charges not included
• Prices & specifications subject to change without notice

10 miles west of Washington, D.C.
Sorry—No COD's
10-5 Tues., Wed., Fri.
10-9 Thursday
10-4 Saturday
Closed Sunday and Monday

Electronic Equipment Bank
516 Mill Street, N.E.
Vienna, Virginia 22180
Order Toll Free 800-368-3270
Virginia 703-938-3350

Listening In: How Sweet It Is!

by Don Jensen

Jackie Gleason. For millions of people around the world, the name conjurs up images of his most popular character, Ralph Kramden. For years, audiences laughed to the antics of the frustrated, beleaguered bus driver, his long-suffering wife Trixie and sewer-cleaner friend Ed Norton. The program was, of course, "The Honeymooners," and through television syndication Jackie Gleason has remained a household name for generations, despite his passing last July at the age of 71.

Offstage, Jackie Gleason the actor lived a life that his character, Ralph, could only dream of. He existed in a constant state of overdrive, hellbent down the fast lane of life. The stories are

> He was looking for a Neanderthal, one of those old fashioned tube monsters

legend—Johnnie Walker by the fifth, lobster by the platterful, brash, swaggering days and nights of carousing, chorus girls, and champagne. And all true, as Gleason freely admitted.

There was, of course, a very private side to this public persona, one that included a near-lifelong love affair with radio. This was the quiet, introspective—he liked to think of it as scholarly—Gleason. He guarded his solitude, especially in his later years, but never lost his zest for living. As always, he had to be a part of anything that was exciting, and world band radio fitted that need perfectly. It was the window on the world that let the actor look out on life by listening in.

Gleason had installed an elaborate listening post on the second floor of his home some years back. Later, it was reportedly destroyed by fire, but the estangement with radio was never complete. Back in the fall of '85, the actor came back to world band radio.

Mike Spivak of Mike's Electronics in Fort Lauderdale remembers the day "The Great One" bounced into his store.

"Jackie Gleason!" said Spivak. "No pal," replied Gleason, "It's Red Buttons."

Attired in a natty golf hat and accompanied by his wife, Marilyn, Gleason told Spivak he wanted to see some world band radios. "I showed him some, but he shook his head." "Show me a *real* set," he said, holding his hands a couple of feet apart to demonstrate the size of the receiver he had in mind.

"He was looking for a real Neanderthal, one of those old fashioned tube monsters," Spivak said. "Like a Hammarlund?" I asked. "Yeah, that's it," he answered. "I told him that those days were long gone!"

Spivak showed Gleason an ICOM IC-R71 receiver, told him how it worked, and left him alone to play with it. Twenty minutes later, the electronics dealer returned to Gleason's side.

"When can you deliver two of these?" the beaming comic asked. Gleason always bought his receivers in pairs. He always had to have a backup just in case—and it never did—one went out on him. Early in 1986, he added a pair of ICOM IC-R7000 VHF/UHF scanning receivers, along with a set of Sony ICF-2010 portables. In fact, a lot of expensive radio gear found its way into Gleason's fourteen-room Lauderhill, Florida, estate—as the world band and 16-element scanner antennas on the roof at 3425 Willowwood attested.

There, in his 20-by-20 foot library—stacked floor to ceiling with bookshelves—was where Gleason would sit, quietly chain-smoking cigarettes and tuning in the world with his battery of radios. Angie Cannon, the *Miami Herald* reporter who managed a lengthy interview with the celebrity not long before his death, describes his monitoring activities.

The Great One hosts a program in 1979

Don Jensen is a reporter for the Kenosha News *and editor of* Numero Uno. *He is a frequent contributor to radio publications.*

THE BEST GETS BETTER AT EEB

THE EEB 2020

FREE LIMITED ONE YEAR WARRANTY!

NEWEST HIGH TECH RECEIVER

$299.95 VALUE
INTRODUCTORY PRICED
$179.95
FREE AC Power Adapter—
Mention This Ad
($14.95 Value)
FREE RADIO STAND
($9.95 value)

Here is the latest in the series of fine receivers from Taiwan (ROC). The last three years have seen big changes from Taiwan. There was the Radio Shack DX 400, the Uniden CR 2021, the Sangean ATS 803 and now the EEB 2020!

- State of the art PLL world band receiver.
- Easy to use with direct digital dialing of frequency.
- 5 tuning functions: direct frequency key-in, auto scanning, manual scanning, preset recall and manual rotary tuning.
- 9 programmable memories.
- Radio, clock and alarm turn on automatically, play preset stations at preset time.
- Wide/narrow bandwidth. You select for better listening in today's crowded band.
- RF gain control to prevent overloading when near strong stations and improve SSB reception.
- Signal strength indicator (5 LED).
- Shortwave button allows user to tune only international shortwave bands. No need to guess start and stop frequency.
- AM button allows full coverage of 150 KHz to 30 MHz.
- FM button allows full coverage of 88-108 MHz.
- Unique speed dialing—the faster the dial is turned the larger the frequency steps.
- Multimode AM-FM-CW-SSB allow full coverage of commercial traffic. Amateur, aircraft, ship at sea, and more.
- 24 hour clock—sleep/alarm selectable switch allows hour & minutes or minutes & seconds.
- Sensitivity and dynamic excellence for a portable radio.
- High stability for good CW-SSB-RTTY reception.
- Excellent audio with separate bass and treble controls, stereo balance.
- Din plug for standard stereo hook-up.
- External speaker jack.
- Built in whip antenna and terminals for external antenna.
- Battery powered 3D & 2AA (not included).
- Same size and weight as Sony ICF 2010

DIPLOMAT 4950

HERE IS THE RECEIVER EVERYONE NEEDS!

$99.95 VALUE
INTRODUCTORY PRICE
$69.95

The perfect radio for the person on the go—The Student or a second shortwave for office or home. Slightly larger than the Sony ICF 4910; otherwise The Diplomat is identical and covers more!

- Medium wave AM 550-1670KHz
- FM 88 to 108 with stereo head set out.
- Shortwave 2.3 to 5 MHz continuous 120, 90, 75 and 60 meter bands (not covered by Sony ICF 4910) and 49, 41, 31, 25, 19, 16 and 13 meter band each expanded for easy tuning.
- LED tuning and stereo indicators.
- Battery portable (3 AA not included).
- AC adapter 120V-4.5V optional—**$9.95**.
- Low battery drain for extended listening.
- Amazing sound for a small package.

TELEX: 62915985
FAX: 703-938-6911
- We ship world-wide
- Shipping charges not included
- Prices & specifications subject to change without notice

10 miles west of Washington, D.C.
Sorry—No COD's
10-5 Tues., Wed., Fri.
10-9 Thursday
10-4 Saturday
Closed Sunday and Monday

**Electronic Equipment Bank
516 Mill Street, N.E.
Vienna, Virginia 22180
Order Toll Free 800-368-3270**
Virginia 703-938-3350

"Gleason spent hours in his library at his shortwave radio set, just listening to the constant traffic of voices drifting through the air. A news broadcast from Moscow . . . a documentary from Spain. Day after day, he manned those machines, just listening."

Spivak agrees. "He listened to everything. He gave me his private number and asked me to call and let him know if anything interesting was on the air. So I'd do that—give him frequencies I'd heard about from some of the guys down here."

The salesman recalls Gleason as a warm and considerate person, not at all technically inclined. But, like many world band listeners, he was an "intelligent and a very precise man." "He'd call up, saying, 'This is Jackie,' or sometimes, even, 'Hello. This is The Great One!' And he'd have questions, especially about his 'R71's." But Gleason was a quick study and mastered their operation without much trouble. A short time later, he ordered the pair of ICOM IC-R7000 radios.

"They were among the first in this country," Spivak says. "I suggested he send an autographed photo to the guy I knew on the order desk at ICOM America in Bellevue, Washington, as a sweetener to speed up delivery. "He did, too," chuckles the dealer, "although the guy still thinks to this day that I forged the name on that picture."

An accurate time standard is important to anyone who monitors the world radio bands. And so it was to the precise Gleason, who kept a Seiko quartz clock on his desk, next to his receivers.

One day he asked Spivak to order him a time piece suitable for traveling. "I got him a Sony cube clock radio with AM and FM." But Gleason had little tolerance for equipment that didn't work the way it was supposed to. Spivak got the call: "Take this thing back, pal. It runs four seconds slow!"

He gave me his private number and asked me to call if anything interesting was on the air

---

On those occasions when he could be cornered by reporters, he was cordial, but not particularly revealing. The answers to questions about past successes came as if by rote, often ended with the same, tired nonenthusiasm. That is until the conversation shifted to the subject that was, to him, always interesting—his world band radios.

"He brightened just discussing it," said reporter Cannon. His usual low-key voice would rise with enthusiasm, sounding, again, like Ralph Kramden on a roll: "You can pick up the *whole world*!"

And so you can. How sweet it is!

Ian McFarland (left) and Larry Magne (right) relax after recording a new program in the "SWL Digest" (see story, p. 22)

Rely on SPECTRONICS — A "REAL RADIO STORE" — Since 1967

FAMOUS "EAVESDROPPER" SWL RECEIVING ANTENNA

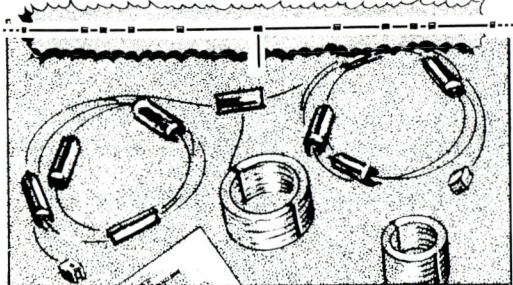

OUR MOST POPULAR SHORTWAVE ANTENNA!

ONLY $59.95
Plus $5 UPS

- Automatic bandswitching
- Completely weatherproof
- 60, 49, 41, 31, 25, 19, 16, 13 & 11 mtrs.
- Complete, no assembly needed

All the world's shortwave broadcast bands are yours with the Eavesdropper All-Band antenna! Individually tuned traps make the Eavesdropper work like seven separate antennas, each tuned to a different international broadcast band. Its 100 foot, 72 ohm balanced feedline provides an exact match to the antenna on every band. Completely assembled, and ready to install with 50 ft. of 450 lb. test nylon rope. Overall length: 42' 10". Wire: #14 copper. Bandswitching: Automatic, impedance to rcvr: 50-75 ohms balanced.

$19.95 POST PAID   **42ND YEAR EDITION**

- Frequency Listings
- Country Listings
- Time Schedules • Reviews

Listen to the world ... hear the voices of people making history ... eavesdrop on local programs from around the globe ... discover music and cultures you never knew existed ... gain a new perspective on international events ... the WORLD RADIO TV HANDBOOK (WRTH) is your guide to an exciting new world — the world of the International Listener! This unique handbook is your personal 24-hour passport to the world's broadcasters and their services listed by country. A special hour-by-hour guide to broadcasts in English directed to your area. Essential station information including frequencies, transmitter powers, operating times, languages, addresses, etc. Listing of stations in frequency order to help you identify them more easily. Maps of principal transmitter sites worldwide. Names and addresses of international radio listeners' clubs. Information on reception conditions. Time Signal Stations and other specialized subjects. Widely acclaimed annual test reports on receivers for the International Listener.

SCANNER ANTENNAS

Get maximum performance from your handheld scanner with these new High Efficiency Antennas from Centurion, Inc.!

(D) Field tunable low band VHF antenna, 30-50 MHz. Tapered helical .050 diameter spring with top loading coil. BNC connector. Approx. length: 10½".
Model A-FT-BN $21.65*

THE TRI-BANDERS ARE HERE! High efficiency flexible helical antennas with tri-resonance design for maximum performance on Low Band, VHF, and UHF. All models are approximately 10½" long, and **DELIVER PERFORMANCE!**

Choose the model that fits your scanner:

(E) Model A-TRI-BN: For all models using BNC connector. **$15.85***
(F) Model A-TRI-RC: For all Bearcat models using a 5/16-32X3/8 threaded connector. **$15.85***
(G) Model A-TRI-BC: For all Bearcat 100 models using a 10-32X3/8 threaded connector. **$15.75***

 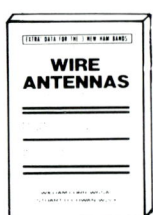 

MONITOR AMERICA — The Communications Travel Guide. Contains information on radio systems used by state & local agencies across the U.S. 98 chapters (50 states, 48 metro areas), covers radio systems and frequencies as well as districts, maps, codes for: State Police/Highway Patrol, Turnpikes, Highways/Transportation Depts., Natural Resources/Conservation, Parks & Recreation, Fish, Game & Wildlife, Forestry-Fire Towers, Civil Defense, Statewide Intersystems, Hotlines, Colleges & Universities, NOAA Weather Broadcasts, Radio Amateur Emergency Nets, and much more. Over 600 pages. .... **$15.95 Ppd.**

WIRE ANTENNAS Simple, low-cost wire antennas for radio amateurs. Easy! How to build efficient antennas that get out — 2 through 160 meters. How to build "invisible" antennas for use in difficult station locations. EXTRA! Data for the 3 new ham bands! .... **$9.95 Ppd.**

BETTER SHORTWAVE RECEPTION New 5th Edition! How your radio receiver works; how to adjust it for best performance. How to eavesdrop on foreign broadcasts, police, fire, aircraft, marine, weather, amateurs, CB, private business radio. All about VHF-FM scanning receivers, long-distance TV and FM reception. Mysterious radio signals from outer space; plans to communicate with other worlds .... **$8.95 Ppd.**

STORE HOURS: Mon.-Fri. — 0900 To 1800; Sat. — 0900 To 1500

(312) 848-6777   **SPECTRONICS, INC**
1009 GARFIELD ST. • OAK PARK, IL 60304

OUR 21ST YEAR

SPECTRONICS — World Class Shortwave Values — Since 1967

WORLD BAND: PASSPORT TO EXCITEMENT!

Super-Compact Direct Access World Band Radio Receiver
SONY ICF-2010

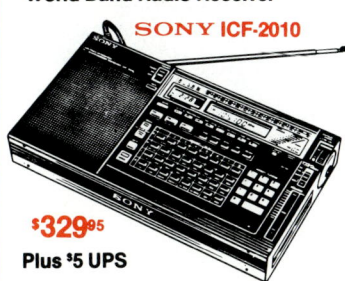

$329.95
Plus $5 UPS

☐ Full AM band coverage (LW, MW, SW), plus FM and Air Band reception ☐ Dual PLL quartz frequency synthesis is the world's most accurate tuning system that virtually abolishes drift and maximizes tuning stability ☐ Direct Access tuning enables you to "key-in" station numbers directly ☐ Band select function for quick access to SW broadcast bands ☐ New synchronous detection circuitry dramatically reduces fading and annoying "beat" frequency interference from adjacent stations ☐ Switchable IF bandwidth with "narrow" mode to select one station out of a crowded band, or "wide" for the lowest distortion ☐ 32 station memory presets for immediate recall at the touch of a button ☐ Memory Scan tuning gives a brief sampling of each preset ☐ Automatic Scan tuning gives a brief sampling of each station on the band ☐ Built-in quartz clock with standby and alarm capability ☐ Convenient programmable timer turns receiver on and off automatically ☐ Sleep timer for 15, 30, or 60 minutes of music as you go to sleep ☐ Switchable 12-hour/24-hour clock indication ☐ Supplied earphone for private listening

Rugged Nine-Band Portable World Receiver In A Paperback Size

$134.95
Plus $5 UPS
ICF-7600A

☐ Exciting 9-band portable world receiver with 7 SW bands/MW and FM ☐ Shortwave bandspread tuning provides you with easy station selection with pinpoint accuracy ☐ LED tuning indicator helps you pinpoint stations you want ☐ 76 to 108 MHz FM band coverage for FM in both the U.S. and abroad ☐ Advanced dual-conversion superheterodyne circuitry on SW for high sensitivity and interference rejection ☐ Separate crystal oscillators for each SW band offer exceptional stability ☐ Ceramic AM filter increases selectivity ☐ Separate FET amplifier on SW ☐ 3-position tone control

ALL PRICES SUBJECT TO CHANGE WITHOUT NOTICE

Ultra-Small, Direct Access World Band Radio
SONY ICF-2003

$249.95
Plus $5 UPS

☐ Full AM band coverage (LW, MW, SW) and FM reception ☐ Quartz frequency synthesis tuning assures optimum stability and reduces drift to an absolute minimum ☐ Direct Access microprocessor tuning enables you to "key-in" station numbers directly ☐ Ten station memory presets ☐ Manual tuning and automatic scan tuning for a total of four tuning options ☐ Dual conversion AM circuitry ☐ Switchable AM sensitivity/RF gain ☐ SSB (Single Side Band/CW capability ☐ Built-in microprocessor quartz clock with standby and alarm capability ☐ Switchable 12 hour/24 hour clock indication ☐ Sleep timer ☐ Multi-function liquid crystal display ☐ LED tuning indicator ☐ Record output jack ☐ Supplied earphone

Exciting World Band Radio With 15 Bands Offers Electronic Rotary Tuning and 15 Station Memory Presets
SONY ICF-7700

$244.95
Plus $5 UPS

☐ Sensational 15 Band Receiver with 12 SW Bands/MW/LW and FM ☐ PLL quartz frequency synthesis tuning is the world's most accurate, drift-free method ☐ Quartz frequency tuning ☐ 15 station memory with 5 for FM, 5 for MW/LW, and 5 for SW ☐ LCD frequency read-out ☐ LCS dial pointer ☐ Electronic rotary tuning control ☐ Electronic Feather-touch for easy operation ☐ Advanced dual conversion superheterodyne circuitry ☐ Tone control ☐ Built-in clock has radio or buzzer wake-up alarm ☐ 65 minute Sleep Timer ☐ Microprocessor control of all clock and tuning functions ☐ 4-way power supply ☐ Complete with earphone, carrying case, and Wave Handbook.

CALL FOR PRICE
ICOM R-71A WORLD HF RECEIVER
• 100 KHz-30MHZ • Keyboard entry • 32 Programmable Memories • SSB/AM/RTTY/CW (FM Optional) • Wide dynamic range • Digital P11 Synthesized Memory scan, Pass Band and Notch Tuning.

Unique, Vertically Designed World Band Radio Offers Ultra-Convenient Portability Plus Full Features
SONY ICF-PRO80

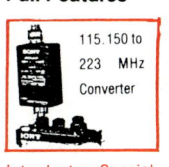

115.150 to 223 MHz Converter

Introductory Special
$359.00
Plus $5 UPS

☐ AM (SW, LW, MW)/Narrow FM/TV 2—6 CH plus Air, PSB, plus 7—13 CH with supplied frequency converter ☐ 150 kHz—108 MHz continuous full coverage ☐ Direct Access 10 key microprocessor tuning lets you key-in station numbers directly ☐ Liquid Crystal Display (LCD) for digital frequency read-out ☐ Versatile professional quality scanning system offers 5 scan options ☐ 3-way scan mode plus precision scan steps for maximum scanning versatility ☐ 40 station memory gives you immediate recall of a tremendous number of stations at the push of a button ☐ PLL quartz frequency synthesis tuning ☐ Quartz frequency for optimum stability and absolute minimum drift ☐ Dual conversion circuitry for AM/narrow FM/SSB ☐ Electronic feather touch functions ☐ Automatic and manual squelch

Compact World Band Radio Offers Nine Band Receiver With Dual Conversion Circuitry
SONY ICF-4920

$99.95
Plus $5 UPS

☐ Sony's 9 band coverage that includes shortwave (1-7), AM, and FM ☐ Sleek, compact design ☐ Dual-conversion superheterodyne circuitry on SW for high sensitivity and interference rejection allows for clear worldwide reception ☐ Bandspread system in shortwave for easy station selection and pinpoint tuning ☐ LED tuning indicator ☐ Soft touch band selectors for effortless selection ☐ Tone selector

CALL FOR PRICE ON ICOM R-7000
UHF/VHF RECEIVER/SCANNER

---

(312) 848-6777
SPECTRONICS, INC
1009 GARFIELD ST • OAK PARK, IL. 60304

OUR **21ST** YEAR

YAESU FRG-8800

150 KHz TO 30 MHz

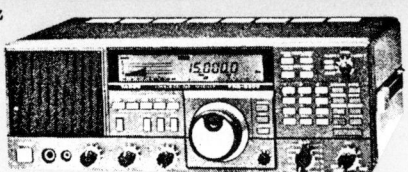

THE FRG-7700 WAS A GREAT RECEIVER. NOW THE NEW GENERATION FRG-8800 TAKES YOU A STEP FORWARD!

* CAT computer compatible
* 12 memories - scan - RIT
* Keyboard frequency entry
* Dual 24 hour clock timer/recorder control
* Noise blanker
* All mode AM-SSB-CW-FM
* Green LCD display
* 150 KHz to 30 MHz
* Weight 11.5 lbs
* Dimensions: 5" x 8 3/4" x 13" W

White Paper Order: RD2

CALL FOR PRICE

OPTIONS

* FRV 8800 VHF Converter (118-174 MHz)
* FM WIDE KIT
* D.C. 8800 (for 13.8 DC operation)
* YH 55 Headphones

CLOSEOUT SPECIALS
SAVE 75% OFF ORIGINIAL COST!!!

* **MU7700** Memory unit for the FRG-7700. NOW 7700 owners can have the convience of 12 channel memory at a fraction of original cost. List price was $119.95 on this unit!
SPECIAL $29.95 + $4 ups

* **FRV7700 VHF** Converters designed to mate with 7700 but works on any radio tuning 20 to 30 MHz (12VDC required-supplied from 7700)
FRV7700F   118-130 & 150-170 MHz
FRV7700A   118-150 MHz
List price was $149.95 on these units!
SPECIAL $29.95 ea + $4 ups.
DON'T MISS THIS OFFER!
(Supplies are limited to stock on hand)

YAESU FRG-9600

60-905 MHz

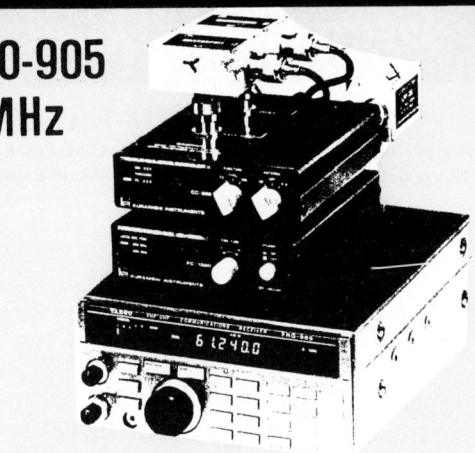

FRG 9600 with both KURANISHI converters.

A PREMIUM VHF/UHF SCANNING COMMUNICATIONS RECEIVER.

* CAT computer compatible
* 60 to 905 MHz
* SSB, CW, AM/wide, narrow FM/wide, narrow
* SCAN steps 5, 10, 12 1/2 & 100 KHz
* 99 Memories   * Scanning
* 24 Hour clock
* Multiplexed output
* Keypad frequency entry
* Watch television programs with Optional VU-9600   CALL

EEB introduces 2 converters to expand frequency coverage of the YAESU FRG-9600. Simple to use and install on your radio.

NEW

KURANISHI CONVERTERS
FC 965   500 KHz to 60 MHz
FC 965DX   20 KHz to 60 MHz
FC 1300   800 to 1300 MHz
WA 965   Pre-amp covers DC to 1.5 MHz
CC 965   Control Console
LPF 05   500 KHz low pass filter
CALL FOR QUOTE AND DETAILS
* Also will work with **ICOM R7000** or any receiver tuning 60 - 170 MHz.

TELEX: 62915985
FAX: 703-938-6911
• We ship world-wide
• Shipping charges not included
• Prices & specifications subject to change without notice

10 miles west of Washington, D.C.
Sorry—No COD's
10-5 Tues., Wed., Fri.
10-9 Thursday
10-4 Saturday
Closed Sunday and Monday

Electronic Equipment Bank
516 Mill Street, N.E.
Vienna, Virginia 22180
Order Toll Free 800-368-3270
Virginia 703-938-3350

Riding the Wave of the Future

With world band radio, an array of events and cultures comes tumbling at you from all over the globe. From the haunting island music of Radio Tahiti to the hourly international newscasts of the BBC, it's a twenty-four hour magazine of the air—one part *Newsweek*, another *National Geographic*.

Affordable and Entertaining

Best of all, anyone can take part. It's neither difficult nor, as you'll see in our 1988 "Buyer's Guide," expensive. Turn on your radio and you'll be treated to lively salsa music from South America or a searching documentary on the Titanic from London. A touch of the keypad adds the *Tour de France* from Paris.

Marriage of Nature and Advanced Technology

Advanced radio technology makes tuning in stations from around the globe almost as easy as turning on the stereo or dialing a pushbutton

---

Advanced technology makes tuning in stations from around the world almost as easy as dialing a pushbutton phone

---

telephone. Still, there are some differences between world band radio and its more familiar AM and FM siblings.

The first, and probably most important, is what makes it so fascinating: the enormous distances these signals can travel. While FM and mediumwave AM signals are limited to local coverage, many world band signals actually girdle the planet, traveling huge distances. This is because the signals are dispersed worldwide by a natural, but invisible, layer surrounding the earth called the *ionosphere*. Because no satellites or other artificial intermediaries are needed, world band radio is the only medium by which programs can be beamed direct—without editing or censorship—from one country to another. This, more than anything else, is why world band programs are so interesting and different from anything you hear or see elsewhere. Nature, as much as man, makes it possible.

All that is great news, but it has its price. Because the ionosphere is a natural entity, it has seasons and temperamental moments just as does, say, the weather. Also, like any traveler of great distances, world band signals tend to get a little weary from the journey.

So, reception of world band stations is different from that of local signals. That's why you'll sometimes hear the same station broadcasting identical programming on more than one channel at the same time. So, the reasoning goes, if one channel doesn't make it to your radio, hopefully one or more of the others will. It's something of a shotgun approach to broadcasting, but for the most part it succeeds in overcoming the day-to-day vagaries of the ionosphere.

Some stations—but not by any means all or even most—find seasonal changes in the ionosphere to be significant enough to prompt them to alter their channel useage from time-to-time. These separate summer and winter schedules are indicated in the "Worldscan" section by a "J" ("June") for summer, "D" ("December") for winter. All other schedules are adhered to year-around, so don't have either a (J) or (D).

Another difference from local stations is that any number of broadcasters—even a dozen or more—may make use of a given channel. Obviously, you can't hear them all at the same time, but you might hear one now, another later. Sometimes you'll not hear what you want because your station's blotted out by more powerful stations, or the ionosphere simply can't carry it to your radio.

How to Listen to Your World

When listening during the local daytime, expect to get the best long-distance reception on the higher frequencies: 11, 13, 15, 17 and 21 MHz. At night, tune around the lower bands having frequencies starting with 5's, 6's, 7's and 9's—plus 3's and 4's during the winter. World band radio is so extensive that there are countless exceptions, but these rules of thumb will get you into the fast lane right away.

You'll find that given regions of the world are

Dominican Republic's Radio Clarin, well-heard on 11700 kHz, airs programs of anti-Castro "La Voz del CID"

more audible during certain times of the year. If you live in North America or Europe, for example, you have a better chance at hearing some of the smaller African and Latin American stations on 4 MHz at night during the winter. Of course, the larger stations tend to be heard year around.

What's Heard Where

In order to be better heard in selected parts of the world, many stations use special antennas that beam most of the signal in a single given direction. *Passport's* "Worldscan" section—the blue pages—indicates where these broadcasts are beamed. But so efficient is the world band medium that you can often hear programs even though they're directed to listeners elsewhere. For example, although Radio France International's English-language program from 1600–1655 World Time (UTC) is targeted to Africa and Europe, listeners in the Americas and most other parts of the world can still hear it . . . especially if an above-average radio is used.

Finally, you'll notice that world band stations operate on a special 24-hour clock called World Time, or UTC. But don't worry about that. It's easy to figure out because so many world band stations announce it over the air. Additionally, there are specialized stations that give you the time. Throughout much of the world, for example, you can hear one or more of the following channels operated by the US National Bureau of Standards: 2500, 5000, 10000 or 15000 kHz. As a bonus, these also provide weather information for the Atlantic and Pacific oceans.

Passport to World Band Radio guides you all the way through the wide world of world band radio. In addition to exhaustive coverage of all schedules, *Passport* includes the "Buyer's Guide" in which you'll find unflinching reviews of the most advanced and popular radios.

Welcome aboard!

Prince Edward Island

Canada by Radio

by Ian McFarland

Canada is one nation that has made good use of world band for reaching its vast territory. Although often unrecognized as such, it is the second largest country in the world, after the Soviet Union. In fact, the entire People's Republic of China could fit neatly within its borders.

But take a look at any map of Canada and you'll see that it looks as if some great hand shook the country, settling the bulk of the population along the eastern part of the border with the United States. That leaves a few hardy souls occupying the remaining expanse of the country. It's clearly not worth it to dot the chilly expanses above the tree line with AM stations. The alternative: world band radio.

In Canada, there are several local commercial stations, all on the mediumwave AM band, that air their programs—mainly for backwoods inhabitants and fishermen—over world band channels. Because these stations are dotted across the country, world band listeners can get a clear view of what's happening in most of Canada's larger cities. Fully half the nation's ten provinces are represented on world band radio in this way.

Wide Variety of Stations Fill the Airwaves

From Quebec, listeners can tune in news and talk shows on Montreal's CFCX, which rebroadcasts the programs of AM-60, CFCF. Like talk radio in the United States, the Canadian version is limited only by good taste. CFCX's world band channel is 6005 kHz and it's on the air fully 24 hours a day.

Want to know what's going on in Toronto? Tune in CFRX. It's Canada's most powerful commercial world band station and the one most often heard by listeners throughout North America and as far away as South America. It relays the programs of CFRB and is found on 6070 kHz.

From Canada's heartland in the province of Alberta comes CFCN on 6030 kHz. Broadcasting from Calgary, its low power transmitter is no match for the big international broadcasters. To give you an idea of the difficulty in hearing some of these local stations, keep in mind that while the Voice of America and Radio Moscow regularly use transmitters of 500,000 watts, CFCN's world band relay—CFVP—is a mere 100 watts, Its channel, too, contributes to its difficulties. 6030 kHz is regularly used by the powerful US Armed Forces Radio and Television Service, which blots out the station's signals for much of the day. Nonetheless, with some patience and persistence—and waiting until AFRTS signs off the channel—you'll have the opportunity to hear it.

Two other stations represent opposite coasts of Canada. CHNX (see photo) on 6130 kHz carries the programs of CHNS in Halifax, Nova Scotia, along the east coast. It benefits from a recently installed new transmitter, but remains a midget among world band signals with only 500 watts.

The other station is CKFX in Vancouver, broadcasting around-the-clock on 6080 kHz. Located on the Pacific coast just across Queen

---
Since it began in 1936, the CBC has been a shining beacon of excellence
---

Charlotte Sound, it uses a scant *ten* watts . . . although they have plans to upgrade to between 15-25 watts in the not-too-distant future. As world band listeners know, even ten teeny watts of world band signal travel quite a distance—much, much farther than on either AM or FM. Indeed, the station reports being heard occasionally even as far as Europe and Australia by radio aficionados who relish being able to hear such a miniscule signal from afar.

Two more low-power world band stations—one along the east coast in St. John's, Newfoundland, the other across the continent in Vancouver—relay the programs of the Canadian Broadcasting Corporation (CBC). Because they share the same frequency—6160 kHz—you have to listen carefully be certain which of the two is being heard at your location.

Since it began in 1936, the CBC has been a shining beacon of excellence in radio programming. It provides Canadians—and now, through world band radio, many others, as well—with a constantly varying fare of music, news, drama, and other entertainment. Sadly, Canadians have tended to look upon the CBC with a certain amount of disdain, viewing it mainly as the producer of elitist and minority-interest programming. But their oversight is the world's gain.

High-Powered Services Heard Worldwide

The CBC also uses local world band radio, with considerably more powerful transmitters—100,000 watts—to serve the population of the northern region of Canada's largest province,

---

Ian McFarland, who prepared this article in cooperation with Larry Miller, is Announcer-Producer of English Weekend Programming at Radio Canada International.

Toronto city scene

Quebec. Here you'll find news, music and some really exotic information not only in English and French, but also in two native Indian dialects, as well as Inuktitut, the language of the Eskimo— or Inuit, as they prefer to be called. World band channels for this CBC Northern Quebec Service include 6065, 6195, 9625 and 11720 kHz. Some of these are heard throughout the Western Hemisphere plus, to a lesser extent, East Asia,

---

It's clearly not worth it to dot the chilly expanses with AM stations

---

Australia/New Zealand and even Europe.

The CBC also has an excellent and easily heard external service, Radio Canada International. Its programs are a mixture of features especially produced for world band radio, such as *North Country*, heard at 1200 World Time (UTC). Also, it rebroadcasts some of the finest of CBC's news programming. Notable among these are *The World at Six* and *As it Happens*, heard Monday through Friday at various times in the local later afternoon and early evening. Non-Canadian listeners and Canadians traveling abroad have made RCI one of the most popular world band stations on the airwaves. Indeed, Gallup surveys have showed it to be running neck-and-neck with the BBC for first place among listeners in the US.

Canada's use of world band radio is, in itself, not all that unusual. There are any number of countries—Brazil, India, Pakistan, Bangladesh, and Nigeria, for example—that have local, regional and international world band stations. But Canada is especially interesting in that its local world band broadcasting is primarily in English. This makes it the perfect target for world band listeners who not only want to eavesdrop on this vast and diverse country, but also wish to hone their listening skills.

Computer Decoding of Radio Signals

Six Digital Modes - Including Weather FAX

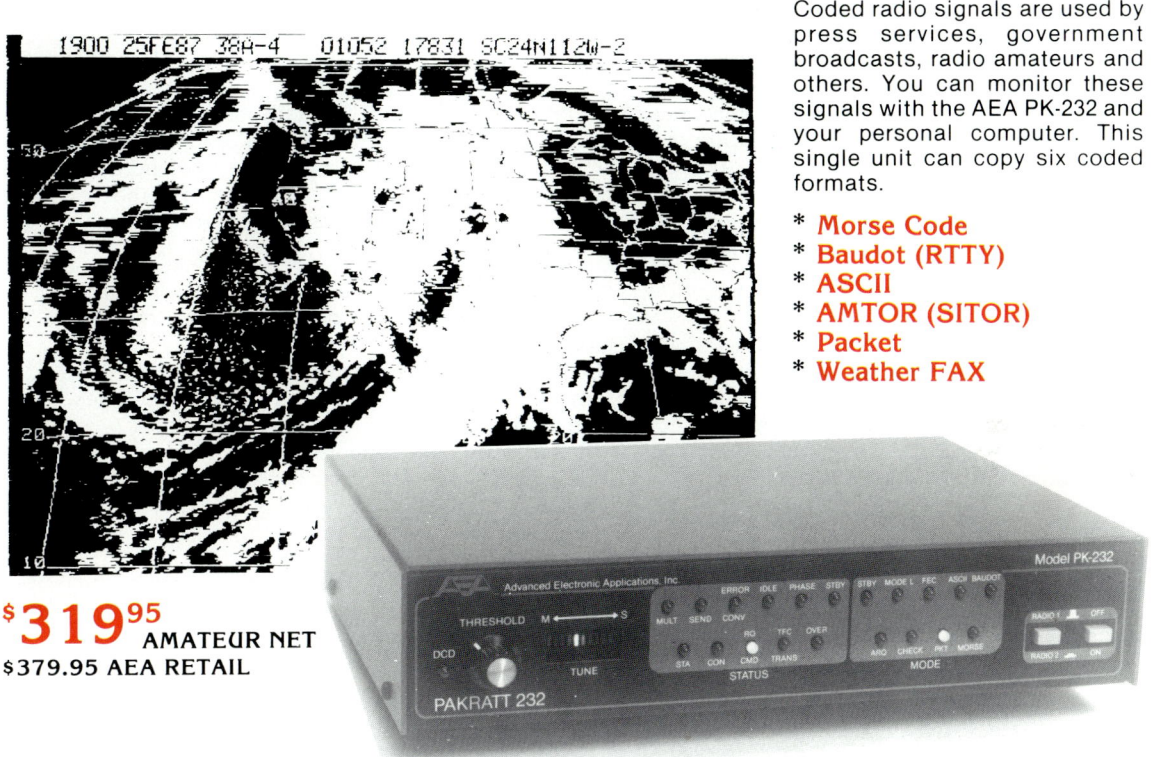

Coded radio signals are used by press services, government broadcasts, radio amateurs and others. You can monitor these signals with the AEA PK-232 and your personal computer. This single unit can copy six coded formats.

* **Morse Code**
* **Baudot (RTTY)**
* **ASCII**
* **AMTOR (SITOR)**
* **Packet**
* **Weather FAX**

$319.95 AMATEUR NET
$379.95 AEA RETAIL

Short-wave listening can be an exciting hobby. But voice broadcasts are just part of the information available to those with short-wave receivers. Digital signals carry everything from foreign news broadcasts to satellite photos like the one pictured in this ad. The special SIAM (Signal Identification and Acquistion Mode) of the PK-232 will even identify the type of signal, transmission rate, and special characteristics of the signal being received.

The PK-232 requires some type of data terminal or personal computer for operation. Any computer or terminal with a serial RS-232 port can connect directly to the PK-232. A terminal program, like those used with telephone modems, is then used to set up the computer for communication. AEA also offers two special PK-232 programs, one for the Commodore 64 and one for IBM PC or compatible computers.

The weather fax mode requires an Epson graphics compatible parallel printer. The special PK-232 cable connects to both the computer and printer, making it easy to tune weather fax signals and output them directly to the printer.

In addition to receiving and decoding the six different modes, the PK-232 is also capable of transmitting in all six modes. This makes the PK-232 the perfect unit for both the short-wave listener and licensed radio amateur. Contact your local AEA dealer today for more information, or call AEA. Then you'll be able to listen in on the world of information you're missing now.

 Brings you the Breakthrough

2006-196th St. SW
Lynnwood, WA 98036
(206) 775-7373

Canada Airs Popular Program for World Band Listeners

Each week Radio Canada International airs "SWL Digest," a program devoted to helping you better understand and enjoy world band radio. Included in this popular program are tests of new world band radios and accessories by *Passport's* resident equipment expert, Lawrence Magne. *Passport* editors Tony Jones and Don Jensen highlight worldwide developments, while other features cover a variety of interesting topics, including new world band schedules.

Hosted by award-winning announcer/producer Ian McFarland, "SWL Digest" is heard in many parts of the world—including the Americas, Europe, the USSR, Africa and Australia/New Zealand—at various times. Although the 0007/0107 broadcast is intended for listeners in North America, 2008/2108 for Europeans, 2137 for Africans and 2307 for residents of the Caribbean, in reality these can be heard simultaneously in many parts of the world. Tune around for the best results and handiest times.

1st Saturday/Sunday of the month	Larry Magne's "Test Report" on new world band radios and accessories, plus news of stations and schedules.
2nd Saturday/Sunday of the month	Official monthly reports from radio umbrella associations, plus news of stations and schedules.
3rd Saturday/Sunday of the month	Either Don Jensen's "Journal" or Tony Jones' "Report from South America."
4th Saturday/Sunday of the month	"Potpourri"—special reports and features to aid in listening to world band radio. Also, news of stations and schedules.

A similar program, "Allô DX" is presented weekly in French by Yvan Paquette, animateur.

World Time (UTC)		
Winter Hiver	Summer Été	Broadcast
	1837	Samedi en français sur 7235 ou 7130, 9555, 11945, 15325 et 17875 kHz
1908	1908	Samedi en français sur 15260, 17820 kHz
1937		Samedi en français sur 5955, 7235 ou 7130, 11945 et 15325 kHz
	2008	Saturday in English on 7235 or 7130, 9555, 11945, 15325, 17820 and 17875 kHz
2108		Saturday in English on 5995, 7235 or 7130, 11945 and 15325 kHz
2137	2137	Saturday in English on 11880 or 11945, 15150 and 17820 kHz
0107	0007	Saturday night in English on 5960 and 9755 kHz
0207	0107	Samedi soir en français sur 5960 et 9755 kHz
2307	2307	Sunday evening in English on 9755 and 11710 kHz (plus 930 kHz AM for the Caribbean)

Parcourir le Canada par les Bandes

par Ian McFarland

Le Canada est l'une des nations qui a le plus profité de la radiodiffusion à ondes courtes pour atteindre les confins de ses vastes territoires. Quoique le fait ne soit pas toujours reconnu, le Canada est le deuxième Etat du monde après l'URSS. En effet, l'on pourrait caser à l'intérieur de ses frontières l'intégralité de la République populaire de Chine.

Mais si l'on regarde la carte du Canada, on dirait qu'une main gigantesque eût éparpillé la majorité de la population le long de la frontière Est des Etats-Unis. Il ne reste donc qu'une minorité d'âmes intrépides qui habitent le reste du pays. A quoi bon, alors, parsemer de stations AM les régions glaciales au nord des zones forestières? L'alternative, c'est l'utilisation des bandes internationales.

Au Canada, en effet, il existe de nombreuses stations locales commerciales émettant sur des fréquences AM, qui rediffusent leurs programmes, surtout à l'intention de bûcherons et de pêcheurs isolés, sur les bandes internationales. Dû au fait que ces stations se trouvent éparpillées à travers le pays, l'auditoire international a la possibilité de s'informer des actualités dès qu'elles se déroulent dans les plus grandes villes canadiennes. La bonne moitié des dix provinces de la nation se fait ainsi entendre sur les bandes internationales.

Une diversite de stations

A Québec les auditeurs peuvent capter des programmes d'informations et de discussions en provenance de CFCX à Montréal, qui réémet les diffusions de AM-60, CFCF. Comme aux Etats-Unis la radio d'opinion canadienne ne connaît aucune censure si ce n'est celle des bienséances usuelles.

Ian McFarland qui a rédigé cet article en collaboration avec Larry Miller, est présentateur-réalisateur des services de fin de semaine en langue anglaise de Radio Canada International. La version française est d'Isabel Walcutt, Ph.D., Directrice de French Instruction and Consulting à New York.

The Rideau Canal, Ottawa

EEB -- THE NATION'S LEADING SHORTWAVE RADIO SUPPLIER
JUST 10 MILES WEST OF WASHINGTON, D.C.

KENWOOD R5000

* PLL Digital tuning.
* AM, FM, SSB, CW, FSK.
* 150 KHz to 30 MHz.
* 100 Programmable memories.
* Optional service manual Order: SMR5000.   White Paper order: RD3
* Optional VHF converter (108-174 MHz). Order: VC 20.
SEE REVIEW IN THIS ISSUE!

KENWOOD R2000
* PLL Digital tuning.
* 150 KHz - 30 MHz.
* AM, FM, SSB, CW.
* 10 Programmable memories.
* Optional VHF converter (108-174 MHz). Order: VC 10.

SONY PRO-80   NEW
* 150 KHz - 232 MHz (with supplied converter).
* Full Featured scanning (Memory, Program, Limit, Priority, Manual!).
* AM, FM, (wide/narrow); SSB.
* HANDHELD RADIO SIZE!
* Optional AC adpt. Order: SPA6.
SEE REVIEW IN THIS ISSUE!

SONY ICF2010
* 150 KHZ to 30 MHz, 76 MHz to 108 MHz FM, 116 to 136 MHz.
* 32 Programmable memories.
* 4 Event timer.
* Scanning.   White Paper order: RD9
* AC adapter included.
FREE RS1 radio stand.

SONY 2003
* 150 KHz to 30 MHz.
* 10 Memories.
* Scanning.
* Keyboard entry.
* Opt. AC adpt. Order: SPA6.
FREE RS1 radio stand!

SONY 4920
* 9 Bands; AM, FM, and 7 Shortwave bands.
* Small size (not much larger than a pack of cigarettes!).
* Opt. AC adpt. Order: SPA3.

SONY ICF7700   NEW
* 15 Bands; LW, AM, FM, and 12 Shortwave bands.
* 15 Memories.
* LCD Frequency readout.
* Clock/sleep timer.
* Opt. AC adpt. Order: SPA6.
SEE REVIEW IN THIS ISSUE.
FREE RS1 radio stand.

SANGEAN ATS801
* LW 155-281 KHz, MW 530-1620, SW 5.8 MHz-15.5, FM 88-108 MHz.
* 25 Programmable memories.
* Digital frequency/Clock readout.
* Manual/auto tune.
* Opt. AC adpt. Order: SPA6.

TOSHIBA RP-F11
* Analog dial.
* AM, FM, and 9 Shortwave bands.
* S meter. * Safety lock switch.
* Carrying strap.
* Opt. AC adpt. Order: TAC64.

PANASONIC RFB60   NEW
* LCD Readout.
* FM, LW, MW, SW.
* 36 Memories.
* Clock/Timer, Sleep/Standby.
SEE REVIEW THIS ISSUE.
FREE RS1 radio stand.

TEN-TEC RX325
* 300 KHz to 30 MHz.
* AM, SSB, CW.
* 25 memories. * Clock/Timer.
* Keyboard/manual frequency entry.
* Noise blanker. * Attenuator.
* Wide/ Narrow filter.

NRD-525

* 90 KHz to 30 MHz.   White Paper order: RD5
* 200 Memory channels.
* 2 Clocks/Timer to control radio and extra equipment (recorders).
* Options for 34-60 MHz, 114-117 MHz, 423-456 MHz.

Look for our FULL page ads in this issue for:
ICOM R71A-R7000
YAESU FRG-8800 FRG-9600
EEB 2020 DIPLOMAT 4950
GRUNDIG 650 AND 400
MARANTZ RECORDERS AND C.R.I.S.

ANTENNAS

DATONG ACTIVE ANTENNA   NEW
* Outdoor model AD370.
* Constant sensitivity from 200KHz to 30MHz.
* No tuning or adjustment required.
* Balanced dipole design for good rejection of local interference.
* Order: AD370.
* Optional DC supply. Order: MPU.
SEE REVIEW IN THIS ISSUE.

DX-SLOPER   NEW
* Covers .5 to 1.5 MHz for optimum AM DXing!
* Covers 11 shortwave bands, 2 - 26 MHz.
* All stainless steel hardware, fully assembled.
* Overall length is only 60 feet.
* You must provide 50 Ohm coax.
* Order: DXSWL.
* Complete review in White Paper. Order: RD8
SEE REVIEW IN THIS ISSUE.

EAVESDROPPER
* Balanced trapped Dipole.
* Maximum performance, Minimum local noise pickup.
* All SW bands, 60 - 11 meters.
* Only 43 feet long, Completely assembled.
* 100 foot feed line included.
* Order: SWL.

AMECO TPA
* Tunable Preamp/Antenna preselector covers 200 KHz to 30 MHz in 5 bands.
* Up to 20dB gain.
* Separate preamp, tuning, band and gain controls.
* Use on external antenna for better match and gain.
* Order: TPA.
* Powered by 9 volt battery or Optional AC adapter. Order: P9T

RADIO TILT STAND   NEW
* Use with any portable radio.
* 47 deg. angle for optimum viewing of display, panel controls and better sound projection.
* Order: RS1

FAX: 703-938-6911
TELEX: 62915985
• We ship world-wide
• Shipping charges not included
• Prices & specifications subject to change without notice

10 miles west of Washington, D.C.
Sorry—No COD's
10-5 Tues., Wed., Fri.
10-9 Thursday
10-4 Saturday
Closed Sunday and Monday

Electronic Equipment Bank
516 Mill Street, N.E.
Vienna, Virginia 22180
Order Toll Free 800-368-3270
Virginia 703-938-3350

NEW FROM VIENNA!

THE ELECTRONIC EQUIPMENT BANK OF VIENNA, VIRGINIA, IS PLEASED TO ANNOUNCE THE ADDITION OF TWO NEW RECEIVERS TO ITS GROWING FAMILY OF SHORTWAVE RADIOS. NOW WITH CLASSIC EUROPEAN STYLING AND UNCOMMON ATTENTION TO DETAIL. THE NEW GRUNDIG SATELLIT 650 AND SATELLIT 400 SWL RECEIVERS!

GRUNDIG®

SATELLIT 650

SATELLIT 400

SATELLIT 650

The Satellit 650 International is the Ultimate in German crafted Portable radios. Along with excellent Audio performance the Satellit 650 also has these fine features: • AM, SSB, CW and FM. • Covers 510—1620 KHz, 1.6—30 MHz AM and 87.5 to 108 MHz FM. • Direct keyboard frequency entry. • PLL Quartz Crystal Synthesized for precise tuning and stability. • 60 Programmable memories (16 FM, 8 MW, 4 LW & 32 SW) • Up to 30 Watts peak audio output power. • Built in Ferrite and Telescopic Antennas. • Dual speaker/switchable. • World Power: 110-127/220-240V 50/60 Hz. • DC power with batteries (not provided) • Dimensions 19¾ × 9½ × 8. • Weight 19 lbs.

SATELLIT 400

The Satellit 400, with its rounded corners and smooth lines is the obvious "style leader" in personal portables. Beautifully crafted, this small portable covers all Shortwave bands plus MW and FM. It's unexcelled audio will surprise you! • Direct keyboard frequency entry. • 24 Programmable memories. • Scanning. • 2 Clocks (turn on/turn off). • World power 120/240 V 50/60Hz. • Battery powered (not supplied) and optional Nicad Batteries (nicads can be recharged while radio is playing in AC mode). • Dimensions 12" × 7" × 3" • Weight 5 Lbs.

ELECTRONIC EQUIPMENT BANK
516 MILL STREET, N.E.
VIENNA, VIRGINIA 22180
TELEPHONE (703) 938-3350
ORDER TOLL FREE: 800-368-3270

IMPRIMÉ: THE WORLD RADIO MARKETPLACE

Philips/Magnavox D2935
$169.95 plus 4.50 UPS
Ships in October!

A great midsize portable at an unusual value. Full shortwave coverage. Digital frequency display, push-button and manual tuning, clock and more. Best value in a portable! Arrives in the US late October '87. Order now!

Clandestine Radio Broadcasting: A Study of Revolutionary and Counterrevolutionary Electronic Broadcasting • The most thorough examination of the topic ever written. Over 350 pages of exciting information on the stations that paved the way for revolution—and failed revolution—in countries all over the globe. Includes an introduction to clandestine broadcasting. Hardback only. $49.95 plus $2.16 UPS.

The toll-free number is for orders only. Questions about prices, inventory and technical information will be refered to the warehouse number.

For same day shipping, use your Mastercard or Visa and call the warehouse direct: 215-383-1150, open Monday through Friday 9:00 AM to 5:00 PM EST.

Listen and Learn Language Tapes • DX smarter! A new streamlined format and low price for the popular Listen and Learn language cassette series. The complete, critically acclaimed course for under ten dollars each! Each edition features 90 minutes of recorded speech by native speakers, spoken first in English and then in the foreign language, plus a convenient manual. Choose from French, German, Italian, Japanese, Modern Greek, Modern Hebrew, Portuguese, Russian, Spanish and Swedish. Each is just $8.95 plus $1.95 UPS.

Confidential Frequency List • A frequency directory of utility stations from 4 to 28 MHz shortwave. $15.95 plus $1.95 UPS.

Tune in the World • A complete course designed to help you get your ham radio license. Includes cassette and book—all you need to get on the air. By the world famous American Radio Relay League. $15.00 plus $1.95 UPS.

How to Tune the Secret Shortwave Spectrum • An easy look at the people and organizations behind some of the more unusual and mysterious broadcasts on shortwave radio. $9.95 plus $1.95.

World Press Service Frequencies • A list of English RTTY press frequencies on shortwave arranged by time, frequency and country. $8.95 plus $1.95 UPS.

Broadcasting Around the World • A country-by-country look at how broadcasting works in over 18 nations. Hardback only. $22.95 plus $2.16 UPS.

DX Edge

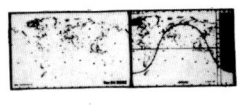

For years, ham radio operators have known that their best chances of getting those hard-to-hear stations is when their location is in sunset and the location of the station they want to hear is in sunrise. Or vice-versa. It's called Greyline DXing. But calculating when this occurs has never been easy until now. With the DX Edge, you can find the perfect time to try for any station. DX Edge puts all this information at your fingertips in an easy to use, 11 inch sliderule device. Get the DX Edge plus 12 overlays for each month for just $19.95 plus $1.95 UPS.

The Map Pack

Keep track of your shortwave listening—and the world—with this handsome pack of nine handy folded Rand McNally maps. You get the world, the U.S., Canada, Mexico, Europe, South America, Africa, and West Indies/Carribbean. Average size: 42" x 28" in full color. $22.50 plus $1.95 UPS.

Sangean ATS-803
$175.95 plus 4.50 UPS
Great Starter Radio

A truly affordable, full feature portable. Perfect for the beginner or anyone looking for a great second radio. Has most of the features of the more expensive Sony ICF-2010 —at over a hundred dollars less.

World on the Air • An exciting, 60 minute cassette, imported from Finland, of radio stations identifications and interval signals from all over the world—some on longer even on the air! $8.95 plus $1.95 UPS.

Latin America on the Air • A 75 minute cassette of radio stations in Latin America, with fascinating program excepts, identification signals and IDs. Complete with guidebook in English. $9.95 plus $1.95 UPS.

Utility QSL Address Guide: The Americas • Enjoy Collecting QSL cards? Here's a book filled with names and addresses of utility stations operating all over the Americas—military, maritime, embassies, aeronautical and much more. $12.95 plus $1.95 UPS.

Language Lab • Write reception reports in a foreign language—and increase your QSL totals—without knowing a single word of the language! An ingenious series of books that gives you hundreds of English phrases and their foreign language equivalents, allowing you to create an endless variety of reception reports. Choose from Spanish, French, Portuguese or Indonesian. $12.95 each plus $1.95 UPS. Or get all four for $45.00 plus $3.09 UPS.

Never received one of our mailings and want to be on our list? Put your name and address on a postcard and mail it to us. We're always adding new and exciting titles and dozens of other shortwave bargains!

Minimum Credit Card Order: $10.00.

How to Repair Old Time Radios • Without any previous experience in radio electronics, you can locate trouble spots, repair defective parts, or find out how to obtain replacements for obsolete tubes and other components—and turn that old time radio into a working radio. $8.95 plus $1.95 UPS.

The World Below 500 kHz • A handy little reference guide to stations on this forgotten—in North America—portion of the radio spectrum: longwave. $4.75 plus $1.95 UPS.

Unos Dos Cuatro: A Guide to the "Numbers" Stations • A rambling discourse on the possible origins and use of these 20 year old—but still identified broadcasts. $13.95 plus $1.95 UPS.

RDI White Papers

In-depth, hands-on and laboratory tests of shortwave listening equipment by the staff of Radio Database International. The following papers are available at $4.00 each, postpaid.
• Gundig 650 • ICOM R71A • Japan Radio NRD-93 • Japan Radio NRD-525 • Kenwood R-5000 • Lowe HF125 • Sony ICF-2010 • Ten Tec RX-325 • Yaesu FRG-8800 • Popular Outdoor Antennas • How to Interpret Receiver Specifications and Lab Tests.

The Azimuth World Time Dual-Zone 24 Hour Station/Travel/Alarm Clock

Digital display, dual world time/local time clock. Flip the switch and get the local time for any of 24 cities around the world. Complete with folding easel stand and black leather-like travel pouch. Just 2" x 4.5" x .5". Comes complete with long life AAA batteries. $25.95 plus $1.95 UPS.

Warehouse Direct: 1-215-383-1150

Box 241 • Radnor Station • Radnor, PA 19087

Eavesdropper Antenna
$59.95 plus 2.62 UPS

America's favorite outside antenna! Ruggedly built and completely assembled and ready to use—just 40 feet long. And its special design gives best reception to whatever shortwave band you're tuned to.

The Sloper Shortwave Antenna
$64.95 plus 2.62 UPS

Another super shortwave antenna. Ruggedly built like the Eavesdropper but with better coverage of the low frequencies (tropical bands). 60 feet long. Completely assembled and ready to connect to your coaxial cable.

McKay Dymek DA-100D Active Antenna
$184.95 plus 4.02 UPS

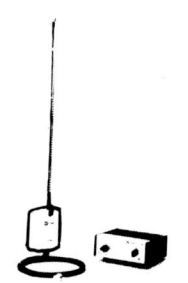

If you can't put up a big outdoor antenna but want great shortwave reception, get the McKay Dymek DA-100D. It's the ultimate outside antenna and perfect for apartment dwellers. Mounts on windowsill or balcony.

World Radio Longwire Antenna
$12.95 plus 2.62 UPS

Experts agree. The simple longwire is best! But not everyone known enough "theory" to build a good one. So, here's a completely assembled, ready to use longwire antenna. Ruggedly made of high quality parts. Everything is included!

Shortwave Books

Shortwave Listening with the Experts • 25 chapters on the fascinating hobby of shortwave listening. An "A to Z" book that takes the reader from how to get started to the fine points of DX-ing. Large format, 500 + page book. Great for beginner and veteran alike. $21.95 plus $2.16 UPS.

Radioteletype Monitoring • There are, between the broadcast stations, dozens of news agencies and press services that transmit on shortwave. And you can tune them in—if you know how. An introductory book. $12.95 plus $1.95 UPS.

U.S. Military Communications • A three volume set of reports detailing the use of shortwave by the US military. Includes frequencies for Army, Air Force, Navy, Marines, and more. $38.85 plus $2.16 UPS.

World Radio TV Handbook • The professional reference book for international broadcasters. Station names, addresses, frequencies, telephone numbers. Current edition $19.95 plus $2.16 UPS.

The Radio Station • An excellent, truly comprehensive book that takes you behind the scenes for a look at how radio stations really operate. $21.95 plust $2.62 UPS.

Guide to Utility Stations • If *Radio Database International* is the "bible" of shortwave broadcasting, then this is the bible of shortwave utility stations: ships at sea, news services, military, aero and much, much more. Almost 500 pages of frequencies, station names and even addresses. $29.95 plus $2.62 UPS.

The World Hamnet Directory • Groups of ham radio operators get together to discuss a whole range of special interests: missionary work, retired people, weather, military and many others. You can hear them! Arranged by frequency, alphabetically and by time. $12.95 plus $1.95 UPS.

Secrets of Successful QSLing • A helpful yet entertaining look at how to college QSL cards and other souvenirs from shortwave stations. $9.95 plus $1.95 UPS.

Shortwave Listener's Antenna Handbook • Robert Traister's classic book on understanding and building shortwave listening antennas. An Imprime best seller. $11.95 plus $1.95 UPS.

Voice of America Calling Finland • An historical look at shortwave broadcasting, focusing on the VOA's Finnish Service —by the former chief. A collector's item. $9.95 plus $1.95 UPS.

International Broadcasting: Limits of the Limitless Medium • The definitive work on international broadcasting. Covers history of the medium, profiles of various stations, and examines their purpose, techniques and successes. A must for any serious student of shortwave radio. Hardcover. $36.90 plus $2.86 UPS.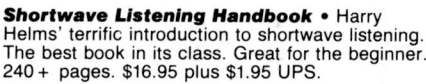

Shortwave Listening Handbook • Harry Helms' terrific introduction to shortwave listening. The best book in its class. Great for the beginner. 240 + pages. $16.95 plus $1.95 UPS.

"Top Secret" Registry of US Government Radio Frequencies • Tom Kneitel's incredible 192 page all new 6th edition of this standard reference book. Covers federal frequencies—FBI, CIA, NASA, Army, Navy, Marine, Coast Guard to Air Force One and more—from 30 to 470 MHz. Almost 120,000 listings. $17.95 plus $2.16 UPS.

Radio Station Treasury: 1900-1940 • A profusion of call signs, old-time station slogans, frequencies, programs and original station and equipment advertisements. 176 pages that archive the birth of radio in the 1900s until 1946. $12.95 plus $2.16 UPS.

So You Bought a Shortwave Radio • The beginner's guide to the world of shortwave radio. Easy to understand with plenty of whimsical illustrations. This should be your first shortwave book! $6.95 plus $1.95 UPS.

World Broadcast Station Address Book • Station names and addresses plus information on getting QSL cards. $8.95 plus $1.95 UPS.

Airscan • A large format, 120 page book loaded with frequencies for listing to all manner of 30-5, 118-136, 138-174 and 400 + MHz aero communications. $10.95 plus $1.95 UPS.

Guide to Embassy and Espionage Communications • Hear news before it happens as it's communicated between the embassies of the world. $10.95 plus $1.95 UPS.

All orders shipped by United Parcel Service wherever possible. Not responsible for items lost or damaged in the mails. Books, tapes and antennas not returnable except in case of damage which must be reported to and documented by shipping agent upon receipt. All others subject to 5% restocking fee. PA resident (only) add 6% sales tax. Prices subject to change without notice.

Order Toll-Free: 1-800-323-1776, ext. 126

Monitor Global Communications with High Technology Equipment from Universal

"HEAR" THE WORLD

Universal offers a complete line of competitively priced communications receivers from $95 to $6495. We feature all major lines including Japan Radio, Kenwood, Icom Yaesu, Ten-Tec, and Sony. The new JRC NRD-525 is shown above.

"READ" THE WORLD

Explore the exciting world of radio teletype. Read messages from marine, press, military and weather stations worldwide! Like to learn more? Send SASE for our **FREE pamphlet, "Listening to RTTY."** Shown above is the new Info-Tech M-6000, the most powerful decoder you can buy at any price.

"SEE" THE WORLD

Receive and print pictures using your shortwave radio! See FAX weather maps, press photos, marine information, military charts, satellite photos and more! Send SASE for our **FREE pamphlet, "How to Receive Facsimile (FAX) On Your Shortwave Radio."** New M-800 FAX converter shown with optional printer.

Contact Universal today for information on these and other exciting communications products. Used equipment also available.

UNIVERSAL SHORTWAVE RADIO
1280 Aida Drive
Reynoldsburg, Ohio 43068
Toll Free 1-800-431-3939

Calgary Stampede, Alberta

CFCX diffuse sur 6005 kHz et s'entend 24 heures par jour.

Et s'il s'agit de se mettre au courant de ce qui se passe à Toronto? On n'a qu'à écouter CFRX. C'est la station commerciale à bandes internationales la plus puissante du Canada et celle qui s'entend le plus souvent par des auditeurs d'un bout à l'autre du continent américain. Elle se charge du relais des programmes de CFRB et émet sur 6070 kHz.

En provenance de l'intérieur de la province d'Alberta s'entend CFCN sur 6030 kHz. Situé à Calgary, son émetteur à faible puissance ne saurait se comparer aux émetteurs des grands diffuseurs internationaux. Pour se faire une idée de la difficulté de capter certaines de ces stations locales, il suffit de songer aux 500,000 watts des émetteurs de la Voix d'Amérique et de Radio Moscou en comparaison avec les quelques 100 watts de CFVP qui transmet vers le monde les programmes de CFCN. D'ailleurs, le canal attribué à cette station ne fait qu'aggraver sa réception car il est utilisé également par la puissante AFRTS (Radio et Télévision des Forces Armées Américaines) sur 6030 kHz, qui recouvre ses signaux pendant une bonne partie de la journée. Avec un peu de patience et beaucoup de ténacité, pourtant, on finit bien par capter ses émissions aux moments où AFRTS quitte le canal.

Les côtes opposées du Canada se représentent par deux stations. CHNX (v. photo) sur 6130 kHz. réémet les programmes de CHNS à Halifax en Nouvelle Ecosse en direction de la côte Est. Munie d'un émetteur neuf, elle reste, en raison de ses signaux qui ne dépassent pas les 500 watts, un nain auprès des géants internationaux.

L'autre station maritime est la CKFX à Vancouver, qui diffuse 24 heures par jour sur 6080 kHz. Située sur le littoral pacifique en face du détroit Reine-Charlotte, elle ne dispose que d'une dizaine de watts. On n'en projette pas moins de les augmenter de 15-25 watts dans le proche avenir. Comme l'apprécient les radiophiles internationaux, même un petit faisceau d'une dizaine de watts d'ondes courtes a une portée remarquable, supérieure à celle des émissions AM ou FM. En effet, la station prétend recevoir de temps à autre, en provenance de l'Europe et de l'Australie, des rapports d'écoute envoyés par des auditeurs enchantés d'avoir capté de si loin des signaux tellement minuscules.

Encore deux stations à faibles émetteurs diffusant sur les bandes internationales, dont l'une à St. Jean, Terre Neuve, et la seconde à l'autre côté du continent, à Vancouver, se chargent du relais des programmes de Radio Canada. Etant donnée leur cohabitation à la même fréquence—6160 kHz—il faut écouter attentivement pour distinguer laquelle des deux on vient de capter.

Depuis sa création en 1936, Radio Canada s'illustre par l'excellence de ses programmes. Conçues pour plaire à l'auditoire canadien et maintenant, grâce aux services extérieurs de Radio Canada International, à des auditeurs étrangers également, ses émissions offrent toute une gamme de musique, d'informations, de théâtre, et d'autres divertissements. Malheureusement, les Canadiens eux-mêmes ont tendance à dédaigner quelque peu les programmes de Radio Canada, les considérant comme visant des minorités ou carrément des auditoires d'élite. Mais s'ils y perdent, les étrangers en retirent un gain certain.

Emissions puissantes a portee universelle

Radio Canada se sert aussi de la radio ondes courtes, cette fois-ci à l'aide d'émetteurs puissants de 100,000 watts, pour atteindre les populations du Nord de la plus grande des provinces canadiennes, le Québec. A leur intention, on prépare des informations et nouvelles, ainsi que de la musique, tout à fait particulières, que l'on diffuse non seulement en français, et en anglais, mais aussi en inuktitut, langue des Esquimaux ou Inuits, comme ils préfèrent s'appeler. Les canaux utilisés par les services de Radio Canada vers le Nord du Québec comportent les 6065, 6195, 9625, et 11720 kHz. Certains d'entre eux s'entendent à travers l'hémisphère Ouest et aussi, à un moindre degré, en Australie/Nouvelle Zélande et même en Europe.

Radio Canada possède un excellent service extérieur qui se capte facilement, Radio Canada International (RCI). Les programmes divers de RCI sont réalisés pour être diffusés sur les bandes internationales, par exemple l'émission "Le Matin des Canadiens", diffusée à 1200 UTC. Elle retransmet également les meilleures actualités de Radio Canada, entre autres, "Radiojournal" et "Présent" que l'on entend tous les jours de lundi à vendredi. Grâce à de nombreux auditeurs étrangers, ainsi qu'aux Canadiens voyageant ou séjournant dans les pays desservis, RCI est devenue l'une des stations internationales les plus populaires. Des sondages Gallup, en effet, la montrent en concurrence avec la BBC pour être la station internationale la plus écoutée par l'auditoire américain.

L'utilisation des bandes internationales que se fait le Canada n'est pas en soi, certes, bien remarquable. Bon nombre de pays, effectivement, tels le Brésil, l'Inde, le Pakistan, le Bangladesh, le Nigeria, diffusent aussi sur les trois plans, local, régional, et international. Mais les émissions canadiennes sont surtout intéressantes par le fait qu'elles se font surtout en français et en anglais. Les radiophiles à l'étranger peuvent donc en profiter non seulement pour connaître davantage un vaste pays très divers, mais aussi pour exercer leurs aptitudes linguistiques.

ALPHA DELTA TRANSI-TRAP® SURGE PROTECTOR

$20.00 Can Save Your Receiver.

Todays's receivers are some of the most sensitive ever made. It's nothing for these state-of-the-art beauties to pull in signals from halfway around the world.

Unfortunately, it's those same solid state components that make them sensitive to something you don't want—lightning and static. And it doesn't have to be a direct strike for lightning to ruin your radio. Even high winds can cause a big enough static build up on your antenna to send your receiver to the repair shop—or worse.

You've spent a lot of good, hard-earned money on your radio. Protect it with an Alpha Delta Transi-Trap Lightning Surge Protector. Transi-traps are in use around the world in thousands of government, commercial and military installations—and ones just like yours.

Ceramic gas tube Transi-Trap lightning surge protectors come in two models. Model LT for receivers from 0.1 to 30 MHz ($19.95) and model RT for receivers with coverage from 0.1 to 500 MHz ($29.95). At your Alpha Delta dealer or direct in the U.S. plus $2.00 shipping and handling. Exports quoted.

ALPHA DELTA
P.O. Box 571 • Centerville, Ohio 45459

SANGEAN

PORTABLE WORLD BAND RECEIVER

ATS-803

ITEM		ATS-803		
Wave Range	FM(MHz)	87.5 - 108		
	SW(MHz)	SW1	2.30-2.50	(120m)
		SW2	3.20-3.40	(90m)
		SW3	3.90-4.00	(75m)
		SW4	4.75-5.06	(60m)
		SW5	5.80-6.20	(49m)
		SW6	7.10-7.50	(41m)
		SW7	9.50-9.90	(31m)
		SW8	11.65-12.05	(25m)
		SW9	15.10-15.60	(19m)
		SW10	17.55-17.90	(16m)
		SW11	21.45-21.85	(13m)
		SW12	25.60-26.10	(11m)
	LW(KHz)	150-281		
	MW(KHz)	520-1620		
Type		FM- Stereo (via headphone)		
Output power		1200mW		
Speaker (inches)		4		
Batteries		6x1.5V "D" = radio power		
		2x1.5V "AA" = u-com back up		
External power		9V = /400mA		
Dimensions (W x H x D)		11.5 x 6.3 x 2.4 (inches)		
Weight (lbs)		3.75		
Accessories		External antenna jack		
		Shoulder Strap		
		Wave hand book		
Features		Sleep (10-90 minutes)		
		BFO		
		LCD		
		PLL-Synthesized		
		SW band spread		
		9 memories		

- Full AM/FM frequency range 150-29999kHz plus FM 87.5-108 MHz
- Fine tuning functions: Direct press button frequency input, Auto scanning, manual scanning, Memory recall and manual tuning knob Built-in 24 hour clock and alarm. Radio turns on automatically at preset time and frequency
- Fourteen memories - nine memory channels for your favorite station frequencies
  - Last setting of mode & waveband stored in 5 memories
- Direct press-button access to all 12 short wave broadcast bands
- Two power sources - battery to AC mains adaptor
- General coverage of all AM bands in LW, MW, SW (dedicated broadcast band coverage on all versions) plus of course the FM band for quality sound broadcasts in headphone stereo
- SLEEP function turns the radio on or off after an adjustable time of 10-90 minutes
- Separate BASS & TREBLE controls for maximum listening pleasure
- External antenna jack for better reception
- Adjustable RF GAIN control to prevent overloading when listening close to other strong stations or if there is interference
- BFO control (Beat Frequency Oscillator) enables reception of SSB (Single Side Band) and CW (Morse Code) transmission
- Illuminated display to facilitate night-time use
- Designed for both portable and desk top use
- Five dot LED signal strength indicator
- Wide and Narrow Switch filter
- Metal speaker grill
- AC adapter 110/220V, shoulder strap and short wave handbook included.
- Manufacturer warranty for 180 days.

SANGEAN AMERICA, INC.

9060 Telstar Avenue, #202, El Monte, CA 91731

Tel. (818) 288-1661 • FAX (818) 288-8231

For further information contact dealers near you.

HOW WE MADE *TUNE IN THE WORLD WITH HAM RADIO* EVEN BETTER!

The **BIG NEWS** about the League's beginner's package is that we have replaced the 60-minute cassette with two 90-minute cassettes in order to give almost three time the Morse Code instruction. Production of these tapes was a team effort. First, selected ARRL registered instructors were asked to review prototype tapes. ARRL staff incorporated the suggested improvements on the master tape. A special keying interface was designed in the ARRL lab.

The result is vastly superior to the code practice material contained in the previous editions. The popular Farnsworth method is used: the letters are sent at 18 WPM with appropriate spacing so that the actual speed is 5 WPM. The code is recorded on both stereo channels, but the voice-over is recorded only on one. Students with a stereo tape player can learn the code as the text is described on the tape, and then switch to the "code only" channel to test themselves as they go along.

The first tape is devoted to teaching the letters of the alphabet, prosigns, and numbers; the knowledge of each is required on the code portion of the Novice exam. Each new letter (or character) is sent several times, then words are sent containing previously learned letters and the new letter before going on the the next. The audio channel explains what is being sent. The first side of the second tape consists of 9 practice sessions; which are described on the tape and in greater detail in Chapter 3 of the *Tune In The World* text. The other side of the second tape consists of six sample Amateur Radio contacts for use as final practice by the student. Sample 10-question tests covering each of the QSOs are also presented in the text in order to give the student a feel for what the code portion of the exam will cover. The new tapes should made learning the code a fun experience.

We've improved the text too! Material has been added to the text to cover what the prospective Novice needs to know in order to pass the new 30-question Novice exam.

Tune In The World With Ham Radio is suitable for individual or classroom instruction. With the expanded text and improved code learning cassettes, this package should be *your* choice for Novice instruction material.

THE AMERICAN RADIO RELAY LEAGUE
225 MAIN ST.
NEWINGTON, CT 06111

HAM RADIO
JUST BECAME MORE FUN!

— More fun to learn
— More fun to operate

Tune In The World With Ham Radio has put the fun back into learning what Amateur Radio is all about. Enhanced Novice class privileges have brought the fun back into operating. Now beginners with their Novice licenses no longer have to spend all of their time on the air using only Morse Code. Novices can now use voice communications on 10-meters and use VHF and UHF repeaters. The new privileges include the use of digital communications so that home computers can be linked through packet radio networks. The FCC requires that Novices know something about their new privileges and that's where the expanded *Tune In The World With Ham Radio* text comes in. You'll find what you need to know explained in clear, concise bite-sized chunks of information. You'll find all 300 possible questions that may appear on the 30-question Novice exam with their distractors and answer key.

Besides improving the text, we've added almost three times the code practice material to the package in the form of two C-90 tape cassettes. One tape teaches the code, the other provides practice. They are recorded in stereo so you can switch off the voice portion for even more practice. These new tapes make learning the code a snap!

Tune In The World With Ham Radio is available at your dealer or from ARRL for $15.00 plus $3.50 for UPS shipping and handling.

THE AMERICAN RADIO RELAY LEAGUE, INC.
225 MAIN STREET
NEWINGTON, CT 06111

...and here is the "RDI" for the remaining 75% of the SW spectrum:

GUIDE TO UTILITY STATIONS

from Klingenfuss Publications, the world's leading publisher in the field of nonbroadcasting stations.

Tired of those crowded broadcasting bands? Of that vicious circle of megawatt power play, interference and jamming? You are invited to discover the exciting world of commercial radiocommunication on SW which is now easily accessible by means of the new generation of general coverage receivers and decoders for facsimile and teleprinter transmissions.

The Klingenfuss GUIDE TO UTILITY STATIONS, including the famous GUIDE TO RADIOTELETYPE STATIONS (13th edition!) published since the early seventies, is the basic international reference book used by both professional and hobbyist radio monitors. It covers all types of utility stations and includes

- numerical frequency list with more than 14000 frequencies from 9 kHz to 30 MHz, each with call sign, name of station and country, modulation type and details;
- alphabetical call sign list with 2800 call signs, each with name of station and country, and corresponding frequency;
- schedules of 75 teleprinter press services on 450 frequencies in both alphabetical and chronological order for easy access;
- schedules of 70 meteorological facsimile services on 250 frequencies;
- schedule of NAVTEX transmissions on 518.0 kHz;
- 950 abbreviations, plus 180 telex services codes;
- SINPO and SIMPFEMO code, complete Q-code and Z-code;
- designation of emissions, with associated examples;
- classes of stations and type of service;
- comprehensive list of ITU terms and definitions;
- ITU frequency allocations from 9 kHz to 150 MHz;
- aeronautical and maritime mobile service regulations and frequency allocations;
- country list with 960 station addresses in 200 countries.

Pulished annually in December, the 1987 edition of the Utility Guide (ISBN 3-924509-87-5) has 473 pages plus 3 world maps. The price of 60 DM (or equivalent in your currency) includes airmail postage to anywhere in the world. Payment can be by cheque (drawn on a German bank), cash, International Money Order, or postgiro (account Stuttgart 2093 75-709). Further publications available are Air and Meteo Code Manual, Guide to Facsimile Stations, Radioteletype Code Manual, etc. All manuals are softbound in the handy 17x24 cm format, and of course written in English. For further information ask for our catalogue (available also in French, German and Spanish), list of recommendations from all over the world, and dealer discount rates. Orders received after 15 NOV 1987 will receive the 1988 edition. Please mail your order to

Klingenfuss Publications, Hageloch, D-7400 Tuebingen, West Germany, Tel. 07071/62830

Radio Canada International

Broadcasting in English and French to the United States, Latin America, the Caribbean, Africa, the Middle East and Europe.

Des émissions en français et en anglais destinées aux États-Unis, à l'Amérique latine, aux Antilles, à l'Afrique, au Moyen-Orient, et à l'Europe.

For a free program schedule, write to:

Programme-horaire gratuit sur demande. S'adresser à:

Radio Canada International (DB)
PO Box/CP 6000
Montréal, Canada
H3C 3A8

ENHANCE PERFORMANCE

IMPROVE RECEPTION WITH THE NEW HIGH FIDELITY SE-3 PHASE-LOCKED AM PRODUCT DETECTOR. Eliminates selective-fade distortion. Provides a new dimension in shortwave and standard broadcast reception. Starts at $299.00. For 455-, 100-, and 50-kHz IFs.

SPECIALTY FILTERS AND MODIFICATIONS FOR DRAKE AND JRC NRD-515.
R-7 (A): 9 bandwidths, such as 10 kHz CD-10K/8, 3 kHz CD-3K/8, CD-1.6K/8, CD-1.0K/8, etc. $80.00.
R- 4C: 14 bandwidths, up to 24 poles!
SPR- 4/SW- 4A: 5 kHz first-IF 8-pole filter, SD-5K/8. $100.00.
NRD-515: 3- to 16-kHz filters and modifications.

Other bandwidth phone and CW filters and accessories available. Filter shipping $3 per order; $6 overseas air.
Europeans: INGOIMPEX, Postfach 24 49, D-8070, Ingolstadt, W. Germany.

Sherwood Engineering Inc.

1268 South Ogden St.
Denver, Colo. 80210
(303) 722-2257

FOR THE BEST IN

(((SWL - DX - HAM RADIO)))

>>> CONTACT <<<

→→→ HOBBYTRONIQUE ←←←

PROFESSIONAL SALES & SERVICE

WE CARRY ALL MAJOR LINES ... AND MORE

```
    ICOM      YAESU     KENWOOD
    JRC       SONY      SANGEAN
         GROVE     TEN-TEC
```

 Hobbytronique **I**nc.

8100 Trans-Canada Hwy. Suite H

Ville St. Laurent, Quebec   H4S 1M5

(514) 336-2423

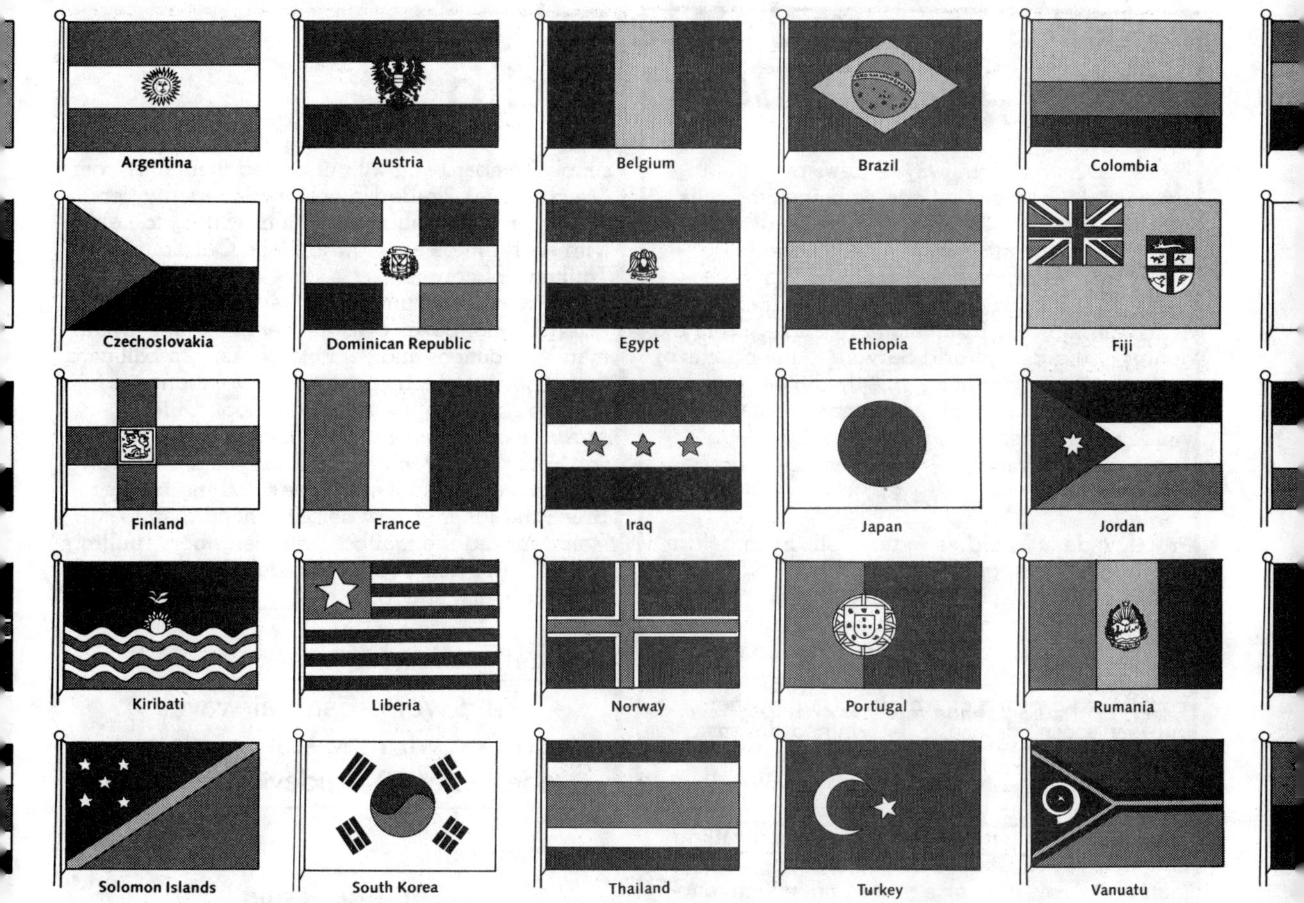

With a Sony World Band Radio, each country clearly has its own voice.

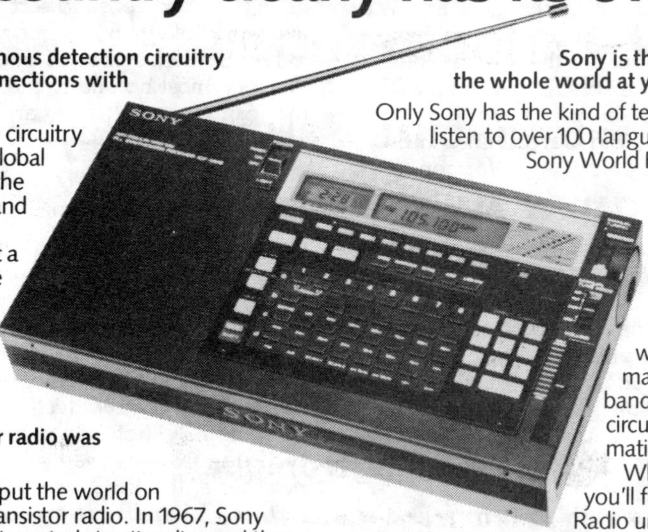

Sony presents synchronous detection circuitry so you can enjoy clear connections with less interference.

Synchronous detection circuitry is a tiny mechanism with global proportions. It locks onto the frequency you've chosen and travels with it, letting you clearly hear one country at a time, with less interference all of the time. Which means if you happen to be listening to Ping-Pong from Peking, São Paulo soccer shouldn't break in.

Inventing the transistor radio was just the beginning.

Thirty years ago, Sony® put the world on its ear with the very first transistor radio. In 1967, Sony unveiled the world's first integrated circuit radio, and then made history again with the shortwave transistor. Ever since then, Sony has been fine-tuning world band radios with world-famous technological advancements.

Sony is the world leader that can put the whole world at your fingertips.

Only Sony has the kind of technology that lets you clearly listen to over 100 languages in 160 countries. Sony World Band Radio® units around the

world are clearly receiving air, marine, longwave and shortwave bands, with synchronous detection circuitry, quartz tuning and automatic search and scan.

When it comes to globe-trotting, you'll find any of our five World Band Radio units handier than a passport. And you'll also find that you understand what's being said around the world more clearly than ever before.

World Band Radio.® SONY. THE ONE AND ONLY.®

© 1987 Sony Corporation of America. Sony, World Band Radio and The One and Only are trademarks of Sony.
No endorsement by any country where flag is displayed is implied.

News... and "News"... Over World Band Radio

There's so much in the way of news over the international airwaves that one commentator calls world band radios "news boxes." With over 160 countries on the air, you can be plugged into nearly any event, great or small, on Planet Earth . . . plus some attempts at news that seem as though they are from another galaxy. From the genius of the BBC World Service to the depths of Radio Pyongyang's "news-ah," the airwaves crackle with the stuff of history one moment, vaudeville the next.

The Hard Sell

Propaganda is as old as radio itself. Even before Tokyo Rose came on the air during World War II, the airwaves have been littered with outrageous claims and incentives for surrender. From Tirana to Tehran, old Rosie lives on.

Some stations have tricks to make you listen. You'd need a prune of a brain to fall for them, but that doesn't deter the determined doyens of disinformation.

For example, Radio Berlin International encourages listeners to join its "club." All that's needed is to write them regular letters detailing how well you're hearing their programs. The object, of course, is to give you some incentive to tune them in every night. Do that for a while and you'll be mumbling about "revanchists"—the station's favorite perjorative—in your sleep.

Broadcasts from other countries along the lines of Berlin's political persuasion might try to stir your emotions with such cliffhangers as, "The Marxist-Leninist Truth Will Triumph Over Revisionist Demagogy," "People of the World Unite and Defeat the US Aggressors and All Their Running Dogs," "Reagan: Destroyer of Worlds," "Democracy: When Will the Curse End?" and Radio Moscow's tireless potboiler, "Warmongers Monthly"—which is, using appropriate logic, weekly.

Libya is in a league of its own when it comes to portrayal of Western countries. Recent shows have told incredulous listeners how the streets of New York are littered with the bodies of people who have jumped from office windows. These people, claim the announcers, have broken under the "intense psychological stress of living in a hellish, computerized capitalistic society." From stations of this ilk, even a weather forecast should be suspect.

Some broadcasters send their listeners goodies if they write. These can vary from a form letter inviting you to participate in a quiz on your knowledge of the statues on Martyr's Square, to offering a free trip to Havana. Some stations will simply bombard you with unwanted literature, from Czechoslovak Trade Union Newsletters to offers of comfort from religious stations willing to help with such things as "How to Make Out Your Will." To them, of course.

Other items rumored to have arrived in listeners' mailboxes over the years include Chairman Mao buttons and coloring books, baseball-card type photos of Latin American revolutionaries and, so some claim, an Iranian book called, *Plastic Explosives and You*. These unexpected "gifts" can follow you around for years and even through several moves. There's nothing better to break the ice in a new neighborhood than to stand around the mailbox with neighbors, pulling out letters "From Your Comradely Friends in Revolutionary Ethiopia."

The world band airwaves crackle with the stuff of history one moment, vaudeville the next

---

The Good Stuff

Only someone suffering from constipation of the brain can possibly sit through all six parts of the series, "Copper Wire Production in Bulgaria." But for those who do enjoy a variety of views and have the good sense to sort through them, there is a world of slants, perspectives and approaches to be sampled. For serious junkies of world events, there's almost no end to the really good newscasts that can be heard. Here are some plums.

BBC World Service

The BBC has long led the field in international news. It is, quite simply, the standard against which all other electronic journalism is measured. The BBC has no axes to grind, and so produces some of the most bias-free reports in the world. It also has a strong sense of perspective, sorting out what's important in the longer run from the fluff that tends to distract lesser news organizations.

The BBC not only is quality, it's also quantity. Aside from its twenty-four daily international newscasts, the BBC takes its responsibility even further. A number of programs—most aired daily and following the regular hourly news—offer in-depth and extensive coverage of the day's events. Arguably the best is "Twenty-Four Hours," which provides on-site analyses of the main news of the day. It is heard daily at 0509, 0709, 1309, and 2009 World Time (UTC).

"Newsdesk" is another, a three-times-daily program—0400, 0600, and 1800 World Time—where correspondents report on a variety of topics. Yet another is "The World Today," heard Mondays through Fridays at 1645 and 2209, plus Tuesdays to Saturdays at 0315, 0545 and 0915 World Time. On "The World Today," which is something a cross between a current events program and a documentary, journalists examine a single topic of interest on the international scene.

More personal commentary than straight news is Alistair Cooke's excellent weekly program, "Letter from America." This pleasurable quarter hour is aired Saturdays at 1015 and Sundays at 0545, 1645 and 2315 World Time.

There are several other similar programs as well, all worthy. With the BBC, the only problem is in finding the time to hear them all.

WCSN

WCSN is the call sign of the World Service of the Christian Science Monitor. The Christian Science Monitor is an internationally respected, Pulitzer Prize-winning newspaper known for its thorough and unbiased coverage of international events. Despite the fact that it's sponsored by a religious organization—The Church of Christ, Scientist—religion is separated from journalism, both in the newspaper and on the air. WCSN is a very recent, and very welcome, addition to the growing lineup of world radio broadcasters. Unfortunately, because it is so new, its schedule is still in a state of flux.

World Service broadcasts are divided into two-hour segments which, Monday through Thursdays, are repeated between 1600 and 1200 World Time, plus Fridays and Saturdays between 1600 and 2400.

There's news on the hour and half hour, followed by in-depth review and analysis of the main stories. Other programs—such as "Roundtable," an informal discussion with Monitor writers and editors; "Kaleidescope," which features conversations with newsmakers; and a selection of columns from the paper itself—round out the broadcasts. For most of the rest of the day and all day Sunday, WCSN devotes itself to programs about the Christian Science religion in time slots separate from the newscasts.

Radio France International

Radio France International has dramatically increased its commitment to world band radio. Its English-language broadcasts now include the one-hour "Paris Calling Africa"—heard worldwide, its title notwithstanding, at 1600 World Time—and a number of short newscasts at, for example, 0200, 0330 and 0415.

The BBC World Service airs authoritative news every hour of the day

Radio France International's English-language newscasts provide a slightly more European outlook at what is important in world events. And, thanks to powerful overseas relay stations, it's well heard, too.

Still, the staple of RFI's broadcasts are those in French. In the past, most programs came from the popular home service, "France-Inter." Now, they're prepared exclusively for a foreign audience. On one hand, this has eliminated the silliness sometimes found on "France-Inter." But, at the same time, much of the local flavor that makes France universally exciting has been lost. At times, it's hard to tell whether you're listening to Radio France or the French service of Radio Sweden.

Armed Forces Radio and Television Service

For the American traveler who wants to stay in touch with home, there is no better source of news than the Armed Forces Radio and Television Service. AFRTS is unusual in that it most closely resembles a domestic US all-news station. The twist is that AFRTS carries all the private and public US radio networks. Twenty-four hours a day, you'll hear newscast after newscast from ABC, CBS, NBC, Mutual, AP and National Public Radio. Interspersed are network features and commentaries, plus live baseball, American football and other sporting events. Even some regional US news and weathercasts are aired.

AFRTS is, in a sense, the ultimate in non-

Make your station really perform

Need to hear the weak ones? No room for an outside long wire? Confused about your frequency? Choose the accessories for the

Use this 54 inch active antenna to receive strong signals from all over-the-world MFJ-1024 . . . $129.95

Receive strong clear signals from all-over-the world with this 54 inch active antenna that rivals long wires hundreds of feet long. The authoritative *World Radio TV Handbook* rates the MFJ-1024 as 'a first-rate easy-to-operate active antenna . . . quiet . . . excellent dynamic range . . . good gain . . . very low noise factor . . . broad frequency coverage . . . excellent choice'.

You'll receive all frequencies 50 KHz to 30 MHz from VLF thru lower VHF - including long wave, medium wave, broadcast and shortwave bands. Mounts anywhere away from electrical noise for maximum signal and minimum noise pickup --mount on houses, buildings, balconies, mobile homes, apartments, on board ships -- anywhere space is a premium.

High dynamic range eliminates intermodulation so you never hear 'phantom' signals.

A 20 dB attenuator and a gain control prevents overloading your receiver. You can select between 2 receivers and an auxiliary antenna. Has weather-proofed electronics. Use 12 VDC or 110 VAC with MFJ-1312, $9.95.

The MFJ-1024 comes complete with a 50 foot coax cable and connector - ready to use!

ANTENNA MATCHER

MFJ-959B . . . $89.95

Did you know you may be missing rare DX because of lost signal power due to impedance mismatch between your receiver and antenna?

The MFJ-959B Antenna Tuner provides proper impedance matching so you transfer maximum power from your antenna to your receiver from 1.8 to 30 MHz. You'll be pleasantly surprised by significant increases in signal strength.

20 dB preamp with gain control boosts weak stations and 20 dB attenuator prevents overload. Select from 2 antennas and 2 receivers. 9x2x6 inches. 9-18 VDC or 110 VAC with MFJ-1312, $9.95.

MOBILE SHORTWAVE CONVERTER

MFJ-308 . . . $99.95
MFJ-304 . . . $79.95

Enjoy new excitement and variety as you listen to the world while driving with these low cost mobile shortwave converters. Choose the MFJ-304 'World Explorer I' that covers 19, 25, 31 and 49 meter bands or the MFJ-308 'World Explorer II' that adds 13, 16, 41 and 60 meter bands for total shortwave broadcast coverage.

World wide coverage brings in Europe, Africa, Asia, Middle East, South Pacific plus North and South America! Just push a button to select your band and tune in stations with your car radio.

PRESELECTING SW/MW/LW TUNER

MFJ-956 . . . 39.95

This MFJ-956 short, medium, long wave preselector/tuner lets you boost your favorite station while rejecting images, intermod and other phantom signals on your shortwave receiver! It greatly improves reception of 150 KHz thru 30 MHz signals. Has tuner bypass and ground receiver positions. 2x3x4 in.

COMPACT SPEAKER

MFJ-280 . . . $18.95

Enjoy crisp, clear audio with this tiny, low cost communications speaker. This rugged unit is conveniently mounted on a tilt bracket attached to a magnetic base. It works with all 8 and 4 ohm impedances and has a 3.5 mm phone plug on the end of a long cord. Handles up to 3 watts of audio. Its dark gray military color matches nearly all rigs. 2 1/2x2x3 inches.

DX'ers FILTER/SPEAKER/AMPLIFIER

MFJ-732 . . . $69.95

Uncover voice signals buried in heterodynes, noise and interference. Signals you thought were unreadable suddenly make sense when you use this MFJ-732 all-in-one audio filter, 2 watt amplifier and speaker combination.

Select lowpass and highpass filter cutoffs at 300, 500, 1500, 2200 and 3000 Hz for best readability.

Use 9-18 VDC or 110 VAC with MFJ-1312, $9.95.

with MFJ shortwave accessories

Troubled by 'phantom' signals? Need to know UTC time and date? kind of performance you need from the many models MFJ offers

24 HOUR CLOCKS

MFJ-108 . . . $19.95
MFJ-107 . . . $9.95

Know the exact 24 hour UTC time and your local 12 hour time at a single glance so you'll tune in your favorite stations on time and keep accurate logs for DXing. And never, never be confused about the correct UTC date because you can read it with a touch of a button. Huge 5/8 inch LCD digits makes glare-free reading easy. MFJ-108, dual 24/12 hour clock, 4 1/2x1x2 in. MFJ-107, single 24 hour clock, 2 1/4x1x2 in. Battery included.

ALL BAND RF PRESELECTOR

MFJ-1045B . . . $69.95

Is your receiver "hearing" all it could? Adding an MFJ-1045B RF preselector can make "lost" signals readable while reducing troublesome images and out-of-band signals. It adds 20 dB of low noise gain with a strong, sharp tuning front end and covers all HF amateur and shortwave bands through lower VHF from 1.8 to 54 MHz.

A gain control prevents overload. 5x2x6 inches. Uses 9-18 VDC or 110 VAC with MFJ-1312, $9.95.

ALL MODE TUNABLE FILTER

MFJ-752C . . . $99.95

Maybe the only filter you'll ever need. Why? Because it lets you zero in am/ssb/rtty/cw/amtor/packet signals and notch out interference at the same time.

The primary filter lets you peak, notch, low or high pass filter out interference.

The auxilary filter gives deep notches and sharp peaks.

Both tune 300 to 3000 Hz with variable bandwidth from 40 Hz to virtually flat. Select 2 receivers. Drive speaker. Use 9-18 VDC or 110 VAC with MFJ-1312, $9.95. 10x2x6 in.

RTTY/ASCII/CW COMPUTER INTERFACE

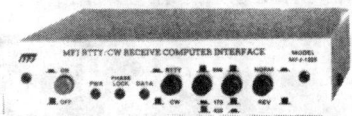

MFJ-1225 . . . $69.95

Open up a whole new and exciting world of shortwave listening with an MFJ-1225 RTTY/ASCII/CW computer interface. Listen to news before it appears on general radio and TV, weather, ship-to-shore communications, hams rag chewing, all kinds of commerical traffic and even the military. You'll be fascinated as traffic scrolls across your home computer screen (some messages may be encrypted).

All you need is a stable shortwave receiver, personal computer and the MFJ-1225 computer interface. Software on tape and cables are supplied for the Commodore 128, 64 and VIC-20 --everything you need. Most other home computers -- with an RS-232 port, suitable software and cable -- can be used, such as IBM PC and clones, Apple, TRS-80C, Tandy, Atari, TI-99. Uses 12-15 VDC or 110 VAC with MFJ-1312, $9.95.

Rival outside long wires with this INDOOR active antenna

Now you'll rival or exceed the reception of outside long wires with the new and improved MFJ-1020A Indoor Tuned Active Antenna with higher gain. Here's what the *World Radio TV Handbook* says about the MFJ-1020: 'Fine value...fair price...best offering to date...**performs very well indeed.**'

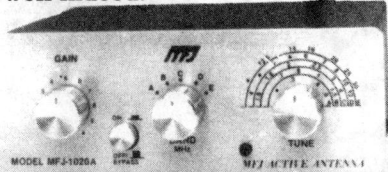

MFJ-1020A . . . $79.95

You get continuous coverage of low, medium and short wave bands from 300 KHz to 30 MHz so you can listen to all your favorite stations. It even functions as a preselector with an external antenna.

Its unique tuned circuitry minimizes intermodulation, improves selectivity and reduces noise so you're less bothered by images, and other out-of-band signals.

It has an adjustable telescoping antenna for maximum signal and minimum noise. There's a full set of controls for tuning, band selection, gain, ON/OFF/Bypass and an LED power 'ON' indicator. It measures just 5x2x6 inches. Use a 9 volt battery, 9-18 VDC or 110 VAC with MFJ-1312, $9.95.

Call toll-free 800-647-1800 and charge to your VISA or Master Card. Order any product from MFJ and try it -- no obligation! If not satisfied for any reason, return it within 30 days for a no hassle refund, less shipping. One-year unconditional guarantee. Add $5 each for shipping and handling. Free catalog. For tech. infor. or outside USA or in Miss. call 601-323-5869 or telex 53-4590 MFJ STKV.

800-647-1800

 MFJ Enterprises, Inc.
921 Louisville Road
Starkville, MS 39759

MFJ . . . making quality affordable

propaganda radio. Instead of airing broadcasts spoon-fed abroad by government functionaries, AFRTS offers an insider's look into what's actually going on in the US. Indeed, it's not only Americans that listen in. In some parts of the world, the AFRTS has been known to pull a larger English-language audience than the US government's official Voice of America.

Because the AFRTS exists solely to reach American armed forces personnel, it's difficult —even impossible—to hear clearly in many parts of the world. In Europe, for example, it's often audible only via a small single-sideband transmitter used, usually on 9334 kHz, to feed programs to the Azores. Lately, for whatever reason, this transmitter—which is fairly well-heard throughout southern Europe—has been on the air only every now and then.

Voice of America

Mention the VOA's news to any group of regular world band listeners or journalists, and you may well start an argument that's best walked away from. The Voice of America is, like the BBC, a government-sponsored organization. But, unlike the BBC, it's clearly a mouthpiece of the government. Another drawback is that, unlike AFRTS, it airs nearly no news about events in the US unless they concern politics.

There is a real concern as to whether the Voice, structured as it currently is, can present truly unbiased news. Still, the talent and resources at the VOA are considerable. Cynics notwithstanding, the Voice does air a number of international news and analysis programs of merit.

Regional Newscasts

Listeners with an interest in a particular part of the world can tune in any of a number of stations with newscasts that include regionally—as opposed to internationally or nationally— oriented newscasts. Sometimes these stations are in the region itself, sometimes they're not. But most are quite good and nearly all are vastly superior to what you'll find in your local media.

For example, for Nordic news try Radio Finland's excellent "Northern Report" or the newscasts of Radio Sweden International. Latin America is not yet well-covered by English-language broadcasts. But if your heart is in Asia, you can try Radio Australia, Japan's NHK or even the BBC's Asian Service. For Africa, try Radio France International or the BBC's African Service, both of which provide excellent coverage.

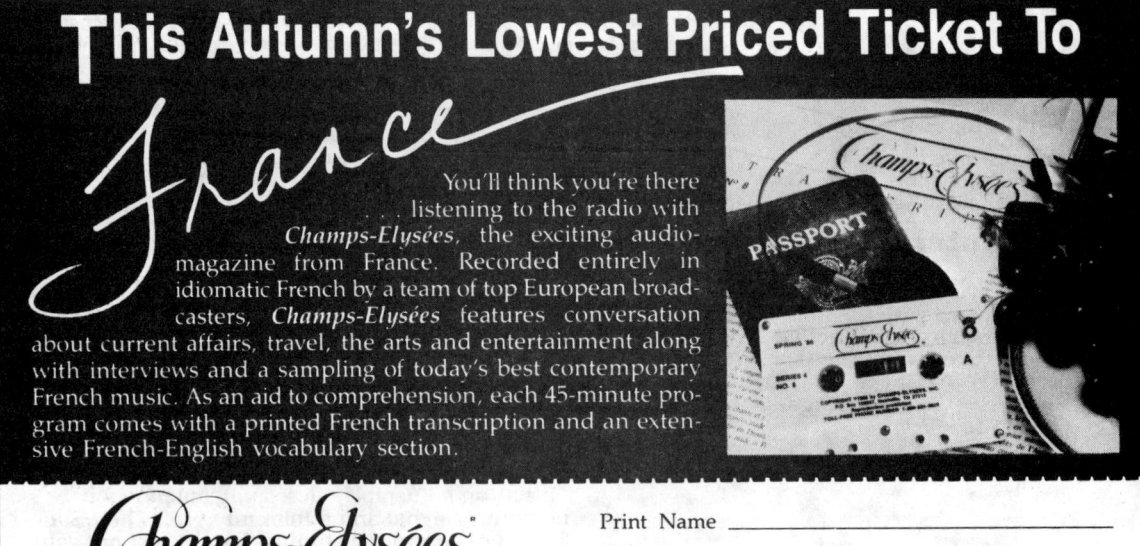

Vladimir Posner: Not Your Average Russian

World band radio listeners have known the name Vladimir Posner for almost two decades now. In the years since 1970 when he first joined the staff of Radio Moscow, Posner quickly developed into one its most visible "personalities" on a station that, until recent years, was not known for personality at all.

What makes this man different from his colleagues is that he understands not only American English but the people who speak it. And in the end, that's what makes him so effective in the Soviet Union's campaign to win hearts and minds in the West. On a station where other announcers tend to sound as if they are shouting from textbooks positioned twenty yards from the microphone, Posner comes off like the fellow next door. Posner is unquestionably likeable, even charming.

Says Alex Beam, former Moscow correspondent for *Business Week*, "He takes advantage of our stock negative image of the USSR as an awful, benighted country where if you ask to go to the bathroom, you're sent to Siberia. [And it's because he presents] such a radically different image . . . that Posner can come to our country and score very big in our public opinion marketplace."

Posner, not surprisingly, agrees. "I feel that most Americans have been bamboozled, to use an old word, to the point where when anything half normal comes out of the Soviet Union, they are surprised . . . I mean, they have made the Soviets such bogeymen, as it where, that when you get someone who is normal, they just go 'uhhh.'"

"He shows himself to be a flesh and blood person," says Beam. "Prick me, and I bleed. It's perhaps a very human image of the Soviet Union that's very desireable . . ."

So what is Posner, a peace missionary or a treacherous propagandist?

Arnold Zenker, one of America's most successful media coaches, compares watching Posner on TV to "being a violinist and watching Isaac Stern." But is Posner a manipulator? Says Zenker, "I think he is. I think he understands very well what he's trying to get across. Is he any more calculating than one of our politicians; say, Ronald Reagan? Probably not. Is he any more calculating than some of the business people I work for? Probably, but only because they don't understand the medium as well."

What is unquestioned is that Posner has become the voice of the Soviet Union in the US.

This article, based on the program "Not Your Average Russian," was prepared in cooperation with WGBH-Boston, Steve Atlas, Producer.

And he's in a unique position for the job because both countries figure prominently in his past.

Posner was born to a French Catholic mother and a Jewish Russian father. His family lived in Paris until 1940 when they fled Nazi-occupied France for the US.

"From a rather early age, I was aware there was another country called the Soviet Union. My father talked to me quite a bit about what socialism was, and why it was a more just society compared to other societies. And gradually that took hold."

When he was in grade school, differences between the US and Russia were put aside as both joined forces to fight the Nazis. It was during Posner's schoolyears that his father really started to emphasize the family's national origins.

"When the Germans invaded the Soviet Union, my father bought a map and he would draw in the German advances in black pencil, And every evening, he would say, "They will never win. They will never take Moscow. They will never take Leningrad." That was the beginning of my political education."

The conclusion of World War II marked the end of hostilities in Europe and the beginning of hostilities toward the thirteen-year-old Posner in the US. It all began after Winston Churchill's Fulton, Missouri, speech, about the Iron Curtain in 1946. "I didn't agree. I defended the Soviet Union. And that led to some problems. I remember when six or seven kids ganged up on me for being a commie and I was bloodied up."

His filmaker father also began experiencing problems at this time. Then he received an offer to work in Soviet-controlled East Germany. Posner was pulled out of classes at New York's Stuyvesant High School and the family immigrated for the second time.

"My father is Russian Jew and you see what the Germans have done to Russia. My mother is French and the French aren't the greatest fans of Germany." In 1952, Posner's father was invited to come to Moscow. But the Soviet Union proved no more hospitable to the Posners than did East Germany.

"We arrived in December of '52 when Stalin was still alive. People are afraid of foreigners. So it turns out we have no apartment, we live in a hotel, and my father can't get a job. Frankly, if Stalin hand't died on March 5th, I think things would have gone very poorly for us. I think my father would have joined those who were sent to some camp in Siberia, And my mother too. And perhaps even myself."

After Stalin's death, however, things changed quickly. "My father got a job. And I enrolled in Moscow U." Five years later, Posner graduated

CANADA'S BEST STOCKED
AMATEUR RADIO AND S.W.L. STORE...
Call or write for best Canadian pricing, immediate delivery, and reliable service...

SANGEAN	ATS-803 • SG-789
ICOM	R71-A • R7000
JRC	NRD-525
GRUNDIG	SATELLIT 650/400
YAESU	FRG-8800 • FRG-9600
SONY	ICF-PRO 80 • ICF-2010 ICF-2003 • ICF-7700

WE STOCK ICOM YAESU JRC SONY SANGEAN INFOTECH REGENCY UNIDEN-BEARCAT TENTEC Hy-Gain CUSHCRAFT MFJ NYE-VIKING ALINCO ASTRON KDK KENPRO BUTTERNUT AMP-SUPPLY AEA BENCHER B&W Daiwa HEIL HANSEN HUSTLER KANTRONICS LARSEN MIRAGE ARRL PALOMAR UNADILLA VIBROPLEX CALLBOOK MICROLOG AVANTI EAVESDROPPER and more to come...
We also take trades. Call us for a list of our current used/reconditioned equipment...

ATLANTIC HAM RADIO LTD.

Tuesdays-Fridays: 10 a.m.-6 p.m.
Saturdays: 10 a.m.-2 p.m.
After 7 p.m. call
(416) 222-2506 for orders.

378 Wilson Ave.
DOWNSVIEW, ONT.
CANADA M3H 1S9
(416) 636-3636

and began working his way up in the Soviet literary establishment—a path that would eventually make him a celebrity both in the Soviet Union and the United States.

Since then, Posner has worked in print, on radio and as a TV commentator. But the medium he seems most committed to is the "Spacebridge," a live TV broadcast between the US and the USSR in which audiences in both countries can talk to one another by satellite.

"It's the image of the Russian. And you've got this squat, fat, bushy eyebrow, large jaw, strong accent. It's part of the picture that's been drawn over the years that reflects the bias and the way Russians are supposed to be."

US television talk show host Phil Donohue figures prominently in the Posner strategy. His program gives gives Posner a direct link to a vast block of middle America. On May 15, 1986, Posner returned to the US after 38 years behind the Iron Curtain. He is now an international media star, set to appear before some eight million American viewers on *Donohue*.

Posner is a success. He is smooth, charming and funny, as this series of exchanges shows.
Donohue: "How do we [America] look after 38 years?"
Posner: "When I walked out on the streets, it was a very emotional thing for me. I've dreamed about it. Because, regardless of being a Soviet

Radio Moscow's Vladimir Posner has become familiar to millions of Americans through his appearances with US talk show host Phil Donohue

LINIPLEX LEADS THE WAY

The Demodulator
The first successful synchronous phase locked demodulator for short wave broadcast reception has been proven in many rebroadcasting applications around the world. Undisturbed by carrier fading owing to its innovative memory, audio distortion is a thing of the past. Automatic synchronisation at the onset of a transmission is a feature of the new Liniplex F2 receiver.

LSB, DSB or USB
With its unique tracking filter consistent SSB performance is provided down to within 50 Hz of the carrier with the result that audio reproduction is excellent, especially music, and in whatever mode.

Crystal Control or Synthesiser
The new F2 receiver based on the F1 offers the best of both worlds. For simple receiving applications such as monitoring the BBC, eight channels suffice and plug-in crystals provide the ultimate in receiver performance at little cost or complexity. (Stocks of crystals for BBC World Service reception are held by Phase Track for this purpose.) The receiver can also be operated from an external synthesiser, but the Liniplex demodulator places severe demands on synthesiser performance since phase jitter becomes an audible noise superimposed on every received signal, especially in SSB modes. What is acceptable for communications receivers now becomes unacceptable for high quality broadcast reception. Low noise synthesisers for LW/MW and short wave will soon be available meeting these new standards.

Signal Strength Measurement
For realistic signal assessment an analogue meter is provided. Temperature compensated and accurate to 2 dB over a linear 80 dB range, provision is also made for an external recorder.

Other Systems
Quadruple frequency diversity systems for unattended operation are also available for high quality domestic HF broadcast links to remote towns and communities.

Available from:
Radio West, 850 Anns Way Drive, Vista, CA 92083, U.S.A. Tel: 619 726 3910

PHASE TRACK LTD., 16 Britten Road, Reading, RG2 0AU, England
Tel: 0734 752666 Tlx: 848888 Tadlex G.

citizen, I love New York."
Donohue: "Are you a Party member? You are, I assume."
Posner: "You assume right."
Donohue [jokingly]: "You're a real 'commie.' "
Posner [pausing, smiling]: "Uh, huh."
Audience: Laughter.
Donohue: "Don't you think this [country] is a great democratic idea. You've got to be impressed."
Posner: "When I walk up Park Avenue and I see people living in the streets, I say to myself, that is not a fair society. It should not be in a society so incredibly rich as the United States. We are a poor society. No one lives in the streets. No one."

It is unquestioned that Posner has become the voice of the Soviet Union in the U.S.

Says Donohue, "We flatter Mr. Posner with our anxiety. Let's listen to the man. If we doubt, we doubt. Let's question, let's challenge, let's even get angry. It's all of those things allowed in our Constitution. But I must not tell a lie. I like him."

Not everyone does.

Reed Irvine, head of the conservative press watchdog organization, "Accuracy in Media," states flatly that "Mr. Posner doesn't express views. For the most part, he tells disinformation or lies."

Marshall Goldman, who runs Harvard's Russian Research Center, has studied the Soviet Union and Vladimir Posner. "When we watch Posner on television, what we have to keep in mind is that he has a line he is trying to peddle. You just have to decide, 'Do I want to buy a car from that man? Am I prepared to take the warranty that he is offering?'"

Says Beam, "He's not a con man any more than Dan Rather or Phil Donohue or Tom Brokaw is a con man. This guy is a spokesperson. He articulates views that are formulated partly by himself and partly by others."

Marshall Goldman ends on a more cautionary note. "When there's an issue that confronts the the two countries, we immediately have someone we can turn to. He misleads us in many respects. But at least we have the chance to enliven and enlarge the debate. And I think that he does this. But I don't think he has changed public policy. I think we're still too sophisticated for that."

In the fine tradition of Quality German Engineering comes the

GRUNDIG SERIES OF WORLD BAND RADIOS.

Satellit 400. Satellit 650.

Two world band radios for people who demand the very best.

The Grundig Satellit 400. A handsome, hightech portable that combines strong performance with superb audio. Truly a world class radio, with AM, FM, LW and full shortwave coverage that can bring the world effortlessly to your home or office. Yours for under $400.

The Grundig Satellit 650. The best engineered and most sophisticated portable ever. Manual and computer controlled tuning. Keypad station selection. And a 30 watt audio amplifier with bass and treble and two built-in speakers. The ultimate listening experience for those who demand the very best. Available for under $1,000.

UNIVERSAL SHORTWAVE RADIO
1280 Aida Drive
Reynoldsburg, Ohio 43068
TOLL FREE 1-800-431-3939
In Ohio 1-614-866-4267

Ghana: Radio Gold Worth Digging For

by Maud Blankson-Mills

Broadcasting in Ghana first saw the light of day under the British Colonial Government, when the country was known as the Gold Coast and was ruled by the then-Governor, Sir Arnold Hodson. The inaugural radio station, ZOY, came on the air in the capital city of Accra in 1935.

Broadcasting Inaugurated with Fleapowered Transmitter

ZOY began its life with a minuscule 10-watt transmitter operating out of a bungalow. As would be expected in a British colony, the station concentrated on relays of the BBC, plus news and music in a few local languages. Only about three hundred nearby "subscribers" were able to hear the broadcasts.

In order to expand the audience, construction of relay stations began the next year in the cities of Cape Coast, Sekondi, Kumasi, and Koforidua. By the time World War II got underway in 1939, the Gold Coast had a total of sixteen stations for transmitting war news in the English, Twi, Fante, Ga and Ewe languages, with Hausa added later on. Programming emphasized the progress of the war and how Ghanaian servicemen were faring at the front.

A pioneering version of world band radio entered the picture that same year. The inauguration took place at Broadcasting House in Accra with the installation of a 1.3 kW transmitter operating on 4915 kHz. The station officially signed on the air in 1940, broadcasting to the Gold Coast and other neighboring countries.

Major Pan-African Broadcasting Follows Independence

By the time Ghana gained its independence in 1957, the country had already learned about the important benefits of world band radio. A new Broadcasting House—still in use—was built, complete with new studios, offices and a new 20 kW transmitter audible throughout West Africa and even beyond. The station continued to rely on the BBC for much of its programming, but now a different emphasis was added: the projecting to Ghanaians and fellow West Africans of an "African Personality," as well as the fanning of national-

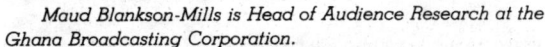

Maud Blankson-Mills is Head of Audience Research at the Ghana Broadcasting Corporation.

Author Maud Blankson-Mills peruses listener mail

ism and President Kwame Nkrumah's brand of Pan-Africanism.

As time went on, programming continued to become more African in nature. There were African news bulletins, programs projecting Ghanaian and African culture, as well as special shows for freedom fighters in various parts of the world.

At the same time, both the Schools' Broadcasts Department and the Rural Broadcasts Department took to the airwaves in an effort to assist in the government's policy of accelerated education. Schools' programs supplemented classroom teaching for young people, while the Rural network concentrated on adult topics, such as agriculture, maternal health care, nutrition and self-help. Shortly after the Rural broadcasts began, a "Rural Radio Forum" was instituted, allowing regular feedback from remote villages.

Mighty Transmitters Present Africa to the World

In 1961, when most world band stations were struggling with transmitters of 100 kilowatts or less, Ghana introduced its External Service, using two new 250 kW units located at Ejura. It was, at that time, one of the most powerful stations on the air. Except for Ghana, only the largest nations—notably China, Russia and the United States—possessed world band transmitters of this magnitude. Programs were beamed to listeners throughout Africa, as well as Europe, North America,

If Our Editorial Staff Were Bigger, Our World Vision Might Be Smaller.

By news magazine standards, WORLD PRESS REVIEW is small. No vast network of foreign correspondents. No team of copy editors in New York or Washington to filter, shape and flavor the news for American tastes.

A drawback? No—a benefit!

We *borrow* our resources—to give you something that the weekly news magazines don't: a *world* view of your world.

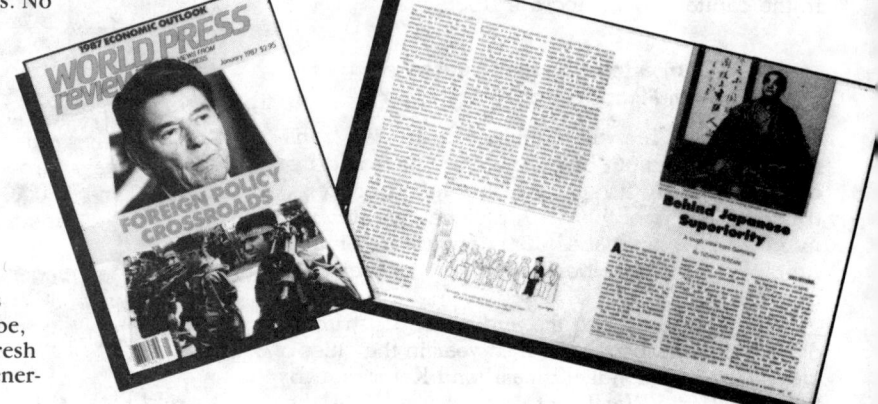

We cull the great newspapers and magazines, all over the globe, to bring you different views. Fresh opinions. Overlooked facts. Generally unnoticed trends.

With WORLD PRESS REVIEW, you get the sense, not just the gist. Because we leave the writing in the authors' own words. Translated, of course, from the French, Finnish, Chinese, Russian, Arabic or whatever. You get the full impact of the original, because you actually get the original: the news as it looks to informed observers close to the scene and familiar with the background.

Nothing between you and the source.

How does the situation look from abroad? What are they thinking, what are they talking about, what concerns them—in Britain, France, West Germany, Australia, Brazil, Mexico? What's being reported, and how, in the Soviet press? The Israeli press? The Syrian press? Now you'll know. Because WORLD PRESS REVIEW never comes between you and the source.

Early warning.

Any signs of political instability? Upheaval in the works? Changes in the wind? *WORLD PRESS REVIEW gives you early warning—months, even years before it breaks out, boils over, impacts on you.*

Global opportunities.

What's the economic outlook? The climate for investment? The new technology? Where are the opportunities in today's global economy and who's in the best position to capitalize? Maybe you! *WORLD PRESS REVIEW could help you profit!*

World awareness.

What's new in books, art, movies, culture, social trends? Where do people in the rest of the world travel when they want to unwind and relax? *Along with everything else, WORLD PRESS REVIEW can add to your pleasure!*

Half price trial.

WORLD PRESS REVIEW can give you what you need in today's world—greater global vision. And we're monthly. Which means you have time to read and absorb each issue before the next arrives.

Not advertised on TV—commercials are too costly—WORLD PRESS REVIEW is available now at the special introductory rate of one year for just $16.97. That's less than half the 12-issue cover cost, and 32% off the regular subscription rate. Satisfaction guaranteed.

To order your subscription, write:

**World Press Review
Department WR
230 Park Avenue #1610
New York, New York 10169**

Japan and Australasia. The rhythms of "Ghana Is Free" mirrored African pride throughout the world.

Today, over 99 percent of the Ghana Broadcasting Corporation's programming is locally produced and the station is manned entirely by Ghanaians. The basic philosophy of broadcasting in the country remains the same, but the Ghana Broadcasting Corporation has, in the meantime, increased to three the number of services available to world band listeners: Radio One, Radio Two and the External Service (see box).

How to Tune in Ghana

Radio One

Ghana has three world band radio networks. However, of these the most commonly heard outside Africa is Radio One. Radio One is on the air Mondays through Fridays from 0525-0905 and 1200-2305, weekends 0525-2305 World Time (UTC), on 4915 kHz. The best chance to hear it outside Africa is during the winter just after sign-on and again right before sign-off.

Radio One carries programs in English and six local languages—Akan, Dagbani, Ga, Ewe, Nzema and Hausa—for both the rural and urban population. While the emphasis is still on agricultural development, other information includes news, commentary, readings from the Ghanaian press, school broadcasts and important addresses by Head of State, Flt. Lt. John Rawlings. Nine million of the 12.5 million people in Ghana reportedly listen to Radio One.

The facilities of Radio One are also used Mondays through Fridays during the school year from 0910-1050 to carry the Schools' Broadcasts.

Radio Two

Radio Two is the Commercial Service, supported by advertising, with programs in the same languages as Radio One. The local audience is estimated at about 9 million.

Radio Two is on the air daily from 0525-0905 and 1705-2305 World Time (UTC) on 3366 kHz, plus 0905-1152 weekends and 1152-1705 daily on 7295 kHz. 3366 kHz is audible in Europe during wintertime, but in the Americas it is somewhat harder to receive. 7295 kHz is usually not well heard outside Africa.

External Service

The External Service, which was suspended in 1982 because of deteriorating equipment, returned to the air at lesser power—100 kW—only last year. Broadcasts are intended for West African audiences, but are also occasionally audible near the hours of darkness in some other parts of the world. English, with a smattering of Portuguese, is aired daily from 0645-0800 and 1845-2000 World Time (UTC) on 6130 kHz. French is broadcast daily from 0800-0900 and 2000-2100 on the same channel.

Good Receiving Equipment Helpful

Ghana is typical of many of the Third World countries that operate on world band radio in that the signals are weak and have to struggle against much stronger signals put out by larger broadcasters.

When listening conditions are right, a number of these smaller stations can be heard on simpler portables. However, the odds improve if you use one of the higher-quality portables or, better, a highly rated tabletop receiver with a properly mounted outboard antenna. Full details on these are given in the "Buyer's Guide" section.

Copy Worldwide Short-wave Radio Signals on Your Computer

Remember the fun of tuning in all those foreign broadcast stations on the short-wave radio? Remember those mysterious sounding coded tone signals that baffled you? Well, most of those beeps & squeals are really digital data transmissions using radioteletype or Morse code. The signals are coming in from weather stations, news services, ships & ham radio operators all over the world. Our short-wave listener cartridge, the "SWL", will bring that data from your radio right to the video screen. You'll see the actual text as it's being sent from those far away transmitters.

The "SWL" contains the program in ROM as well as radio interface circuit to copy Morse code and all speeds/shifts of radioteletype. It comes with a cable to connect to your radio's speaker/earphone jack, demo cassette, and an excellent manual that contains a wealth of information on how to get the most out of short-wave digital DXing, even if you're brand new at it.

For about the price of another "Pac-Zapper" game, you can tie your Commodore 64 or 128 into the exciting world of digital communications with the Microlog SWL. $64. Postpaid, U.S.
MICROLOG CORPORATION,
20270 Goldenrod Lane
Germantown, Maryland 20874.
Telephone: 301 428-3227.

MICROLOG
INNOVATORS IN DIGITAL COMMUNICATION

We have the finest in Communications Receivers

R-2000 KENWOOD R-5000

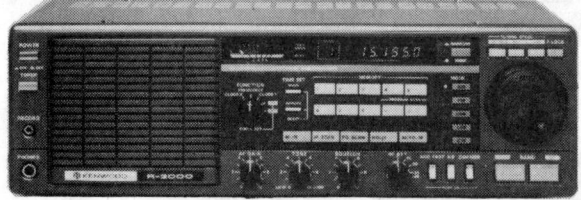

JRC NRD-525

These, and many other fine Ham radio products are detailed in our latest mail-order catalogue.

GLENWOOD TRADING COMPANY LTD.
278 East 1st St., North Vancouver, B.C. V7L 1B3

ORDER DESK
(604) 984-0405

Tuning in the Middle East Powder Keg

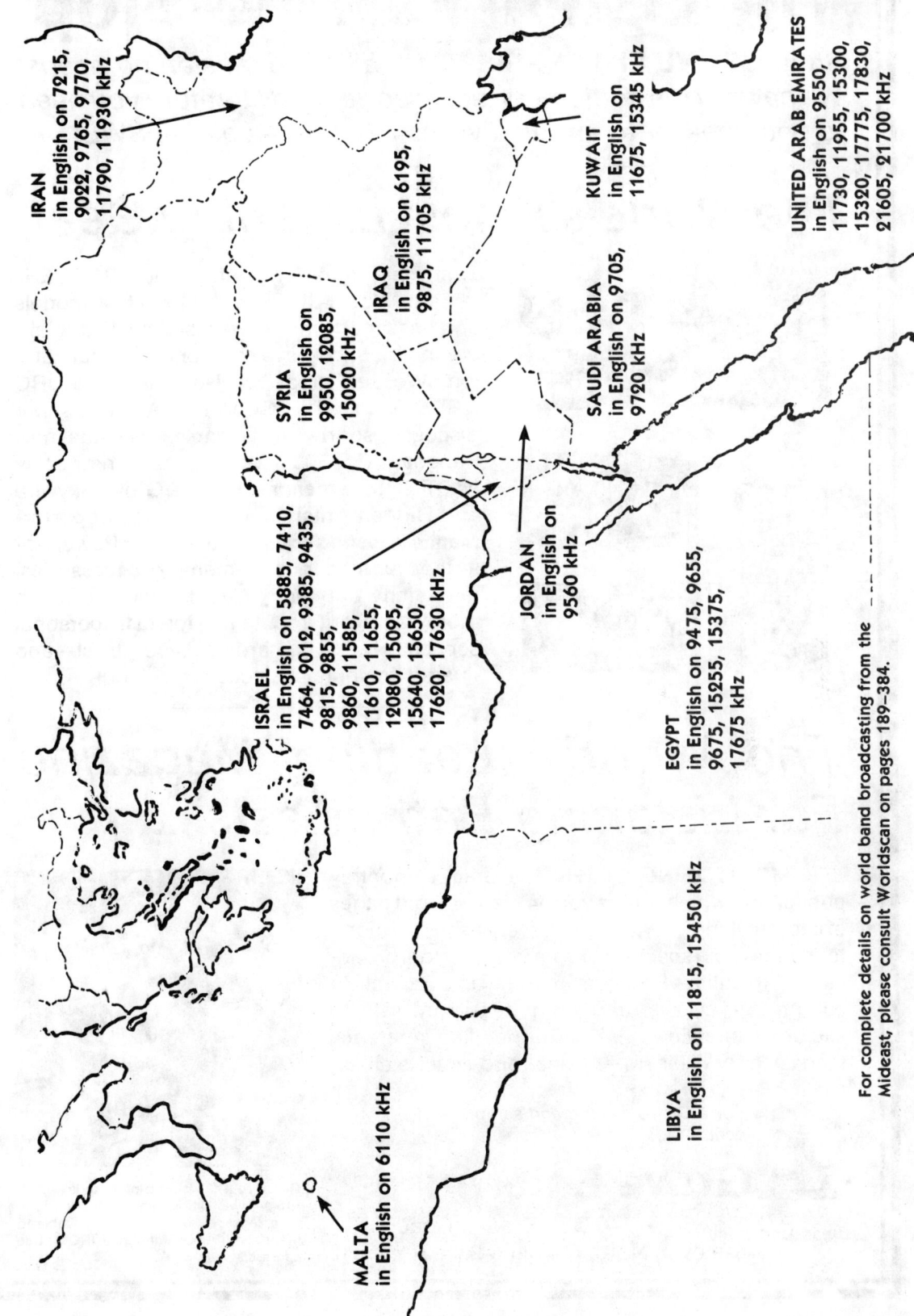

Cartography: Robin Miller

The Listener's 'One-Stop'

Shop **GROVE ENTERPRISES** for all of your listening needs. We have all the equipment, accessories and literature you need to become—or to improve your status as—a global SWL!

The World's Finest Listening Gear

... at the bargain prices for which **GROVE** is famous. You will find only the latest models and finest performers (we test them in our lab) among the shortwave components in our catalog. We sell great **receivers**, like the JRC NRD525 and the ICOM R-71A, at a super discount; shortwave **portables**, like the new Sangean ATS-803 and the amazing new Sony ICF-PRO80; **antennas**, like the Grove Skywire and Hidden Antenna; our own line of **performance boosting equipment**, like our Power Ant and Minituner; not to mention **accessories**, **microfiche frequency files**, **books**, and much more! With our reputation for fast, personal service, we work hard for your trust—and guarantee your satisfaction in writing!

The World's Leading Magazine For Listening Enthusiasts ...

... is **MONITORING TIMES**, the unique monthly publication which brings you 64 giant tabloid pages of late-breaking information on every aspect of monitoring the radio spectrum. From international broadcasting to utilities DXing, from electronic projects to equipment reviews, and from frequency lists to exclusive interviews, **MT** is the leading reference magazine for listeners—novices and experts alike!

For a Grove catalog and sample issue of MT, send $1 to:

Grove Enterprises

P.O. Box 98 • 140 Dog Branch Road
Brasstown, North Carolina 28902

Order your MT subscription today! Only $15 per year U.S & Canada ($22 foreign). Sample issue: send $1 check payable to Monitoring Times (foreign, 2 IRC's). Use address at left.

China's Potpourri of World Band Stations

by Robert J. Hill

China is a nation with one foot in the 21st century, the other firmly rooted in its historic past. The clash of past and present is often obvious, even in the field of communication. It's not unusual, for example, to see a worker on a bicycle hauling a satellite dish to a work site.

Still, when cheap and efficient communication is needed for over one billion people spread over 3.6 million square miles, the Chinese do what many others do: turn to world band radio. It's in constant use throughout the People's Republic and allows foreign listeners a very special peek inside this unusual land.

Numerous World Band Outlets

In many parts of the world, China's official external service, Radio Beijing, is a favorite among those seeking to gain a more meaningful insight into this venerable land of the Han and other peoples. Its programs are broadcast virtually around the clock in dozens of languages to all parts of the globe. But there's more to Chinese world band radio than Radio Beijing.

Domestic world band radio proliferates throughout China. The main headquarters for domestic radio in the People's Republic of China is the Central People's Broadcasting Station (CPBS). And while it sounds as if it is just one station, it's really an umbrella organization for a number of different networks.

Based in the capital city, Beijing (formerly called Peking), CPBS operates two national networks, "CPBS-1" and "CPBS-2," as well as two networks beamed to Taiwan. In addition, there is a fifth service aimed at the so-called "domestic minorities"—residents of the country whose language is other than Standard Chinese. These programs are relayed to outlying areas by regional and provincial stations scattered throughout the country. They are also broadcast directly from Beijing.

World band radio in China does not end with Radio Beijing or the CPBS networks. The People's Republic of China comprises fully twenty-one provinces, five "autonomous regions" and three municipalities. Of these subdivisions, all but five of the provinces and the municipalities operate local world band stations.

Bob Hill, currently a promotional writer, was formerly a radio monitor for the US government.

What to Look For: Languages

Except for the domestic minorities services, most CPBS and provincial stations broadcast in Standard Chinese (essentially Mandarin)—although if you listen long enough, you may catch English lessons on CPBS-1 and CPBS-2 channels from time-to-time. Programs beamed to Taiwan—which the government in Beijing considers merely an errant province of the People's Republic—carry Amoy, Hakka and other dialects in addition to Standard Chinese. Minority service programs are in Kazakh, Korean, Mongolian, Tibetan, and Uighur.

What to Look For: Programs

Ten years ago, world band programs from the People's Republic were characterized by strident rhetoric, revolutionary Chinese music, and generally, a heavy-handed approach to its audience.

Today, programming is based more on the need to entertain and inform than to indoctrinate. In some places—notably southern China, where Western-sounding programs roll in over the air-

A Trade Ministry somewhere outside Peking

> **Today, world band radio in China is designed to entertain rather than indoctrinate**

waves from Hong Kong—local stations work extra hard to try and build an audience. Still, some traditions die slowly. So you can still hear such programming relics as vigorous wake-up exercises, complete with insistent one-two-three-four cadences, at the crack of Chinese dawn. Overall, though, Chinese world band radio has changed in a way that both Chinese and overseas listeners seem to enjoy.

What to Look For: Identification

With a multiplicity of world band stations on the air and the fact that many nations are intent on reaching China with their own broadcasts, it is not always easy to be sure whether you're actually listening to a station in China or, say, the Voice of America's Chinese programs. Fortunately, there are several characteristics of the Chinese domestic stations that make them easy to identify.

- When the top of the hour arrives, you'll hear six "time pips" with the final pip being exactly on the hour and an octave higher than the first five. Many stations around the world use time pips, but no other has exactly this combination of sounds.
- After the pips, the announcer will say several words, which are followed by "Renmin Guangbo Dientai." That phrase means "People's Broadcasting Station," and it's the last part of virtually every CPBS regional and provincial station identification. But that's only half the identification puzzle, because the words preceding those will reveal specifically which station you're hearing.

  If you're listening to the national network itself, the "Renmin Guangbo Diantai" will be preceded by the word "Zhongyang," which means "Central."

  If it's a provincial station—from, say, Gansu—the word "Gansu" will precede the "Renmin Guangbo Dientai."

  As for the minority services, remember that they are not in Standard Chinese. So listen, instead, for a rendition of the Chinese Communist hymn, *The East is Red*, at the beginning of each broadcast.
- Many local stations relay news from CPBS on the hour, much as local stations in North America relay ABC or the CBC. Don't be fooled! Stick around for a *local* identification at the conclusion of the CPBS newscast.
- Unlike stations in a number of other countries, the majority of domestic stations in the People's Republic of China remain on the same channel or channels for many years at a clip. Too, most Chinese domestic outlets operate outside the internationally allocated world radio broadcasting bands. This makes them easier to pinpoint as being located in China.

When and Where to Look

A number of generalizations can be made about the best times and places on the dial to find Chinese domestic stations. CPBS outlets tend to utilize lower frequencies—under 10 MHz during their hours of local darkness—moving to higher frequencies during the day. This is to make them easier to receive within China.

> **Domestic world band radio proliferates throughout China**

As a rule, you're not limited to a single spot on the dial for a particular station, as almost all use more than one frequency at the same time. And most domestic transmitters are of modest power by international standards. Although 100 kW transmitters are sometimes used, 10, 15 or 50 kW is more common. Nevertheless, these are strong enough to generate a respectable signal continents away.

NEED A RELAY?

PROGRAMMING AIR TIME?

WWCR has patterns for AFRICA, NORTH AND SOUTH AMERICA, EUROPE, MEXICO, and THE CARIBBEAN.

AIR TIME AVAILABLE NOW. REASONABLE RATES.

CALL TOLL FREE:
1-800-238-5576

OR CALL COLLECT:
(901) 278-7878

George McClintock

WWCR
3314 West End Avenue
Nashville, TN (USA) 37203

BUYER BEWARE - Insist on New SANGEAN AMERICA'S Receivers Backed by Manufacturer 180 Day U.S.A. Warranty

ATS-803
JUST ARRIVED
ADVANCED UPDATED MODEL

$299.95 VALUE
INTRODUCTORY PRICED
$189.95
INCLUDES AC Power Adapter

EVERY UNIT PACKED WITH SANGEAN AMERICA'S WARRANTY

- State of the art PLL world band receiver.
- Easy to use with direct digital dialing of frequency.
- 5 tuning functions: direct frequency key-in, auto scanning, manual scanning, preset recall and manual rotary tuning.
- 9 programmable memories.
- Radio, clock and alarm turn on automatically, play preset stations at preset time.
- Wide/narrow bandwidth. You select for better listening in today's crowded band.
- RF gain control to prevent overloading when near strong stations and improve SSB reception.
- Signal strength indicator (5 LED).
- Shortwave button allows user to tune only international shortwave bands. No need to guess start and stop frequency.
- AM button allows full coverage of 150 kHz to 30 MHz.
- FM button allows full coverage of 88-108 MHz.

MANY NEW FEATURES

- Unique speed dialing—the faster the dial is turned the larger the frequency steps.
- Multimode AM-FM-CW-SSB allow full coverage of commercial traffic. Amateur, aircraft, ship at sea, and more.
- 24 hour clock—sleep/alarm selectable switch allows hour & minutes or minutes & seconds.
- Sensitivity and dynamic excellence for a portable radio.
- High stability for good CW-SSB-RTTY reception.
- Excellent audio with separate bass and treble controls, stereo balance.
- Din plug for standard stereo hook-up.
- External speaker jack.
- Built in whip antenna and adaptor for external antenna.
- Battery powered 6D & 2AA (not included).
- Optional AC wall power pack switchable 120/240V 50/60 Hz.
- Same size and weight as Sony ICF-2010
- Metal speaker grill

ATS 801 Computerized Radio Is Now Here

Operate with touch of your finger, automatic program searching and 25-memory preseting just a part of its function. Note its perfect design and compact size!

- 4 Band Receiver
- SW/LW/MW/FM Stereo (using headset)
- SW 5.8-15.5 MHz-MW 530-1620 kHz FM 87.5-108 MHz LW 155-281 kHz
- Digital frequency readout
- Scanning and manual tuning
- 25 program memories
- Built-in telescope ant.

- SW/FM Ferrite bar on MW/LW
- Alarm-Sleep function
- Fully synthesized gives excellent stability
- Battery power 4 "AA" AC adapter optional $9.95
- *Free* 100 page S.W. frequency book

Introductory Special **$99.95**
+ $4.00 U.P.S. List $179.95

FULL LINE OF SHORTWAVE RECEIVERS—ASK FOR DETAILS

AT AN AUTHORIZED DEALER NEAR YOU: Universal Shortwave 800-431-3939
EGE, Inc. 800-336-4799

DEALER INQUIRIES - Call Manufacturer's Rep: BLAZE INTERNATIONAL INC.
65 Oser Ave., Hauppauge, NY 11788 • (516) 434-9399

Latin America, Land of Traditional Music

by Don Jensen

According to Peruvian mythology, Viracocha, the god of creation, rose from the foam of the sea and gave to his people the gift of music.

Viracocha's ear wasn't attuned to the music of the West, so he taught his people a pentatonic musical scale: E-G-A-B-D. The Incas responded by using their native instruments—the quena, pinkillo and drums—to create an eerie, almost Asian-sounding music. These strains, loosely called "Andean," are heard today over world band stations through most of western South America. Mixed with the Spanish guitar of the conquistadores and some European rhythms, these have evolved further into the spirited *huayno*.

Simon and Garfunkel Create Andean Hit

It was the huayno that inspired Peruvian folklorist Daniel Alomia Robles. Back at the turn of the century, Alomia traveled through the mountains, across the *altiplanos* and into the valleys of his country to study the music of the people. In 1912, drawing on traditional themes he had learned, he wrote the most famous huayno of all, "El Condor Pasa." The tune, in memory of the fierce bird of the Andean heights, was later recorded by the internationally known duo of Simon and Garfunkel. It was thus that the huayno had its first million-selling hit.

Today, the memory of Simon and Garfunkel's tune has faded. But back in the Peruvian Andes where the huayno began, it's very much alive and kicking. Sung in either Spanish or Quechua, it continues to delight listeners of all ages.

The huayno isn't limited to Peru, though. It's also a hit across the Bolivian border, where it's often spelled *huaino*. In this version an instrument called the *siku* predominates.

The siku consists of about six bamboo tubes of varying lengths, stuck together. The musician plays his instrument much as a youngster would blow across the mouth of a Coke bottle. The result, especially on the low notes, is a haunting, breathless sound. The omission of certain notes from the descending scale imparts an inherent sadness to the music.

Troupes That Snore in Ecuador

Although you might not notice it right off, there is a difference between the music of Peru and its

Don Jensen, whose interest in Andean world band radio goes back several years, has traveled widely throughout Latin America.

Folkloric musical group from Imbabura, Ecuador

northern neighbor, Ecuador. The two types of Ecuadorian music most commonly heard on world band radio are the *pasacalle* and the *pasillo*. The pasacalle features both traditional Indian influences and a local native flute. This flute's distinctive low sound has led it to be dubbed *el roncador* . . . "the snorer." Fortunately, it sounds better than the spousal variety!

Pasillos are lamentations, sadness over love gone sour. The titles don't let you forget. Try "*Lagrimas*," "*Dos Lagrimas*," "*Cuatro Lagrimas*," and "*Vaso de Lagrimas*"—literally "Tears," "Two Tears," "Four Tears" and, for a good cry in your *cerveza*, "A Glassful of Tears."

Come Back to the Five and Dime, Carmen Miranda

The music of Argentina is no stranger to Europe and America, although the tango is now considered a bit *passé*. No less romantic, however, is the cool Brazilian samba. Also from that vast land—Brazil is the world's fifth-largest country—comes the bossa nova. It's a relatively new form of music which mixes the samba with jazz.

Moving northward into Colombia brings the world band listener to the tropical rhythm of the cumbia. Tropical suggests "hot," and the rhythm of the cumbia is as hot as a horse's gallop: rat-tat-tat, rat-tat-*tat*! The popular cumbia has emigrated

TGVN, "La Voz de Nahuala," is a friendly station specializing in traditional Indian music

throughout Latin America, so you're as likely to hear this sound from stations in Guayaquil, Ecuador, as you are from Bogotá, Colombia.

Traditional Instruments Thrive

The harp plays an important role in the music of two South American nations, Paraguay and Vene-

How to Listen to Traditional Latin American Music

There are hundreds of world band radio stations in South and Central America. Many, sometime during their daily schedules, play traditional national music. The trick is how to hear them. Few are intended to be heard worldwide, so most operate at very low power, making them a challenge to receive half a world away. And there are so many signals on the air that they often interfere with each other.

So the first requirement is that you have good receiving equipment. It doesn't have to be a superset . . . just something above average, and preferably with a good antenna, to boot. Look for models with at least a three-star rating in our "Buyer's Guide." And, among these, look for those with superior ability to bring in tough signals. Antennas are also reviewed in the "Buyer's Guide."

The second tip to succeed is to listen during the months of maximum darkness—winter and the equinoctial months of autumn and early spring. The signals you're seeking travel best at that time of year, and static from thunderstorms is down.

The third "must" is timing. Just as you wouldn't show up for a dinner banquet at 10:30 AM, so you shouldn't expect much if you poke around for Latin signals at high noon. The night time is the right time. Just make sure it's night—or at least dusk or dawn—not only at your location, but also in the country you're trying to receive.

For example, in eastern North America the fun starts in the pitch dark of early morning around 0800 World Time (UTC), when the Brazilians begin to roll in. From then until shortly after your local sunrise the full complement of Latin countries makes its appearance. In the evening a variety of signals begins to roll in just before sunset and continues until they bid *adios* to their listeners some hours later.

In Europe, mornings are duds. But from late afternoon until you nod off, you can hear excellent examples of traditional Latin American music.

Latin stations, like peanuts, tend to be addictive. Some listeners spend hour after hour tuning listening to as many as they can unearth. But, so you can get started, here are some potential catches.

zuela. Because it has only one regularly operating world band station, Paraguay is not the easiest place in the world to hear. It's a pity, too, as the Paraguayan *polca*—a lively variation of the traditional polka, but played with a large harp—is immensely popular at beer-swilling festivities. The haunting Paraguayan *guarania*, the music of lovers longing for each other's presence, is no slouch, either. It's considered the most beautiful of its type in South America and is one of the musical gems of world band radio.

In Venezuela, the tempo is up and a smaller harp is heard. The best example of this is the national toe-tapping rhythm called the *joropo*. The beat—well heard over the air—is fast and sharp.

North, along the isthmus of Central America, up-tempo music also prevails. In Guatemala, and even El Salvador, the music is easily recognized by the instrument rather than the rhythm. The unmistakable marimba, the famous wooden xylophone, dominates.

Finally, there's familiar Mexico. Here two types of music dominate, the *mariachi*—popularized by Herb Alpert and the Tijuana Brass—and the *ranchera*.

Mariachi includes the violin, bass guitar and, of course, those indispensible trumpets, playing in "thirds." Ranchera is the untamed cowboy music of the northern Mexican states. With the ranchera you'll be treated to wailing accordions, guitars and vocalists. Once heard—it can be daily fare on Spanish-speaking AM stations in some US cities—it's easily recognized.

Latin America represents only one part of the worldwide musical mosaic heard over world band radio. But it's also one of the most fascinating. If you have a first-class listening setup, listen in and enjoy!

Author Jensen strolls through El Ejido Park on a drizzly Quito day

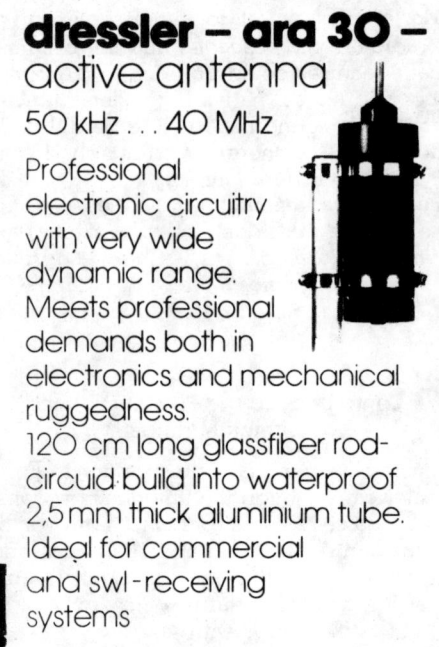

dressler – ara 30 –
active antenna
50 kHz ... 40 MHz

Professional electronic circuitry with very wide dynamic range. Meets professional demands both in electronics and mechanical ruggedness. 120 cm long glassfiber rod-circuid build into waterproof 2,5 mm thick aluminium tube. Ideal for commercial and swl-receiving systems

dressler – ara 900 –
active antenna
50 ... 650 MHz

Vertical wide band antenna system with superior dynamic range and low noise figure of 1 – 3,5 dB.

Professional thick-film hybrid amplifier.

Ideal for vhf-uhf-monitoring and scanner applications.

dressler
hochfrequenztechnik gmbh
WERTHER STR. 14-16 · D-5190 STOLBERG
TEL. (0 24 02) 7 10 91 · TLX 832 287

Where to Find Traditional Latin American Music

Country	Station	Channel
Argentina	★★Radio Nacional, Buenos Aires	9690 kHz
	Radio Nacional, Mendoza	6179 kHz
Bolivia	Radio Fides, La Paz	4845, 6155 kHz
	Radio Illimani, La Paz	4945 kHz
	Radio Panamericana, La Paz	6105 kHz
Brazil	★Radiodifusora do Maranhão, Saõ Luís	4755 kHz
	★Radio Clube do Pará, Belem	4885 kHz
	Radio Brasil Central, Goiânia	4985 kHz
	★Radio Rio Mar, Manaus	9695 kHz
	★Radio Guaíba, Pôrto Alegre	11785 kHz
	★★Radio Clube Ribeirão Preto, Ribeirão Preto	15414 kHz
	★Radiodifusora do Amazonas, Manaus	4805 kHz
	★★Radio Nacional, Manaus	4845 kHz
	★★Radio Nacional, Brasília	6180, 11780 kHz
	★Radio Inconfidência, Belo Horizonte	15190 kHz
Colombia	★Radio Cadena Nacional, Bogotá	6160 kHz
	★La Voz de la Selva, Florencia	6170 kHz
	Ondas del Orteguaza, Florencia	4975 kHz
	La Voz de Yopal, Yopal	5050 kHz
	La Voz de los Centauros, Villavicencio	5955 kHz
Ecuador	★Radio Quito, Quito	4920 kHz
	★Radio Rio Amazonas, Macuma	4870 kHz
	La Voz del Napo, Tena	3280 kHz
	Radio Centinela del Sur, Loja	4890 kHz
	Radio Federación, Sucúa	4960 kHz
	★★HCJB, Quito ("Musica del Ecuador" program)	9870, 11910, 15155 kHz
Guatemala	Radio Maya de Barillas, Huehuetenango	3325 kHz
	★La Voz de Nahualá, Nahualá	3360 kHz
	Radio Tezulutlán, Cobán	3370 kHz
	★Radio Chortís, Jocotán	3380 kHz
Mexico	XEQQ, "La Q," Mexico City	9680 kHz
	Radio Huayacocotla, Huayacocotla	2390 kHz
	XEUJ, Linares	5980 kHz
	XEQM, Mérida	6105 kHz
Paraguay	★★Radio Nacional, Asunción	9735 kHz
	Radio Encarnación, Encarnación (irregular)	11941 kHz
Peru	Radio San Martín, Tarapoto	4810 kHz
	Radio Huancavelica, Huancavelica	4885 kHz
	★Radio Andina, Huancayo	4996 kHz
	Radio Los Andes, Huamachuco	5030 kHz
	Radio Unión, Lima	6115 kHz
	★Radio Amistad, Soritor	8515 kHz
	Radio Continental, Arequipa	6055 kHz
	Radio Tacna, Tacna	9486 kHz
Venezuela	★Radio Nacional, Caracas	5020 kHz
	★Radio Rumbos, Caracas	4969 kHz
	★★Ecos del Torbes, San Cristóbal	4980 kHz
	Radio Valera, Valera	4840 kHz
	Radio Táchira, San Cristóbal	4830 kHz
	Radio 980, El Tigre	3255 kHz
	Radio Mara, Maracaibo	3275 kHz

★★Reception may be fairly good except during summer.
★Fair reception possible except during summer.
No stars = Reception difficult but possible except during summer.

Barry Electronics Corp.

WE SHIP WORLDWIDE
WORLD WIDE AMATEUR RADIO SINCE 1950
Your one source for all Radio Equipment!

For the best buys in town call:
212-925-7000
Los Precios Mas Bajos en Nueva York...

KITTY SAYS: WE ARE NOW OPEN 7 DAYS A WEEK.
Saturday & Sunday 10 to 5 P.M.
Monday-Friday 9 to 6:30 PM Thurs. to 8 PM
Come to Barry's for the best buys in town.

MAY We Help You With the Best in Commercial and Amateur Radios? Jan KB2RV, Toni, Kitty WA2BAP, Mark K2CON

ONV Safety belts-in stock

IC-R71A, 751A, 745, 28A/H, 38A, 48A, Micro2/4, R-7000, IC-761, IC-375A, 275A/H, 3200A, 475A/H, 735, IC-900

KENWOOD

TS440S/AT, R-5000, R-2000, TS-940 S/AT,
TM 221A/421A, TM-2570A/50A/30A, TR-751A,
Kewood Service /Repair, TH 21/31/41 BT,
TS-711A811A, TM3530A, TH205AT, TH215A,
TW-4100A , TM-321A

NEL-TECH DVK-100 Digital Voice Keyer

Media Mentors—
Amateur Radio Course $99.95

VoCom/Mirage/Alinco
Tokyo Hy-Power/TE SYSTEMS
Amplifiers &
5/8λ HT Gain
Antennas IN STOCK

Soldering Station,
48 Watts, $68
MICROLOG-ART 1, Air Disk, SWL, Morse Coach

KANTRONICS
UTU, KAM, UTU-XT,
KPC 2400, KPC IV

EIMAC
3-500Z
572B, 6JS6C
12BY7A &
4-400A

AEA 144 MHz
AEA 220 MHz
AEA 440 MHz
ANTENNAS

BIRD
Wattmeters & Elements
In Stock

Antennas
A-S
Cushcraft
Hy-Gain
Hustler
KLM
METZ
Mini-Products
MULTIBAND
Mosley
MODUBLOX
TONNA

YAESU
FT-23/73/727R
FT-2/1/709R/H
FT-1903/1123

ICOM
IC2AT/12AT
ICO2AT
IC-03/04AT
IC-A2/U16

FT-767GX, FT-757GX II,
FRG-8800, FT-726, FRG-9600,
FT-211/711RH, FT-2700RH

YAESU

Land-Mobile H/T
Midland/Standard
Wilson   Maxon
Yaesu FTC 1123, FTC 1143
ICOM IC-M5 (Marine) M700
Tempo M-1

RF Concepts

AMERITRON AMPLIFIER AUTHORIZED DEALER

Yaesu FTR-2410, Wilson
ICOM IC-RP 3010 (440 MHz)
ICOM IC-RP 1210 (1.2 GHz)

Computer Interfaces
Stocked: MFJ-1270B,
MFJ-1272, MFJ-1224, AEA
PK-87, PK-64A, PK-64, PM-1,
PK-232 W/FAX

ETO
ALPHA AMPLIFIERS

Complete Butternut Antenna Inventory In Stock!

DIGITAL FREQUENCY COUNTERS
Trionyx, Model TR-1000, 0-600 MHz
AMP SUPPLY STOCKED
Long-range Wireless
Telephone for export in stock

BENCHER PADDLES,
BALUNS,
IN STOCK
MIRAGE AMPLIFIERS
ASTRON POWER SUPPLIES
Saxton Wire & Cable

SMART PATCH
CES-Simplex Autopatch 510-SA Will Patch FM Transceiver To Your Telephone. Great For Telephone Calls From Mobile To Base. Simple To Use.
PRIVATE PATCH III,
Duplex 8000 in stock

Budwig ANT. Products
FLUKE 77 Multimeter

NYE MBV-A 3 Kilowatt Tuner

SANTEC
ST-222/UP
ST-20T
ST-442/UP
HT-7

MFJ-989B

Ten-Tec Tuner 229B

HEIL EQUIPMENT IN STOCK

SANGEAN Portable Shortwave Radios

Tri-Ex Towers
Hy-Gain Towers & Antennas, and Roters will be shipped direct to you FREE of shipping cost.

New TEN-TEC
Corsair II, PARAGON, Century 22, RX-325

MAIL ALL ORDERS TO: BARRY ELECTRONICS CORP., 512 BROADWAY, NEW YORK CITY, NY 10012 (FOUR BLOCKS NORTH OF CANAL ST.)

New York City's LARGEST STOCKING HAM DEALER
COMPLETE REPAIR LAB ON PREMISES

"Aqui Se Habla Espanol"
BARRY INTERNATIONAL TELEX 12-7670
MERCHANDISE TAKEN ON CONSIGNMENT
FOR TOP PRICES
Monday-Friday 9 A.M. to 6:30 P.M.   Thursday to 8 P.M.
Saturday & Sunday 10 A.M. to 5 P.M. (Free Parking)
AUTHORIZED DISTS. MCKAY DYMEK FOR
SHORTWAVE ANTENNAS & RECEIVERS.
    IRT/LEX-"Spring St. Station"
Subways: BMT-"Prince St. Station"
    IND-"F" Train-Bwy. Station"
Bus: Broadway #6 to Spring St.
    Path—9th St./6th Ave. Station.

Commercial Equipment Stocked: ICOM, MAXON, Midland, Standard, Wilson, Yaesu. We serve municipalities, businesses, Civil Defense, etc. Portables, mobiles, bases, repeaters...

We Stock: AEA, ARRL, Alpha, Ameco, Antenna Specialists, Astatic, Astron, B & K, B & W, Bencher, Bird, Butternut, CDE, CES, Collins, Communications Spec. Connectors, Covercraft, Cushcraft, Daiwa, Dentron, Digimax, Drake, ETO (Alpha), Eimac, Encomm, HeilSound, Henry, Hustler (Newtronics), Hy-Gain, Icom, KLM, Kantronics, Larsen, MCM (Daiwa), MFJ, J.W. Miller, Mini-Products, Mirage, Newtronics, Nye Viking, Palomar, RF Products, Radio Amateur Callbook, Rockwell Collins, Saxton, Shure, Telex, Tempo, Ten-Tec, Tokyo Hi Power, Trionyx TUBES, W2AU, Waber, Wilson, Yaesu Ham and Commercial Radios, Vocom, Vibroplex, Curtis, Tri-Ex, Wacom Duplexers, Repeaters, Phelps Dodge, Fanon Intercoms, Scanners, Crystals, Radio Publications.

WE NOW STOCK COMMERCIAL COMMUNICATIONS SYSTEMS
HAM DEALER INQUIRES INVITED   PHONE IN YOUR ORDER & BE REIMBURSED
COMMERCIAL RADIOS stocked & serviced on premises.
Amateur Radio Courses Given On Our Premises, Call
Export Orders Shipped Immediately. TELEX 12-7670

ALL SALES FINAL

How to Listen to Worldwide Weather Information

Up-to-the-minute weather information is so much a part of our daily routine that we tend to take it for granted. A day at the beach, a weekend in the mountains or just deciding when to mow the grass—they're all dependent on Old Sol and the rest of nature's temperamental weather system.

It's the best data modern weather technology can provide

But what do you do if you're flying to New York from London and want to know the weather on the other side of the Pond? You won't have any luck with the Voice of America. It will only tell you about the political winds out of Washington.

In fact, often the best bet is not to tune in normal world band broadcasts at all. Instead, hone in on the *aeronautical weather* stations—called "VOLMET"—that share the world radio spectrum with broadcasters. These are the same English-language transmissions airline crews listen to when they're out over the ocean.

You know that these pros, hurtling along above the earth in aluminum-foil cylinders, are going to get the best data modern weather technology can provide. And now, with your world band radio, you can, too.

To receive VOLMET weather information, you only need a radio that can tune the world band spectrum continuously from at least 5-14 MHz . . . 2-14 MHz if you're really fussy. Many of these stations are audible on an ordinary world band radio, but a number use special *lower-sideband* signals that can be understood only on radios designed to cope with them. Fortunately, most larger and midsize portables—and virtually all tabletops—can receive lower-sideband signals. Even some compacts, such as the Sony ICF-2003 and ICF-PRO80, are designed to pick these up properly.

Following, extracted from the *Shortwave Directory* (Grove Enterprises), are some of the

Rideau Canal, Ottawa, Ontario

Buenos Aires's cosmopolitan Avenida 9 de Julio by night

major VOLMET stations currently on the air. Weather data for several cities is usually aired sequentially on the same channel, with the "cycle period" for any given city being thirty minutes. Once you're heard what you want, you can be assured of hearing it repeated thirty minutes later, hour-after-hour, day-after-day.

Eastern/Central North America

2905, 3485, 5592, 6604, 8870, 10051, 13270 and 13276 kHz

United States: New York, Detroit, Chicago, Cleveland, Niagara Falls, Milwaukee, Indianapolis, West Palm Beach, Tampa, Bangor, Pittsburgh, St. Louis, Syracuse, Minneapolis, Newark, Boston, Baltimore, Philadelphia and Washington. Miami weather is also heard on these channels, plus the additional channels of 2950, 5580 and 11315 kHz.
Canada: Windsor Locks and Gander.

Western/Central North America and Nearby Pacific

2863, 8828, 6679 and 13282 kHz

Continental US: Los Angeles, San Francisco, Sacramento, Las Vegas, Seattle/Tacoma, Portland and Alaska.
Pacific US: Hawaii and Guam.
Canada: Vancouver and Ontario.

Carribean

2905, 3485, 5592, 6604, 8870, 10051, 13270 and 13276 kHz

Bermuda, Nassau and Freeport.

2950, 5580 and 11315 kHz

Port-of-Spain and Merida.

NOW! DX the World!
WHEREVER YOU ARE

In a recent review in a leading shortwave listening magazine, this new shortwave receiving antenna by Com-Rad Industries, besides its superior performance, was so easy to use and so portable the author dubbed it the "FUNtenna".

You too can enjoy the advantages of using the only truly effective **passive** antenna in a portable, desk-top package.

FUNtenna™

Frequency range from 6.8 to beyond 50 MHz.
ONLY $79.95 plus $4 shipping.
(Attention Hams. Call or write about our portable transmitting version of the FUNtenna, the UNtenna HI-Rizer.)

COM-RAD INDUSTRIES
1635 West River Road
P.O. Box 554
Grand Island, NY
14072-0554
716-773-1445

ONLY $79.95
plus $4.00 Shipping
(New York State Residents add appropriate Sales Tax)

Yaesu has serious listeners for the serious listener.

Yaesu's serious about giving you better ways to tune in the world around you.

And whether it's for local action or worldwide DX, you'll find our VHF/UHF and HF receivers are the superior match for all your listening needs.

The FRG-9600. A premium VHF/UHF scanning communications receiver. The 9600 is no typical scanner. And it's easy to see why.

You won't miss any local action with continuous coverage from 60 to 905 MHz.

You have more operating modes to listen in on: upper or lower sideband, CW, AM wide or narrow, and FM wide or narrow.

You can even watch television programs by plugging in a video monitor into the optional video output.

Scan in steps of 5, 10, 12½, 25 and 100 KHz. Store any frequency and related operating mode into any of the 99 memories. Scan the memories. Or in between them. Or simply "dial up" any frequency with the frequency entry pad.

Plus there's more, including a 24-hour clock, multiplexed output, fluorescent readout, signal strength graph, and an AC power adapter.

The FRG-8800 HF communications receiver. A better way to listen to the world. If you want a complete communications package, the FRG-8800 is just right for you.

You get continuous worldwide coverage from 150 KHz to 30 MHz. And local coverage from 118 to 174 MHz with an optional VHF converter.

Listen in on any mode: upper and lower sideband, CW, AM wide or narrow, and FM.

Store frequencies and operating modes into any of the twelve channels for instant recall.

Scan the airwaves with a number of programmable scanning functions.

Plus you get keyboard frequency entry. An LCD display for easy readout. A SINPO signal graph. Computer interface capability for advanced listening functions. Two 24 hour clocks. Recording functions. And much more to make your listening station complete.

Listen in. When you want more from your VHF/UHF or HF receivers, just look to Yaesu. We take your listening seriously.

YAESU

Yaesu USA
17210 Edwards Road, Cerritos, CA 90701
(213) 404-2700

Yaesu Cincinnati Service Center
9070 Gold Park Drive, Hamilton, OH 45011
(513) 874-3100

Dealer inquiries invited.

Prices and specifications subject to change without notice.
FRG-9600 SSB coverage: 60 to 460 MHz.

Europe and Near East

3413 (night), 5640, 8957 and
13264 (day) kHz

Europe: Shannon, London, Prestwick, Paris, Madrid, Amsterdam, Oslo, Copenhagen and Athens.

2998, 3413, 5640, 6580, 8957, 11378,
13264 kHz

Europe: Prague, Bratislava, Brno, Vienna, Frankfurt, Berlin, Warsaw, Budapest, Moscow.
Near East: Tel Aviv, Haifa, Elat, Jerusalem, Larnaca, Athens, Ankara, Istanbul.

Asia and Australia

2965, 3458, 5673, 6676, 8849, 11387
and 13285 kHz

Asia: Karachi, Nawabshah, Lahore, Calcutta, Bombay, Delhi, Dhaka, Mingaladon, Singapore, Lebar, Kuala Lumpur, Jakarta, Brunei, Kota Kinabalu, Bali, Penang, Bangkok, Mingaladon, Tan son Nhut, U-Tapeo, Colombo, Madras and Ahmadabad.
Australia: Sydney, Brisbane, Melbourne, Adelaide, Alice Springs and Perth.

We Ship around the WORLD

Dealers for:

Icom • Yaesu • Ten-Tec

KLM • Cushcraft • MFJ

Bencher • B&W • AEA • Hustler

Palomar • Grundig

ARRL Publications • Kantronics

Com-West Radio Systems Ltd.

(604) 321-1833

8179 Main Street
Vancouver, B.C.
V5X 3L2   Canada

Unloading Crab, Newfoundland

Atlantic and Pacific Weather Forecasts

2500, 5000, 10000 and 15000 kHz

In addition to VOLMET weather information, there are oceanic weather forecasts prepared by the US National Oceanic and Atmospheric Administration (NOAA). These are extremely helpful for seafarers, especially yachtspersons.

Forecasts for the Atlantic, aired over radio station WWV in Colorado, are heard from 8-11 minutes after the hour. These are often audible in the Americas, Europe, West Africa and the Atlantic Ocean.

Forecasts for the Pacific, aired over radio station WWVH in Hawaii, are heard from 48-52 minutes after the hour. These tend to be audible in Australasia and throughout much of the Pacific Ocean region.

AUTOMATE YOUR CASSETTE RECORDER WITH THE CC-2020 CASSETTE CONTROLLER.

Attention Sony, Panasonic and Sangean Shortwave Radio Owners!!!

Automate your cassette recorder with the CC-2020 Cassette Controller. Now operate your cassette recorder via the Timer/Alarm function on board your shortwave radio. Using the CC-2020 gives you the flexibility of automated SW cassette recordings.

How?

The CC-2020 links the Timer/Alarm on your reviever by submini plug to the remote control jack of a cassette recorder, enabling you to make cassette recordings of shortwave programs. No matter what time or frequency they may be broadcast just by setting the on board timer. Following is a list of S.W. receivers with a Timer/Alarm function that operate with the CC-2020 Cassette Controller. Sony: ICF-2010/2001D, ICF-2002/7600D, ICF-7700/7600DA, ICF-2003. Panasonic: RF-B60. Sangean: ATS-801, ATS-803.

No disassembly of receiver is necessary.

Requires a cassette recorder with a remote 2.5 mm submini jack (Fig.1), an AC adapter or equivalent transformer, and an audio patch cord between the radios' tape output jack and the cassette mic. input jack (Fig. 2).

The price is $37.50 plus $3.00 for shipping ($5.00 for overseas shipments). Includes twelve month warranty. Audio patch cords, mini plug to mini plug are available at time of order for $2.00. Payment by check in U.S. funds with U.S. bank code imprints or by money order. For customers outside of North America; payment may also be made by International Postal Money Order, in U.S. funds. Please make all payments to *"Soltronic"*.

SOLTRONIC 6720 N. 11th ST. Philadelphia, Penna., 19126 U.S.A. TEL: 215-548-4747. Formerly SOLAR LIGHT CO.

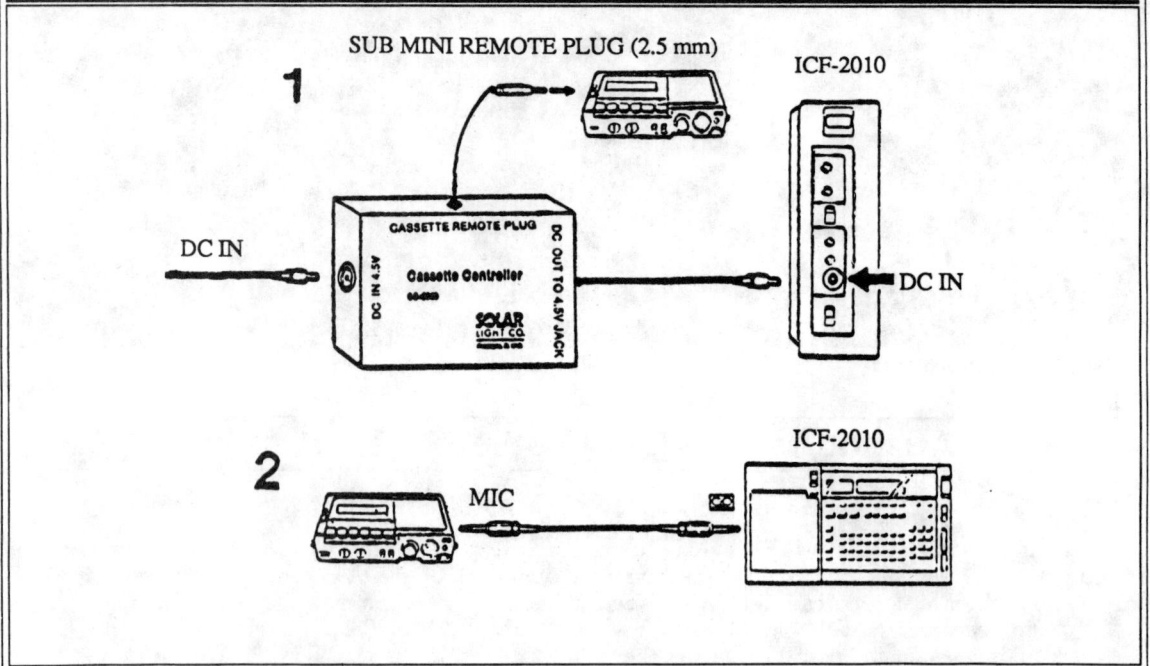

THE JAPAN RADIO NRD-525

UNPARALLELED PERFORMANCE AND SOPHISTICATED FEATURES!

Rated a full FIVE STARS by Larry Magne in RADIO DATABASE INTERNATIONAL Whitepaper.

"...it must be said that the NRD-525 is as close to the optimum shortwave listener's receiver as is in existence."

"Japan Radio has taken the features shortwave listeners have always sought...and packaged the lot into what is unquestionably the best overall shortwave listener's receiver on the market today."

"The NRD-525 exemplifies once again that Japan Radio receivers are for the connoisseur."

Larry Magne, International Broadcast Services

The JRC NRD-525 truly stands alone in performance and features! Enjoy exceptional sensitivity and selectivity coupled with rock-solid stability. Continuous coverage from 90.00 to 34000.00 Khz with readout to 10 Hz! Razor-sharp notch filter and passband tuning for digging out that weak DX! All modes are standard including FM and FAX! 24 hour digital clock timer with relay contacts. Incredible 200 channel scanning sweeping memory stores frequency, mode, bandwidth, AGC and ATT settings for each channel. Other standard features include keypad, RIT, MONITOR, AGC, ATT, BFO and dual NB. Available options include VHF/UHF, RS-232, RTTY Demodulator and a wide variety of filters. Operates from 110/220 VAC or 13.8 VDC! Write today for your full color brochure!

Japan Radio Co., Ltd.

Japan Radio Co., Ltd.
USA Branch Office   T. Yamaguchi
405 Park Avenue
New York, New York 10022
(212) 355-1180   Telex 961114

Universal Shortwave Radio
1280 Aida Drive
Reynoldsburg, Ohio 43068
Toll free: 1-800-431-3939
In Ohio: 614-866-4267

Gilfer Associates
52 Park Avenue
Park Ridge, New Jersey 07656
201-391-7887

HUGE
70 PAGE
SHORTWAVE CATALOG

SEE WHAT'S NEW IN...
Communications Receivers
Portable Shortwave Radios • Antennas & Supplies
Radioteletype Equipment
Facsimile (FAX) Equipment • Communications Books
Part & Accessories

SEND $1.00 (OR 3 IRCS) • REFUNDABLE

UNIVERSAL SHORTWAVE RADIO
1280 Aida Drive
Reynoldsburg, Ohio 43068
(614) 866-4267

SPECIAL!! BIG "D" Sale Prices

Regularly $389.95

SONY ICF-2010

SALE PRICE

CALL FOR SPECIAL PRICING

ICOM R7000A
KENWOOD R-2000
KENWOOD R-5000
YAESU FRG 8800
YAESU FRG 9500

ICOM R71A

Prices ON REQUEST

- 150 kHz to 20 MHz AM, CW, SSB
- 76 MHz to 108 MHz FM
- 116 MHz to 136 MHz AM Air Band
- 32 Programmable memories
- 4 Event Timer
- Synchronous Detector
- Wide/Narrow Bandwidth

1-800-441-0145 (IN TEXAS) 1-800-527-2156 • ask for Ham Dept.

ALSO ASK ABOUT super saving prices on Kenwood, Icom, Encomm, Yaesu and all ham items.

electronic center, inc.
ROSS AT CENTRAL EXPRESSWAY, DALLAS, TX 75201

MasterCard VISA

Photo page 67: Advanced receivers are made in many countries. Japanese listener Masayuki Uchiike uses German, Japanese, and American receivers to hear the world

1988 BUYER'S GUIDE TO WORLD BAND RADIO

HOW TO BUY
A WORLD BAND RADIO

World band radios have to receive signals from afar, so it's not surprising that some better models are on the leading edge of technology. That's the good news. Less encouraging is that there is no shortage of low-technology clunkers on the market.

Specialized Tests for Evaluating Radios

That's why, starting ten years ago, International Broadcasting Services conceived and carried out the concept of specialized systematic testing of world band radios. These tests have grown over the years to involve a number of individuals, including "hands on" test panelists and specialized laboratory personnel. The reports, which have nothing to do with advertising, form the basis of our "Buyer's Guide," and even more detailed data on sophisticated world band receivers is available via *RDI White Papers*.

Taken together, these findings indicate that while really cheap world band radios serve little good purpose except for occasional use on trips, it's not necessary to spend hefty sums to obtain a really pleasant radio. Quality advanced-technology portables list from $189, and sometimes sell for less.

Best-Value Performers Start at $189

The best values are usually found among portables in the $189–280 range that are given a two-star rating in the "Buyer's Guide." These perform surprisingly well and are more than adequate for most listeners' needs. Too, they can be used both in the home and on trips, and they don't require a special antenna.

Higher-priced portables can have advantages, especially if you are a seasoned listener or inclined to go for the best. Even costlier tabletop models are available, but these are primarily for ferreting out the weakest and most obstinate stations. Tabletop models usually require a separate antenna, too.

What Do You Want to Hear?

In the final analysis, it's what you're trying to hear that should determine what you purchase. If you're interested only in the strong, clear stations, then there's no need to spend a king's ransom on equipment designed to hear a fly scratch its back on a Pacific Island. Look for a reasonably performing portable with high-quality audio.

Ratings of Overall Performance

★★★★★ Superb
★★★★ Excellent
★★★ Very Good
★★ Good
★ Fairly Good
☐ Fair
◪ Fair to Poor
■ Poor

Classement Général

★★★★★ Superbe
★★★★ Excellent
★★★ Très Bon
★★ Bon
★ Assez Bon
☐ Moyen
◪ Moyen à Médiocre
■ Médiocre

Clasificación General

★★★★★ Magnífico
★★★★ Sobresaliente
★★★ Muy Bueno
★★ Bueno
★ Bastante Bueno
☐ Regular
◪ Regular asta Mediocre
■ Mediocre

Einteilung in Klassen

★★★★★ Ausgezeichnet
★★★★ Vorzüglich
★★★ Sehr Gut
★★ Gut
★ Ziemlich Gut
☐ Nicht Sehr Gut
◪ Nicht Sehr Gut zu Schlecht
■ Schlecht

Comparative Ratings of Portable World Band Radios

Portables these days are handy, affordable and widely distributed—not to mention fun. Trouble is, there are a number of models that are so awful as to be essentially useless. For example, some credit card companies and department stores have been known to prey on newcomers by offering "multiband portables," usually for around $100–130. These sets are so terrible that their disgusted owners almost invariably turn away from world band radio, blaming the medium instead of the outfit that sold them the radio in the first place. Had they spent only another $50 or so, they could have had a world class radio.

The main rule of the road is to remember that world band portables —unlike, say, TV sets—vary considerably in performance from model-to-model. We've tested models performing anywhere from poor (■) to very good (three stars on a scale of five). Only costly tabletop models earn an "excellent"— four stars—or "superb"—five stars.

Obviously, best results among portables are found among the two top-rated models. However, high-performance bargains can be found among such two-star models as the Philips/Magnavox D2935 and the various incarnations of the redesigned Sangean ATS-803. You can get an awful lot of performance for under $200!

Travel Specialty Sets Available

Most radios tested are compact enough for travel. For those who place a premium on traveling light, the Panasonic RF-B60, followed by the Sony ICF-2003/ICF-7600D, perform nicely and come equipped with a wide range of travel-oriented features. For bikers and others who must consider weight above all else, the Sony ICF-4920/ICF-4900 Mark II, followed by the Panasonic RF-B20, are the obvious choices.

Models are listed in order of suitability for listening to world band radio broadcasts. To facilitate the comparison of prices among competing models, US list prices are given whenever possible. Thus, once adjustments are made for local commercial, currency and tariff considerations, these prices serve as an effective worldwide index.

★★★ Sony ICF-2010
★★★ Sony ICF-2001D
$389.95

Advantages: Innovative synchronous detector circuit results in superior reception in world, longwave and mediumwave AM bands. Numerous tuning features. Separately displayed 24-hour World Time (UTC) clock. Alarm/sleep/timer facilities.
Disadvantages: Audio quality slightly below par for price class. Proliferation of controls and features may intimidate or confuse some. Minimally larger and heavier than desirable for airline travel.
Overall: The best. Reviewed in more detail elsewhere in this "Buyer's Guide."

★★★ Grundig Satellit 650
$1,149.00

Advantages: Above-average performance in all bands. Numerous helpful tuning features. Unusually pleasant audio quality. 24-hour World Time (UTC) clock. Alarm/timer facilities. Superior FM reception.
Disadvantages: Rather large and heavy for portable use. Relatively costly. Motorized preselector tuning creates unnecessary complexity.
Overall: A favorite for hour-after-hour listening, the '650 is arguably the best-sounding world band radio available new. Reviewed in more detail elsewhere in this "Buyer's Guide."

★★ Philips D2999
★★ Magnavox D2999
$399.00

Advantages: Unusually pleasant audio quality. Numerous helpful tuning features. Performance above average in most bands. 24-hour World Time (UTC) clock. Alarm facility.
Disadvantages: Somewhat large and heavy for air travel.
Overall: A fine-sounding world band receiver for the price. A complete test report on the Philips/Magnavox D2999 may be found in last year's edition of this book.

D-2999, sold by Philips and Magnavox alike, is larger and heavier than the cheaper D-2935

★★ Grundig Satellit 400
$449.00

Advantages: Pleasant audio quality. Numerous helpful tuning features. World band performance at least average. 24-hour World Time (UTC) clock. Alarm facility. Superior FM reception. Unusually handy carrying handle.
Disadvantages: Unstable synthesizer produces a "whooshing" sound in some world radio bands. Slightly larger and heavier than desirable for airline travel.
Overall: A pleasant all-around receiver. Reviewed in more detail elsewhere in this "Buyer's Guide."

★★ Panasonic RF-B60
$279.95

Advantages: Very easy-to-operate state-of-the-art radio. Light and compact for travel. Pleasant audio quality. Numerous helpful tuning features. 24-hour World Time (UTC) clock. Alarm/sleep facilities. Two-year warranty.
Disadvantages: Tuning knob cumbersome to turn rapidly.
Overall: A very nice all-around portable for listening to favorite programs. Reviewed in more detail elsewhere in this "Buyer's Guide."

★★ Philips D2935
★★ Magnavox D2935
$249.99

Advantages: Pleasant audio quality. Numerous helpful tuning features. Performance good in most bands.
Disadvantages: Pushbuttons lack "feel." Slightly larger and heavier than desirable for airline travel.
Overall: A simplified and more compact version of the Philips/Magnavox D2999, the D2935 is an exceptional value. Reviewed in more detail elsewhere in this "Buyer's Guide."

Panasonic RF-3100 is one of Panasonic's early synthesized designs

---

★★ Sony ICF-2003

★★ Sony ICF-7600DS

$269.95

Advantages: Light and compact for travel. Numerous helpful tuning features. Separately displayed 24-hour World Time (UTC) clock. Alarm/sleep facilities.
Disadvantages: Only modest sensitivity with built-in antenna. Pedestrian audio quality. Lacks tuning knob.
Overall: A high-quality portable for air travel. Re-Reviewed in more detail elsewhere in this "Buyer's Guide."

★★ Panasonic RF-3100

★★ National DR-31

$499.95

Advantages: Fairly pleasant audio quality. Two-year warranty.
Disadvantages: Overall performance slightly below average for price class. High battery consumption. Somewhat large and heavy for air travel.
Overall: A pleasant, but not distinguished, performer.

★★ Sangean ATS-803

★★ EEB 2020

★★ Realistic DX-440

★★ Eskab RX33

$189.00–199.95

Advantages: Numerous helpful tuning features. Performance slightly above average.
Disadvantages: None.
Overall: An unusually worthy value. Reviewed in more detail elsewhere in this "Buyer's Guide."

Radio Shack's new DX-440 "Voice of the World" portable

★★ Sony ICF-PRO80

★★ Sony ICF-PRO70

$419.95

Advantages: Comes equipped with versatile VHF scanner. Fairly light and compact for travel. Good at bringing in weak world band stations.
Disadvantages: Cumbersome to operate. Mediocre audio quality. Lacks tuning knob and most travel features.
Overall: Of interest mainly to "DXer's" in need of something smaller than an ICF-2010, plus VHF scanner listeners who also tune in world band broadcasts. Reviewed in more detail elsewhere in this "Buyer's Guide."

★ Sony ICF-7700
★ Sony ICF7600DA
$269.95

Advantages: Very easy to use. Light and compact for travel. 24-hour World Time (UTC) clock. Alarm/sleep functions.
Disadvantages: Coarse tuning increments create performance shortcomings. Limited coverage of world band spectrum.
Overall: For the technically timid. Reviewed in more detail elsewhere in this "Buyer's Guide."

☐ Panasonic RF-B20
$109.95

Advantages: Very light and compact for worldwide air travel. Unusually good audio for such a small radio.
Disadvantages: Limited coverage of world band spectrum. Mediocre ability to reject adjacent-channel interference. Lacks digital tuning display.
Overall: Would be the top mini were it able to sort stations out more successfully.

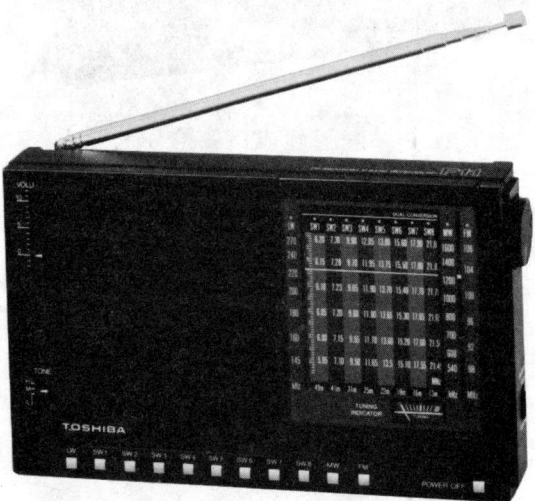

Toshiba's RP-F11 is one of the few compact portables with an analog signal strength meter

☐ Toshiba RP-F11
$129.95

Advantages: Light and compact for travel.
Disadvantages: Limited coverage of world band spectrum. Lacks digital tuning display. Modest overall performance.
Overall: Reasonable performer for the price.

★ Sony ICF-4920
★ Sony ICF-4900 Mark II
$99.95

Advantages: Exceptionally light and compact for worldwide air travel.
Disadvantages: Limited coverage of world band spectrum. Lacks digital tuning display. Mediocre audio quality.
Overall: The best of the minis, the '4920—new for 1988—is nonetheless of interest almost exclusively to the weight-conscious traveler.

Sony introduced the ICF-4920 for 1988

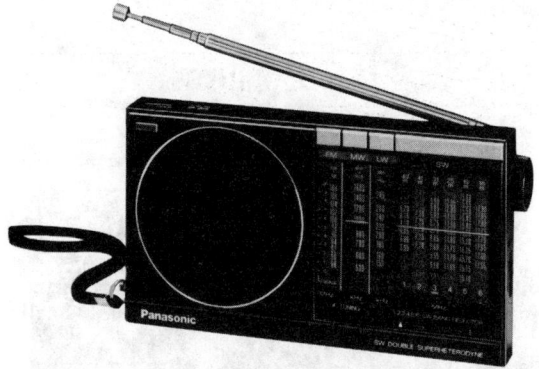
Panasonic's miniature RF-B20 travel portable

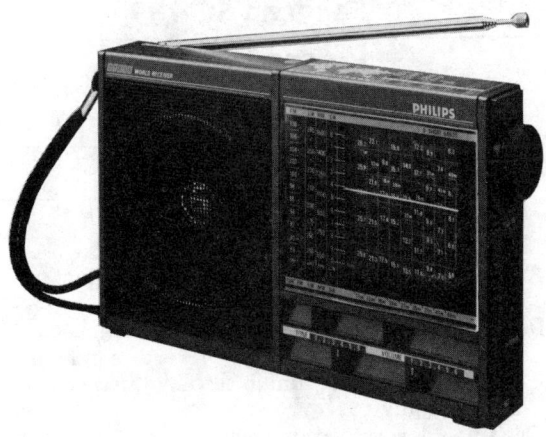
Model D-1835 compact portable is sold under both Philips and Magnavox names

☐ Silver XF1900

About $150.00 (Europe only)

Advantages: Fairly pleasant audio quality. 24-hour World Time (UTC) clock. Sleep/timer functions.
Disadvantages: Mediocre overall performance. Does not receive 120, 90 and 75 meter world bands.
Overall: An out-of-date design, but interesting because of its moderate price.

◪ Philips D1835
◪ Magnavox D1835

$69.99

Advantages: 24-hour World Time (UTC) clock. Sleep/timer functions.
Disadvantages: Mediocre overall performance.
Overall: Better choices are available near this price range.

Silver XF-1900 provides only fair performance

☐ Grundig Yacht Boy 700

About $250.00 (Europe only)

Advantages: Light and compact for travel. Inexpensive.
Disadvantages: Substandard overall performance. Limited coverage of world band spectrum. Lacks digital tuning display.
Overall: A low-priced, modest performer.

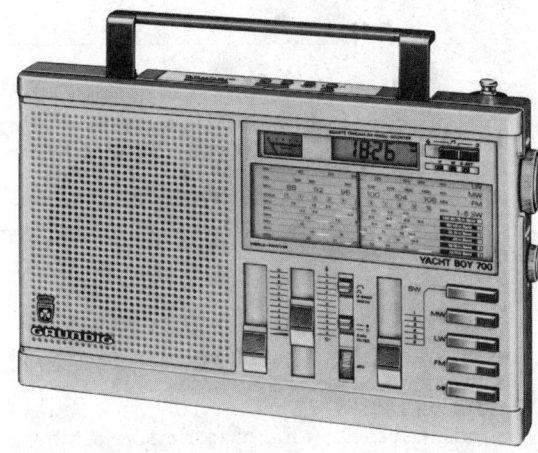
Grundig's Yacht Boy series performs far less well than its Satellit models

■ Sangean SG-789
■ EEB 4950
■ Emerson PSW4010
$69.95

Advantages: Exceptionally light and compact for worldwide air travel. Inexpensive.
Disadvantages: Mediocre overall performance. Somewhat limited coverage of world band spectrum. Lacks digital tuning display. Mediocre audio quality.
Overall: A low-priced, modest mini portable of interest almost exclusively to the weight- and price-conscious traveler.

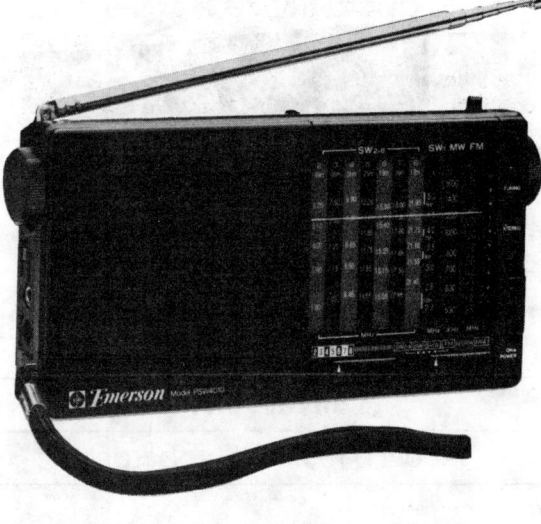

Sangean's SG-789 is sold under various names, including Emerson

Sangean ATS-801

■ Sangean ATS-801
$99.00

Advantages: Light and compact for travel.
Disadvantages: Dreadful overall performance. Tortoise-slow tuning. Limited coverage of world band spectrum. High battery consumption for compact radio.
Overall: A model with little to commend it.

■ Realistic DX-360
$99.95

Advantages: Fairly pleasant audio quality.
Disadvantages: Mediocre overall performance. Does not receive 120, 90 and 75 meter world bands. Lacks any form of reasonably accurate tuning display.
Overall: Barely acceptable. Aficionados of Radio Shack products should look, instead, to the new DX-440.

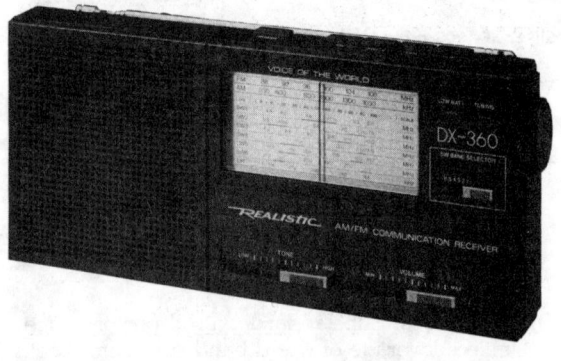

Realistic DX-360 is stripped-down version of Silver XF-1900

Discontinued Models

The following models have reportedly been discontinued, but may still be available new at some retail outlets. Prices are in the range of actual or estimated sale prices as of the time this book went to press. Because these are discontinued models, prices vary widely and interesting bargains can sometimes be struck with retailers seeking to unload outdated inventory.

★★★★ Sony CRF-1
About $1,000.00

The CRF-1 is a moderately large semi-professional communications receiver suited to field portable applications. Nearly every aspect of performance is top-drawer—dynamic range, selectivity, sensitivity, image rejection . . . the works. Only the "wide" bandwidth and audio quality disappoint, plus tuning features are limited.

★★★ Panasonic RF-9000
★★★ National DR-90
Under $3,800.00

With excellent audio quality and overall performance in all bands, plus numerous helpful tuning features, this one-time pride of Panasonic's fleet is nonetheless much too large and heavy for portable use. Grossly overpriced when in Panasonic's line, it may be available new here and there at steep discounts. If you can find it for under $1,000, it could be worth considering.

★★★ Grundig Satellit 600
Under $800.00

Nearly identical to Grundig Satellit 650.

★★ Sony ICF-2002
★★ Sony ICF-7600D
Under $250.00

Essentially identical to Sony ICF-2003.

★★ Panasonic RF-2900
★★ National DR-29
About $175.00

Very similar to the General Electric World Monitor II. Especially interesting is a high-performance version sold in the United Kingdom some years back. For more details, see the description of the GE World Monitor, below.

★★ General Electric World Monitor
Under $235.00 (US only)

This model has modest frequency stability and a tendency to produce unwanted signals 6 MHz down. For air travel, the World Monitor is larger and heavier than many other portables. Lacks some of the snazzier "bells and whistles" found on a number of newer models. However, in most other respects—notably audio quality—it produces results on a par with some more costly world band receivers.

★★ Sony ICF-2001
Around $170.00

The '2001 is unselective, letting in interference from adjacent stations, and its keypad is prone to failure. Battery consumption is excessive. Otherwise, in

Sony's established ICF-2002 has been replaced for 1988 by the ICF-2003

most respects it's like the smaller and lighter ICF-2003/ICF-7600DS, except that the '2003 is a bit less sensitive and much more selective.

tions. Selectivity mediocre. In fact, neither its sensitivity nor dynamic range is appropriate to a set in this price class.

★★ Panasonic RF-2200
★★ National DR-22
Under $180.00

Overall, performance is only slightly inferior to that of the General Electric World Monitor, above, to which the '2200/'22 adds something unique: a fully rotatable mediumwave AM antenna. The rub is that in lieu of a straightforward digital frequency readout, the '2200/'22 utilizes a Rube Goldberg arrangement to tune in stations. For most listeners, this complexity is too much to bear.

★ Panasonic RF-B50
★ National DR-B50
About $150.00

The Panasonic RF-B50 has two bandwidths and above average audio quality for a compact portable. The overall FM and mediumwave AM performance are also good for such a small set. The 'B50 covers most, but not all, of the world radio spectrum. It utilizes an analog, rather than digital, frequency readout. It's also not very sensitive to weak signals.

★ Supertech SR-16H
Around $125.00 (Europe only)

The Supertech SR-16—an early and unselective version of the Sangean ATS-803—is a low-cost buy for world band listening in Europe. For more details, refer to the test report, "Affordable Portable II," on the early ATS-803 in last year's edition of this book.

★ Sony ICF-4910
★ Sony ICF-4900
Under $100.00

Identical in all but styling to Sony ICF-4920/ICF-4900 Mark II.

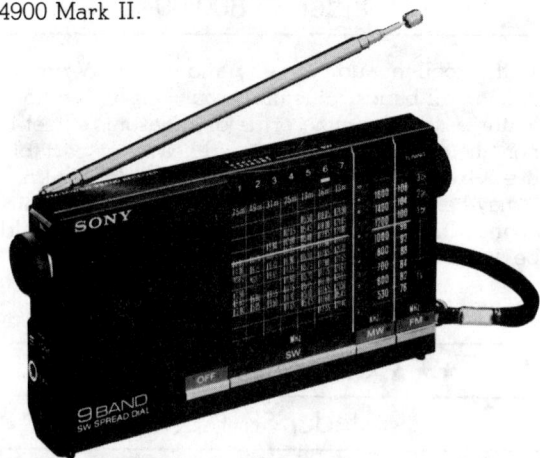

Sony's mini-portable ICF-4910 has been replaced for 1988 by the ICF-4920.

★ Sharp FV-610GB
Under $170.00

The FV-610GB—sold in various parts of the world with different suffix letters—is notable mainly for its relatively high sensitivity and concomitant propensity to "overload." Coverage of the world band spectrum is incomplete and selectivity is only fair.

★ Panasonic RF-6300
★ National DR-63
Around $350.00

The '6300/'63 has pleasant audio for listening to world band and local broadcasts alike. Includes twelve "memory" buttons for storing favorite sta-

---
☐ Kenwood R-11
☐ Trio R-11
Under $125.00

Virtually identical to the Toshiba RP-F11.

☐ Sony ICF-7600A
☐ Sony ICF-7600AW
About $150.00

Very similar to the Sony ICF-7700/ICF-7600DA in size, shape and performance, the '7600A covers most, but not all, of the world radio spectrum with an analog frequency readout. Overall performance, except for the frequency readout, resembles that of the newer '7700.

☐ Grundig Satellit 300
Under $250.00 (Europe only)

The Grundig Satellit 300 is longer on features than on performance. World band performance is uninspiring in nearly every respect, with its major shortcoming being insensitivity to weak signals.

☐ Sharp FV-310GB
Under $140.00

Identical to the Sharp FV-610GB, above, except that it lacks digital frequency readout, clock/timer and stereo FM.

☐ Grundig Yacht Boy 650
Under $200.00

The Grundig Yacht Boy 650 covers most, but not all, of the world radio spectrum used for broadcasting. Frequency readout is both analog and digital, with the latter sharing the display with a 24-hour World Time (UTC) clock/timer. Audio quality is above average in quality and strength, and FM performance is similarly commendable. However, world band radio performance is only so-so, overall.

☐ Panasonic RF-799
Under $300.00

The '799 incorporates features in profusion. However, it doesn't cover the entire world radio spectrum. Overall performance is mediocre, especially for a set in this price class.

■ Panasonic RF-9
■ Panasonic RF-9L
■ National Micro 00
Under $100.00

This miniature radio, similar in appearance to Sony's respectable ICF-4920/ICF-4900 Mark II, has only size working in its favor. Unselective, with false signals aplenty—plus incomplete coverage of the world radio spectrum and tinny audio—the RF-9 is more a toy than a serious radio.

WHY YOU SHOULD—OR SHOULD NOT—
BUY A TABLETOP RECEIVER

Most of us who have discovered world band radio are delighted with what we hear on our portables. But if you want to have a greater chance of unearthing the fleapowered local stations that proliferate in the lower "tropical" radio bands—60, 90 and 120 meters—you may wish to consider a tabletop receiver. These elite models aren't cheap, plus they require a good outboard antenna and electrically quiet location to strut their stuff. But when all the pieces fall into place, these supersets can really perform.

Hearing What the Home Folks Hear

For most listeners, the point of spending a lot of money on a world band radio is to eavesdrop on what the home folks are hearing. In the exotic countries to the south, world band is often used instead of FM or mediumwave AM for local radio broadcasts. These domestic broadcasts aren't intended to be heard outside the home turf. But at night in the winter they are, anyway, thanks to world band radio's exceptional ability to propagate signals around the globe.

Local Signals Often Heard Far Away

The rub is that these stations are intended for local audiences. This means that they're usually far less powerful than the international giants that come in so nicely on $250 portables. You need a highly sensitive set to flush these low-powered signals out. And, because there are so many stations on the air, you need a set that is unusually selective to lasso any given one of these little radio heifers from the thundering herd.

Sets that are unusually well-equipped to pick up these sorts of signals qualify as "DX receivers," DX being an old telegraph term to indicate great distance. Technically, almost all world band radio signals come from far away, so the term "DX" has evolved to mean weak signals—not just any signals—from afar.

That's not all. The world radio bands used for these local tropical broadcasts in equatotorial countries are also used for other and specialized types of transmissions in countries to the north. Some of these make "dih-dah" sounds which can be eliminated only by special circuitry within a radio.

It's a different radio world within the tropical broadcasting bands. If you listen to Albania's external services, targeted to foreigners, in the higher world radio bands, you'll hear lectures on Marxism-Leninism-Stalinism that only a commissar could love. But if you listen to Albania's home service in the tropical bands, you'll hear a stream of haunting Balkan music. Radio Australia beamed to the world outside Australia will tell you what officialdom wants foreigners to hear. But if you listen to the local ABC outlet in Perth, you'll be treated to such tidbits as shark alerts at the beaches.

Not all countries broadcast both external and domestic services via world band radio. Some, particularly in Latin America, broadcast domestic services only. There's no other way to hear these countries.

Either you like this kind of eavesdropping or you don't. If you do, you'll need the right equipment if you want to hear them as well and as often as possible. This is where high-quality tabletop receivers—at least those with good audio quality—and outboard antennas come into the picture.

Local Broadcasts Often Beamed Abroad

Fortunately, it's not necessary to purchase a costly tabletop receiver to hear a good number of the exotic domestic stations operating within the world radio bands. In many cases local broadcasts within these bands are targeted abroad, rather than to the home audience. Usually these are via powerful transmitters and are intended to be heard by expatriates, mariners and diplomats abroad. But, with a good radio, anybody can listen.

These dozens of outlets include "Radio Luxembourg" in English, French and German; Iran's "Familiar Voice" in Farsi; Israel's "Reshet Bet" in Hebrew and "Reshet Aleph" in English, French and others; Turkey's "TRT"; Sweden's "Sveriges Riksradio" (also in upper sideband); and numerous transmissions from the Arab World. Some of the most interesting music on the airwaves is heard on these stations, so language is not a make-or-break consideration. And nearly all are easier to hear than signals intended strictly for domestic consumption.

Comparative Ratings of Tabletop World Band Receivers

Results with state-of-the-art portables can be truly satisfying. But for some, satisfying isn't good enough. They seek the best the marketplace has to offer, and so turn to tabletop world band receivers.

What Tabletops Will and Won't Do

The best tabletop models will always outperform the best portables—but only if you're trying to hear the really weak and difficult signals. It's something like a Ferrari, which can outrun the pack on a winding mountain road but does no better than a Toyota van on a crowded freeway. Tabletop sets also tend to be better made and easier to repair than portables.

Although there are exceptions, performance among tabletops tends to rise with price, so good models aren't cheap. Additionally, tabletop models usually aren't self-contained. This means you have to go through the rigmarole of setting up an outboard antenna.

Another point is that audio quality is only fair with some models. If you're fond of listening to music over powerful world band radio stations, those models can be a poorer choice than some of the better portables. Fortunately, there are some tabletop models that sound quite good, even if none quite equals the aural quality of the Grundig Satellit 650 portable.

Because they are targeted to an elite of world band radio listeners, tabletops tend to be harder to find than portables. But the good news is that many of the electronics specialty firms that carry them also provide advice and service after the sale.

The bottom line is that we have found only four tabletop models—all costly—that clearly outperform the best of the available portables. If you're a dedicated world band listener, one of these high-quality tabletops may be ideal. But if you're a relative newcomer to the medium, or are looking for your first set, stick to the portables. If you decide later on to get a tabletop receiver, you can always use the portable for trips and listening on the balcony or backyard.

How They Perform

Among the three tabletops—two from Japan Radio, one from Kenwood—given our top five-star rating, differences in performance are relatively minor. However, differences in price can be considerable. Any of these three top sets should be a delight to own, with the only disappointment being the collective absence of fidelity-enhancing synchronous detection.

Those who are interested mainly in tuning in weak, hard-to-hear signals may also wish to consider the ICOM IC-R71. Note, though, that this otherwise-outstanding model suffers from audio that some find tiring to hear hour-after-hour.

If these models are too pricey for you, the Yaesu FRG-8800 or Lowe HF-125 may be appealing. Although neither is a star performer, both provide reception quality that is quite pleasing.

Receivers are listed in descending order of suitability as devices for listening to world band broadcasts. Keep in mind that if you listen more to weaker stations—or speech, such as newscasts—audio quality will be of less importance than if you listen mainly to musical programs over strong, clear stations.

★★★★★ **Japan Radio NRD-93**

$7995.00

Advantages: Superb reception of all types of world band signals. Exceptional durability and quality of construction. Unusually easy to repair. Excellent ergonomics. Above-average audio for a tabletop model. Sophisticated optional scanner allows for a certain degree of automated listening.

Disadvantages: Dreadfully expensive! Lacks certain advanced-technology controls.

Overall: A sterling professional-quality receiver for those to whom money is no object. Nonetheless, performance is not appreciably different from that of less-costly models, such as the Kenwood R-5000 and Japan Radio NRD-525.

Japan Radio NRD-93 with NDH-93 "intelligent" scanner, for when money is no object

---

★★★★★ **Kenwood R-5000** (except UK)

★★★★★ **Trio R-5000** (UK only)

$899.00

Advantages: Superb reception of all types of world band signals. Unusually good audio for a tabletop, provided a suitable outboard speaker is used.

Exceptionally flexible operating controls.
Disadvantages: Ergonomics only fair. Replacement of standard, but mediocre, wide bandwidth filter with high-quality substitute adds $70 to cost.
Overall: Superb performance at a non-stratoshperic price. Reviewed in more detail elsewhere in this "Buyer's Guide."

★ ★ ★ ★ ★ Japan Radio NRD-525
$1,285.00

Advantages: Superb reception of all types of world band signals. Highly flexible operating controls. Very good ergonomics. Quality of construction slightly above average. Modular plug-in circuit boards enhance ease of repair. Sophisticated scanner allows for a certain degree of automated listening.
Disadvantages: Audio quality only fair. Surface-mounted devices make replacement of discrete components very difficult.
Overall: A superb performer that's well put together. Reviewed in more detail elsewhere in this "Buyer's Guide."

★ ★ ★ ★ ICOM IC-R71
$949.00

Advantages: Superb reception of weak, hard-to-hear signals. Highly flexible operating controls.
Disadvantages: Mediocre audio quality. Substandard ergonomics. Operating system software erases, requiring reprogramming by ICOM service center, if backup battery dies.
Overall: A DXer's treat, but not equal to most comcompeting models for quality-audio program listening. Reviewed in more detail elsewhere in this "Buyer's Guide."

★ ★ ★ Yaesu FRG-8800
$699.95

Advantages: Flexible operating controls. Audio quality slightly above average for tabletop model. Fairly good ergonomics. 24-hour World Time (UTC) clock.
Disadvantages: Bandwidths too broad as compared to most other tabletop models. Reportedly above-average rate of microprocessor failure, usually user-correctable.
Overall: A well-balanced performer. Reviewed in more detail elsewhere in this "Buyer's Guide."

★ ★ ★ Lowe HF-125
£375 in UK
$599.95

Advantages: Unusually straightforward to operate for tabletop model. Generous bandwidth flexibility. Most physically rugged receiver tested, except for Japan Radio NRD-93. Very good audio for tabletop model. Optional battery pack allows for field portable use.
Disadvantages: Some operating controls provide limited flexibility. Reception of world band signals suffers from slight insensitivity to weak signals, digital-circuit "whine" and excessively slow automatic-gain control recovery.
Overall: A pleasant, tough, easy-to-operate little unit. Note that circuit improvements may be forthcoming in later production units. Reviewed in more detail elsewhere in this "Buyer's Guide."

★ ★ Ten-Tec RX325
$699.00

Advantages: Fairly straightforward to operate for tabletop model. Physically rugged. 24-hour World Time (UTC) clock.
Disadvantages: Overall performance somewhat below par.
Overall: A decent performer. Reviewed in more detail elsewhere in this "Buyer's Guide."

★★ Kenwood R-2000 (except UK)
★★ Trio R-2000 (UK only)
$649.95

Advantages: Straightforward to operate, with good ergonomics. Audio quality slightly above average for tabletop model. 24-hour World Time (UTC) clock.

Disadvantages: Overall performance materially hampered by "overloading." Bandwidths too broad as compared to other tabletop models.

Overall: Generally pleasurable for listening to most world radio programs, but not much more. Aficionados of Kenwood products should look, instead, to the '2000's superb new sibling: the R-5000.

Kenwood's lowest-priced receiver, the R-2000

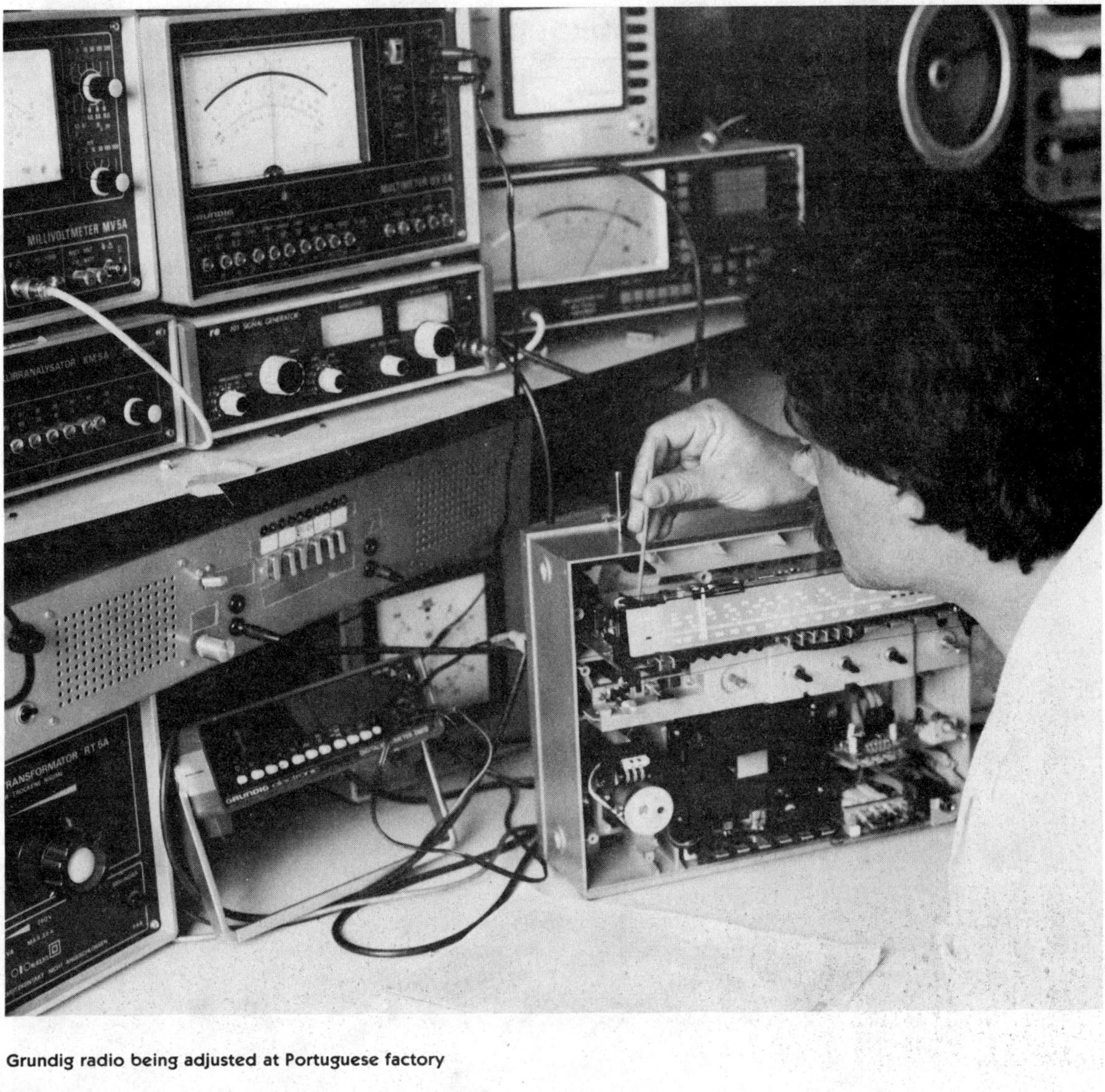

Grundig radio being adjusted at Portuguese factory

☐ Heathkit SW-7800
$349.95

Advantages: Lowest-priced tabletop model tested.
Disadvantages: Poorest performance of any tabletop tested. Available only in kit form, requiring dozens of hours to construct.
Overall: With the marriage of such names as "Heathkit" and "Zenith"—now the parent company of Heathkit—you'd think the SW-7800 would be a world-class receiver. It isn't. Unless kits are your special passion, you'd be better off spending the same money on a much-superior Sony ICF-2010 or ICF-2001D portable.

Heath, now part of Zenith, offers the only world band radio available in kit form

KILOBUCK RADIO CAN'T BE TUNED

Liniplex F1 non-tunable radio, to be joined by tunable F2 in 1988

. . . But Help Is on the Way

You assume that when you take a radio home, you'll be able to sit down and tune it to whatever channel you want. But with the circa $1,000 Liniplex F1 this isn't so. You get nine factory-preset channels, and no more. If you want to change any, you're out of luck unless you buy some parts from the manufacturer in Reading, England.

This makes the Liniplex F1 about as popular with the public as castor oil. Nevertheless, tuning aside, the F1 performs quite well, even though it has only one bandwidth and a tiny speaker. Its high point is a synchronous detector—the same type of high-tech circuit that makes the Sony ICF-2010/ICF-2001D so special. It performs flawlessly, even better than that of the '2010.

All of this hasn't escaped the attention of the F1's manufacturer, Phase Track. It's announced that it will shortly be unveiling a model, the F2, which—*mirabile dictu*—can be equipped to be tuned. The new F2, due out sometime in 1988, will be based on the F1 and will include an analog signal-strength indicator. Add to this the synchronous detection feature of the F1 and this should be a radio of exceptional interest to serious world band listeners.

Discontinued Models

A number of discontinued models are still available new on dealer shelves. However, unlike discontinued portables, discontinued tabletop models are rarely heavily discounted. Indeed, in the case of some prized models prices may be higher than they were when the sets were still in the manufacturer's catalog.

Here's a glance at some of the more recently discontinued models you might come across in your wanderings through electronics stores and catalogs. Prices are estimated or recent actual *selling* prices for new units.

★★★★★ Japan Radio NRD-515
Under $1,500.00

The Japan Radio NRD-515 is built like a fortress and provides intelligible signal reception under all but the most excruciating situations. Operation is straightforward and pleasant, but the '515's woolly audio is not well-suited to the reproduction of music.

★★★ Sony ICF-6800W
Under $600.00

Lacking the "bells and whistles" found on most other tabletop models, the Sony ICF-6800W provides unusually well-rounded and pleasant results, along with uncomplicated operation. In particular, its audio quality is well above average. Another unusual plus with the '6800 is that it operates off either ac power or batteries. Its chief shortcoming lies in its modest dynamic range, which can result in "overloading," with stations sounding as though they're piled atop each other when they really aren't.

★★★ McKay Dymek DR 101-6
Under $1,000.00

The best of the McKay Dymek line, the DR 101-6 manages to incorporate McKay's innovative ideas, such as a variable-rate slewing knob, without some of the drawbacks evidenced on earlier versions of the DR 22 and DR 33 series. As with other models in the McKay line, the DR 101-6 comes equipped with a substandard speaker and mediocre front-end selectivity. But with a worthy external speaker the end result can be quite pleasant.

★★★ Yaesu FRG-7700
About $550.00

Similar in performance to the newer FRG-8800, but lacking many of the '8800's "bells and whistles," the '7700—which has three bandwidths, instead of the '8800's two—can be an attractive alternative when it is priced to move.

★★ Kenwood R-1000
★★ Trio R-1000
Under $500.00

The first of Kenwood's modern world radios to come on the market, the R-1000 was an early favorite among radio aficionados and electronics experimenters. Although the current Kenwood R-2000 is in many ways a more modern receiver than the older R-1000, the '1000 is less prone to "overloading," and thus continues to be preferred where long antennas are used.

★★ McKay Dymek DR 33-C
Under $1,000.00

Some years back, McKay Dymek decided to produce world band radio receivers that would look like and interface with conventional stereo hi-fi components. The earliest models, the DR 22 and DR 33 series, were not entirely successful performers. A small quantity of the final version of the DR 33-C— the DR 33-C/6—may still be on hand, new, at Stoner Communications' warehouse.

★ Kenwood R-600
★ Trio R-600
About $400.00

Kenwood's low-cost receiver, the R-600, provides decent, uncomplicated world radio reception with a minimum of operating controls and features. Introduced on the market long after the appearance of the R-1000, and just as advanced-technology portables were presenting serious competition, the R-600 was too little, too late.

★ Bearcat DX1000
Under $600.00

When Bearcat, long known as a pioneer in the manufacture of scanning devices, announced its first world band receiver, it was expected that something advanced and quite special would emerge. Instead, what appeared was a thoroughly servicable, but uninspiring, radio seemingly designed to be produced at rock-bottom cost. Given the several better alternatives, there is precious little reason to select this model.

★ Panasonic RF-4900
★ National DR-49
Under $500.00

An otherwise-modest performer with surprisingly pleasant audio quality, the Panasonic RF-4900 is notable because it operates off both ac power and batteries.

The Perfect Starter Radio

SANGEAN ATS-803
EEB 2020
REALISTIC DX-440
ESKAB RX33

All the best portable world band radios come from Japan. Well, that's how it used to be. Today, it's simply not true.

Japanese domination of the field is declining rapidly. The upward spiral of the yen has sent prices through the roof and confusion over trade restrictions in some countries has cast a shadow over the market. Too, in some parts of the world, distribution channels for such in-demand models as the Sony ICF-2010 have fallen into disarray, making it difficult for some dealers to obtain receivers for their customers.

As a result, many people have been turning to other countries for their world band radios. Fortunately, there appears to be no shortage of places willing to jump into this growing void. Chief among these is Taiwan.

The hub of activity in this country is centered on a company known as Sangean Electronics, Inc. Until recently, Sangean (pronounced "san-GEE-uhn") was hardly a household name, best known only in that shadowy subculture of department stores buyers and the like. All over the world, thousands of Sangeans masqueraded as Supertech SH-16H's and goodness only knows what else.

Special Versions Found Worldwide

Today, the Sangean line is in the process of stepping out of the closet, and with good results. The pride of the fleet, the ATS-803, is often called the "poor man's '2010"—a reference to the $300 + price tag on the Sony ICF-2010. The radio has a lot going for it, most notably the fact that it retails for under $200.

Still, to be an affordable portable, a radio's got to be worth its price. And the ATS-803 basically is. In addition to its affordability, it boasts a keypad for easy frequency selection, as well as the traditional knob. A digital display shows where you are tuned, and good stability prevents the radio from drifting away from that spot on the dial. There's also a modest assortment of gadgets, like memory buttons for storing your favorite stations, and so forth. The radio even looks roughly like the '2010.

Early models of the '803 weren't without their drawbacks, however. Most appalling was the lack of selectivity. The filter was so wide that when you tuned in one frequency, you'd hear a number of others at the same time. So while you might be trying to listen to the BBC by tuning in 6175 kHz, you'd also hear Radio Interference on nearby 6170 and its sister station, Radio Whino, on 6180 kHz. Add to that a package of lesser annoyances, and you had a radio that was worth the price—but only vis-à-vis the competition.

Enter Dick Robinson of the Electronic Equipment Bank. Robinson, a Sangean dealer, noted that, with a very simple modification, he could correct the ATS-803's most annoying problem, the lack of selectivity. Pleased with the results, he ordered a number of the radios, with modification, made for his own company, which he now sells under the name "EEB 2020."

But the people at Sangean are no fools. And herein lies the greatest promise the radio has for world band listeners: evolution. Every time someone—à la Robinson—tells the people at Sangean about a way to improve their radios, they incorporate the design changes into the Sangean line. As a result, the '803 has become a constantly improving radio.

Numerous Features Add to Versatility

This year, the ATS-803 continues its climb up the evolutionary ladder. With a wide range of features, reasonably good performance and pleasant audio quality, it is an even better value for its price tag than last year's. It's also surfaced anew, at $199.95, as Radio Shack's "Realistic DX-440", and continues to be sold as the EEB 2020 and the Eskab RX33. All these are essentially identical to the '803 with Sangean's label.

In order to correct the earlier selectivity problem, the 1988 version has two selectivity positions: full fidelity, or "wide," for stations in the clear; and "narrow" for stations hemmed in by interference. In this respect, it's now unusually flexible for an affordable portable. A number of other changes—improved sensitivity and dynamic range, for example—have also been made for 1988, as has the elimination of a strange "thooping" sound that used to appear as the radio was being tuned.

The '803 remains a highly stable receiver, making retuning unnecessary. Sensitivity with the built-in telescopic antenna is quite good, too. This allows the set to bring in a variety of both weak and strong stations without having to be hooked up to an outdoor antenna. And it works well in just about any part of the world, covering mediumwave AM, FM stereo, plus the entire world band radio spectrum.

The '803 is fully portable and self-contained, being well-suited for both travel and in-home use. It's powered not only by batteries, but also by a dual-voltage ac converter that works in most countries. There's also a digital readout so you can tell immediately what channel you're tuned to, plus an LED signal-strength indicator. A dial light, separate bass and treble tone controls, sleep-delay timer, clock-alarm/timer, and power lock—to keep the set from accidentally switching on while stuffed in a suitcase—round out the '803's list of features. In all, it's a fine showing and well worth consideration.

WANT TO TUNE INTO MORE STATIONS?

WE HAVE THE FREQUENCIES!

Whether you want to monitor Interpol, Aircraft, News Agencies, the Military, or just listen to the BBC World Service we have the books to show you how.

Tens of thousands of frequencies are listed. Whatever your interest we will show you how to find the stations you want. What's more we will tell you which books are suitable for your country.

So send today for our free catalogue.

INTERBOOKS

RD2, Stanley, Perth, PH1 4QQ, Scotland.
Tel: (0738)-828575. International
+44-738-828575 (24 Hrs)

High-Tech Radio for the Technically Timid

SONY ICF-7700
SONY ICF-7600DA

The old Sony ICF-7600/7600A series was one of the great success stories of world band radio. In fact, its sales managed to pass the million-unit mark some time back—no small feat in the world band radio market. The main ingredients in the '7600's recipe were good performance, combined with compactness and ease of operation. As a result, for years the '7600 was the radio of choice for foreign correspondents and overseas government personnel. Even the globe-trotting U.S. Secretary of State Henry Kissinger had one for keeping in touch with the world during his myriad trips abroad.

High Tech Made Easy

It's in this spirit that Sony has replaced the '7600A with the totally new ICF-7700, sold outside North America as the ICF-7600DA. In a lot of ways, Sony has decided not to upset its successful applecart. While the 1988 '7700 is a high-tech redesign inside, outside it looks and operates very much like the '7600—with a few rather unusual exceptions.

The first clue to this is the dial. At first blush, it looks just like a regular analog dial, with numbers painted on for the various frequencies and a needle that goes up and down to show where you're tuned.

But here's the catch. This isn't an ordinary dial and needle at all. Instead, it's a huge LCD that's rigged up to look like a traditional dial face and needle. Turn on the radio and the needle appears; turn it off and it's gone.

All dial-and-needle frequency readouts are inherently imprecise, and the ersatz version on the '7700 is no exception. It moves jerkily, only every 20 to 25 kHz or so, functioning more as a pacifier for the technically timid than as a frequency indicator. So Sony went ahead and added an exacting digital frequency display to provide true accuracy.

That's not all about the '7700 that's designed to gladden the hearts of the technologically faint-of-heart. There is no keypad for frequency selection, only a standard tuning knob. Again, Sony has indicated that they don't want to intimidate customers.

Yet another simplification is that the set tunes only in 5 kHz single-channel increments, and there's no control to fine-tune between these 5 kHz points. To compensate for this, the '7700 uses a single wide bandwidth filter, one wide enough to let in stations a bit away from where you're tuned. As a result, the set is not terribly selective and gives out more than the usual assortment of howls and squeals. And since you can't detune, there's absolutely nothing that can be done about it.

Too, the '7700 segments the world radio spectrum into a dozen separate broadcasting bands. And, sure enough, the radio will tune only within those bands. So, if you want to hear, say, Iran on 9022 kHz, you're out of luck. Still, the coverage is quite generous, so the number of missed stations is small.

There are, however, five easy-to-use push-tuning buttons, like on a car radio. When you're listening to FM, you get five buttons for FM. When you're listening to world band radio, you get five for that, and so on. In all, a total of fifteen channels can be stored in the radio's "memory," which is enough for most people.

Features for Traveling

Because the '7700 is meant to be simple to operate, its list of features is largely limited to those of use to the traveler. These include a 24-hour digital clock, a sleep switch to allow you to doze off with the radio going, an alarm/clock-radio function to rouse you from your slumber, and a power safety switch to prevent the set from going on accidentally in your suitcase. There's also a tuning knob, a slider-type volume control, a two-step tone control, a single-LED "glow light" to aid in tuning, and an elevation leg to angle the radio towards the listener. Both the longwave and extended—to 1700 kHz—mediumwave AM bands are covered. FM includes both the usual 88–108 MHz band and the lower Japanese FM band that starts at 76 MHz.

For whatever reason, no dial light is provided. For $270—yen rise or no yen rise—it hardly seems reasonable to expect listeners to squint or light matches to see where they're tuned. The '7700, as you would expect from a set designed to be very simple to operate, doesn't handle single-sideband signals, either.

The '7700's audio quality is clean and crisp for pleasant listening to voice-oriented programs. Music sounds a bit "tinny," but it's no worse than on most other compact portables.

The new ICF-7700 is not the receiver of choice for active radio aficionados or high-tech types. It should, however, appeal to the noncritical world band radio listener with an aversion to anything smacking of high technology.

Free Clothesline

One "why-hasn't-somebody-thought-of-that-before" feature of the '7700 is that it comes with a 22½', or 6.8 meter, reel-in antenna—the AN-6—that looks for all the world like a tape measure that uses wire instead of tape. You simply clamp the end of the wire onto the set's telescopic antenna, then unreel it to wherever you want it hung. It's a great idea, and it can even second as a makeshift traveler's clothesline for underwear . . . although the case on ours tends to pop open accidentally—and it's no fun to reassemble. A piece of tape cures the problem.

Because the '7700 is reasonably sensitive, the AN-6 antenna is of use mostly for daytime listening and tuning in low-powered broadcasters within the tropical bands. At night on the lower world radio bands, the set tends to overload at times merely with the telescopic antenna. So more input from the antenna is just about the last thing you'd want. This tendency to overload is why the AN-6 is coupled inductively, rather than directly, to the set. Inductive coupling acts, in effect, like an attenuator to keep signal strength within acceptable bounds.

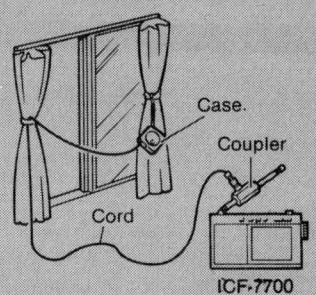

Best of the Easy-to-Operate Portables

PANASONIC RF-B60

Back in the late Seventies and early Eighties, Panasonic was a name to be reckoned with in world band radio. Their pioneering RF-2800 was the first digital portable to hit the market, and it was so successful that until early last year a variant of it—the GE World Monitor—was still being sold. History has left us other Panasonic gems: the RF-2200, which came with its own rotatable antenna and, later, an improved version of the '2800, the RF-2900.

These radios sold so well, in fact, that Panasonic promoted its world band radio design team into the computer hardware division. Unfortunately, the team that replaced them didn't do as well and Panasonic world band radios faded into the background.

Unusually Easy to Operate

With the RF-B60 AM/FM/world band radio, however, Panasonic has stepped back into the market with a vengeance. What makes the 'B60 so special is the niche it has carved out for itself. In a world where most radios seem to be striving to cram so many buttons, gadgets and displays on their front panels that they could double as flight trainers for the space shuttle, the RF-B60 is refreshingly simple and easy to operate. If you want a world band radio that's all but foolproof to operate—yet which doesn't sacrifice performance for simplicity of operation—then this set is for you.

Variety of Tuning Options

The 'B60's tuning is state-of-the-art, while its size, weight and features combine to make it ideal for traveling. Tuning is in 1 kHz increments by either a conventional tuning knob or elevator-like up/down slewing buttons. There's also a handy push-button arrangement to zip from one band to another, along with a calculator-like keypad for instant access to desired stations.

The set also has 36 programmable memory channels, nine of which cover the world band radio spectrum. A nice touch is a flip-up chart to indicate which stations are in each memory.

Designers of portables try to keep protrusions to a minimum, as these tend to get hit and sometimes even knocked off while the set is being toted about. In this regard, the 'B60's tuning knob is interesting for two reasons. First, it appears as a conventional—yet almost flush—round knob on the front panel. But, in addition, it operates as a knurled control from the side of the set.

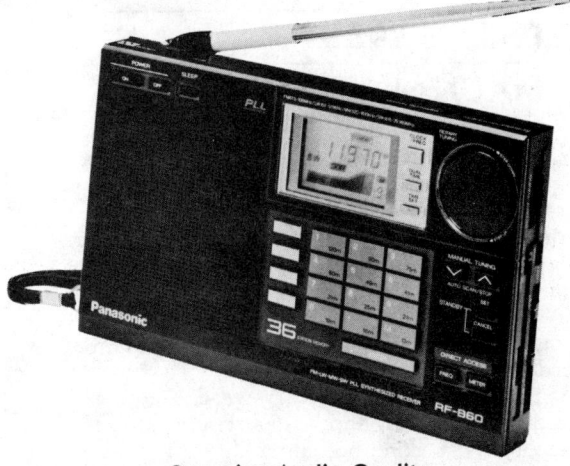

Superior Audio Quality

The 'B60 is a very compact radio. It's roughly the size of a paperback book and weighs in at just under two pounds—just the thing for air travel. And while most compact sets of this type provide only mediocre audio quality, it's clear that the people at Panasonic weren't about to make that mistake. The 'B60 sounds quite pleasant on all bands.

Limited Number of Features

There are other niceties, too. There are twin clocks, for example, and a lengthy two-year warranty. The 'B60 also has a somewhat optimistic digital signal-strength indicator with no less than fifteen increments.

The 'B60, unlike Sony's similar ICF-2003, doesn't handle single-sideband signals. For now, this doesn't mean much, as few, if any, broadcasters operate this way.

The 'B60 has only one selectivity position. It's well-suited to reception of ordinary international broadcasters, which tend to be quite strong. But it isn't so effective for rejecting adjacent-channel interference as, for example, the '2003.

Like most compact portables, the 'B60 has only fair field sensitivity with its built-in telescopic antenna and it's not designed to operate with an outdoor antenna. Also, alas, it doesn't have a dial light to aid in nighttime operation.

Logical Choice for Uncomplicated Operation

The bottom line is that Panasonic's compact RF-B60, which lists in the US at $269.95, is an especially appropriate choice for the world band radio listener who seeks worthwhile performance without a phalanx of high-tech controls. It's one of the easiest advanced-performance world band radios to operate, and its audio quality is clearly superior for such a small set.

Best Value in a Portable

PHILIPS 2935
MAGNAVOX D2935

While manufacturers in the Land of The Rising Yen have been dreaming of price rise after price rise, yet another manufacturer has landed in the world band marketplace—on all fours. This one—based in Europe—is the giant Dutch electronics conglomerate of Philips, also known in America as Magnavox. Their D2935 is truly a bargain. In fact, it's just about all you could ask for in an affordable portable.

The D2935's price is very reasonable, its performance good, and it uses state-of-the-art technology, making operation all but foolproof. It's also large enough to provide good audio quality without being too hefty to toss into your baggage. And, unlike most world band portables, it comes with a built-in ac power supply that works in just about any country of the world—a battery-saving bonus when you're traveling abroad.

As with the Sangean ATS-803, the makers of the '2935 have recently made improvements to their product. (You can identify the latest version by its large baby-blue 1⅛ inch, or 3 cm, power button.) It's more selective than the original model and it works better on mediumwave AM, too.

Easy to Operate

The D2935 is both flexible and easy to operate. It comes with a calculator-like keypad for pushbutton station selection, plus the customary tuning knob. A nice touch for nighttime listening is that the digital frequency display can be lit at the touch of a button. The strength of a station's signal is shown by a bank of five LED's.

Performance Above Average

The '2935 includes a relatively powerful audio stage with a single tone control and a fairly good speaker. The end result is above-average audio quality, although it's not quite in the same league as that of the larger but considerably more expensive Philips/Magnavox D2999; much less the top-sounding, but even more pricey, Grundig Satellit 650.

Other aspects of performance are also surprisingly good for a portable. Selectivity—or adjacent-channel rejection—is effective, although only one position is provided. Sensitivity with the built-in antenna is fairly good.

The '2935 is fairly stable, too—once a station is tuned in, you don't have to fiddle around retuning it. Dynamic range is reasonable by portable standards, but front-end selectivity is mediocre. What this means is that so long as the built-in antenna is used, most listeners won't hear other stations mixed in with the one they're trying to hear. But it also means that you shouldn't connect the '2935 to a fancy outdoor antenna. As is the case with most other portables, that could send it into a confusing fit of noises you don't want to hear.

FM performance is pleasant, although not quite equal to that of the similarly sized Grundig Satellit 400.

Now Widely Distributed

Until early 1987, the '2935 was distributed almost exclusively within Western Europe. Philips' North American subsidiary, Magnavox, didn't carry the receiver, and the parent firm wouldn't let other firms in North America handle it.

Now, all that's in the process of being changed. Officials at Magnavox have announced that by early 1988 they'll be carrying all three Philips world band portables, including the '2935, under the "Magnavox" label. In the US, the Magnavox D2935 lists for $249.99, and in Europe the Philips version is even cheaper.

In all, the Philips/Magnavox D2935 is clearly one of the best buys on the market today. It lacks some of the sophisticated features of the top-rated, but more pricey, Sony ICF-2010/ICF-2001D portable. For most listeners, however, the '2935's combination of price and performance makes it hard to pass up.

Travel Favorite in New Package

SONY ICF-2003
SONY ICF-7600DS

In the beginning, there was the Sony IFC-2001. In 1980, this radio almost single-handedly revolutionized shortwave listening with its LCD frequency display, keypad tuning and compact size. Three years later, the '2001 was updated and improved, becoming the ICF-2002. Back then, we were so excited about the '2002 that we called it "The most exciting entry into the shortwave market since the introduction of the ICOM R-70."

It's been some time since we wrote that review and once again, Sony has unveiled the latest generation of its revolutionary ICF-2001, this time dubbed the ICF-2003. The problem is that this time the review's not so gushy. What was once revolutionary is now routine.

Old Cognac in New Bottle

The new '2003 is virtually identical to the '2002 except that it comes with a newly designed cabinet and special accessory antenna. Smaller than a copy of this book, the '2003 is light enough even for those who trot the globe with nothing more than a sandwich in a paper bag.

Many Features for Compact Portable

Unlike the simpler-to-use sibling ICF-7700, the '2003 is equipped with a wide array of features. There's a scanner, for starters, and a keypad that not only serves to enter frequencies, but also to activate the ten memory channels. For tuning around, the '2003 uses elevator-like up/down slewing buttons instead of a conventional tuning knob, but their movement is limited to 5 kHz increments, although a little knob at the side allows you to tune more finely.

A separate clock/timer display is positioned above the frequency readout. This means that you can see what station you're listening to—as well as the time—at the same moment. The '2003, unlike either the '7700 or Panasonic's RF-B60, also handles single-sideband signals, although there is only limited interest in this type of transmission. There's a useful power lock, too, to help prevent the radio from turning on accidentally in your luggage.

Uncorrected from the '2002 is the radio's essentially useless, single-digit LED signal strength indicator. This is about as useful as an igloo in the Sahara. Even more baffling is the continued absence of a dial light.

Generally Superior Performance for Small Radio

The '2003 does have very good skirt selectivity (rejection of interference from stations on adjacent channels). For those who enjoy hearing the more difficult world band stations, this gives the '2003 an advantage over such competing models as the '7700 and 'B60. Dynamic range and spurious-signal rejection are both at least average for a portable.

On the other hand, sensitivity with the built-in telescopic antenna may disappoint some. To get around this, Sony has equipped each '2003 with the AN-6 reel-in antenna described in our review of the '7700. This helps, and works even better when the inductive connector that hooks onto the radio is snipped off and replaced with a direct-connecting alligator clip, available at nearly any electonics parts store. Another easy trick that sometimes helps boost sensitivity is to run the radio off ac instead of batteries. For this, you need to purchase a separate power supply accessory.

Audio quality is, unfortunately, nothing to write home about. As with previous versions of the model, it's weak and tinny with music, although for listening to voice broadcasts in quiet rooms it's more than adequate.

The '2003's FM high sensitivity is typical of Sony's world band radios. This is great if you live in the remote countryside, where stations are invariably weak. but urbanites have to put up with the '2003's propensity to suffer from spurious signals when there's a strong station in the area. If you don't live in the hinterland and want superior FM on a world band portable, look to Grundig or Panasonic, instead.

No other compact portable quite equals the '2003's combination of reasonable price, very good selectivity, flexibility of operation and compactness. For the sophisticated listener who doesn't want or can't afford Sony's excellent but complicated ICF-2010 portable, but who finds the '7700 simply too restrictive, the '2003 is an excellent choice.

Superior Audio in a Portable

GRUNDIG SATELLIT 400

The final non-Japanese entry into the range of affordable portables has a multifaceted national background. The Satellit 400, made in Portugal by the West German firm of Grundig, is a subsidiary of the Dutch firm of Philips. And while Grundig is indeed well known in Europe, radios carrying this brand name have been virtually absent from North America since the Seventies. But, like the Taiwanese and the Dutch, the West Germans have sensed both the declining participation of the Japanese and the increasing popularity of world band listening in North America. So, in they have come . . . with vigor.

Actually, Grundig resurfaced in North America only last year when the American firm of Lextronix obtained exclusive rights to distribute the West German company's audio and video products. Included in Lextronix's line of Grundig audio products are two world band radios, one of which is the $449 Satellit 400.

The '400 covers the longwave, mediumwave AM, FM and world radio bands. It uses, of course, a knob to tune stations. But, additionally, it has an easy-to-use keypad for direct frequency entry, plus a scanner and 24 programmable channel memories to store—like on a car radio—your most listened-to stations. Additionally, there are up/down band slewing buttons for world band radio that second as a signal seeker in other bands.

Numerous Travel Features

For traveling, there's a nice, solid flip-up handle that's extremely handy. The fused ac power supply is built-in and dual-voltage for worldwide use. In North America, the '400 also comes with an ac plug adapter that allows the set to be operated from the most popular types of wall sockets in most countries.

The '400 incorporates two clocks and a timer, which is handy for making sure you get up on time when the hotel desk clerk forgets your wakeup call. But the clocks and frequency share the same digital display, so you can see only one of these three at any given moment. If you're listening at night, you'll be pleased to find that the '400 has a light that illuminates both the digital frequency readout and the large analog signal-strength/battery check meter.

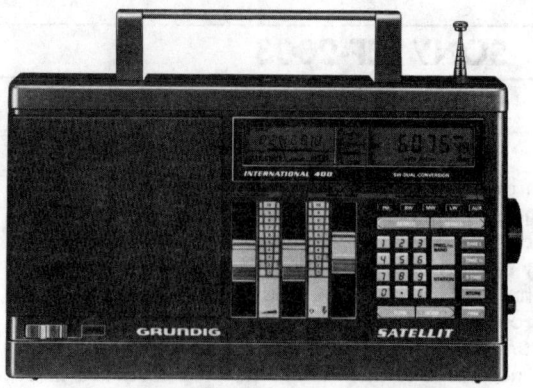

Generally Worthy Performance Except for Single Sideband

In all, for listening to world band radio, the '400 performs fairly well. Distortion in the AM mode is uncommonly low, and ultimate rejection is superb. While only a single bandwidth is used, it's well chosen and performs nicely. Dynamic range is acceptable, but this travel receiver is obviously not designed to be connected to a lengthy outdoor antenna unless you want to encounter the mishmash of stations heard all jumbled together that's known as "overloading."

However, spurious signal rejection is substandard and the '400's tuning circuitry is slightly unstable below 9 MHz. This latter characteristic results in a slight wavering in voices and music heard, plus seriously deficient single-sideband reception.

Grundig radios are known for their superb audio quality, and the '400 is no exception. It has special high-frequency audio filtering for when world band stations are being received, separate bass and treble controls, plus a high-quality audio stage and speaker. So it sounds quite good . . . although it's definitely not in the exceptional league of its bigger, and more expensive, brother, the Satellit 650. The '400's FM performance—especially its capture ratio—is also top drawer. At our *Radio Database International* listening post, located between New York and Philadelphia, you can place the '400 on a single FM channel and find two or three stations popping in separately and clearly simply by rotating the antenna.

In all, the '400, being in the Grundig tradition, is designed not for "DXing" but, rather, for unusually pleasant listening to the more easily heard radio broadcasts.

Handheld Scanner Doubles as World Band Radio

SONY ICF-PRO80
SONY ICF-PRO70

Most advanced-technology radios are specialty products designed to do one thing well: bring in the world by radio. Some also excel at receiving FM broadcasts, and yet others succeed at receiving specialized radio signals—ships at sea and the like.

Sony's new ICF-PRO80, however, tries to do all of these . . . and more. With this $419.95 portable, you can tune in all manner of police, fire and aircraft communications, along with such arcane chitchat as the back-and-forth of security agents. In short, this is what's referred to as a "handheld scanner." And it's a pretty sharp-looking one, at that. Hold one in your hand and you can easily picture yourself directing dozens of hooks and ladders at a seven-alarm blaze.

Few World Band Features

What makes Sony's scanner different is that it also picks up world band broadcasts. However, the 'PRO80 lacks many of the features found on competing world band portables. There's no tuning knob, no signal-strength indicator, no clock and no timer or sleep control. If you want to receive single-sideband signals, you have to settle for just that: single sideband. There's no way to separate lower from upper sideband as there is on better world band radios. And the excellent synchronous detector found on Sony's less-costly ICF-2010/ICF-2001D portable is nowhere to be found.

Too, the 'PRO80 tunes only in 5 kHz increments. There's a supplementary fine-tuning control, just as on the Sony ICF-2003/ICF-7600DS that costs half as much. Also like the '2003, the 'PRO80 has a set of up/down slewing buttons for bandscanning. In addition to the keypad, there are no less than 40 memories to store your favorite stations.

Mixed Performance

World band performance varies from the mediocre to the outstanding. On one hand, the 'PRO80 is very stable, so it doesn't need retuning from time-to-time. Additionally, it's sensitive throughout the world band spectrum, so even weak stations tend to be heard. Its dynamic range is superior for a portable, too, so it's likely to function nicely even in such high-signal-strength parts of the world as Europe and North Africa.

Selectivity, however, is a mixed bag. On one hand, ultimate selectivity is remarkable for a portable—or even a costly tabletop communications receiver. Too, there are fully three bandwidths, two of which are useful for listening to world band broadcasts. These bandwidths are well chosen. Nevertheless, skirt selectivity is inferior to that found on less-costly Sony portables, such as the ICF-2010/ICF-2001D.

At the other end of the quality scale, spurious-signal rejection is mediocre. As a result, unwanted signals tend to pop up in various parts of the world radio bands where they don't belong. The problem with this is that these "repeats" can cause unnecessary interference to stations you're trying to hear.

Modest Audio Quality

The 'PRO80 might be OK if it stopped here, but its audio quality—because of moderately high distortion and a tiny speaker—is mediocre. Listening to chitchat from the local fire house with this device is one thing, but trying to enjoy a clear broadcast from, say, France with the 'PRO80 requires some degree of aural forbearance.

Good Overall Scanner Performance

According to scanner expert Bob Grove of *Monitoring Times*, the 'PRO80 performs fairly well as a handheld VHF scanner. It's sensitive, reasonably selective, its squelch works well, and its scan rate is not too slow. The problem is in trying to operate it, particularly within the 151-174 MHz "high band" and 174-223 MHz TV band (channels 7-13) found on the 'PRO80, but not on the 'PRO70.

Peculiar Ergonomics

The major ergonomic characteristic is that the 'PRO80's role as a handheld scanner has resulted in a set that is unusually complicated to operate. For example, its keypad tuning scheme requires more button-pushing than do those of most other portables. And changing bandwidth calls for pushing two buttons, rather than the usual one, simultaneously.

Too, the volume control is shared with the two-step tone control, and it's located right next to the screw-on telescopic antenna. As a result, this often-used control is awkward to operate, especially when the button control is depressed to the "low" tone position.

But when the 'PRO80 used as a high-band VHF scanner—the 'PRO70 only receives the 30–50 MHz "low band"—the ergonomics become ludicrous. Simply to be able to *tune* the high band, you first have to remove the battery pack, reach into the battery cavity to flip a minuscule switch secreted within, reinsert the battery pack, unscrew the telescopic antenna, screw on the high-band VHF converter module, reattach the antenna, press the "Program" and "Direct" buttons simultaneously, press "11500" to command the receiver to function with the high-band converter, then press "Execute." If your patience hasn't been exhausted, you're now ready to try tuning the high band.

Of course, the entire procedure has to be reversed when you wish to leave the high band and tune the low band . . . or, God forbid, listen to world band or local broadcasts!

All ergonomic characteristics aren't negative, however. The 'PRO80 comes equipped with an easy-to-use display light that fades out automatically after several seconds. This makes nighttime listening much handier than on such unilluminated portables as the Sony ICF-7700/ICF-7600DA.

In short, the Sony ICF-PRO80 occupies, with mixed success, a niche for radio buffs who wish to have a handheld portable that "does it all."

RDI REFERENCE VOLUMES NOW AVAILABLE AT EXCEPTIONAL SAVINGS

Now you can complete your world radio reference library at a fraction of the usual cost! Prior volumes of the authoritative *Radio Database International*—the "new bible" of world band radio—may be ordered, while supplies last, at exceptional savings.

_____**ORDER 1:** 1984 Radio Database International Tropical Bands Edition. Formerly $3.95, now only **$1.95**

_____**ORDER 2:** 1985–86 Radio Database International Tropical Bands Edition. Formerly $4.95, now only **$2.95**

_____**ORDER 3:** 1987 Radio Database International Combined Tropical and International Bands Edition, including extensive world band equipment reviews and comparative ratings. Formerly $12.95, now only **$6.95**

_____**ORDER 4-SPECIAL COMBINATION OFFER:** Save even more! All three volumes—a $21.85 value—now only **$9.95**

Please add $1.95 per order for surface shipping and handling worldwide. Send check or money order, in US funds only, to

**RDI Special Offer
P.O. Box 300
Penn's Park, PA 18943 USA**

Europe's Finest Portable

GRUNDIG SATELLIT 650

The Grundig Satellit 650 is the best of the breed in Germany's world band stable. Although it weighs in at a bicep-building 20 pounds, or 9 kilograms, it is—nominally, at any rate—a portable. Priced around $1,000 in the US, the '650 provides not only worthy overall performance, but also exceptionally pleasant audio quality and a wide array of useful features. FM performance is also top-drawer.

However, the '650—like virtually all other current portables—is not fully equal to the rigors of fishing for faint catches from within congested bands. It's oriented, instead, to allowing you to hear programs pleasurably, hour-after-hour. For serious "DXing," one looks to tabletop models.

Wide Variety of Features

In addition to the usual knob tuning, the '650 also has keypad tuning and no less than 60 memories, 32 of which can be used to store favorite world band stations. The large digital display, which provides readings to the nearest kHz, is easy to read even if your contact lenses are soaking.

The '650 also comes equipped with a tuning knob incorporating "variable-rate incremental tuning" (VRIT). With this, the faster you spin the knob, the more quickly the stations zip by. That's nice, but it can be tricky if you turn the knob too fast.

Helpful features include a digital clock, which displays seconds and uses the 24-hour clock format appropriate to World Time, and a programmable timer with up to three straightforward on/off cycles per day. A large, easy-to-read analog signal-strength meter—which doubles as a battery strength indicator—is included, too. The '650 also comes with a dual-voltage ac power supply suitable for use worldwide.

There's a motorized "tracking" preselector that automatically peaks the set's circuitry to exactly where you're tuned. It's a great idea that's something of a throwback to higher-quality designs of an earlier era. It doesn't quite live up to its performance potential, but it does add to the receiver's complexity and potential for malfunction.

Overall Performance

Sensitivity, using either the built-in telescopic antenna or an external antenna, is very good. Blocking, a related measurement, isn't, though.

On the face of it, the '650, which has no less than three bandwidths, should have top-notch selectivity to cope with strong competing signals on adjacent channels. However, the intermediate bandwidth uses only a simple audio filter. Shape factors—another measure of selectivity—are also not what they could be. Still, selectivity is adequate for most world band listening purposes.

Portables usually have mediocre dynamic range, and the '650 is no exception. It's fine for when you're listening with the built-in telescopic antenna, and if you live in the Americas you'll find it to be adequate. But listeners in Europe and other high-signal-strength parts of the world should be conservative when using the '650 with an outboard antenna. Otherwise, "overloading"—a mishmash of false signals all jumbled together—can be heard.

Top-Notch Audio Quality and FM Performance

Compared with other world band radios, the '650 has audio that is *par excellence*. Included are effective separate bass and treble tone controls, plus excellent speakers and even a tweeter switch. This is where the German engineers really did their homework. No other world band radio, regardless of price, sounds quite this good.

Pleasant Listening

For pleasant listening to foreign stations—but not serious "DXing"—the '650 is hard to beat. Its size notwithstanding, it also functions as a fully self-contained field portable for use on automobile trips and around your property. As a bonus, it also works unusually well in receiving FM and other closer-by broadcasts.

The Prince of Portables

SONY ICF-2010
SONY ICF-2001D

Let's get right to the point. No other world-band portable currently on the market quite equals the overall performance of Sony's ICF-2010, sold outside North America as the ICF-2001D. It utilizes exceptionally advanced circuitry to provide sophisticated and effective reduction of interference; yet it can be operated by most laymen.

The '2010 thus represents a major advance over conventional world band radios. This state-of-the-art portable is midsize, yet small enough for most travel. It covers the entire world band radio spectrum, plus longwave, mediumwave AM (including the proposed extension in the Americas), FM (both the Japanese and Western FM bands), plus the 166–136 MHz VHF aeronautical band to allow globetrotters to eavesdrop on airline traffic. A single-voltage outboard ac converter comes with each set.

Wealth of Features

The front panel of the '2010 literally bristles with pushbuttons, making it, at first glance, look like something dreamed up by a mad scientist. Remarkably enough, most of these pushbuttons each have *two* functions, allowing the radio to do all manner of handstands, such as leaping from one band to another. It takes the better part of an hour with the instruction manual to get even a rudimentary grasp of the set's 76 switches, knobs, buttons and sliders.

Housed within this radio is a scanner, 32 programmable channel memories, control-locking facilities, keypad tuning . . . plus on/off and a power lock so the radio won't go on accidentally in your suitcase.

The dial light, by the way, has a nice little battery-saving feature of its own: it turns off automatically after fifteen seconds. Of course, this same feature can drive you nuts if you're busily scanning about the bands at night and every fifteen seconds the light keeps turning off. In any case, battery consumption is above average for a midsize portable, but isn't out of line for a set of this performance caliber.

The timer can turn on the receiver automatically up to four times a day in order to catch favorite broadcasts stored in up to four separate memories, then turn it off again 15, 30 or 60 minutes later—a bit annoying if what you wish to hear doesn't last 15, 30 or 60 minutes. The sleep switch can be used to turn off the set automatically. You can even go to sleep listening to one station, yet be awakened to another.

You can also use this capability to tape broadcasts automatically, using the '2010's timer, much as with a VCR. The '2010 doesn't come equipped to do this on its own, but one American firm, Solar Light Co., produces a radio-to-recorder interface accessory—the CC-2020—for this purpose.

The '2010's thirty-two memories are handy in that they allow not only the frequency, but also the bandwidth and mode—including synchronous detection, if desired—to be preprogrammed by the listener for future access. This is the sort of advanced technology heretofore found only on costly tabletop communications receivers. Yet, even with all this flexibility, the memories are straightforward to "program."

The '2010 is laden with other useful features, including a clock that operates in either the 24-hour (World Time) or 12-hour (local time) mode. Another feature is the '2010's 10-LED signal-strength indicator, which seconds as a battery-strength indicator.

Overall, the '2010's audio certainly is acceptable, even fairly good. But this is faint praise, indeed, considering how good the '2010 can sound when its record output is fed through a good audio/speaker system.

Worthy Overall Performance

The '2010—except for within the VHF aeronautical band—is adequately sensitive, provided the batteries are not run down. Weak batteries tend to cause a number of problems in the ICF-2010. Both bandwidths provide good selectivity, as well. In addition, dynamic range and other earmarks of a set's ability not to generate spurious signals are above average for a portable.

The '2010 is sensitive on FM—a plus if you live in a rural area far removed from any FM stations. But in urban areas the '2010 tends to produce spurious signals and to overload. As with overloading on the world bands, that on the FM band may be alleviated, at the expense of reduced sensitivity, by reducing the length of the telescopic antenna. Similar shortcomings, coupled with low sensitivity, are found within the VHF aeronautical band. In some locations FM broadcast signals also penetrate into the '2010's shortwave circuitry, causing distorted interference to world band signals.

The initial version of the '2010, which has long since disappeared from dealers' shelves, suffered from various problems with digital-circuit noise and the battery cavity. Sample-to-sample variations in performance were also distressingly high. Fortunately, most of these difficulties have now been overcome.

Some years from now, all shortwave—and mediumwave AM—radios will have advanced detection/interference-rejection circuitry such as is found on Sony's innovative ICF-2010. Although the $389.95 '2010 shouldn't be mistaken for a serious tabletop receiver, it is the obvious choice if you are seeking a world-class travel portable.

What Makes the '2010 So Special?

What puts the '2010 a generation ahead of the pack is a feature called synchronous-detection. Here's how it works. World band and mediumwave AM broadcasters alike operate in what's known as the "AM mode." That's why the mediumwave band is usually referred to as the "AM band," even though the AM mode is also used within other bands.

The AM mode consists of three parts: two "sidebands" and one "carrier." The sidebands contain the information—speech, music and so on—while the carrier is the "vehicle" that carries that information. Perhaps it's easiest to think of the carrier as being like a sheet of paper, with the sidebands being the "ink"—information—on that sheet of paper.

Why two sidebands? How do they differ? The answer is, quite simply, they don't, unless a broadcast is in stereo. So, in monaural, it would seem as if only one would be needed. The problem is that if you eliminate one of these sidebands, distortion rises

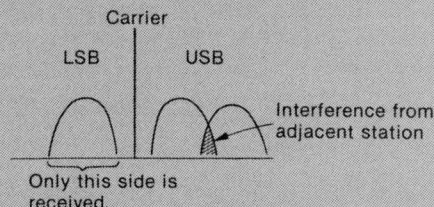

enough to make listening unpleasant. But with synchronous detection, you can listen to just one sideband by itself without adding distortion. In fact, synchronous detection tuned to only one sideband actually reduces distorion over what you'd hear on an ordinary radio tuned to both sidebands!

But the real advantage is that synchronous detection allows you to select the sideband that is less bothered by interference from competing stations on adjacent channels. For example, if you're listening to the BBC on 5975 kHz, and it's being bothered by the USSR on 5980 kHz, you can tune into the BBC's lower sideband and knock out the Soviet interference. This improves world band reception considerably at times, and is also helpful in improving twilight reception on the longwave and mediumwave AM bands.

The '2010's synchronous detector is remarkably straightforward to use once you know how. Unfortunately, the ICF-2010 owner's manual treats the whole subject of synchronous detection too briefly and in arcane language which inadequately explains this major plus. However, our *RDI White Paper* on the '2010 remedies this by explaining how it should be used.

Superior Audio from a Top Performer

KENWOOD R-5000
TRIO R-5000

Because moderately priced world band portables are now so good, fancy tabletop sets might seem to be superfluous. But just as there are monitor TV's to complement the ordinary variety, so there are exceptional tabletop receivers that stand out from the common herd. Among these are a handful of elite models that excels at flushing out tough catches. The newest—and in some ways, the best—of these is the recently introduced Kenwood R-5000, known in the United Kingdom as the Trio R-5000.

Exceptional Receiver

Is the world ready for yet another gilt-edged performer? For at least some perfectionistic listeners, the answer is a resounding "Yes!" Costly tabletop receivers usually have disappointing audio quality, but what makes the R-5000 so interesting is its audio, which is clearly superior by world band standards. It sounds great.

The '5000 provides one of the best combinations of performance, features and price of any tabletop model tested to date. While its ergonomics are only fair, the '5000 is a superb performer. Its only real shortcoming is the lack of a synchronous detector, such as is found on the Sony ICF-2010/ICF-2001D or the forthcoming Liniplex F2.

The '5000, as you would expect, covers the longwave, mediumwave AM and world radio bands in nearly every mode imaginable. It comes in three versions, depending on where you buy it: a US version that works only from 108–132V ac; a Japanese version for 90–110V ac; and—best—a multivoltage worldwide version that operates from 108–132, 198–242 or 216–264V ac. In Canada both the worldwide and US versions have appeared for sale thus far.

Features Aplenty

The '5000 is all but dripping with just about every feature imaginable. Among these goodies are two switchable antenna inputs; 100 memories that store frequency, mode, and antenna; keypad and multi-speed knob tuning; Megahertz up/down slewing controls to allow for large tuning leaps; a scanner of sorts; true passband tuning ("IF shift"); selectable fast/slow automatic gain control decay; a tunable audio notch filter to eliminate howls and squeals; two adjustable noise blankers; two 24-hour clocks; a simple on/off timer; a three-level switchable attenuator; an analog signal-strength meter; a front-panel-display dimmer; plus RF gain and squelch controls.

The only thing this receiver lacks is synchronous detection and a sophisticated scanner along the lines of that of the Japan Radio NRD-525.

The synthesizer tunes in tiny 10 Hertz segments, and the frequency actually displays to this precise degree of resolution. Two voice bandwidths—6.3 kHz and 2.5 kHz—come standard, with a third—1.9 kHz—being available as an option. Additionally, some specialty vendors offer additional bandwidths.

The '5000's timer and recording socket, in conjunction with a DIN socket on the rear panel, provide for full on/off control of a tape recorder. An optional interface is also available to allow the computer to interact with the receiver.

Another plus is that the all-important microprocessor software that is responsible for operation of the receiver is permanent on the '5000. You don't have to worry, as you do with the ICOM IC-R71, that your radio will roll over dead, like a cockroach, and need factory repair because some battery inside the set conked out.

Ergonomics Acceptable

Ergonomics determine how handy and restful your receiver will be to operate. Overall, the ergonomics of the '5000 are reasonable. Most controls and displays are where they should be and operate smoothly and fairly logically. And while keypad operation could be simpler, it is as user-friendly as one of these receivers gets.

The digital readout consists of a multicolored fluorescent display that tells you just about all you'll ever want to know about what the set is up to. A plus is that any one of the two 24-hour clocks displays independently of the frequency readout so it can always be seen. Light-emitting diodes (LED's) on the front panel also add to the sense of knowing what's what. As if this weren't enough, when the mode (USB, AM, etc.) buttons are depressed, the speaker chirps out the selected mode in Morse code. This chirping can be a gimmicky annoyance, but if your vision is impaired it can also be useful.

Superb Overall Performance

One of the main criteria for worthy performance concerns how well a receiver copes with signals from adjacent or other nearby frequencies. Here, there is both good news and bad.

The bad news is that the 6.3 kHz "wide" bandwidth filter that comes standard with the set has skirt selectivity far worse than that of a simple $99.95 Sony portable. Indeed, at −60 dB, this filter—actually, a pair of cheap little filters in tandem—measures a whopping 24.8 kHz!

The good news is that this is easily remediable. Kenwood offers the optional high-quality YK-88A-1 6.5 kHz replacement "wide" crystal filter. It performs superbly, measuring only 10.5 kHz at −60 dB—quite an improvement, indeed! It's easy to install and requires no soldering.

Kenwood deserves a collective onion for pulling this stunt. A "wide" bandwidth filter isn't a luxury feature—it's one of the most-needed parts of a world band receiver. To sell the set with a filter so obviously mediocre, then offer the correct filter as an extra-cost option, is reminiscent of the days, years back, when Chevrolet used to offer its oil filter as an option.

The '5000 also comes with a 2.5 kHz "medium 2" standard bandwidth and a 1.9 kHz "medium 1" bandwidth. Both—thankfully—are excellent, and the overall effect of these three voice bandwidths is one of real flexibility.

Sensitivity and the related "blocking" measurement are both top-drawer. If something is there to be heard—even if it's feeble—the '5000 will snatch it from the tireless grip of the airwaves. However, dynamic range, although adequate, is only fair—slightly worse than that of the Japan Radio NRD-525.

Superior Audio Quality . . .

Ridiculous as it might seem, there are costly world band receivers being turned out that distort worse than cheap telephones. This doesn't necessarily mean that your ears will immediately rebel. In fact, to some listeners a radio with moderate distortion sounds fine. But regardless of how finely tuned your ears are, even low levels of distortion eventually produce aural fatigue. Listening over long stretches of time tends to become tiring, even producing headaches.

The '5000, except when used to make tape recordings, sounds much better than most tabletops. Overall distortion is low, allowing listening to be a pleasure—not just tolerable—over extended periods of time. Additionally, less "hiss" is heard with the '5000 than with some other world band radios.

But the audio's not all Fat City. The '5000 has no tone controls, although the passband tuning circuit can be used to some extent to shape the tone. Additionally, with strong sounds you can sometimes hear "breakup" distortion. The noise blankers can also cause distortion, and the small upward-facing speaker is no prize. A high-quality outboard speaker—other than Kenwood's pedestrian SP-430 optional unit—is needed.

In all, the performance of the '5000 is top drawer. At last, there is a serious world band receiver that excels not only at ferreting out Radio Flyspeck, but also at listening to hour-long concerts from London.

One of the Best Is Now Even Better

JAPAN RADIO NRD-525

For those who demand the very best, there is a handful of elite communications receivers that are the BMW's and Jaguars of world band radio. One such model is Japan Radio's $1,285 NRD-525.

If you're looking for a quality constructed receiver with nearly every feature imaginable, the '525 is all but in a class unto itself. For one thing, it comes with an unusually sophisticated scanner that allows the set to tune itself while unattended. For another, its modular plug-in board construction considerably reduces the complexity of repairs.

Current Versions

There are three versions of the '525. There's really no difference among these except that they come preset from the factory to the correct local ac voltage. All have the same multivoltage capability.

The "U" version is intended for the United States and various other countries with a 120V ac power standard. The "E" version is intended for use in Europe and most other parts of the world where 220-240V ac is the norm. The "J" version is intended for use in Japan, where 100V ac is used.

Do you need to buy the special "U" version if you live in the United States? Or a "J" version if you live in Japan? Hardly. All you have to do is adjust a control on the back of the radio to your correct voltage and put in the proper fuse.

Handy to Operate

Operating controls should be laid out so they are logically located and easy to use. In order to provide the operator with a high degree of operating flexibility, Japan Radio did not take the usual expedient route and clutter up the '525's front panel with tiny controls suitable for an elfin. Instead, creative engineering and design have allowed the '525 to be operated comfortably and logically. Too, most of the controls—keypad buttons excepted—have above-average "feel."

Flexible Tuning

The '525 gives the user numerous ways to tune in a signal. In addition to the traditional tuning knob—the mainstay of any operator-controlled receiver—there are also up-down slewing buttons to aid in rapid bandscanning, a keypad for direct frequency selection, the scanner, and no less than 200 programmable channel memories. Additionally, the receiver is potentially well suited to whatever computer-controlled mischief world band radio hackers may come up with.

The '525's built-in timer can be used to turn the set on and off each day, adding further to the set's robotic capabilities. Too, the timer is designed to control a tape recorder at the same time.

So, this state-of-the-art radio will select channels, tune and retune them for high-quality reception, turn itself on and off, and even record your favorite programs when you're not home . . . just like a good VCR.

Rock-Solid Stability

The Japan Radio NRD-525, like the Kenwood R-5000 and ICOM IC-R71, is an exceptionally stable receiver. This makes it well-suited to unattended reception of stations by allowing you to tune in only one side of the signal—rather than both, as is the usual case—thus avoiding interference from any station alongside the one you're trying to hear.

The reason for this is that world band radio signals consist of two identical Siamese-twin-like parts. With certain radios, you can listen to just one of those two parts—the quieter one.

Why Siamese-twin transmissions? This "twofer" approach allows signals to be received without heavy distortion on inexpensive radios. But for pricier sets—and even the Sony ICF-2010 portable—you can actually select the desired sideband.

Good as the '525 is in this regard, it's hardly ideal. It doesn't include synchronous detection, which is found on the far-cheaper '2010. Synchronous detection can allow the desired sideband to be selected even better.

Special Features Help Improve Reception

The '525 uses a passband tuning control to help reduce interference and improve audio fidelity. For example, if a signal is "boomy," the passband tuning control can be adjusted to an outer edge of the signal to incorporate more high, or "treble," audio frequencies and fewer low, or "bass," frequencies. This sort of flexibility is a real aid not only in hearing difficult stations, but also in listening to music and voice programs from major world broadcasters.

Another useful feature of the '525 is its tunable notch filter. This helps reduce the howls and squeals often associated with world band radio listening when there's lots of interference from competing stations. The '525 also comes with a noise blanker that's unusually effective in coping with the *tap-tap-tap* of Soviet over-the-horizon radar that sometimes interferes with world band radio listening.

Up to Three Bandwidths

Every receiver needs at least one bandwidth filter, preferably more, to select the desired station out of the morass of competing stations on the air. A bandwidth filter acts as does a cattle chute to let through only one bull—or station—at a time.

For voice and music, the '525 comes with filters that provide two bandwidths: 5.7 and 2.2 kHz. Both bandwidth filters have very fine performance characteristics, but in a radio of this caliber a bandwidth roughly midpoint between these is called for to give optimum flexibility for the wide variety of reception situations encountered on the world radio bands.

There are provisions for just such a third bandwidth filter. Although no suitable filter is presently available from Japan Radio, some dealers will install, at extra cost, a high-quality Collins 3.8 kHz filter that fits the bill almost perfectly.

Audio Quality

Most communications receivers tend to have mediocre audio reproduction. This is not so illogical as it might seem at first blush, as typically they are purchased by the military and the like. These outfits could care less about hearing music clearly. All they want is good voice reproduction, like on a telephone.

The '525 is no exception, plus there's a slight persistent "hiss" audible in the background. Additionally, if an unshielded antenna lead-in is used, digital-circuit whine can be heard, like the "hiss," at low level.

The bottom line is that for listening to both voice and music programs, the aural quality of the '525 doesn't equal, for example, that of the new rival Kenwood R-5000 or the Grundig Satellit 650.

Improvements for 1988

Originally, the '525 could only be tuned very slowly. Now, units with serial numbers BR36771 and higher allow you to select between a tuning rate of either 2 or 20 kHz per revolution, as well as a slewing rate of either 1 or 10 kHz.

As the '525 comes from the factory, it's programmed such that the user has to fiddle with the tuning controls to receive lower or upper sideband. With early units, this was the only way the set could be tuned—a pain if you're trying to select a sideband. Now, units as of serial number 36471 don't require this. By initiating a simple keypad command, you can bypass this annoyance altogether.

Fortunately, Japan Radio didn't forget its existing customers. Owners of older versions of the '525 may bring their existing units fully up to date by installing a "New-ROM IC Kit," about $40. This requires only that the new plug-in ROM be substituted for the old and a single wire be cut. No soldering is needed.

Overall, the Japan Radio NRD-525 is an unusually well-constructed and easy-to-repair receiver that provides superb performance. It lacks synchronous detection and its audio quality is not all it could be. Nevertheless, it's one of the few truly superb tabletop world band receivers available. And it's likely to have an unusually long operating life, to boot.

A Favorite for Hearing the Tough Ones

ICOM IC-R71

World band tabletop radios used to be so esoteric and unfriendly to operate that only the most dedicated took them to heart. Now, advanced technology has changed all that. Most of today's high-performance receivers are not much harder to operate than, say, a Sony ICF-2010/ICF-2001D portable or a good VCR.

There are three models priced in the vicinity of $1,000 that qualify as first-rate receivers: Kenwood's R-5000, Japan Radio's NRD-525 and ICOM's IC-R71. The chief difference among these is audio quality, with the '5000 being the best in this regard and the 'R71 being only fair—although adding a good external speaker helps. However, beyond this there are other differences that can make one model a favorite over others.

Unique Variable-Bandwidth Control

The $949 'R71 has the expected array of sophisticated operating controls—keypad tuning, a digitalized tuning knob, 32 memories for favorite stations, and so on. The narrower bandwidths provide breathtaking overall tough-signal performance, while the widest is at least average for a worthy receiver.

But there is a tradeoff between the ability of a receiver to eliminate adjacent-channel interference, on one hand, and to maintain acceptable audio quality, on the other. As the bandwidth is narrowed to eliminate interference, the audio quality of the station you are trying to hear gets worse. So the idea is to narrow the bandwidth no more than necessary to eliminate annoying interference.

On most receivers, you have only some fixed bandwidths from which to choose. If these are, say, 7 and 3 kHz, you may find that at 7 kHz you get good audio quality, but too much interference. At 3 kHz the interference may be gone, but the audio quality will be degraded.

The 'R71 has fixed bandwidths, like the rest. But, in addition, it has a unique variable-bandwidth feature—called, misleadingly, "passband tuning"—that allows you to adjust the bandwidth *exactly* to the "sweet spot" for any given station. As you turn this control, you can set it to, say, 4.7 kHz for one station, 2.4 kHz for another, 3.5 kHz for another, and so on. It's an excellent feature totally absent from competing models, and it can be made even better by the addition of ICOM's $178 FL44A filter accessory or a similar filter from independent sources.

You don't have to use the continuously variable bandwidth feature, either. If you don't need it, you can simply use the fixed bandwidths just as you would with any other receiver.

Whistle Filter of Limited Use

The 'R71, like the R-5000 and the NRD-525, comes with a tunable notch filter that can reduce some of the whistles and howls heard in the world radio bands—particularly the tropical bands. Unfortunately, unlike those of the '5000 and '525, that of the 'R71 does not operate in the AM mode. This is no small drawback, inasmuch as nearly all world band broadcasters operate in that mode. If you want to use the notch, you have to tune a broadcaster as though it were single-sideband.

One thing the 'R71 doesn't have is synchronous detection. It's a pity, but none of the three elite world band receivers currently on the market incorporates this fidelity-enhancing feature already available on the far-cheaper Sony ICF-2010/ICF-2001D portable reviewed elsewhere in this Buyer's Guide.

Dead Battery Means Inoperative Radio

A unique drawback of the 'R71 is that its operating software, needed to make the set work, must be reprogrammed by an ICOM service center should an inboard lithium battery go dead. The chances are that the radio will be headed for the scrap heap before its battery dies, but with the '5000 and '525 you don't have to take this gamble. Fortunately, ICOM has indicated they will be making this repair available indefinitely.

Outstanding Tough-Signal Performance

There are several variables that can help predict how well a radio will pick up a weak station, especially when there are powerful competing stations nearby. High sensitivity, low noise, a "clean" synthesizer, adequate dynamic range, good automatic-gain control action and other factors all work together to provide an overall index of weak-signal performance.

In these respects, the 'R71 is, overall, right up there with the very best. Indeed, in some respects the 'R71 outperforms any other model tested. As a result, it has become a favorite among radio enthusiasts concentrating on the really tough "catches."

Great Performance, Fair Audio

What to buy?

If there were an Olympics of world band receivers, the Kenwood R-5000, Japan Radio NRD-525 and ICOM IC-R71 would all be given gold medals for performance. What really separates them—aside from price, personal preference and the 'R71's unique variable bandwidth—is audio quality.

For many listening to signals that are already weak and of pedestrian fidelity, the audio quality differences among the three are beside the point. But if you also use your receiver to hear the clearer world band radio superstations, you'll either have to come to terms with the R71's audio quality, which is only fair, or to look elsewhere. For most, it's adequate. For some, it's not.

An Affordable, Well-Balanced Performer

YAESU FRG-8800

The Japanese firm of Yaesu has always had a policy of making only one model of world band radio at a time. It started in 1977 with the introduction of the FRG-7—or "Frog," as aficionados call Yaesu sets—and it continues today with the current model, the $699.95 FRG-8800.

For the active world band program listener, the '8800 makes a good choice. The standard bandwidths are reasonably adequate, the set is handy to operate, and its overall audio quality is suitable for listening to favorite programs hour-after-hour. Its design is well balanced, and the set has few drawbacks.

Varied Features

Stations are selected either by keypad entry or via a conventional tuning knob. The '8800 also has a dozen programmable channel memories for favorite stations, plus a scanner.

A digital LCD, rather than an analog meter, is used to indicate signal strength. That LCD also includes a 24-hour World Time (UTC) clock with daily on-off cycles, as well as a "sleep" switch to shut off the set after a period of anywhere up to an hour. Alas, the clock/timer's digital display is shared with the frequency readout; you can read one or the other, but not both at the same time.

Worthy Overall Performance

The '8800's dynamic range is good, as is blocking performance. Also top-drawer is the '8800's sensitivity. These measurements, taken together, help us understand why this model sifts out weaker stations with relative ease.

A major plus is the '8800's audio quality, which is above average. Perhaps because the set performs relatively well with its inboard speaker, Yaesu does not offer an optional outboard speaker.

The '8800 is particularly easy to use for what's known as "manual exalted-carrier selectable-sideband (ECSS)" reception. This useful interference-reducing technique—a less-advanced version of the synchronous detection found on Sony's ICF-2010/ICF-2001D portable—is carefully explained in the comprehensive and well-written owner's manual.

Suboptimal Bandwidths

Two bandwidths—as opposed, alas, to three on the discontinued FRG-7700 that was replaced by the '8800—are provided. Unfortunately, the 8.3 kHz "wide" bandwidth is excessively wide for reception of most world band broadcasts, while the 3.8 kHz "narrow" bandwidth is too skinny to serve as a "wide" and too broad to be a good "narrow." The '8800 thus is effectively left with only one regularly useful bandwidth. This is the only significant shortcoming of this model, although unwanted signals also sometimes pop up when they shouldn't.

Microprocessor Failures Reported

Some '8800's reportedly have had microprocessor failures, brought about by voltage "spikes" or static charges, that cause the set to operate improperly. It's an annoyance that can often be dealt with by simply removing, then reinserting, the three "AA" cells in the microprocessor. Another cure, replacing the chip, requires professional attention. Fortunately, disconnecting the set during thunderstorms and not letting the backup batteries run down should help prevent the problem from arising.

Overall, however, the FRG-8800's combination of price and performance, plus its user friendliness and good audio, make it a worthy choice for world band listening.

Britain's New Bulldog

LOWE HF-125

One of the most interesting of the new tabletop offerings is the HF-125 which is designed and manufactured not in Japan, but in Europe. The manufacturer, Lowe Electronics Limited, clearly does not fit the image of a giant electronics conglomerate. If you visit their headquarters, you won't find a sixty-story chrome and glass skyscraper filled with secretaries. There are no mile-long factories belching smoke into the air. In fact, Lowe is primarily a distributor of radio equipment tucked away in the rustic town of Matlock in Derbyshire, England.

The fact that Lowe doesn't have an office tower or a factory shouldn't be disturbing. Removed from the pressure-cooker environment of a large industrial complex, the people at Lowe have been able to think clearly about their product. Too, this compactness allows Lowe to make small production runs so the set can be improved from time-to-time. The result is a unique and quite British set.

The first thing you notice about the '125 is its sensible price: It's the lowest of any preassembled tabletop model tested.

The second thing is that it's quite compact and has few controls. There are only four knobs on the set—including the tuning knob—and five pushbuttons, plus the signal-strength meter. That's a total of only nine in all, as compared with phalanxes found on many Japanese tabletop models.

If you're wondering where the keypad is hidden, it's not. The keypad, which performs superbly, is optional and comes as an outboard device. If you want to hear, say, Radio Canada International on 5960 kHz, you just tap in 5-9-6-0, and it appears. There's no fooling around with "enter" keys, leading zeroes, decimals, or any of the other business found on most other keypads. In fact, it's made even easier to use—and handier, too—because it lies flat on the table instead of sitting vertically on the front panel.

The next thing you'll notice with the '125 is that it sounds unusually good for a communications receiver. Lab measurements show uncommonly low distortion—almost on the level of hi-fi equipment. When you hook up a good speaker to this receiver, it really does sound pleasing, even though there is some "whine" from the set's digital circuitry than can sometimes be heard faintly in the background.

Another thing that helps this set sound so good is that it comes with no less than *four* bandwidths from which to choose. Most receivers come with only two, so this is a real plus. And they work quite well, thanks to some really innovative engineering on Lowe's part. Of the four bandwidths, the widest is good for listening to local medium-wave AM stations, while the other three work nicely for world band channels.

The third unusual characteristic of the '125 becomes apparent when you take one of the covers off. This set is built more like a tank than a radio. Inside is a pair of thick, cast-aluminum side panels that look as if they could survive a nuclear blast. Other panels and covers are no less impressive in their construction, being made from heavy-gauge stamped aluminum. This is the sort of bulldog construction you find on costly professional communications receivers designed for battlefield applications . . . not a commercially available $600 receiver.

Another plus is that the front panel is protected by a sheet of plastic laminate. You'd have to work hard to scratch up this radio.

The only problem is that if you like gadgets, the '125 will be a disappointment. There is no notch filter, passband tuning, or scanner. And the frequency readout is only to the nearest kilohertz. What's ironic is that in an attempt to simplify this radio, the people at Lowe have actually complicated some things, especially if you're constantly switching modes and the like.

Those regularly tuning stations in the tropical bands won't care for the automatic gain control (AGC) decay rate in the AM mode. It's awfully slow, and there's no faster rate you can switch in. Also, the '125 isn't an ultrasensitive set. Sensitive, yes, but not like some of the Japanese supersets, such as the Kenwood R-5000. And there is a trace of digital-circuit "whine" sometimes audible in the background.

One very unusual aspect of the '125 is that you can get it with an optional synchronous detector, like you find on Sony's excellent ICF-2010/ICF-2001D portable. Alas, this is one thing Lowe didn't do well. This option is hard to operate, doesn't lock properly, and you can't even select sidebands with it. The manufacturer has indicated that its engineer is working to correct at least some aspects of it.

Another option allows the '125 to be used as a portable. We haven't been able to test this, but the manufacturer states that it consists of a rechargeable NiCad battery pack and built-in active antenna. Eventually, they'll offer a carrying case with a shoulder strap, too. This should make the compact '125 appropriate for high-quality field portable applications, especially since the radio is so rugged.

So, in all the '125 is an agreeable choice for listening to world band broadcasts. And if the construction of this tough little set is any indication, it should last a long time, too.

A Tough American-Made Receiver

TEN-TEC RX-325

Towards the very end of 1986, the American firm of Ten-Tec, located in Tennessee, introduced its first world band radio, the $699 RX-325. After a period of test-marketing in which it wasn't clear whether the '325 would be made a permanent part of the line, in late 1987 Ten-Tec went ahead with plans for regular production.

High-Quality Construction

What the RX-325 has that you don't find in most other receivers is compactness and high quality of construction. In fact, the '325 is such a small unit that you'd almost think it was meant to be a portable. As for that quality of construction, the '325 looks as though it was designed for use in Beirut. The cabinet and chassis are almost exclusively of steel and aluminium, with plastic only moderately in evidence.

Usual Array of Controls

The '325 comes with the kinds of controls we've all come to expect from newly designed world band radios. Take tuning, for example. The '325 isn't tuned only by the usual knob, but also by a keypad—plus a scanner and memories to store your favorite stations. Another plus is that it's an easy set to operate. In fact, the only real problem is that the keypad is too small for comfortable day-to-day use. It also uses soft-rubber keys that have all the response of unroasted marshmallows.

There are some other ergonomic shortcomings, too—especially with the fluorescent display. But, overall, the controls are well laid out and the set is straightforward to operate. A World Time (UTC) clock is included, as well.

Reasonable Performance

The '325 performs fairly well, overall. For one thing, it's very sensitive, which is helpful if you're trying to receive some faint signal from afar.

Its audio quality has the potential of being above average, too. "Potential" because it's not anything to write home about with the set's dinky little built-in speaker, which is on the bottom of the set. Ten-Tec doesn't offer an optional external speaker, but with a good outboard speaker the sound is quite pleasant.

The real problem with this radio is that it doesn't do much else all that well. For example, for half the price of the '325 you can get a Sony ICF-2010 or ICF-2001D with synchronous detection. Synchronous detection helps in reducing interference, as well as in eliminating certain types of fading and distortion. But there is no synchronous detection on the '325. Even if you try selecting sidebands manually, the results are mediocre because some of the circuitry is drifty and the set doesn't tune closer than to the nearest 50 Hz. Also, there's no passband tuning, no notch circuit and no tone controls.

Dynamic range, front-end selectivity and certain other laboratory measurements of performance are not encouraging, either. For difficult listening situations, especially in such high-field-strength locations as Europe, the '325 is going to have some tough sledding. Even in North America you may encounter difficulties with the '325 if you live near a local FM or mediumwave AM station.

In all, the Ten-Tec RX-325 is a pleasant, easy-to-use little set with only reasonable performance. It's awfully well made—much better than nearly anything else on the market under $1,000 except the Lowe HF-125. In that respect, Ten-Tec has showed that when American manufacturers choose to do so, they can compete successfully . . . at least when it comes to quality of construction.

Antennas to Hear the World By

High-tech radios usually come equipped with foolproof little telescopic antennas that make an outdoor antenna unnecessary. But so do TV's. Yet, most of us hook up our televisions to an outdoor antenna or cable system for better reception.

World Band Portables Rarely Need Extra Antenna

The same isn't always true of world band radios, though. Nearly all portables—even the better ones—are designed to be operated off the telescopic antenna that comes with the set . . . or perhaps a bit of wire, at best. If you try to be too fancy and hook a portable up to a whiz-bang antenna, you may overload the radio's circuits. This won't damage the radio, but you could end up with a mishmash of signals . . . Radio Canada International might pop up in the middle of the wrong band, things like that. Remove the fancy antenna and, *voilà*! The radio works properly again.

But if you're using a truly excellent receiver, you need to feed it with something other than a meager diet of signals grasped from the air by a short telescopic antenna. That's why most tabletop world band receivers don't come with a built-in antenna: You don't feed table scraps to a thorobred.

The main reason for using an outboard antenna is that it can be mounted away from local electrical noises and things that might absorb radio signals before they reach your antenna. As a result, more radio signals and fewer noises work their way into your set, so you increase what's called the *signal-to-noise ratio*. More signal and less noise—those are the main things to keep in mind. A so-so antenna that's mounted where signals are strong and noises are weak will do better than a fancy antenna mounted next to a microwave oven or hidden behind thick walls.

So, Rule 1: The most important thing about an outboard antenna is where you put it. Get your antenna out into the fresh air. Best is to mount it outdoors, well away from buildings and utility lines. However, if you're in an apartment at least try to place your antenna near a window or out on the balcony—or even have it protruding inconspicuously from the building, flagpole style.

Outdoor Wire Antennas: Two Winners

In an *RDI White Paper* testing exercise running several months, we put popular outdoor wire antennas through the hoops and twists of receiving faint world band radio signals. The best were Alpha-Delta's "Sloper" and Antenna Supermarket's "Eavesdropper." The Sloper, although longer, is a bit more versatile in that it is designed to cover even the lowest tropical bands, as well as the popular world bands. Otherwise, both perform very well.

Which brings us to Rule 2: All outdoor wire antennas should be routed through a high-quality surge protector, such as the Alpha Delta Transi-Trap or similar devices produced by the R.L. Drake Company. Keep in mind that your antenna doesn't have to suffer a direct strike for damage to occur. Such awesome energy is released by a lightning strike that serious electrical charges can be picked up some distance away. These—and similar charges from windy snow and sand storms—are more than enough to damage your receiver's sensitive solid-state innards.

Surge protectors are a bit like burglar alarms. Most people don't install them until after their first burglary, and most don't purchase a surge protector until after they've found their Pride and Joy world band radio mute following a thunderstorm. 'Nuff said.

Antennas for Apartment Dwellers

Not everyone has the room to erect an outside antenna. The Eavesdrooper is 42 feet, or 12.8 meters, long; the Sloper 60 feet, or 23 meters. So if you live on the twelfth floor of an apartment building, you'll need something a bit more discreet. One solution is something called an "active" antenna. What distinguishes an active antenna from

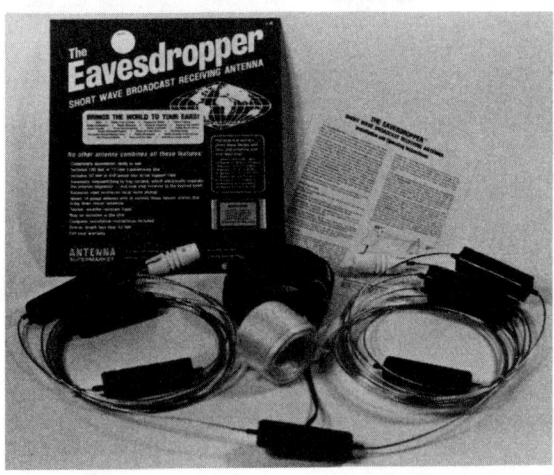

Eavesdropper outdoor antenna performs very well with tabletop radios

* Stero location recording
* Dbx noise reduction (75 db dynamic range)
* 3 Head off the tape monitoring while recording
* Dolby B noise reduction
* Built-in speaker/Channel select
* Built-in power
* Astonishing High frequency performance
* Illuminated VU meters
* 3 position mic attenuation
* Light weight/impact resistant

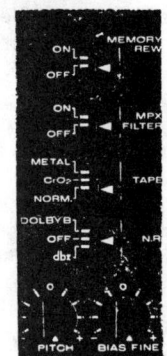

The PMD 430 has many useful features!

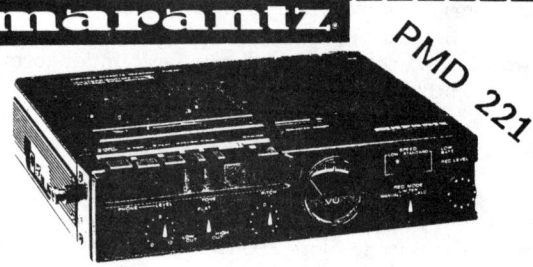

* Monaural location recording
* 3 Head off-the-tape monitoring while recording
* Direct telephone connective jack
* 2-Speed (1-7/8 & 15/16 IPS) allows up to 2 hours of unattended recording (using C-120 tapes)
* Built-in speaker with volume control
* 3 position mic attenuator
* Line IN/OUT

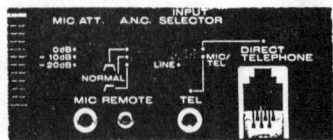

Direct telephone connective jack on the PMD 221

The PMD 430 comes with its own soft case.

COMPUTER CONTROL
YOUR RADIO WITH C.R.I.S.
(Computer/Radio Interface System)

All you need is **C.R.I.S.**, your computer and one of these radios:
<u>ICOM:</u> R71A, R7000, 271, 275, 375, 471, 475, 575, 735, 751, 761, 1271. <u>KENWOOD:</u> R5000, TS-440S, TS-940S, 811, 711. <u>YAESU:</u> FT757GX, 757GX II, FRG8800, 9600, FT-980, 767, 727.

CALL FOR PRICE AND DETAILS

FUNCTIONS:
* <u>AUTO LOG</u>: Log frequency, mode, signal strength[1], date, time and your comments to a file.
* <u>AUTO SCAN</u>: Scan up to 800 channels or between upper/lower limits (programmable). Scan terminates on strong signal and resumes on signal loss[1], operator intervention, or programmable delay.
* <u>AUTO TRACKING</u>: Tracks dual VFO's for splits.
* <u>EDIT MODE</u>: Read disk file or radio memory into computer RAM; edit computer RAM; write computer RAM to disk file or radio memory.

OPTIONS:
* EEB-SDA: Data acquisition for spectral analysis, store to memory/disk and display on CRT or printer.
* EEB-DBS: Shortwave Broadcast Database.
* EEB-DBU: Database of Utility stations/channels.
* EEB-RCM: Remote control via modem.

<u>UNDER DEVELOPMENT:</u> Multi receiver control, **C.R.I.S.** for: AR2002, NRD-525, PARAGON

HARDWARE REQUIREMENTS:
IBM PC/XT, AT or compatible with MS-DOS 2.11 or later, serial port, a DSDD floppy disk drive, 256K bytes of user memory and a monochrome or color monitor. Interface requires 13.8 Vdc @ 1 Amp. Optional D.C. adapter #571512 @ $19.95 + $4 UPS.
[1] Some radios may require modification.

TELEX: 62915985
FAX: 703-938-6911
• We ship world-wide
• Shipping charges not included
• Prices & specifications subject to change without notice

10 miles west of Washington, D.C.
Sorry—No COD's
10-5 Tues., Wed., Fri.
10-9 Thursday
10-4 Saturday
Closed Sunday and Monday

Electronic Equipment Bank
516 Mill Street, N.E.
Vienna, Virginia 22180
Order Toll Free 800-368-3270
Virginia 703-938-3350

something like the Eavesdropper—or even a simple hank of wire—is amplification. An active antenna contains electronic circuitry that actually amplifies the strength of signals it receives.

Active Antennas: Midgets of the Airwaves

Active antennas, however, are mixed blessings. True, their smaller size allows them to be affixed to windowsills and the like, plus they're inconspicuous. So they're a favorite of apartment dwellers. The drawback is that their electrical circuitry is prone to the same foibles as any other electrical circuitry. If the antenna has a so-so circuit and you're connecting it to a superset, you're almost certain to be disappointed, because the antenna's mediocre amplifier will provide a weak link in the signal chain. Also, too much antenna amplification can strain—"overload"—most receivers . . . especially portables.

No antenna—active or otherwise—is worth its salt if its pickup wire or rod has to be mounted close to the receiver, where it will be exposed to electrical noises emanating from the receiver itself. So we restricted this year's tests of active models to those having separate pickup and control modules. These allow signals to be picked up well away from the receiver, so the signal-to-noise ratio tends to be improved.

Our tests of five leading active world band antennas indicates that whether any given model will succeed depends on the specific conditions you face. For example, the Palomar PA-351 contains a special filter to help keep local medium-

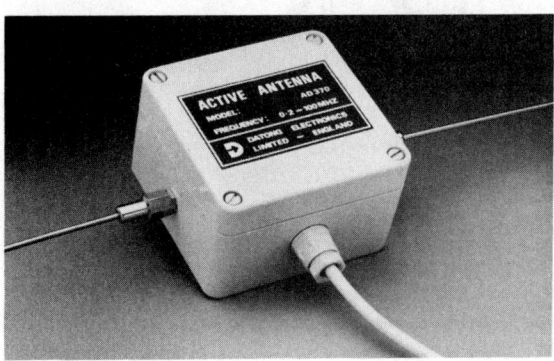

Datong's well-built AD370 active antenna, suitable for use outdoors

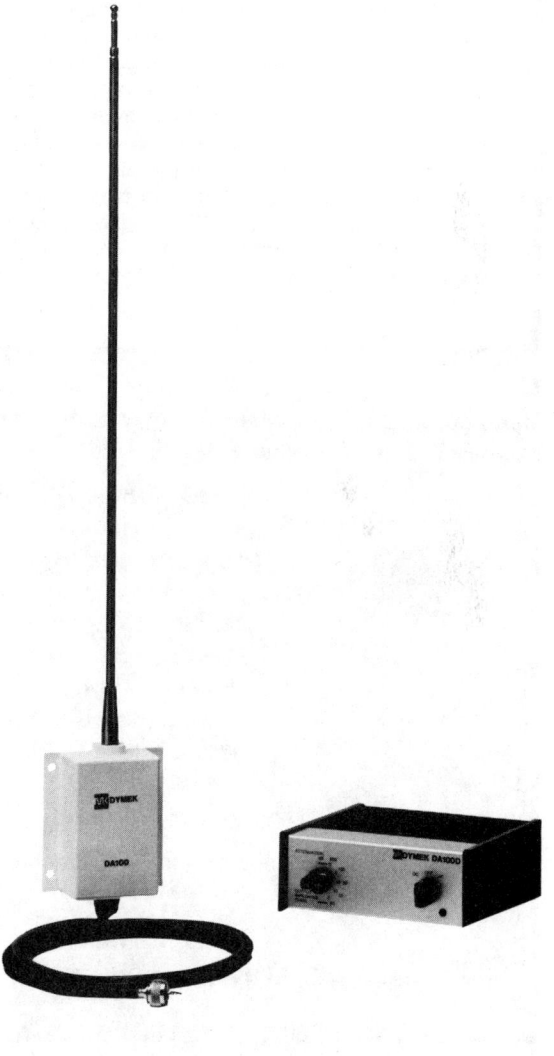

Dymek DA100D active antenna can be weatherproofed for use on the high seas

RDI WHITE PAPERS

EVERYTHING YOU NEED TO KNOW BEFORE YOU BUY

World band radios can be a real pleasure to use. But today's best models are also advanced-technology devices with a wide range of performance possibilities... and potential pitfalls.

Now, RDI's private library of in-depth test reports—the *RDI White Paper* series—is being made available to you. Within the pages of each report is the full range of measurements and opinions prepared by IBS' award-winning "hands-on" and laboratory experts. Findings are conveniently organized for ready reference and the results—good or ill—are revealed without fear or favor. Nothing is held back, whether for reasons of space or editorial policy.

RDI White Papers are thoroughly up-to-date. Because each report covers only one model, revised editions can be issued from our Database as soon as a manufacturer alters a previously tested model. Only the freshest material appears, and you buy only the reports you need.

The latest editions of these *RDI White Papers* are now available from participating dealers worldwide, or direct from RDI postpaid for US$4 each in North America, $6 each worldwide:

RDI Evaluates the Grundig Satellit 650 Receiver
RDI Evaluates the ICOM IC-R71 Receiver
RDI Evaluates the Japan Radio NRD-93 Receiver
RDI Evaluates the Japan Radio NRD-525U/525E/525J Receiver
RDI Evaluates the Kenwood R-5000 Receiver
RDI Evaluates the Lowe HF-125 Receiver
RDI Evaluates the Sony ICF-2010/ICF-2001D Receiver
RDI Evaluates the Ten-Tec RX325 Receiver
RDI Evaluates the Yaesu FRG-8800 Receiver
RDI Evaluates Popular Outdoor Antennas
How to Interpret Receiver Specifications and Lab Tests

The following reports are in process and will be available in the very near future:

RDI Evaluates Popular Indoor Antennas
RDI Evaluates the Grundig Satellit 400 Receiver
RDI Evaluates the Sony ICF-PRO80/ICF-PRO70 Receiver

Get all the facts before you buy. For a complete, up-to-date list of **RDI White Papers** and other RDI publications, please write, enclosing a self-addressed envelope, to

RDI Publications Information
Box 300
Penn's Park, PA 18943 USA

wave AM stations from intruding into the world radio bands. This makes it an obvious candidate for use in urban areas or by those living near mediumwave AM stations. On the other hand, the Grove Power Ant III—by itself, more a preamplifier than an active antenna, as you have to purchase separate antenna components—lacks such a filter and has uninspiring dynamic range, to boot. This makes it a dubious choice for use in urban locations. But it's quiet at full gain, which allows it to be considered for use with high-quality receivers in outlying areas where mediumwave AM signals are weak.

The Datong AD270—for indoors—and AD370—for outdoors—are the most balanced performers. Their amplification is strong without being excessive, dynamic range is impressive, and they're reasonably quiet. However, they have no filter to reject local stations. If you live near local mediumwave AM stations, this could make the Palomar a better choice.

The MFJ-1024 has the best dynamic range of the group. This, plus its moderate level of amplification, means that it is the least likely to cause "overloading" within either your antenna or radio. However, both it and the Dymek, although not really noisy, are also the least quiet models tested—a consideration if you own a particularly "hiss" free receiver.

The Dymek DA100D performs similarly to the MFJ unit, except that the Dymek's amplification rises with frequency, whereas that of the MFJ falls. Regular listeners to the tropical bands thus will prefer the MFJ. However, listeners to the popular international world bands should lean to the Dymek—especially if they tune around during the daytime, when the higher bands are most active.

Finally, one disappointing finding is that virtually all the active antennas tested tend to cause audible hum from the speakers of the receivers to which they are connected. For the most part, this is caused by the antenna's own outboard ac power supply. The only practical way to cope with this, short of using a battery power supply, is to mount your antenna's pickup head a good distance from the control box and power supply.

The Bottom Line

If you are using a portable, you probably won't need an outboard antenna of any sort. However, if you have a tabletop receiver, you will almost certainly need one. The best choice, if it's feasible, is a passive antenna, such as the "Sloper" or "Eavesdropper." Otherwise, choose the most suitable active antenna based on the conditions at your location.

Outdoor antennas can have unexpected consequences. At twilight, a hot-air balloon becomes entangled in antenna at **Passport's** monitoring site in Pennsylvania

"The Best Results throughout the Shortwave Spectrum."

— Larry Magne, Radio Database International White Paper

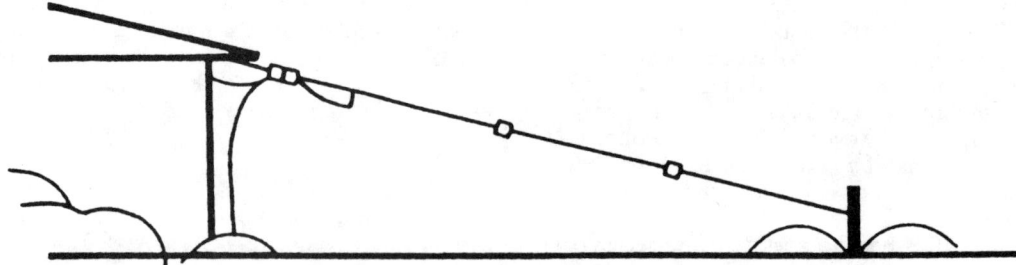

Get world-class, multi-band reception with

ALPHA DELTA DX–SWL SLOPER ANTENNA

Just $69.95 plus shipping from your Alpha Delta dealer!

- Fully assembled, ready to use and built for long life. So strong, it can even be used to transmit — up tp 2 kW!
- Superior multi-band performance on 13, 16, 19, 21, 25, 31, 41, 49, 60, 90, 120 meters plus the AM broadcast band (.5-1.6 MHz). All in a single compact antenna. Alpha Delta first!
- Efficient multi-band frequency selection by means of special RF choke-resonators — instead of lossy, narrow band traps.
- Overall length just 60 feet. Requires only a single elevated support — easier to install than a dipole.
- 50 ohm feedpoint at apex of antenna for maximum DX reception. A UHF connector is provided on the mounting bracket for easy connection to your coax.
- A top overall rating in Radio Database International's hard-hitting White Paper, "RDI Evaluates the Popular Outdoor Antennas."

There's a lot happening on the shortwave broadcast bands. Don't miss a thing by skimping on your antenna. Get world class, multi-band DX reception with the Alpha Delta model DX-SWL Sloper. Just $69.95 plus shipping from your local Alpha Delta dealer.

ALPHA DELTA COMMUNICATIONS, INC.
P.O. Box 571 • Centerville, Ohio 45459

Photo page 113: Radio Nederland, like many world band broadcasters, uses directional antennas to improve signal strength.
Courtesy Radio Nederland

WORLDSCAN
- Worldwide Broadcasts
- The Blue Pages

Receive the world's short wave bands with only one antenna and one feed line!

11 Meters (25.6 to 26.1 MHz)
13 Meters (21.45 to 21.75 MHz)
16 Meters (17.7 to 17.9 MHz)
19 Meters (15.1 to 15.45 MHz)
25 Meters (11.7 to 11.975 MHz)
31 Meters (9.5 to 9.775 MHz)
41 Meters (7.1 to 7.3 MHz)
49 Meters (5.95 to 6.2 MHz)
60 Meters (4.75 to 5.06 MHz)

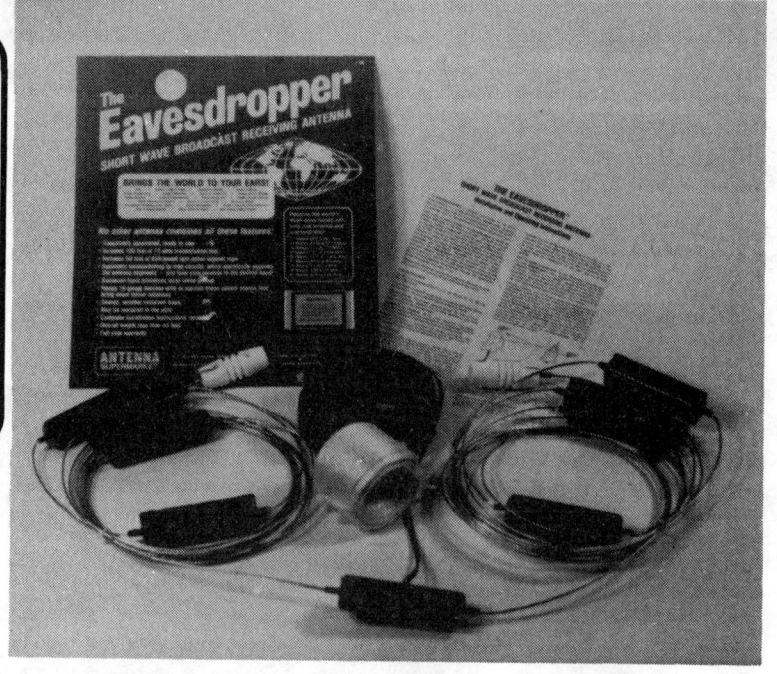

Only
$64.50
plus shipping

No other antenna combines all these features:

- **Completely assembled, ready to use**
- **Includes 100 feet of 72-ohm transmission line**
- **Includes 50 feet of 450-pound test nylon support rope**
- **Automatic bandswitching by trap circuits, which electrically separate the antenna segments — just tune your receiver to the desired band!**
- **Balanced input minimizes local noise pickup**
- **Heavy 14-gauge antenna wire to survive those severe storms that bring down lesser antennas**
- **Sealed, weather-resistant traps**
- **May be installed in the attic**
- **Complete installation instructions included**
- **Overall length less than 43 feet**
- **Full year warranty**

Carried in stock by many progressive dealers
Used in 45 countries we know of as of 1983
Large SASE or 3 IRC's (foreign) for brochure

TO ORDER:
Please patronize your stocking dealer. We both need him to survive. If no luck there, you may order direct, certified funds or C.O.D. Add $3.50 for UPS Cont. U.S. C.O.D.'s cash only. Dealer inquiries invited.

ANTENNA SUPERMARKET

P. O. Box 563, Palatine, Illinois 60067 U.S.A.   MADE IN U.S.A.
Manufacturers of Quality Short Wave Listening Antenna Systems

WORLDWIDE BROADCASTS

You can tune in world band radio broadcasts in either of two ways: by country or by channel.

This quick-access "Worldwide Broadcasts" section tells you right away the best channels to hear in English, French, German, Japanese and Spanish from each country. It's perfect for getting right to the station you want.

Readers tell us that they also especially enjoy "traveling" around the bands . . . dialing about to discover what they like to hear, rather than simply tuning directly to a given country. The at-a-glance "Worldscan" section within the blue pages of this book allows you to do just that. It also includes a wealth of details, including time and target zone, for every station on the air.

Channels in **bold** type often come in best, as they are from special relay transmission facilities located near the countries where the broadcasts are intended to be heard.

AFGHANISTAN
RADIO AFGHANISTAN
English 9635 kHz

ALBANIA
RADIO TIRANA
English 6200, 7065, 7075, 7080, 7120, 7205, 9480, 9500, 9760, 11835, 11855, 11960, 11985, 15185 kHz
French 6170, 7075, 7080, 7120, 7135, 7165, 7205, 9480, 9500, 11855, 11905, 11960 kHz
German 6080, 7310, 9375, 11985 kHz
Spanish 6200, 7300, 9430, 9500, 9750, 9760, 11985 kHz

ALGERIA
"VOICE OF FREE SAHARA"
Multilingual **9640**, 15215 kHz
RTV ALGERIENNE
English 9509, 9640, 15215, 17745 kHz
French 9509, 9640, 15160, 15215, 17745 kHz
Spanish 9509, 9640, 15160, 15215 kHz

ANGOLA
RADIO NACIONAL
Multilingual 7245, 9535, 9720, 11955 kHz

ARGENTINA
RADIO ARGENTINA—RAE
English 9690, 11710, 15345 kHz
French 15345 kHz
German 15345 kHz
Portuguese 9690, 11710 kHz
Spanish 6060, 9690, 11710, 15345 kHz
Japanese 11710 kHz
RADIO NACIONAL
Spanish 6060, 9690 kHz

AUSTRALIA
AUSTRALIAN BROADCASTING CORP
English 5025, 9610, 15425 kHz
Multilingual 4835, 4910 kHz
RADIO AUSTRALIA
English 5995, 6035, 6060, 6080, 7205, 7215, 9580, 9620, 9645, 9655, 9710, 9770, 11705, 11720, 11800, 11910, 11945, 15140, 15160, 15180, 15240, 15315, 15320, 15395, 15415, 17715, 17750, 17795 kHz
French 11945, 15160, 15315, 15320, 15395 kHz
Japanese 9710, 11800 kHz

AUSTRIA
RADIO AUSTRIA INTERNATIONAL
English 5945, 6000, 6155, 7115, 7210, 7245, 7260, 9550, 9585, 9600, 9620, 9635, 9685, 9725, 11760, 11805, 11825, 11830, 11840, 11855, 11915, 11920, 12015, 15320, 15410 kHz
French 5945, 6000, 6155, 7115, 7260, 9550, 9600, 9725, 11760, 11805, 11830, 11920, 12015, 15410 kHz
German 5945, 6000, 6155, 7115, 7210, 7245, 7260, 9550, 9580, 9585, 9600, 9620, 9635, 9655, 9660, 9685, 9720, 9725, 11660, 11720, 11760, 11805, 11825, 11830, 11840, 11855, 11915, 11920, 12015, 15270, 15320, 15410 kHz
Spanish 5945, 6000, 6155, 9580, 9585, 9655, 9660, 9720, 11660, 11720, 12015 kHz
Other 5945, 6000, 6155, 7210, 9600, 9635, 9685, 11915, 15410 kHz

BANGLADESH
RADIO BANGLADESH
English 4880, 6240, 7505, 11645, 12030, 15525, 17653 kHz
Multilingual 4880, 4890, 6195 kHz

BELGIUM
BELGISCHE RADIO & TV
English 5910, 9790, 9830, 9880, 9905, 9925, 11695, 11985, 15515, 15580, 15590, 17595 kHz
French 5910, 6035, 9905, 11695, 11985, 15515, 15580, 15590, 17595 kHz
German 5910 kHz
Spanish 5910, 9790, 9830, 9905, 9925, 11985 kHz
RADIO TV BELGE FRANCAISE
French 6050, 7140, 9900, 9925, 11660, 11855, 15540, 17580, 17675 kHz
German 7140, 17675 kHz

BOTSWANA
RADIO BOTSWANA
Multilingual 4820, 5955, 7255 kHz

BRAZIL
RADIO NACIONAL
English 11745, 15265, 15335 kHz
French 11765 kHz
German 15265, 15335 kHz
Spanish 9655, 11745 kHz

BULGARIA
RADIO SOFIA
English 6070, 7115, 7155, 7280, 9560, 9595, 9700, 9740, 11720, 11735, 11750, 11765, 11835, 11840, 15140, 15310, 15330 kHz
French 6070, 7155, 7280, 9560, 9595, 9700, 9740, 11720, 11735, 11765, 11835, 11840, 15140, 15310 kHz
German 6035, 6070, 7155, 9700, 11720, 15140 kHz
Spanish 6115, 9560, 9745, 9750, 9755, 11710, 11765, 11840, 11860, 11870, 15370 kHz

BURKINA FASO
RTV BURKINA
English 7230 kHz
French 4815, 7230 kHz
Multilingual 4815, 7230, 9515 kHz

BURMA
BURMA BROADCASTING SERVICE
English 5985, 7185, 9730 kHz

CAMEROON
RADIO CAMEROON
French 4795 kHz
Multilingual 4795, 4850, 5010, 7150, 7240, 9746 kHz

CANADA
CANADIAN BROADCASTING CORP
English 6065, 6195, 9625, 11720 kHz
French 6065, 6195, 9625, 11720 kHz
Multilingual 6195, 9625, 11720 kHz
RADIO CANADA INTERNATIONAL
English 5960, **5995**, **6050**, 6140, **7130**, **7155**, **7185**, **7235**, **7295**, 9535, **9555**, 9625, **9740**, 9750, 9755, 9760, 11710, **11775**, 11790, **11840**, 11845, **11915**, **11935**, 11940, 11945, 11955, 11960, 15140, 15150, **15160**, **15180**, **15235**, 15260, 15325, 15440, 17820, 17875 kHz
French 5960, **5995**, **6050**, 6140, **7130**, **7155**, **7185**, **7230**, **7235**, **7295**, 9535, **9555**, 9650, **9740**, 9750, 9755, 9760, 11710, **11775**, **11840**, 11845, **11915**, **11935**, 11940, 11945, 11960, 15140, 15150, **15160**, **15180**, **15235**, 15260, 15325, 15440, 17820, 17875 kHz
German **5995**, **7235**, 15325, 17820 kHz
Spanish 9535, 11845, 11940 kHz
Japanese **3925**, **6055**, **9595** kHz

CENTRAL AFRICAN REPUBLIC
RTV CENTRAFRICAINE
Multilingual 5035 kHz

CHAD
RADIODIFFUSION NATIONALE
Multilingual 4904, 7120 kHz

CHILE
RADIO SISTEMA NACIONAL
Spanish 15140 kHz

CHINA (PR)
NEI MONGOL PEOPLE'S BS
Chinese 7105, 9750 kHz
Other 6974 kHz
RADIO BEIJING
English 4130, 4200, 5250, 6825, 7480, 9440, 9535, 9550, 9570, 9645, 9700, 9730, 9860, 11755, 11860, 11905, 11980, 15165, 15280, 15440, 15445 kHz
French 4020, 7055, 7185, 7800, 9880 kHz
German 4130, 5250, 9860 kHz
Spanish 7375, 9570, 9590, 9645, 9860, 9945, 11445, 11650, 11685, 11695, 11980, 15100, 15120, 15180, 15200, 15445, 17650, 17680 kHz
Japanese 4960, 7295, 7480 kHz

CHINA (TAIWAN)
VOICE OF FREE CHINA
English **5985**, **6155**, 7445, **9555**, **9680**, 9685, 9955, **11740**, 15345.2, 15370, **15440**, **17845** kHz
French 9765, **11805**, 15370, **15440**, **17845** kHz
German 9765, 9845, **11805**, **15440**, **17845** kHz
Spanish **9555**, 9765, 9955, **11740**, **11855**, **11885**, **15130**, **15215**, **17805**, **17845** kHz
Japanese 7130, 15345.2 kHz
VOICE OF ASIA
English 7285, 7445 kHz

COLOMBIA
RADIO SUTATENZA
Spanish 5095 kHz

CONGO
RTV CONGOLAISE
French 6115, 9715 kHz
Multilingual 6115, 9715 kHz

COSTA RICA
FARO DEL CARIBE
Spanish 9645 kHz

CUBA
RADIO HABANA
English 6090, 6120, 6140, **6165**, 9525, **9695**, 9740, 11725, 11795, 15300 kHz
French **6165**, **7165**, **9695**, **9730**, **11755**, 11795, 11950, **17705** kHz
Spanish 6060, 6090, 9550, **9695**, **9720**, **9730**, 9770, 11725, 11760, 11815, 11950, 11970, **15230**, 15300, **17705** kHz

CZECHOSLOVAKIA
RADIO PRAGUE
English 5930, 6015, 6055, 7290, 7345, 9605, 9630, 9740, 11855, 11990, 13715, 15110, 15155, 17730, 17840, 21505, 21705 kHz
French 5930, 6055, 7290, 7345, 9505, 9600, 11990, 13715 kHz
German 6055 kHz
Spanish 5930, 6015, 6055, 7345, 9630, 9740, 11990 kHz
Multilingual 6055, 7345, 9505, 9630 kHz

DOMINICAN REPUBLIC
RADIO CLARIN
Spanish 11700 kHz

ECUADOR
HCJB—VOICE OF THE ANDES
English 6075, 6130, 6205, 6230, 9745, 9845, 9860, 9870, 11740, 11910, 11925, 15115, 15155, 15270, 17790, 17890 kHz
French 9860, 11740, 11835, 11910, 15155, 15220, 15270, 17790 kHz
German 6075, 6205, 9715, 9845, 9860, 11740, 11835, 15250, 15270, 17790 kHz
Spanish 6050, 9765, 9860, 11835, 11910, 11960, 15160, 15270, 17790, 17890 kHz
Japanese 6075, 9715, 11835, 15295 kHz
RADIO NACIONAL
Spanish 15270, 17790 kHz

EGYPT
RADIO CAIRO
English 9475, 9655, 9675, 15255, 15375, 17675 kHz
French 9900, 15335 kHz
German 9900 kHz
Spanish 9475, 9740, 11715 kHz

EQUATORIAL GUINEA
RADIO NACIONAL
English 9553 kHz
Multilingual 4925.6 kHz

ETHIOPIA
"RADIO FREEDOM"
Multilingual 9595 kHz
"VOICE OF NAMIBIA"
Multilingual 9595 kHz
VOICE OF REVOLUTIONARY ETHIOPIA
English 9560 kHz
French 9560 kHz

FINLAND
RADIO FINLAND
English 6120, 9560, 9605, 9610, 9655, 11715, 11755, 11850, 11935, 11945, 15185, 15245, 15400 kHz
French 6120, 9610, 11755, 11850, 11945, 15185, 15400 kHz
German 6120, 9610, 11755 kHz

Aktuelles, Kultur und Gesang

PRESENTING

Schau ins Land, the new biweekly German language magazine on cassette. Recorded entirely in idiomatic German by a team of top European broadcasters and journalists, *Schau ins Land* offers unique insights into the diverse landscapes and cultures of German speaking Europe: West and East Germany, Austria, and Switzerland. Each 45-minute edition features conversation about current affairs, travel, the arts and entertainment, as well as interviews and a sampling of today's best German popular music. As an aid to comprehension, each program comes with a complete printed German transcription and an extensive German-English vocabulary section.

Schau ins Land

Subscribe now and receive the Autumn '87-Spring '88 series (a total of 14 editions) for just $118. Or, if you prefer, order only the Autumn '87 series for $69 (7 editions). You'll receive a new issue twice-monthly beginning in early September.

Please send me ☐ Autumn and Spring ☐ Autumn only

Print Name _____
Address _____
City/State/Zip _____

Tennessee residents add 7.75% sales tax.

TO: *Schau ins Land*, Dept. SD1
P.O. Box 158067
Nashville, TN 37215-8067

Do not sent cash. © 1987, CHAMPS-ELYSEES, INC.

**To order by VISA or MasterCard or for more information call: 1-800-824-0829.
In Tennessee and Alaska call collect: 615-383-8534**

FRANCE
RADIO FRANCE INTERNATIONAL
English 4890, **5990**, 6045, **6055**, 6115, 6175, 7135, 7235, 7280, 9535, 9550, 9715, 9790, **9800**, 11700, 11705, 11805, 11995, 15315, 17620, 17795 kHz
French **4890**, 5950, **5990**, 5995, 6045, **6055**, 6115, 6175, 7120, **7135**, 7160, 7235, 7280, 9535, 9550, 9575, 9605, 9715, 9745, **9790**, **9800**, 9805, 9810, **11670**, 11690, **11700**, 11705, 11790, 11800, 11805, 11845, 11930, 11955, 11965, **11995**, 15135, 15155, 15180, 15190, 15195, **15300**, 15315, 15365, 15425, **15435**, 15460, 17620, **17720**, 17785, 17795, 17800, 17845, 17850, **17860**, 21620 kHz
German 6150, 7145 kHz
Spanish 5950, 5995, 6040, **6055**, **6175**, 9715, **9790**, **9800**, **11670**, **11700**, 11965, **11995**, **15200**, **15435** kHz
Multilingual **9790**, 9805, 11670, 11690, **11700**,

GABON
ADVENTIST WORLD RADIO
French 9630 kHz
AFRIQUE NUMERO UN
French 4830, 7200, 11940, 15200, 15475, 17870 kHz
Multilingual 7200, 15200, 15475 kHz
RTV GABONAISE
Multilingual 4777, 7270 kHz

GERMANY (DR)
RADIO BERLIN INTERNATIONAL
English 5965, 6010, 6040, 6070, 6080, 6115, 6125, 6165, 7185, 7260, 7295, 9505, 9560, 9620, 9665, 9730, 11705, 11750, 11785, 11795, 11810, 11890, 11920, 11970, 15145, 15170, 15240, 15255, 15440, 15445, 17705, 17755, 17880, 21465, 21540 kHz
French 5965, 6040, 6115, 7170, 7185, 7260, 7295, 9505, 9620, 9665, 9730, 11750, 11785, 11810, 11890, 11970, 15145, 15170, 15255, 17755, 21465 kHz
German 5965, 6010, 6040, 6070, 6080, 6105, 6115, 6125, 7105, 7170, 7185, 7295, 9505, 9560, 9600, 9620, 9645, 9665, 9730, 9770, 11705, 11750, 11785, 11795, 11810, 11890, 11920, 15170, 15240, 17705, 17755, 17880, 21465, 21540 kHz
Spanish 6010, 6040, 6070, 6125, 6165, 7185, 7260, 7295, 9505, 9600, 9620, 9645, 9665, 9730, 11750, 11785 kHz
STIMME DER DDR
German 6115 kHz

GERMANY (FR)
BAYERISCHER RUNDFUNK
German 6085 kHz
DEUTSCHE WELLE
English 5960, **5995**, 6010, 6035, **6040**, **6085**, **6120**, 6130, 6145, **6170**, **6185**, 7110, 7130, 7150, 7195, **7200**, **7225**, 7285, **9545**, 9565, 9585, **9600**, 9610, **9615**, 9625, **9640**, **9650**, **9690**, **9700**, **9735**, 9745, 9765, 11765, **11785**, **11945**, **15105**, 15135, 15160, 15185, 15205, 15320, 15330, **15410**, 17765, 17780, **17800**, 17825, **21560**, 21600, 21680 kHz
French 7150, 7195, **7225**, **9565**, 9610, 9625, 9700, **9735**, 9765, 11765, 11945, 15135, 15185, **15410**, 17765, **17800**, 17825, 21600 kHz
German 3995, 6040, **6045**, **6065**, 6075, **6085**, 6100, 6145, **6160**, 7130, 7145, 7175, **7225**, 7285, **9545**, 9570, **9605**, 9615, **9640**, 9645, 9650, **9690**, 9700, 9715, **9735**, **11705**, **11730**, **11765**, **11785**, **11795**, 11820, 11855, **11945**, **11955**, **15105**, **15210**, **15245**, 15260, **15270**, **15275**, 15320, **15355**, **15410**, **17715**, 17810, 17845, **17860**, **17875**, 21560, **21680** kHz
Spanish 6010, 6045, 6120, 6130, 6145, 7235, **9545**, **9565**, **9605**, **9690**, 9700, **9735**, **11705**, **11785**, **11795**, 11810, **11865**, **15105**, **15205** kHz
Japanese **7170**, **7265**, **9505**, **9640**, 9670, 9680, 11810, **11845**, **11865**, 15185, 15400 kHz
RADIO IN THE AMERICAN SECTOR
German 6005 kHz

GHANA
GHANA BROADCASTING CORP
English 7295 kHz
Multilingual 4915 kHz
RADIO GHANA
English 6130 kHz
French 6130 kHz

GREECE
FONI TIS HELLADAS
English 7430, 9395, 9420, 9425, 9855, 11615, 11645, 15630, 17565 kHz
Spanish 9395, 9420 kHz
Multilingual 7430, 9395, 9425 kHz

GUAM
ADVENTIST WORLD RADIO
English 9830, 11980 kHz
Japanese 11965 kHz
KTWR—TRANS WORLD RADIO
English 9840, 11715, 11840 kHz
Japanese 9675, 9820, 11840 kHz

GUINEA
RTV GUINEENNE
Multilingual 7125, 9650, 15310 kHz

HOLLAND
RADIO NEDERLAND
English 5955, **6020**, **6165**, 7175, **9515**, **9540**, **9590**, **9630**, **9650**, **9715**, 9895, **11735**, 11740, 11930, 13770, 15560, **15570**, **17575**, 17605, **21480**, **21485**, 21685 kHz
French **9540**, 9840, 9895, 11730, **11740**, 15280, 15560, **17605**, **21685** kHz
Spanish 6020, 6110, **6165**, **9590**, 9610, 9895, **11715**, 11930, **15315**, **17605** kHz

HUNGARY
RADIO BUDAPEST
English 6025, 6110, 7220, 7225, 9520, 9585, 9835, 11910, 15220, 17710 kHz
German 7220, 7225, 9585, 9835, 11910 kHz
Spanish 6025, 6110, 7225, 9520, 9585, 9835, 11910 kHz

INDIA
ALL INDIA RADIO
English 6035, 7150, 7215, 9545, 9550, 9595, 9665, 9755, 9910, 11620, 11715, 11765, 11795, 11810, 11845, 11865, 11870, 15110, 15130, 15175, 15230, 15320, 15335, 17387, 17705, 17875 kHz
French 9755, 11865, 15365, 17830 kHz
Multilingual 6045, 6050, 6160, 7110, 7160, 7280, 9545, 9610, 9615, 9675, 10335, 11620, 11730, 11815, 11830, 11850, 11895, 15160, 15245, 15250, 15305, 15320, 17705 kHz

INDONESIA
VOICE OF INDONESIA
English 11788, 15150 kHz
French 11788, 15150 kHz
German 11788, 15150 kHz
Spanish 11788, 15150 kHz
Japanese 11788, 15150 kHz

IRAN
VOICE OF THE ISLAMIC REP
English 7215, 9022, 9765, 9770, 11790, 11930 kHz
French 9022, 9575, 9765, 9770, 11930, 15084 kHz
German 9022, 9765, 9770, 11930 kHz
Spanish 9022, 9575, 15084 kHz

IRAQ
RADIO BAGHDAD
English 6195, 9875, 11705 kHz
French 9875 kHz
German 9875 kHz

ISRAEL
KOL ISRAEL – VOICE OF ISRAEL
English 5885, 7410, 7464, 9012, 9385, 9435, 9815, 9855, 9860, 11585, 11610, 11655, 12080, 15095, 15485, 15640, 15650, 17620, 17630 kHz
French 5885, 7410, 7464, 9012, 9385, 9435, 9815, 9855, 9860, 11585, 11610, 11655, 12080, 15095, 15485, 15640, 15650, 17620, 17630 kHz
Spanish 5885, 7410, 7464, 9012, 9435, 9815, 9855, 9860, 11585, 11610, 11655, 12080 kHz

ITALY
RTV ITALIANA
English 5990, 6165, 7235, 7275, 7290, 9575, 9710, 9775, 11800, 11905, 15330 kHz
French 5990, 7235, 7290, 9575, 9710, 11800, 11905 kHz
German 5990, 7275, 7290, 9575 kHz
Spanish 5990, 7275, 9575, 9710, 11800, 11905, 15245 kHz
Multilingual 6060 kHz

UTILITY QSL ADDRESS GUIDE

VOLUME 1:   THE AMERICAS
VOLUME 2:   THE REST OF THE WORLD

The authority for shortwave utility (non-broadcast) station addresses for the serious QSL collector.

Contains addresses for Military facilities, ships, & MARS; USA & Canadian-flag merchant marine vessels; all air carriers; public & private coastal stations; air-route traffic control centers; cruiseships; federal radio agencies; and many more. Plus hot tips on how to obtain those utility QSL's!

Order yours today. Available at $12.95 each plus $2 for USA & Canadian shipping or $4 to other countries. Volume 2 available Fall of 1987. Payable in USA funds only.

Radio InfoSystems
P.O. Box 399
Holland, OH 43528
USA

Dealer Inquiries Invited.

IVORY COAST
RTV IVOIRIENNE
English 11920 kHz
French 4940, 6015, 11920 kHz

JAPAN
NIHON SHORTWAVE BROADCASTING
Japanese 3925, 6055, 6115, 9595, 9760 kHz
RADIO JAPAN/NHK
English **5960**, 5965, 5990, 5995, **6120**, 7105, 7140, 7240, 7280, 9505, 9525, **9645**, 9695, 9725, 11840, 11875, 11950, 11955, 15195, **15230**, 15235, 15350, **17785**, 17810, 17845, **21625** kHz
French 5965, 6070, 7105, 7180, **9570**, 11955 kHz
German 6070, 7105, 7180, **9570**, 11955 kHz
Spanish 5990, 6065, 9525, 9725, 11950, 15195 kHz
Japanese **5960**, 5965, 5990, 5995, 6065, **6120**, 7140, 7240, 7280, 9505, 9525, **9645**, 9695, 9725, 11840, 11950, 11955, 15195, **15230**, 15235, 15350, 17755, **17785**, 17810, 17845, **21625** kHz

JORDAN
RADIO JORDAN
English 9560 kHz

KAMPUCHEA (CAMBODIA)
VOICE OF THE PEOPLE
English 11937.7 kHz
French 11937.7 kHz

KENYA
VOICE OF KENYA
English 6045, 6100, 7270 kHz

KOREA (DPR)
RADIO PYONGYANG
English 6576, 7290, 9325, 9530, 9600, 9650, 9715, 9940, 9977, 11735, 11830, 13650, 13750, 15120, 15140, 15160, 15180, 15340 kHz
French 6576, 7290, 9325, 9530, 9600, 9940, 9977, 11735, 11830, 13750, 15340 kHz
German 6576, 7290, 9325 kHz
Spanish 6576, 7290, 9325, 9600, 9777, 11735, 13650, 15120, 15140, 15160 kHz
Japanese 6540, 9505, 11780 kHz

KOREA (REPUBLIC)
RADIO KOREA
English 5975, 6060, 6480, 7275, 7550, 9570, 9640, 9750, 9870, 11740, 13670, 15375, 15395, 15575 kHz
French 5975, 6480, 7275, 7550, 9515, 9870, 13670, 15575 kHz
German 6480, 7275, 7550, 9515, 9870, 13670, 15575 kHz
Spanish 6480, 7550, 9570, 9870, 11725, 13670, 15575 kHz
Japanese 5975, 6165, 7275, 9640 kHz

KUWAIT
RADIO KUWAIT
English 11675, 15345 kHz

LAOS
LAO NATIONAL RADIO
French **11870**, **11960**, **15190**, **15420** kHz

LESOTHO
RADIO LESOTHO
Multilingual 4800 kHz

LIBERIA
LIBERIAN BROADCASTING SYSTEM
Multilingual 6090 kHz
RADIO ELWA
English 11830 kHz
French 6070, 9550, 11830 kHz

LIBYA
RADIO JAMAHIRIYA
English 11815, 15450 kHz

LUXEMBOURG
RADIO LUXEMBOURG
English 6090 kHz
German 6090 kHz

MADAGASCAR
RADIO MADAGASIKARA
Multilingual 6135 kHz
RADIO MADAGASIKARA
French 5010 kHz
Multilingual 5010 kHz

MALAWI
MALAWI BROADCASTING CORP
Multilingual 5995 kHz

MALAYSIA
RADIO MALAYSIA
English 7295, 9665 kHz
Multilingual 7295, 9665 kHz
VOICE OF MALAYSIA
English 6175, 9750, 15295 kHz

MALI
RTV MALIENNE
Multilingual 5995, 7285, 9635, 11960 kHz

MALTA
RADIO MEDITERRANEAN
English 6110 kHz
French 6110 kHz

MAURITANIA
ORT DE MAURITANIE
Multilingual 4845, 7245 kHz

MEXICO
RADIO MEXICO INTERNATIONAL
Spanish 15430 kHz

MONACO
TRANS WORLD RADIO
English 7105, 7205 kHz
French 5945 kHz
German 6230, 7205 kHz
Multilingual 7205 kHz

MONGOLIA
RADIO ULAN BATOR
English 7235, 9575, 9595, 9616, 12015, 15305 kHz
French 9677, 11790 kHz

MOROCCO
RADIO MEDI UN
Multilingual 9575 kHz

MOZAMBIQUE
RADIO MOCAMBIQUE
English 9525, 11818, 11835 kHz

NAMIBIA
SOUTHWEST AFRICA BC CORP
Multilingual 4965 kHz

NEPAL
RADIO NEPAL
English 5005 kHz

NETHERLANDS ANTILLES
TRANS WORLD RADIO
English 9535, 11815 kHz
German 9535, 15355 kHz
Spanish 6180, 9535, 9665, 15355, 15385 kHz

NICARAGUA
LA VOZ DE NICARAGUA
English 6015 kHz
Spanish 6015, 6100 kHz

NIGER
LA VOIX DU SAHEL
English 5020 kHz
French 5020, 6060, 7155, 9705 kHz
Multilingual 5020, 6060, 7155, 9705 kHz

NIGERIA
FEDERAL RADIO CORP
Multilingual 6050 kHz
RADIO NIGERIA
English 4931.8 kHz
Multilingual 4770, 4990, 6089, 7285 kHz
VOICE OF NIGERIA
English 7255, 11770, 15120, 15185 kHz
French 7255, 15120 kHz
German 15120 kHz

NORTHERN MARIANA IS
KFBS—FAR EAST BC
Japanese 11930 kHz
Multilingual 9545, 9685 kHz
KYOI
Multilingual 9465, 9670, 11900, 15405, 17775, 17780 kHz

NORWAY
RADIO NORWAY INTERNATIONAL
English 6015, 6040, 7125, 7235, 7255, 7265, 9525, 9530, 9580, 9590, 9605, 9650, 9655, 11735, 11850, 11865, 11870, 11925, 11930, 15165, 15175, 15180, 15185, 15230, 15250, 15300, 15310, 15315, 17770, 17840 kHz
Multilingual 6015, 6040, 7125, 7235, 7255, 7265, 9525, 9530, 9580, 9590, 9605, 9650, 9655, 11735, 11850, 11865, 11870, 11925, 11930, 15165, 15175, 15180, 15185, 15230, 15250, 15300, 15310, 15315, 17770, 17840 kHz

PAKISTAN
RADIO PAKISTAN
English 7315, 9465, 9885, 11675, 11740, 15114, 15595, 15605, 17660 kHz
French 9465, 11635 kHz
Multilingual 15605, 17660 kHz

PARAGUAY
RADIO NACIONAL
Spanish 9735 kHz
Multilingual 9735 kHz

PHILIPPINES
FAR EAST BROADCASTING CO
English 9510, 11850, 15350, 15445 kHz
RADIO VERITAS ASIA
English 9710, 15135, 15215, 15270 kHz
Japanese 7190, 9505, 9730, 15275 kHz

POLAND
RADIO POLONIA
English 6095, 6135, 7125, 7145, 7270, 7285, 9525, 9540, 9675, 11815, 11840, 15120 kHz
French 6095, 6135, 7125, 7270, 7285, 9525, 9540, 9675, 11840, 15120 kHz
German 6095, 6135, 7125, 7270, 7285, 9540 kHz
Spanish 6135, 7145, 7270, 9525, 9675, 11840, 15120 kHz
Multilingual 6135, 7125, 7145, 7270 kHz

PORTUGAL
ADVENTIST WORLD RADIO
English 9670 kHz
German 9670 kHz
RADIO PORTUGUESA
English 6100, 9565, 9605, 9680, 9705, 9740, 11915, 15105, 15250 kHz
French 6100, 9605, 9740, 11915, 15250 kHz
German 6100, 9605, 9740 kHz

ROMANIA
RADIO BUCHAREST
English 5990, 6055, 6155, 7105, 7135, 7145,

7195, 9510, 9530, 9570, 9640, 9685, 9690,
9750, 11740, 11775, 11790, 11810, 11830,
11840, 11940, 15250, 15270, 15335, 15340,
15380, 17720, 17790, 17805, 21665 kHz
French 5990, 6105, 7105, 7195, 7225, 9690,
11775, 11790, 11810, 11885, 11940, 15250,
15335, 15365, 15380, 17720 kHz
German 5990, 6105, 6150, 6190, 7195, 7235,
9690, 11940, 15250 kHz
Spanish 5990, 6150, 6155, 6190, 7225, 9510,
9570, 11790, 11810, 11940 kHz

RWANDA
RADIO REPUBLIQUE RWANDAISE
Multilingual 6055 kHz

SAUDI ARABIA
BC SERVICE OF THE KINGDOM
English 9705, 9720 kHz
French 9705, 9720 kHz

SENEGAL
ORT DU SENEGAL
Multilingual 4890, 7173 kHz

SEYCHELLES
FAR EAST BROADCASTING ASS'N
English 9590, 11865, 15120, 15325, 17795 kHz
French 11900 kHz

SINGAPORE
SINGAPORE BROADCASTING CORP
English 5010, 5052, 11940 kHz

SOMALIA
RADIO MOGADISHU
English 6095 kHz
French 6095 kHz
Multilingual 7200 kHz

SOUTH AFRICA
RADIO RSA
English 4990, 5980, 6010, 6120, 7270, 7295,
9585, 9615, 9695, 11900, 15185, 15220, 15245,
17705, 17780, 21590 kHz
French 5980, 7270, 9585, 9615, 9675, 11810,
11900, 15185, 15220, 15245, 17780, 21590 kHz
German 11900, 15185, 15240, 17790,
17850 kHz
Spanish 6065, 6140, 6160, 9540, 9580 kHz
SOUTH AFRICAN BROADCASTING CORP
English 4880, 9665 kHz
Multilingual 3955, 4880, 6105 kHz

SPAIN
RADIO EXTERIOR DE ESPANA
English 6020, 6125, 7275, 9630, 9745, 9765,
11820, 11880, 15375 kHz
French 6020, 6125, 7275, 9745, 9765, 11820,
11880, 15375 kHz
Spanish 5970, 6020, 6125, 7450, 9360, 9570,
9630, 9650, 9685, 9745, 9875, 11730, 11775,
11790, 11815, 11920, 11945, 15125, **15365**,
15380, 15395, 17770, 17845, 17890, 21575 kHz
Multilingual 7450, 9570, 9875, 11790, 11815,
11920, 15380, 15395, 17770, 17845, 21575 kHz

SRI LANKA
SRI LANKA BROADCASTING CORP
English 9720, 11800 kHz

SURINAME
RADIO SURINAME INTERNATIONAL
Multilingual **17755** kHz

SWAZILAND
SWAZI COMMERCIAL RADIO
English 4980 kHz
TRANS WORLD RADIO
English 7120, 9550 kHz
French 7195, 9540, 9550 kHz
German 4760 kHz

SWEDEN
IBRA RADIO
French **9720** kHz
German **6110** kHz
Multilingual **9590**, **9685** kHz
RADIO SWEDEN INTERNATIONAL
English 6045, 6065, 7175, 9565, 9630,
9655, 9660, 9695, 9700, 9715, 11705, 11735,
11785, 11845, 11950, 15110, 15190, 15245,
15345, 15390, 17785, 17840, 21690 kHz
French 6065, 9615, 9630, 11845, 15240,
15245, 15345 kHz
German 6065, 7265, 9615, 9655 kHz
Spanish 6065, 9605, 9615, 9655, 9695,
11705 kHz

SWITZERLAND
RED CROSS BROADCASTING SERVICE
English 6135, 7210, 9560, 9665, 9725, 9730,
9870, 9885, 11745, 11795, 11905, 11955,
17830 kHz
French 7210 kHz
German 7210 kHz
Spanish 6135, 7210, 9625, 9725, 11925 kHz
Multilingual 9535, 9670, 9885, 11955,
17830 kHz
SWISS BROADCASTING CORP
French 3985, 6165, 9535 kHz
German 3985, 6165, 9535 kHz
SWISS RADIO INTERNATIONAL
English 3985, 6135, 6165, 6190, 9535, 9560,
9590, 9625, 9665, 9725, 9730, 9870, 9885,
11745, 11795, 11840, 11905, 11925, 11935,
11955, 12030, 15430, 17830 kHz
French 3985, 5965, 6135, 6165, 9535, 9560,
9590, 9625, 9665, 9670, 9725, 9730, 9870,
9885, 11745, 11795, 11840, 11905, 11935,
11955, 12030, 15430, 17830 kHz
German 3985, 5965, 6135, 6165, 9535, 9560,
9590, 9625, 9665, 9670, 9725, 9730, 9870,
9885, 11745, 11795, 11840, 11905, 11925,

11955, 12030, 15430, 17830 kHz
Spanish 5965, 6035, 6135, 9590, 9625, 9725,
9885, 11925, 11955 kHz

SYRIA
SYRIAN BROADCASTING SERVICE
English 9950, 12085, 15020 kHz
French 9950, 12085, 15020 kHz
German 9950, 12085, 15020 kHz
Spanish 9950, 12085 kHz

TANZANIA
RADIO TANZANIA
English 9684 kHz

THAILAND
RADIO THAILAND
English 9655, 11905 kHz
French 9655, 11905 kHz
Japanese 9655, 11905 kHz

TOGO
RADIO LOME
English 5047, 7265 kHz
Multilingual 5047, 7265 kHz

TURKEY
TURKISH RADIO TV CORP
English 7135, 7215, 7225, 9505, 9560, 15260, 17760 kHz
French 7215 kHz
German 7215 kHz

UNITED ARAB EMIRATES
UAE RADIO
English 9550, 11730, 11955, 15300, 15320, 17775, 17830, 21605, 21700 kHz

UNITED KINGDOM
BBC
English 3915, 3955, 3975, 5965, 5975, 6005, 6010, 6030, 6045, 6050, 6065, 6080, 6120, 6125, 6140, 6150, 6175, 6190, 6195, 7105, 7120, 7135, 7150, 7160, 7165, 7180, 7185, 7210, 7230, 7260, 7320, 7325, 9410, 9510, 9515, 9580, 9590, 9600, 9610, 9640, 9660, 9715, 9725, 9750, 9760, 9915, 11745, 11750, 11775, 11780, 11820, 11830, 11860, 11955, 12040, 12095, 15070, 15105, 15215, 15260, 15270, 15310, 15360, 15390, 15400, 15420, 15440, 17695, 17705, 17710, 17740, 17790, 17810, 17880, 17885, 18080, 21470, 21710 kHz
French 3975, 5975, 6010, 6125, 6195, 7105, 7150, 7165, 7210, 7230, 9600, 9610, 9670, 9915, 11720, 11780, 11825, 15105, 15115, 15150, 17810, 21640 kHz
German 3975, 6030, 6195, 9530, 9565 kHz
Spanish 6055, 6110, 7140, 9765, 11820, 15285, 17830 kHz
Japanese 6080, 7180, 9580, 9725, 11955, 15360 kHz

USA
AFRTS—US MILITARY
English 6030, 6125, 9530, 9700, 11730, 15265, 15330, 15345, 15430 kHz
KCBI
English 11735, 11910 kHz
KGEI—VOICE OF FRIENDSHIP
English 15280 kHz
Spanish 9615, 15280 kHz
KNLS—NEW LIFE STATION
English 7355, 11700, 11820, 11860, 11930 kHz
KUSW
English 6005, 9680, 9755, 11930, 17720 kHz
KVOH—VOICE OF HOPE
English 9495, 17775 kHz
Spanish 17775 kHz
RADIO EARTH
English 7355 kHz
RADIO MARTI
Spanish 6075, 9525, 9570, 9590, 11930 kHz
VOICE OF AMERICA
English 3990, 5965, 5995, 6035, 6040, 6045, 6060, 6080, 6095, 6110, 6125, 6130, 6180, 7170, 7195, 7200, 7205, 7275, 7280, 7325, 8110, 9350, 9455, 9530, 9540, 9550, 9575, 9635, 9640, 9650, 9670, 9700, 9715, 9740, 9760, 9770, 9775, 9815, 10869, 11090, 11580, 11695, 11715, 11720, 11740, 11760, 11775, 11805, 11835, 11915, 11920, 11925, 14398, 15120, 15155, 15160, 15185, 15195, 15205, 15215, 15260, 15290, 15305, 15410, 15425, 15445, 15580, 15600, 17735, 17740, 17775, 17785, 17800, 17820, 17850, 17870, 18137.5, 19480, 21540 kHz
French 6020, 6180, 7135, 7265, 9565, 9605, 10869, 11840, 11850, 11875, 11890, 11920, 15195, 15315, 15400, 15600, 15650, 17640, 17705, 17730, 19261.5, 19480, 21550, 21680 kHz
Spanish 5745, 6040, 6140, 6155, 6190, 6873, 9465, 9525, 9540, 9640, 9670, 9840, 11740, 11890, 11895, 11950, 15160, 15185, 15195, 15265, 15285, 15375, 15400, 17710, 17715, 17730, 17765, 17810, 17830, 21560, 21580, 21590, 21610 kHz
Multilingual 7175, 9575, 9750, 10869, 11710, 11915, 15600, 17715, 21500 kHz
VOICE OF THE OAS
Spanish 9565, 11830, 15160 kHz
WCSN—CHRISTIAN SCIENCE MONITOR
English 7365, 9465, 9815, 11945, 15270, 15280, 15300, 15390, 15395, 17640 kHz
French 7365, 9465, 9815, 11945, 15270, 15280, 15300, 15390, 15395, 17640 kHz
German 7365, 9465, 9815, 11945, 15270, 15280, 15300, 15390, 15395, 17640 kHz
WINB—WORLD INTERNATIONAL BC
English 15400 kHz
Multilingual 15145 kHz
WMLK—ASSEMBLIES OF YAWEH
English 9455 kHz

WORLD HARVEST RADIO
English 5995, 6000, 6010, 6100, 6155, 7355, 7400, 9580, 9770, 11770, 11790, 11980, 15105, 17835 kHz

WRNO WORLDWIDE
English 6155, 6185, 7355, 9852.5, 11705, 15420 kHz
Multilingual 6185, 9715, 11965 kHz

WYFR—FAMILY RADIO
English 6065, 6100, 7355, 9510, 9535, 9660, 9852.5, **11550**, 11580, 11805, 11830, 11875, **15055**, 15170, 15355, 15440, 15566, 17640, 17750, 17845, 21525, 21615 kHz
French 5960, 5985, 6015, 6085, 6175, 9605, 9815, 9852.5, 11580, 11830, 15170, 15365, 15440, 17750, 17845, 21525 kHz
German 6015, 7355, 9815, 9852.5, 11580, 15440, 17845 kHz
Spanish 6065, 6085, 6105, 6175, 9510, 9550, 9605, 9660, 9680, 9705, 9715, 11580, 11715, 11855, 11885, 15170, 15215, 15225, 15355, 15365, 15440, 15566, 17640, 17730, 17750, 17785, 17805 kHz

USSR

"RADIO MAGALLANES"
Spanish **7240**, 7440, **9470**, **9490**, **9640**, **9650**, **9715**, **11630**, **11650**, **11745**, **11800**, **11860**, **11900**, **12020** kHz

KAZAKH RADIO
German 4545, 5970, 6180, 9780, 11950 kHz

RADIO KIEV
English 6020, 6165, 7165, 7195, 7205, 7280, 7330, 9640, 9710, 9765, 9800, 11720, 11790, 11860, 13605 kHz
German 5980, 6165, 7195, 7330 kHz

RADIO MOSCOW/RADIO PEACE & PROGRESS
English 5900, 5905, 5920, 5940, 7100, **7115**, 7305, 7310, 7315, 7320, 7330, 7335, 7355, 7400, 7440, 9450, 9470, 9480, **9600**, 9775, 9780, 9785, 9795, 9800, 9810, 9865, 11675, **11750**, **11840**, 11975, 12000, 12005, 12010, 12015, 12030, 12045, 12050, 12055, 12060, 12070, 12075, 13605, 13645, 13655, 13660, 13680, 13705, 15110, **15135**, 15460, 15480, 15490, 15500, 15515, 15530, 15540, 15585, 17680, 21585, 21690, 21725, 21740 kHz
French 5950, 6010, 7160, 7175, 7240, 7400, 7440, 9480, 9490, 9610, 9650, 9710, 9745, 9810, 11745, 11805, 11880, 12010, 12020, 12050, 12055, 15530 kHz
German 5950, 6130, 6145, 7360, 12020, kHz
Spanish 6010, 6105, 6115, 6145, 7160, 7240, 7280, 7350, 7360, 7420, 7440, 9470, 9490, 9520, 9610, 9640, 9650, 9665, 9670, 9675, 9710, 9785, 9795 kHz
Japanese 4825, 5940, 5950, 5960, 6020, 6050, 6060, 6065, 7170, 7175, 7185, 7260, 7280, 7340, 9520, 9540, 9565, 9795, 9865, 9895, 11730, 11915, 11960, 12030, 15560 kHz
Multilingual 9470, 15140, 15330, 15490, 15500 kHz

RADIO TASHKENT
English 5945, 5985, 7325, 9540, 9600, 9715, 11785, 15460 kHz

RADIO TIKHIY OKEAN
English 4485, 5900, 5950, 5980, 6020, 6035, 6080, 6190, 7210, 7260, 7370, 9580, 9620, 9635, 9795, 9810, 11730, 11815, 11900, 11910, 11950, 12030, 12050, 12070, 17775 kHz

RADIO VILNIUS
English 6020, 6100, 6200, 7165, 9640, 9765, 11720, 11790, 11860, 13605 kHz

RADIO YEREVAN
English 11790, 11860, 13605 kHz
French 15280 kHz
Spanish 9480 kHz

RADIO STATION SOVIET BELORUSSIA
German 5980, 6165, 6185, 7175, 7205, 7330, 7420, 9560, 11960 kHz

VATICAN STATE

VATICAN RADIO
English 6015, 6030, 6145, 6150, 6185, 6190, 6248, 7125, 7250, 9605, 9610, 9615, 9625, 9645, 9650, 11700, 11725, 11740, 11760, 11775, 11780, 11810, 11830, 11845, 15115, 15120, 15190, 17730, 17840, 17865, 21485 kHz
French 6030, 6150, 6185, 6190, 6248, 7250, 9605, 9625, 9645, 11700, 11725, 11740, 11760, 11780, 11810, 11845, 15120, 15190, 17730, 17840, 21485 kHz
German 6185, 6190, 6248, 7250, 9645, 11740 kHz
Spanish 6030, 6035, 6150, 6190, 6248, 7250, 9615, 9645, 9705, 11740, 11780, 11845, 15405, 17740, 17870, 21725 kHz
Japanese 6015, 9615, 11830, 15190, 17865 kHz
Multilingual 6190, 6248, 7250, 9625, 9645, 11700, 11740, 15120 kHz

VENEZUELA

RADIO NACIONAL
Spanish 9500, 9540 kHz

YUGOSLAVIA

RADIO LJUBLJANA
Multilingual 6100, 7240, 9620 kHz

RADIO YUGOSLAVIA
English 5980, 6100, 7240, 9620, 11735, 15240 kHz
French 5980, 6100, 7240, 9620, 15240 kHz
German 5980, 6100, 7240, 9620 kHz
Spanish 6100, 7240, 9620, 11735 kHz

ZAIRE

LA VOIX DU ZAIRE
Multilingual 15245 kHz

ZAMBIA
RADIO ZAMBIA
Multilingual 4910 kHz
RADIO ZAMBIA—ZBS
English 9505, 11880 kHz
Multilingual 6165, 7220, 7235 kHz

ZIMBABWE
ZIMBABWE BROADCASTING CORP
English 6020 kHz
Multilingual 5975 kHz

Andy's Having A Ball...
and you can too!

Andy is a Ham Radio operator and he's having the time of his life talking to new and old friends in this country and around the world.

You can do it too! Join Andy as he communicates with the world. Enjoy the many unique and exclusive amateur bands ... the millions of frequencies that Hams are allowed to use. Choose the frequency and time of day that are just right to talk to anywhere you wish. Only Amateur Radio operators get this kind of freedom of choice. And if it's friends you're looking to meet and talk with, Amateur radio is the hobby for you. The world is waiting for you.

If you'd like to be part of the fun ... if you'd like to feel the excitement ... we can help you. We've got all the information you'll need to get your Ham license. Let us help you join more than a million other Hams around the world and here at home. Who are we? We're the American Radio Relay League, a non-profit representative organization of Amateur Radio operators.

For information on becoming a Ham operator write to:

AMERICAN RADIO RELAY LEAGUE   Dept. RI, 225 Main Street
Newington, Conn. 06111.

This space donated by this publication in cooperation with the American Radio Relay League.

SUMMARY OF WORLD BAND BROADCASTING ACTIVITY

Fully 161 of the world's countries, from the giant USSR to tiny Bhutan—plus various extralegal stations—broadcast within the world radio spectrum, which is divided into two major parts: the *Tropical bands* from 2.2–3.9 and 4.0–5.73 MHz, plus the *International bands* from 3.9–4.0 and 5.7–26.1 MHz.

Although the Tropical bands consist mainly of broadcasts intended for the home audience, the nature of world band radio is such that they are often heard thousands of kilometers away. The International bands, in contrast, consist mainly of powerful stations intended for audiences abroad. Even then, many domestic stations operate here, as well, and also can be heard for great distances.

Jamming consists of deliberate interference directed against a broadcaster by those who are trying to keep the programs from being heard. Jamming takes place within the Tropical and International bands alike, but in practice is much more common within the International bands. Jammed hours given here are listed alongside the "victim" country—not the country actually doing the jamming.

Countries listed are those responsible for the material being broadcast. So, Radio Moscow, for example, is listed under "USSR" regardless of whether it is aired from transmitters within the USSR or from relay sites abroad.

Summary of World Broadcasting Activity—Tropical Broadcasting
(Total Spectrum Occupancy in Hours per Week)

#	Country	Total Hours	Total Jammed	#	Country	Total Hours	Total Jammed	#	Country	Total Hours	Total Jammed
1	BRAZIL	10071	None	33	NEPAL	198	None	65	MALAWI	83	None
2	PERU	5554	None	34	KENYA	198	None	66	DJIBOUTI (JIBUTI)	83	None
3	USSR	5379	None	35	GHANA	186	None	67	SOLOMON IS	75	None
4	CHINA (PR)	5219	None	36	BENIN	185	None	68	FALKLAND ISLANDS	74	None
5	INDONESIA	2977	None	37	BOTSWANA	182	None	69	BURKINA FASO	74	None
6	ECUADOR	2669	None	38	PAKISTAN	180	14	70	ZIMBABWE	73	None
7	BOLIVIA	2471	None	39	FRENCH GUIANA	175	None	71	PAKISTAN (AZAD K)	72	None
8	USA	1507	None	40	LIBERIA	169	None	72	CENTRAL AFRICAN REP	72	None
9	PAPUA NEW GUINEA	1332	None	41	NEW CALEDONIA	168	None	73	BURUNDI	70	None
10	VENEZUELA	1323	None	42	COLOMBIA	168	None	74	AUSTRIA	60	None
11	COLOMBIA	1302	None	43	SWAZILAND	157	None	75	MAURITANIA	58	None
12	KOREA (DPR)	1125	None	44	TOGO	155	None	76	CLANDESTINE (AFRICA)	55	22
13	INDIA	1014	None	45	SURINAME	147	None	77	SENEGAL	55	None
14	MALAYSIA	834	None	46	NIGER	134	None	78	EQUATORIAL GUINEA	55	None
15	ANGOLA	745	None	47	BELIZE	133	None	79	KAMPUCHEA (CAMBODIA)	53	None
16	MONGOLIA	667	None	48	LESOTHO	133	None	80	MEXICO	49	None
17	AUSTRALIA	594	None	49	ALBANIA	133	None	81	ARGENTINA	44	None
18	GUATEMALA	553	None	50	TANZANIA	128	None	82	MAURITIUS	42	None
19	CLANDESTINE (M EAST)	453	369	51	ZAMBIA	126	None	83	MARSHALL ISLANDS	36	None
20	MOZAMBIQUE	396	None	52	CHAD	126	None	84	COMOROS	35	None
21	NIGERIA	354	None	53	THAILAND	125	None	85	IVORY COAST	35	None
22	NAMIBIA	344	None	54	HAITI	125	None	86	BURMA	29	None
23	SRI LANKA	334	None	55	GABON	122	None	87	CLANDESTINE (C AMER)	24	24
24	CAMEROON	294	None	56	ZAIRE	115	None	88	UNIDENTIFIED	24	None
25	COSTA RICA	292	None	57	YEMEN (PDR)	113	None	89	TRISTAN DA CUNHA	15	None
26	SOUTH AFRICA	282	None	58	MALI	110	None	90	IRAN	14	None
27	HONDURAS	271	None	59	RWANDA	105	None	91	FRANCE	10	None
28	AFGHANISTAN	266	None	60	MADAGASCAR	104	None	92	ICELAND	9	None
29	VIETNAM	257	None	61	GUINEA	104	None		Total Tropical Jamming	518 Hours per Week	
30	SINGAPORE	252	None	62	BANGLADESH	96	None		TOTAL TROPICAL BROADCASTING	55400 Hours per Week	
31	LAOS	217	None	63	CHINA (TAIWAN)	89	89				
32	CLANDESTINE (ASIA)	200	None	64	UGANDA	85	None				

Summary of World Broadcasting Activity—International Broadcasting
(Total Spectrum Occupancy in Hours per Week)

#	Country	Total Hours	Total Jammed	#	Country	Total Hours	Total Jammed	#	Country	Total Hours	Total Jammed
1	USSR	28981	None	19	KOREA (REPUBLIC)	1466	None	37	VIETNAM	1064	None
2	USA	18478	7185	20	SWITZERLAND	1378	None	38	COLOMBIA	1060	None
3	CHINA (PR)	10674	6	21	HOLLAND	1327	40	39	ANGOLA	1047	88
4	BRAZIL	5358	None	22	ALBANIA	1322	51	40	IRAN	1043	375
5	UNITED KINGDOM	4608	461	23	INDONESIA	1307	None	41	VATICAN STATE	1021	6
6	GERMANY (FR)	4570	425	24	IRAQ	1284	14	42	ROMANIA	951	None
7	PERU	3258	None	25	ECUADOR	1274	None	43	SOUTH AFRICA	932	None
8	CHINA (TAIWAN)	2589	993	26	ITALY	1263	None	44	KUWAIT	886	None
9	FRANCE	2346	None	27	BOLIVIA	1198	None	45	PHILIPPINES	878	None
10	CANADA	2090	None	28	ALGERIA	1184	None	46	AUSTRIA	874	None
11	AUSTRALIA	2025	None	29	POLAND	1164	None	47	SINGAPORE	798	None
12	INDIA	1920	None	30	SPAIN	1161	None	48	PAKISTAN	773	None
13	KOREA (DPR)	1909	None	31	CZECHOSLOVAKIA	1144	None	49	ARGENTINA	746	None
14	GERMANY (DR)	1904	None	32	BULGARIA	1142	None	50	CLANDESTINE (M EAST)	743	628
15	MALAYSIA	1860	None	33	NIGERIA	1127	None	51	AFGHANISTAN	683	None
16	ISRAEL	1751	723	34	SAUDI ARABIA	1103	None	52	SWEDEN	617	None
17	EGYPT	1667	None	35	CUBA	1087	None	53	CHILE	592	None
18	JAPAN	1466	None	36	HUNGARY	1083	None	54	TURKEY	578	1

Country	Total Hours	Total Jammed	Country	Total Hours	Total Jammed	Country	Total Hours	Total Jammed
55 UNITED ARAB EMIRATES	569	None	89 SENEGAL	238	None	123 EQUATORIAL GUINEA	119	None
56 LIBYA	531	None	90 NORWAY	233	None	124 MALAWI	118	None
57 COSTA RICA	520	None	91 YEMEN (REPUBLIC)	226	None	125 CLANDESTINE (EUROPE)	118	None
58 MOZAMBIQUE	506	None	92 NIGER	225	None	126 QATAR	114	None
59 YUGOSLAVIA	494	None	93 VANUATU	225	None	127 CYPRUS	108	None
60 SOCIETY ISLANDS	492	None	94 NORTHERN MARIANA IS	217	None	128 RWANDA	105	None
61 SRI LANKA	491	None	95 SOMALIA	214	None	129 GUINEA	104	None
62 GREECE	476	None	96 ETHIOPIA	206	None	130 GHANA	100	None
63 CLANDESTINE (AFRICA)	469	238	97 YEMEN (PDR)	198	None	131 TOGO	100	None
64 LEBANON	469	60	98 SEYCHELLES	192	None	132 MAURITIUS	84	None
65 THAILAND	468	None	99 MALI	191	None	133 UGANDA	81	None
66 PORTUGAL	461	None	100 PAPUA NEW GUINEA	189	None	134 DENMARK	79	None
67 GUAM	448	None	101 NICARAGUA	186	None	135 FRENCH GUIANA	77	None
68 MOROCCO	434	None	102 GUATEMALA	178	None	136 MARSHALL ISLANDS	77	None
69 PARAGUAY	410	None	103 LIBERIA	178	None	137 FALKLAND ISLANDS	74	None
70 TUNISIA	404	None	104 IVORY COAST	175	None	138 CAPE VERDE	71	None
71 LAOS	403	None	105 SWAZILAND	171	None	139 UNIDENTIFIED	67	17
72 ZAMBIA	398	None	106 NEW CALEDONIA	168	None	140 BENIN	56	None
73 CLANDESTINE (ASIA)	396	92	107 COOK ISLANDS	168	None	141 BURUNDI	50	None
74 BELGIUM	393	None	108 ZAIRE	165	None	142 COMOROS	46	None
75 FINLAND	382	None	109 BANGLADESH	162	None	143 CHAD	45	None
76 TANZANIA	333	None	110 NEW ZEALAND	160	None	144 BURKINA FASO	43	None
77 CAMEROON	325	None	111 MADAGASCAR	154	None	145 SOLOMON IS	38	None
78 CLANDESTINE (C AMER)	321	47	112 MEXICO	154	None	146 MAURITANIA	38	None
79 OMAN	321	None	113 ANTARCTICA	149	None	147 ICELAND	34	None
80 LUXEMBOURG	319	None	114 BOTSWANA	147	None	148 BHUTAN	25	None
81 SYRIA	313	190	115 KAMPUCHEA (CAMBODIA)	138	None	149 PAKISTAN (AZAD K)	24	None
82 KENYA	309	None	116 GUYANA	134	None	150 MALTA	21	None
83 MONGOLIA	306	None	117 SIERRA LEONE	130	None	151 PIRATE (EUROPE)	20	None
84 VENEZUELA	305	None	118 MONACO	126	None	152 KIRIBATI	16	None
85 GABON	279	None	119 BURMA	121	None	153 SURINAME	3	None
86 ZIMBABWE	249	None	120 NETHERLANDS ANTILLES	121	None			
87 URUGUAY	246	None	121 GREENLAND	120	None	Total International Jamming	11640 Hours per Week	
88 DOMINICAN REPUBLIC	245	None	122 HONDURAS	119	None	TOTAL INTERNATIONAL BROADCASTING	156170 Hours per Week	

RADIO WEST

"The Best In The West," To Serve All Of Your Shortwave And AM Radio Equipment Needs!

We are proud to offer the leaders in shortwave and mediumwave receivers, associated equipment, along with the best receiver modifications available anywhere, to improve your radio enjoyment and listening abilities.

ICOM R-71A, Japan Radio Co. NRD-525, YAESU FRG-8800, LINIPLEX F-1, Sony receivers, Grundig, Kenwood, Panasonic, Eskab & others.

Radio West offers the largest mediumwave loop antenna selection in the world, along with SWL antennas, improvement of your stock receiver with our selectivity filter modifications; Collins & Kokusai Mechanical, and a very large selection of quality ceramic filters. Miscellaneous receiver modifications (all makes and models, please call), Eskab P.L.A.M. (ECCS) mods for the NRD-525/515 & ICOM R-71A/70, books, and many other equipment accessories.

We are very pleased to offer the *friendly service, experience and expertise* that has satisfied our many customers and friends over the last *nine* years.

Call or write for a free catalog . . . TODAY!!
Please enclose $1 for postage and handling.

**850 ANNS WAY DRIVE
VISTA, CALIFORNIA 92083
(619) 726-3910**

"Sail Around The World With Radio West"

YAESU FRG-8800

LINIPLEX F-1

ICOM R-71A

WORLDSCAN: THE BLUE PAGES

Quick-Access Guide to World Band Schedules

There are over 1,100 channels of news and entertainment available on world band radio. As if this weren't enough, these channels are often shared among several stations. With such an abundance from which to choose, it can take some doing to figure out what's on the air.

Ordinary listings of what's on world band radio are unwieldy because there's so much material. This is why *Passport to World Band Radio* includes this quick-access Worldscan section—the "blue pages." With Worldscan's comprehensive charts, all you need—station, times, languages, targets and more—is there at a glance. If something's not clear, there's a lexicon in the back of the book to explain it. There's even a handy key to languages and the like at the bottom of each Worldscan page.

For example, if you're in North America or Europe listening to the channel of 5975 kHz at 2300 World Time, a glance at Worldscan shows that the BBC's World Service is broadcast in English to both these areas at that time.

World Time Simplifies Listening

World Time—a handy concept also known as UTC or GMT—is used to eliminate the potential complication of so many time zones throughout the world. It treats the entire world as a single zone and is announced regularly over the air by many world band stations. For example, if you're in New York and it's 6:00 AM EST, you'll hear the time announced as "11 hours UTC" . . . that is, five hours ahead of your local time. You can either keep this "add five hours" figure in your head or keep a separate clock for World Time. Too, as detailed in the "Buyer's Guide," some world band radios come with World Time clocks built-in.

World Band Stations Widely Heard

In the "Introduction to World Band Radio" found elsewhere in this book, it's mentioned that with so many stations on the air at the same time you can't begin to pick them all up. Nevertheless, world band radio is uniquely efficient at traveling great distances. Just because a station targets its programs to another part of the world doesn't mean that you can't hear it where you are. Tune around with "Worldscan" at your side and you'll find hundreds of stations that can be enjoyed even though they're beamed elsewhere.

GUIDE TO WORLDSCAN FORMAT

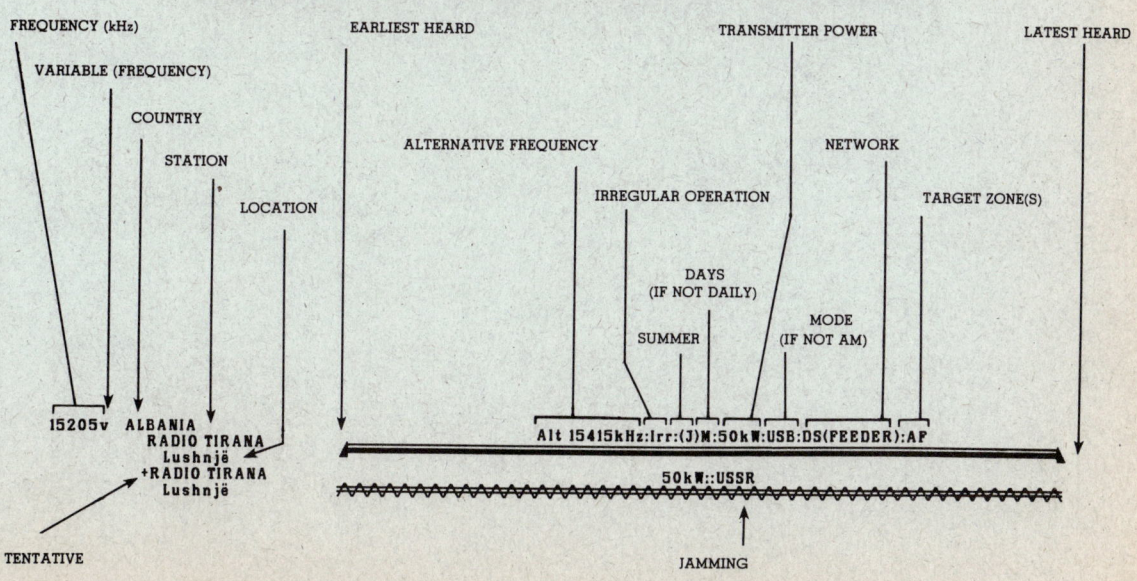

Guides to Worldscan in French, German, and Spanish are included with the lexicons.

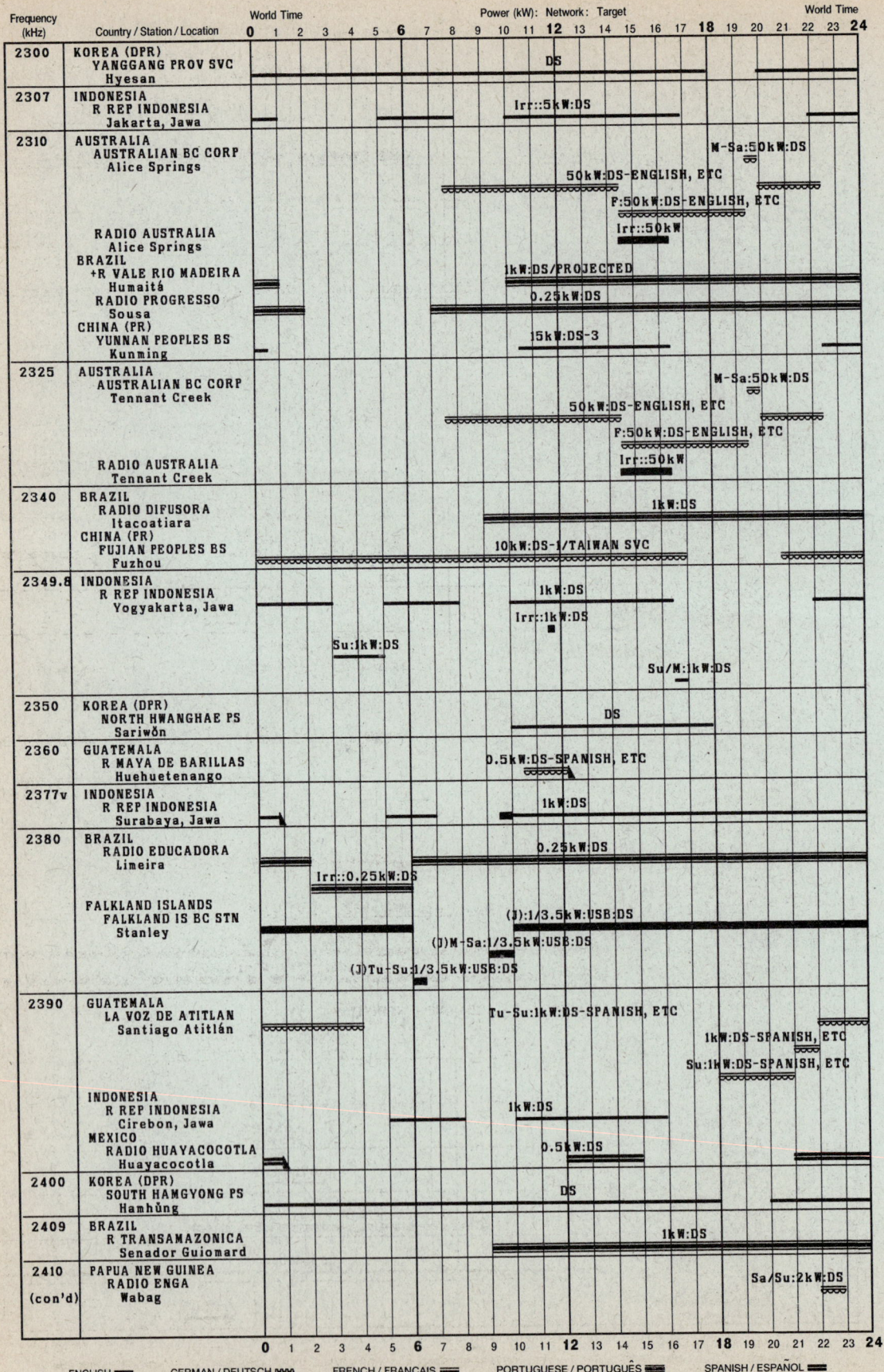

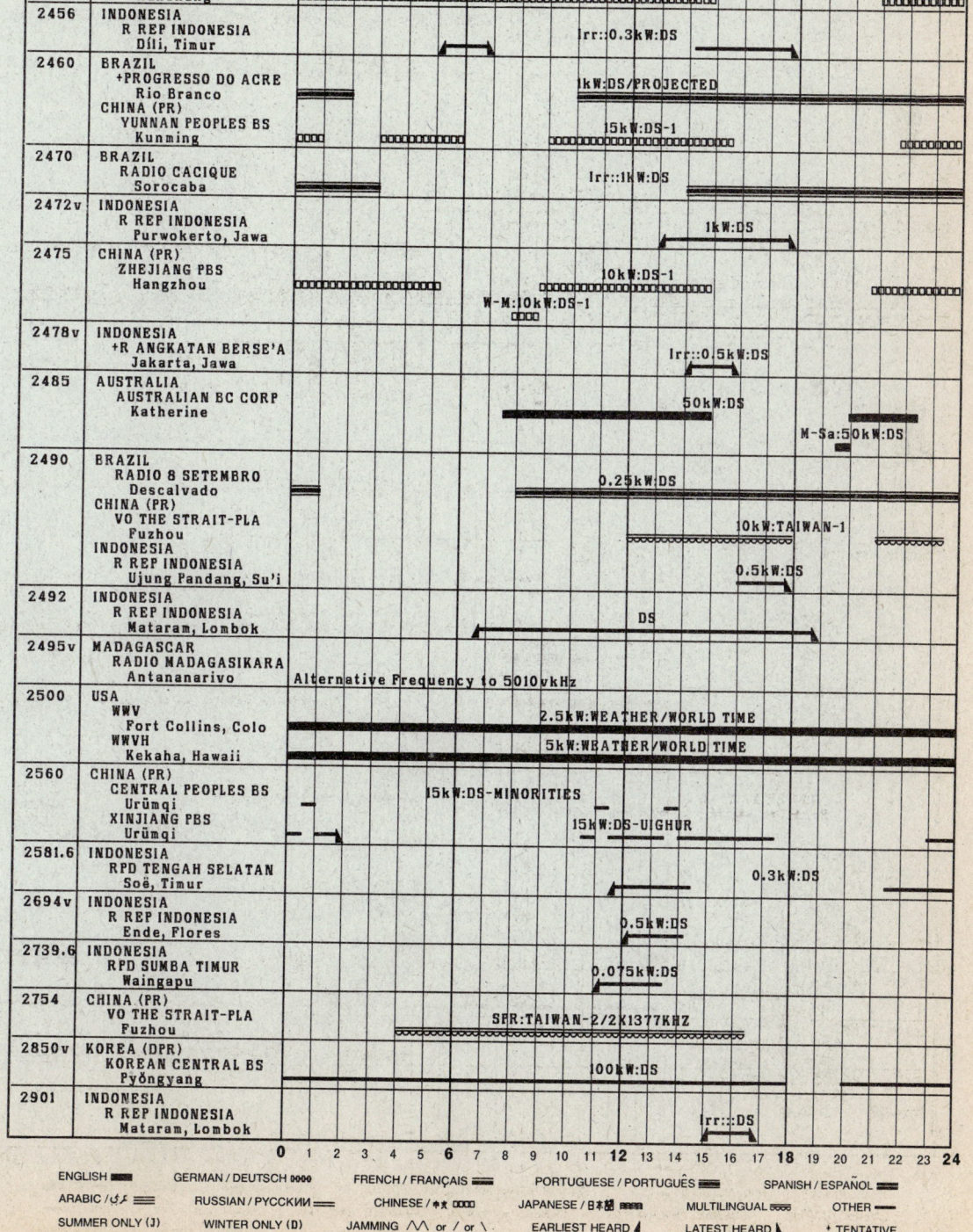

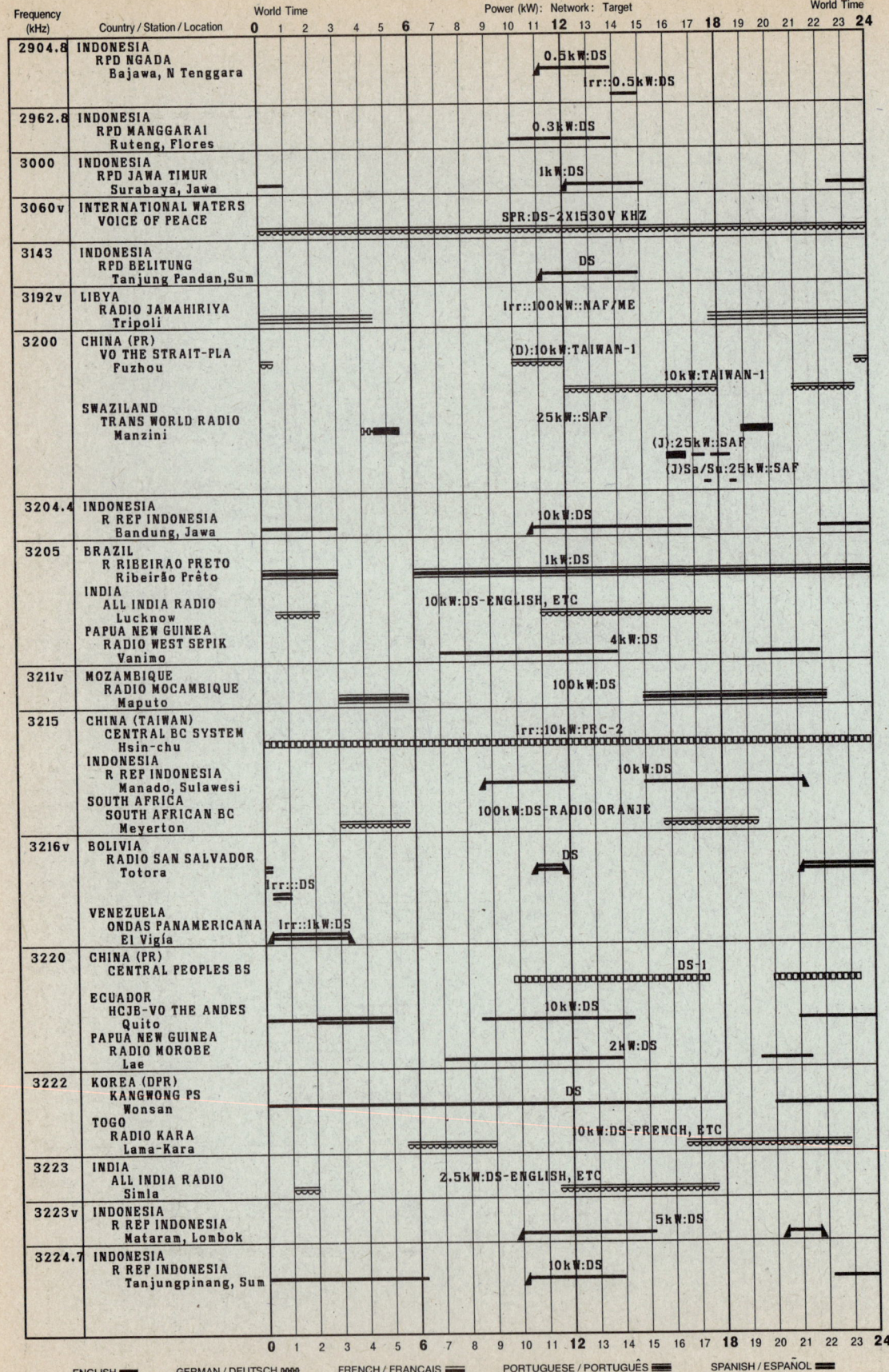

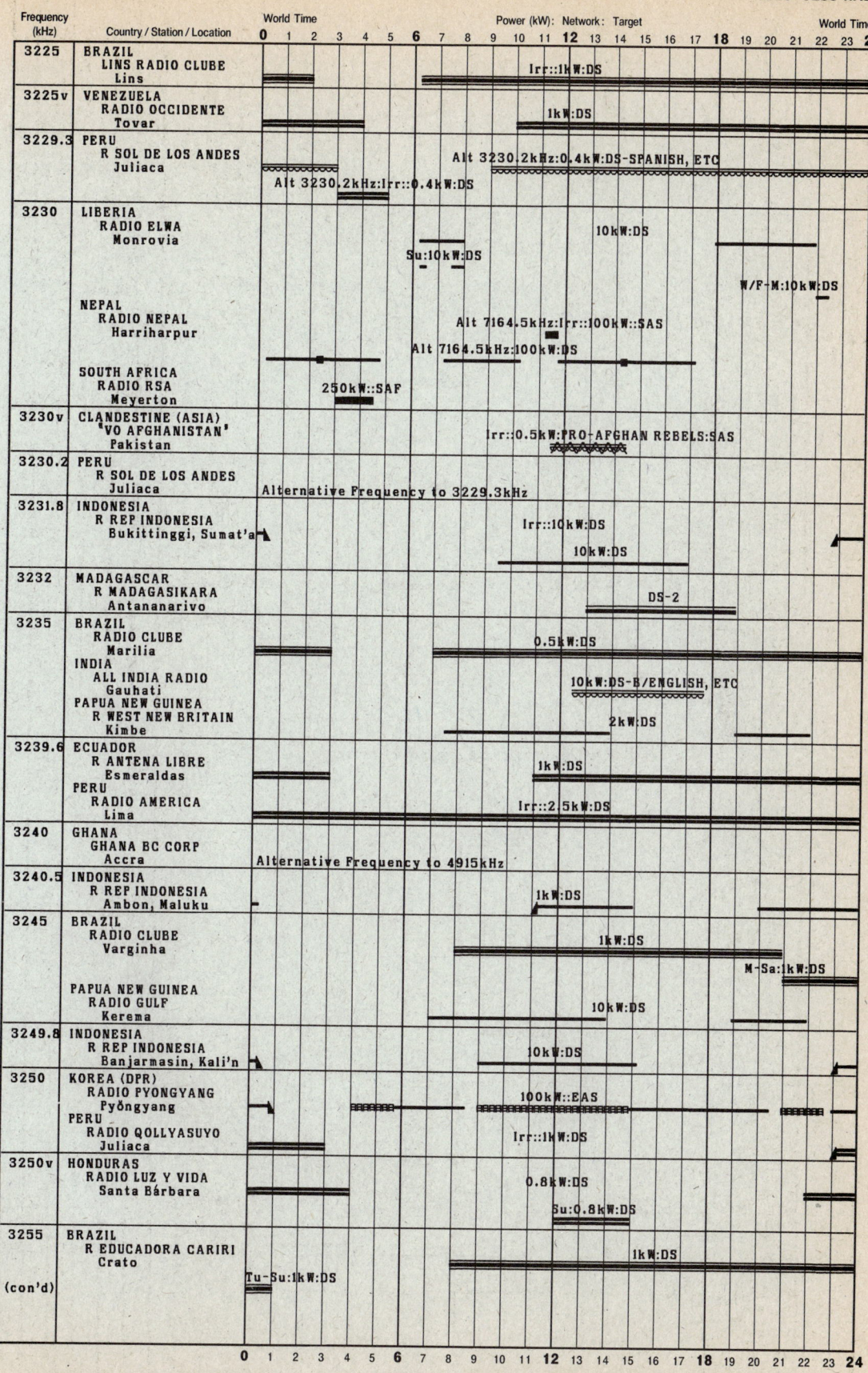

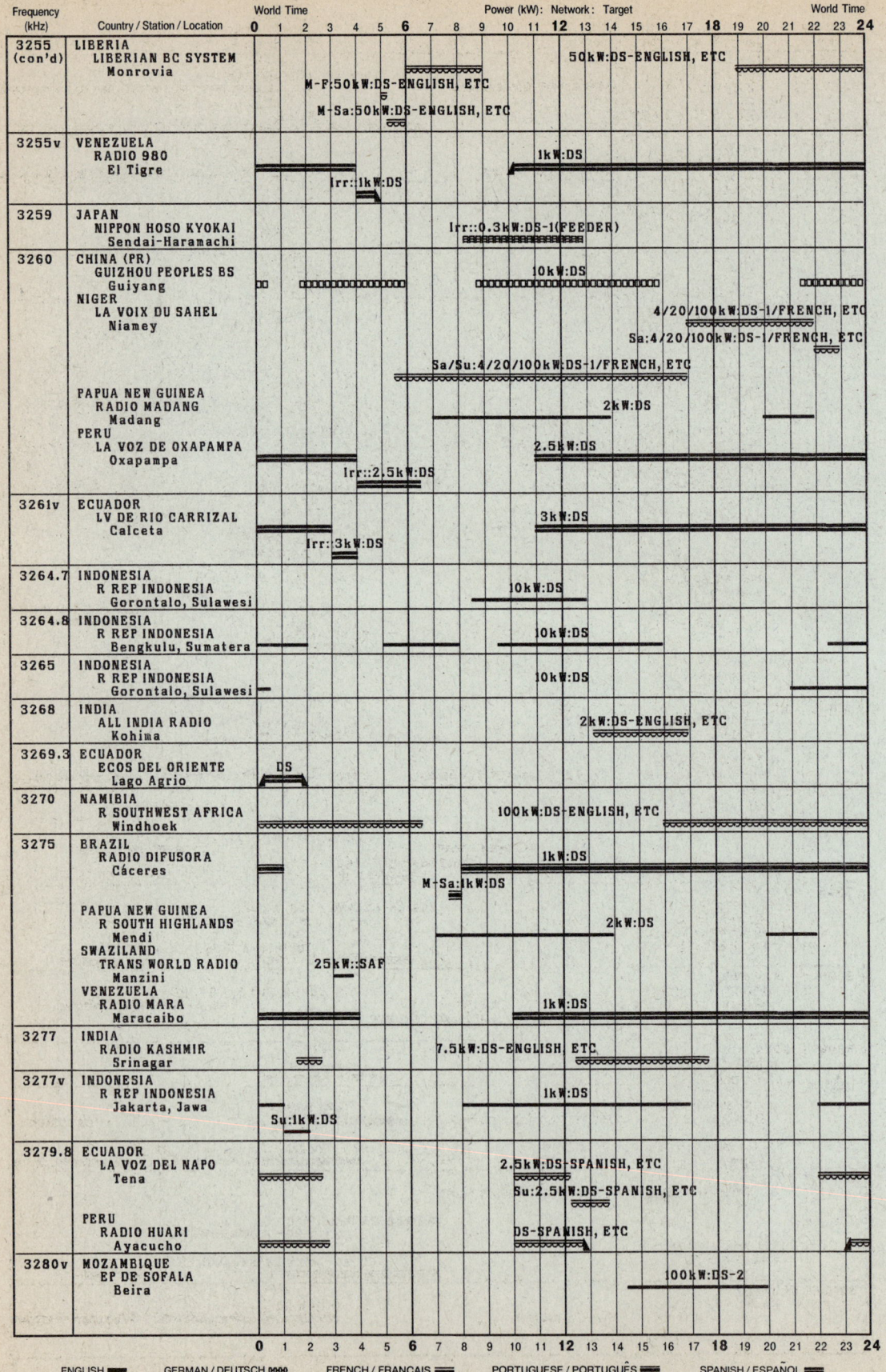

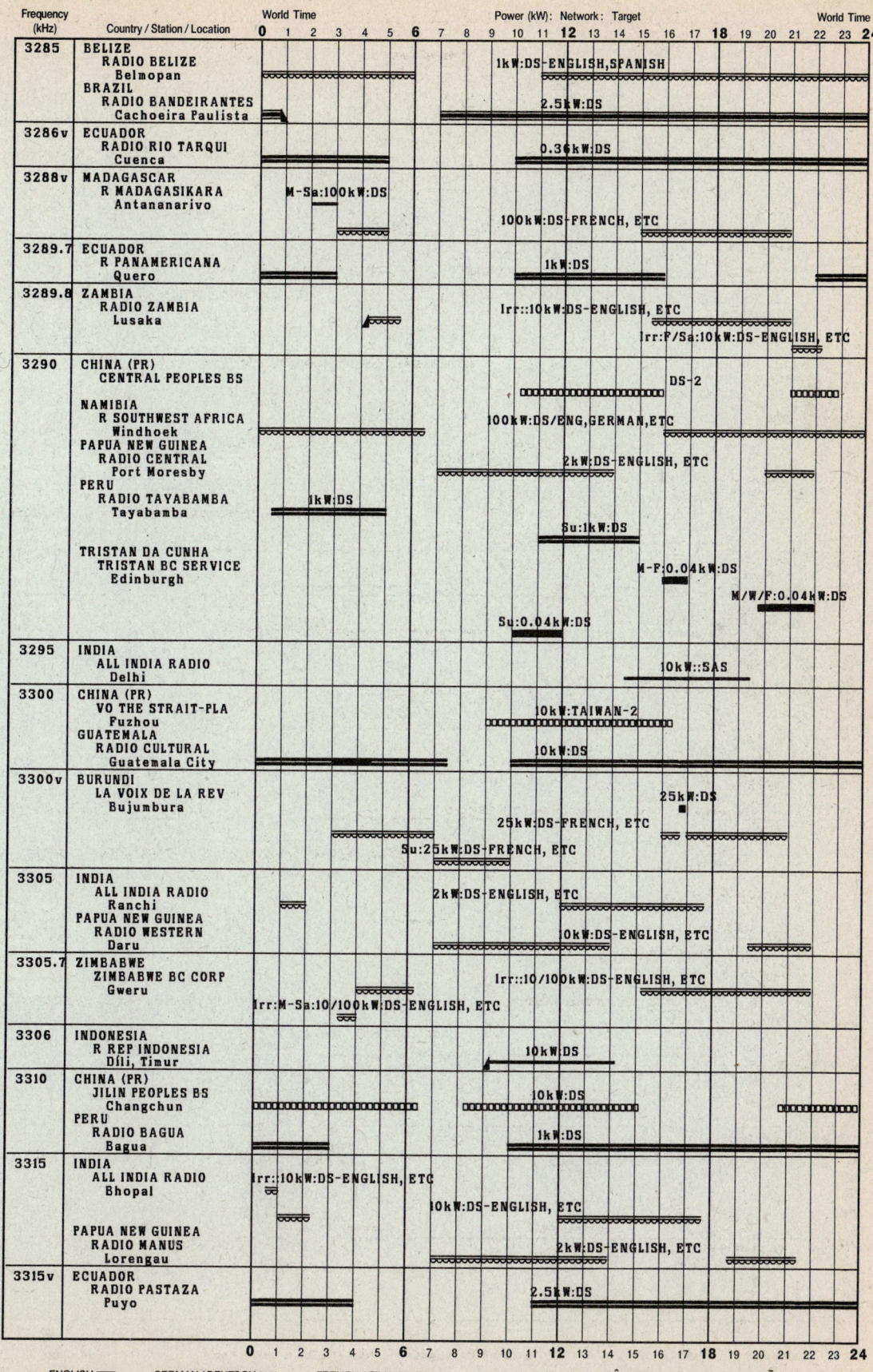

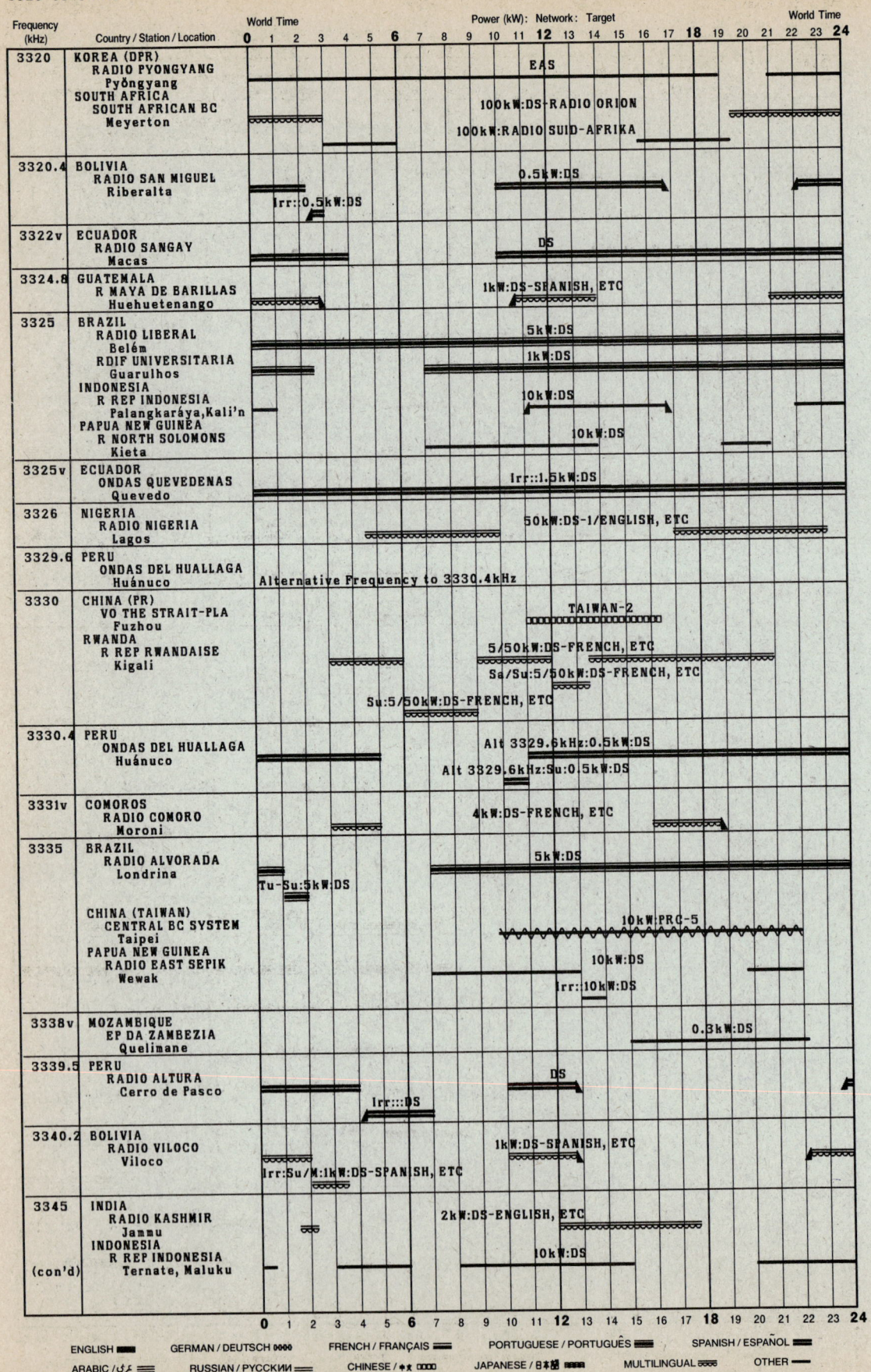

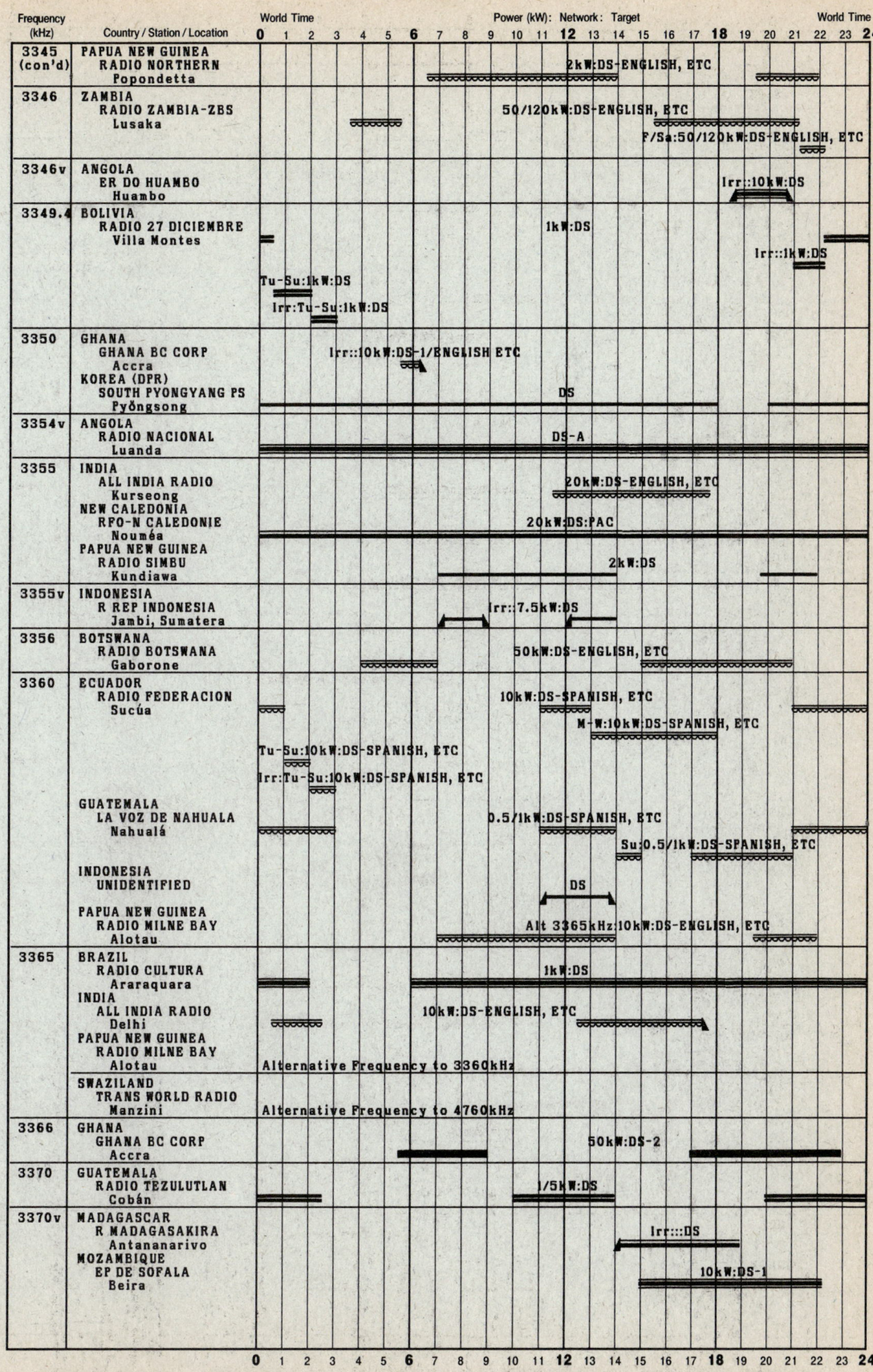

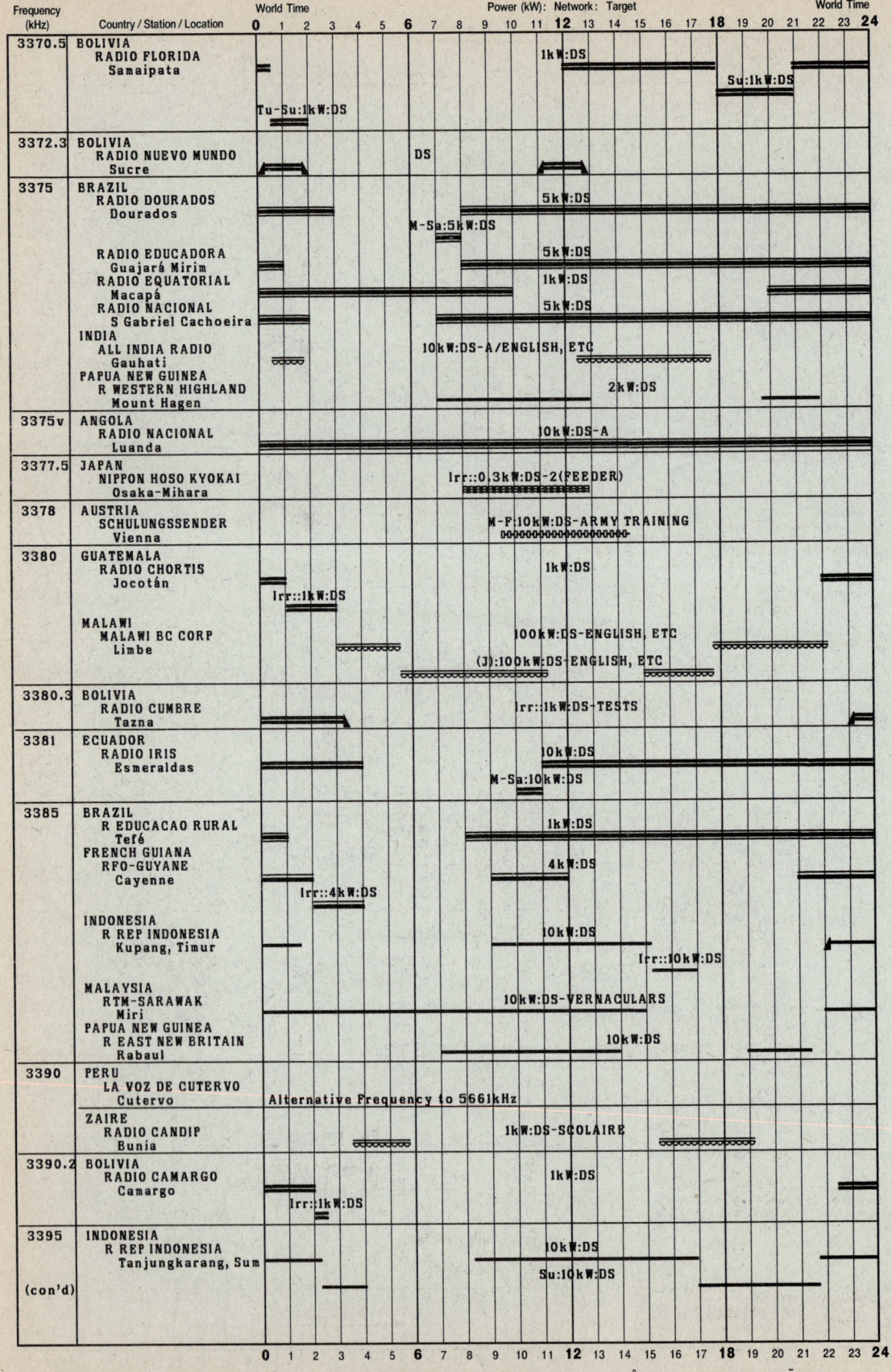

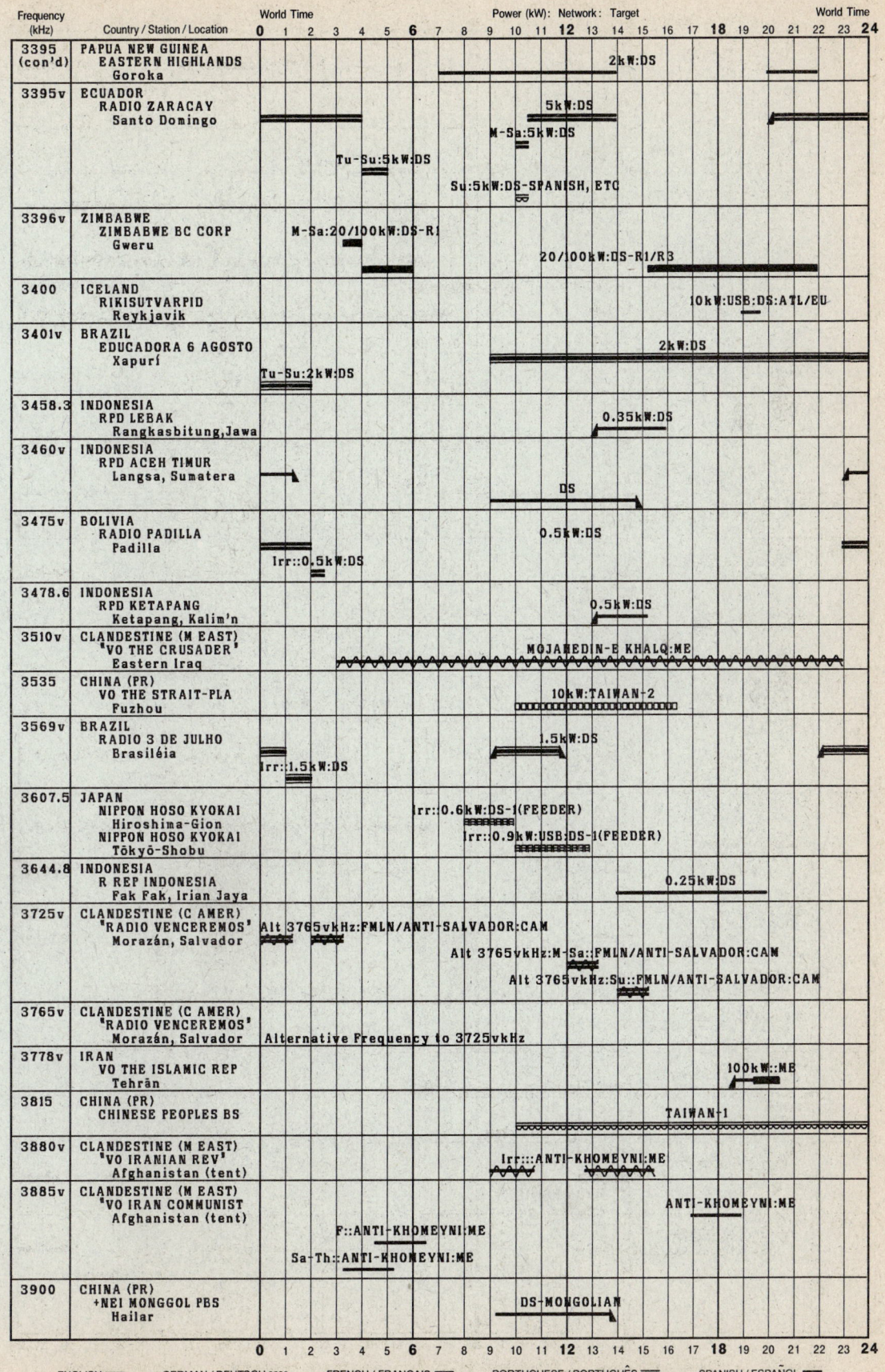

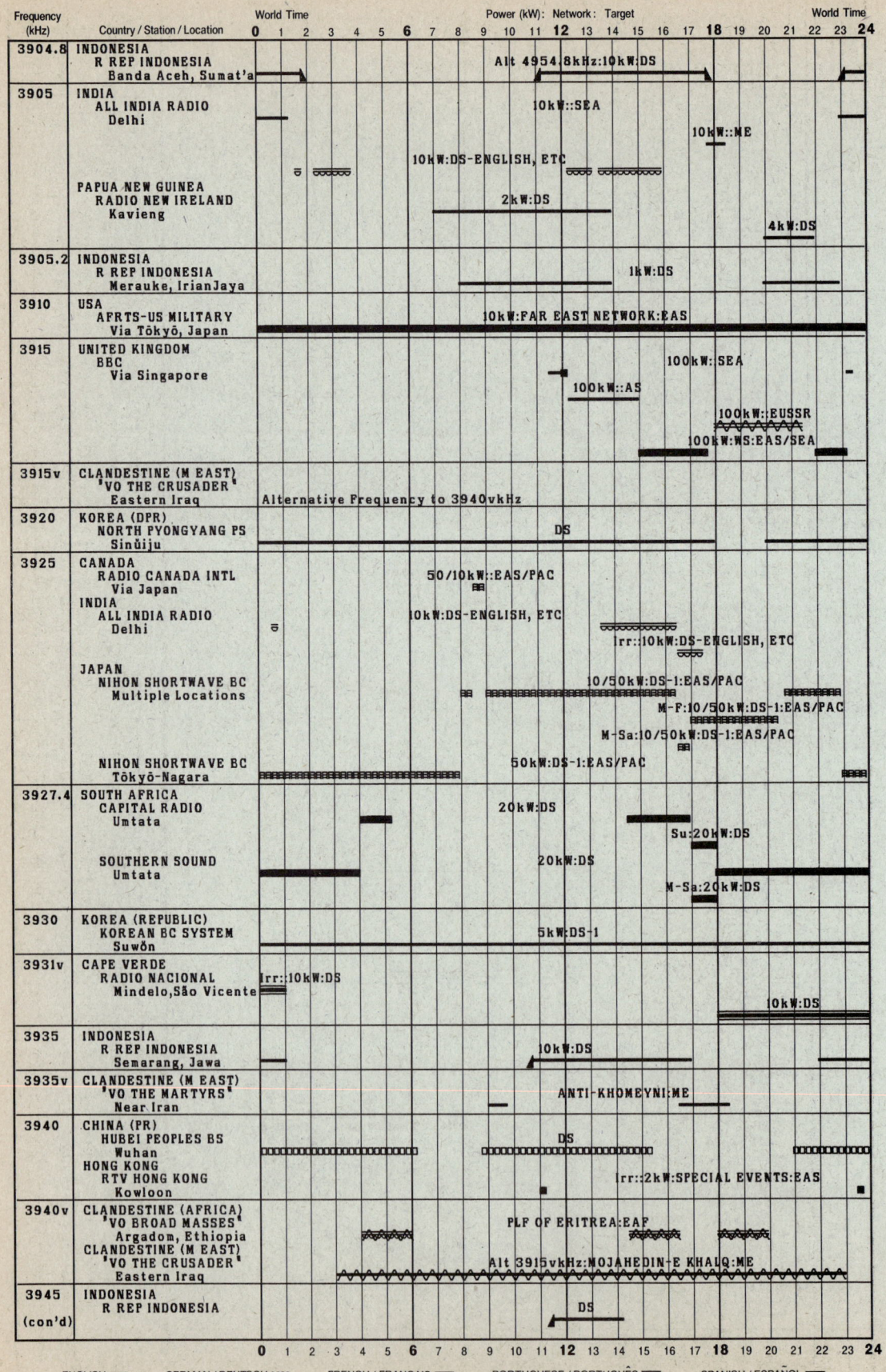

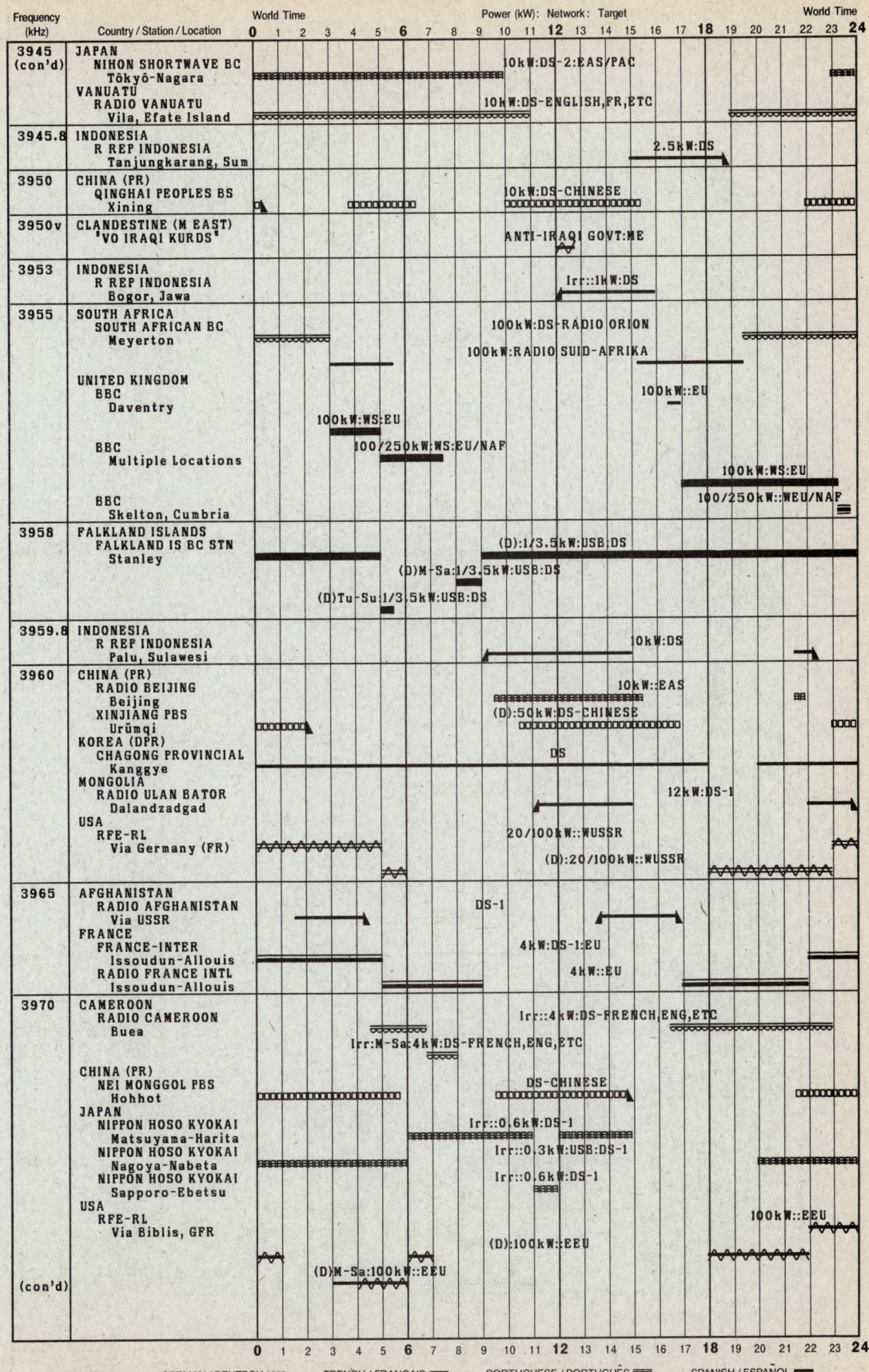

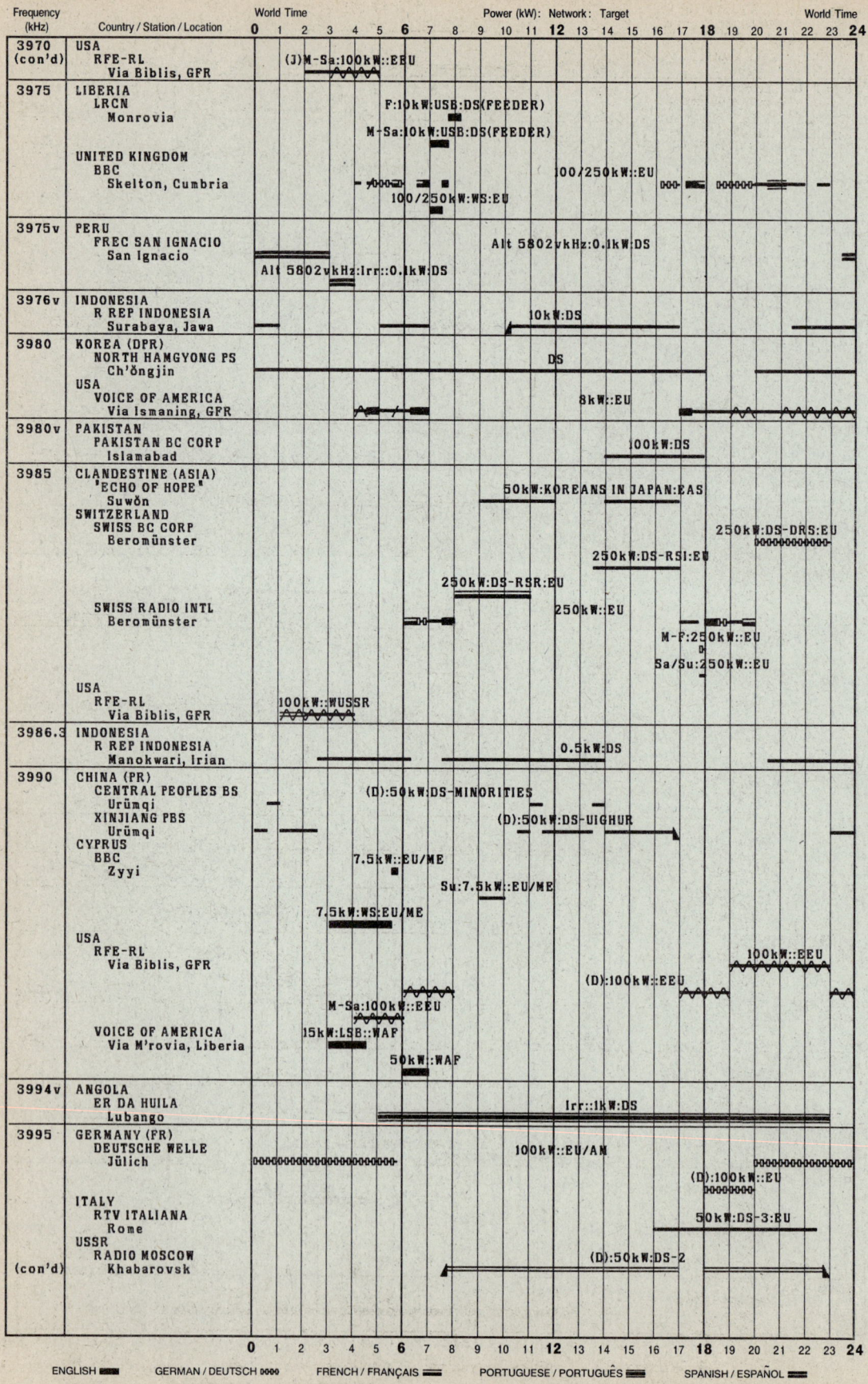

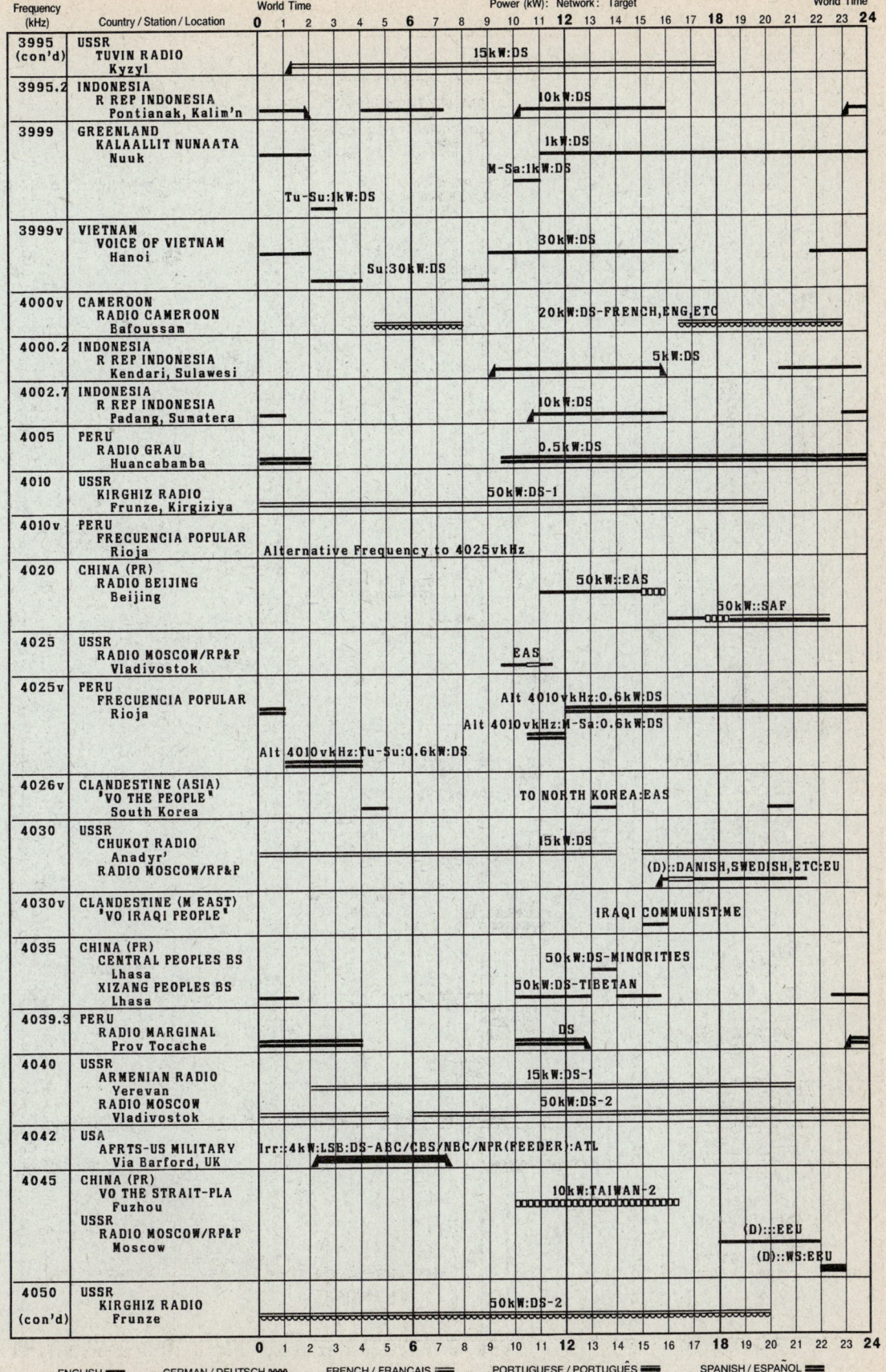

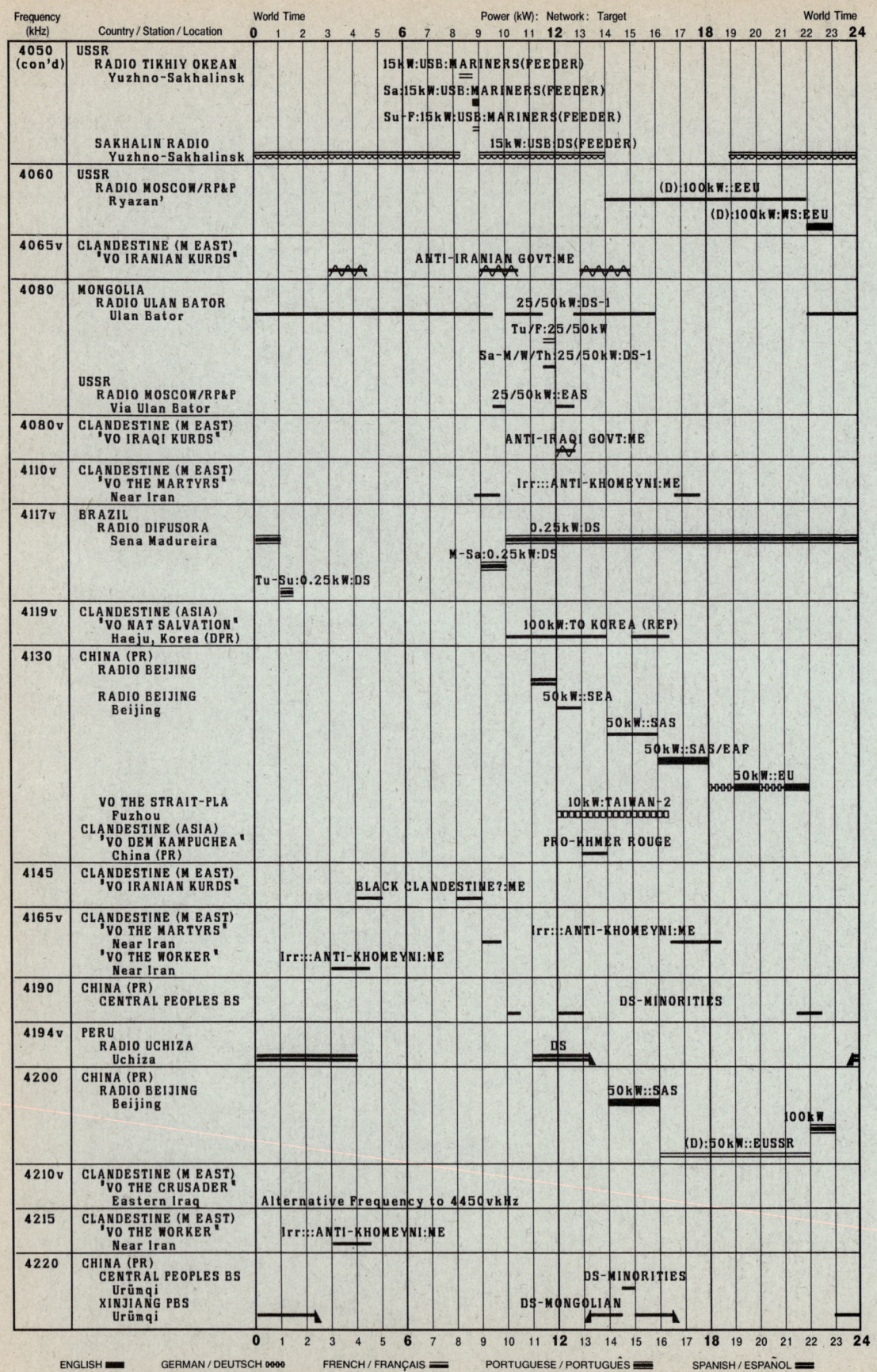

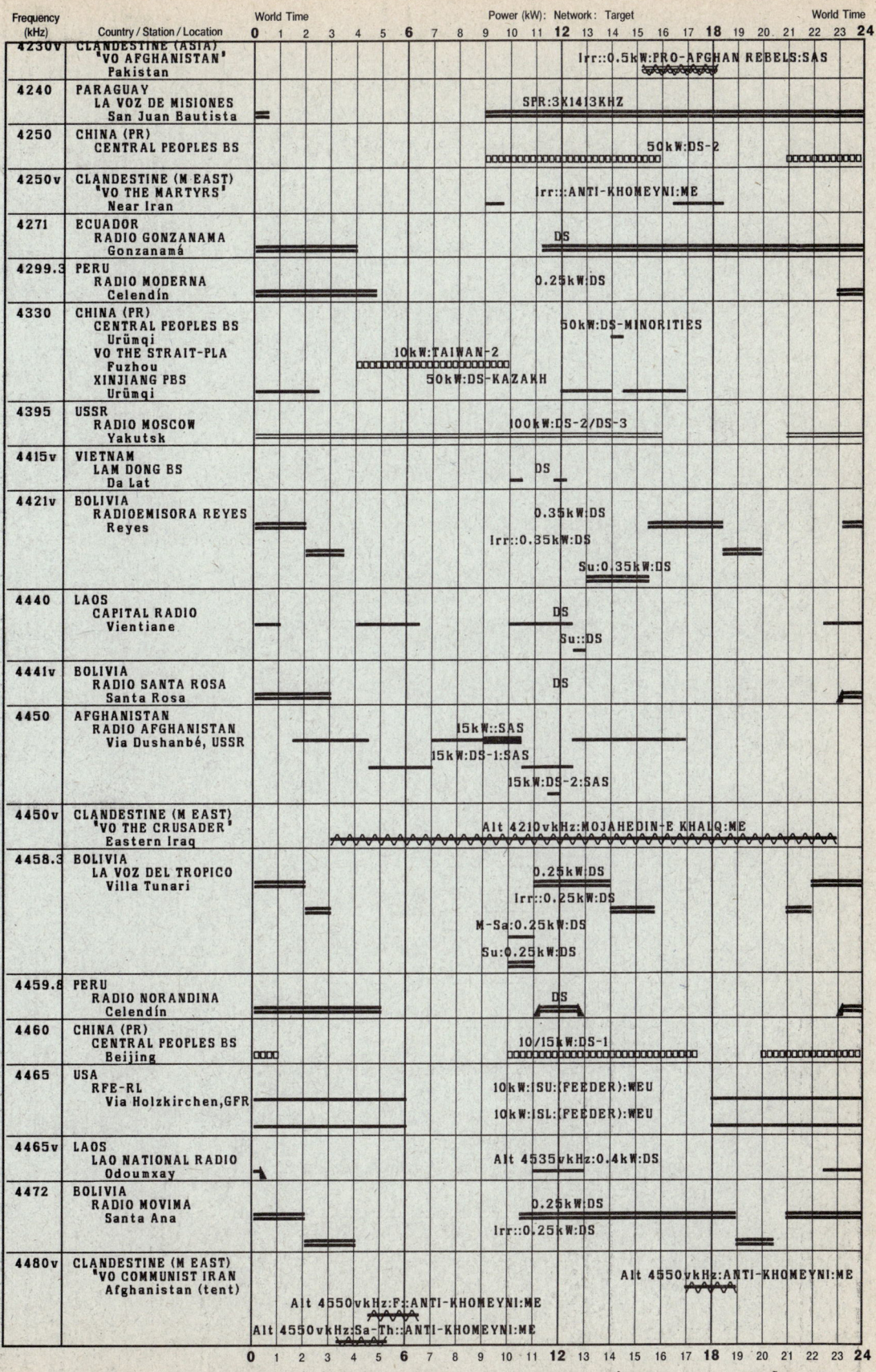

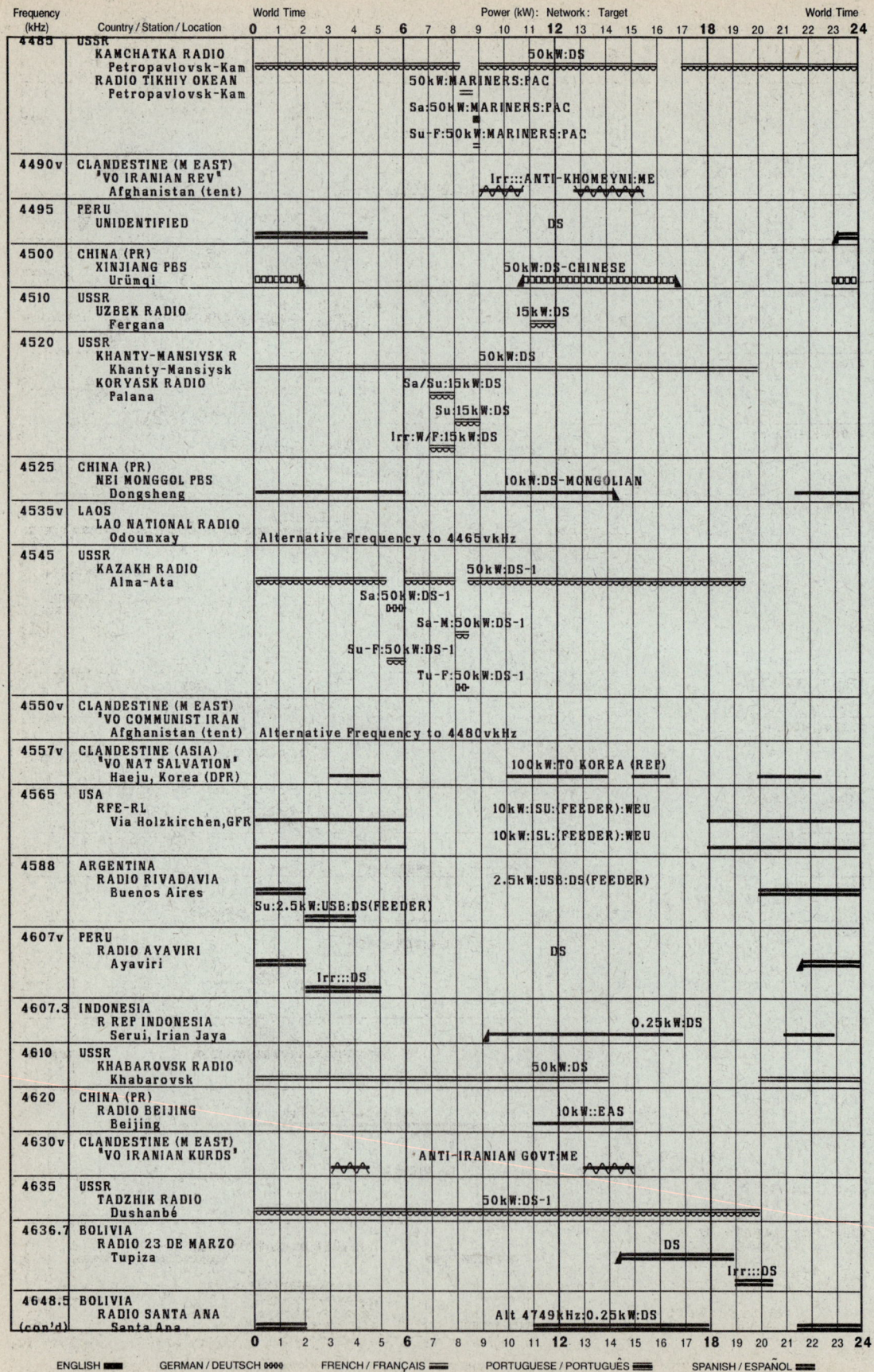

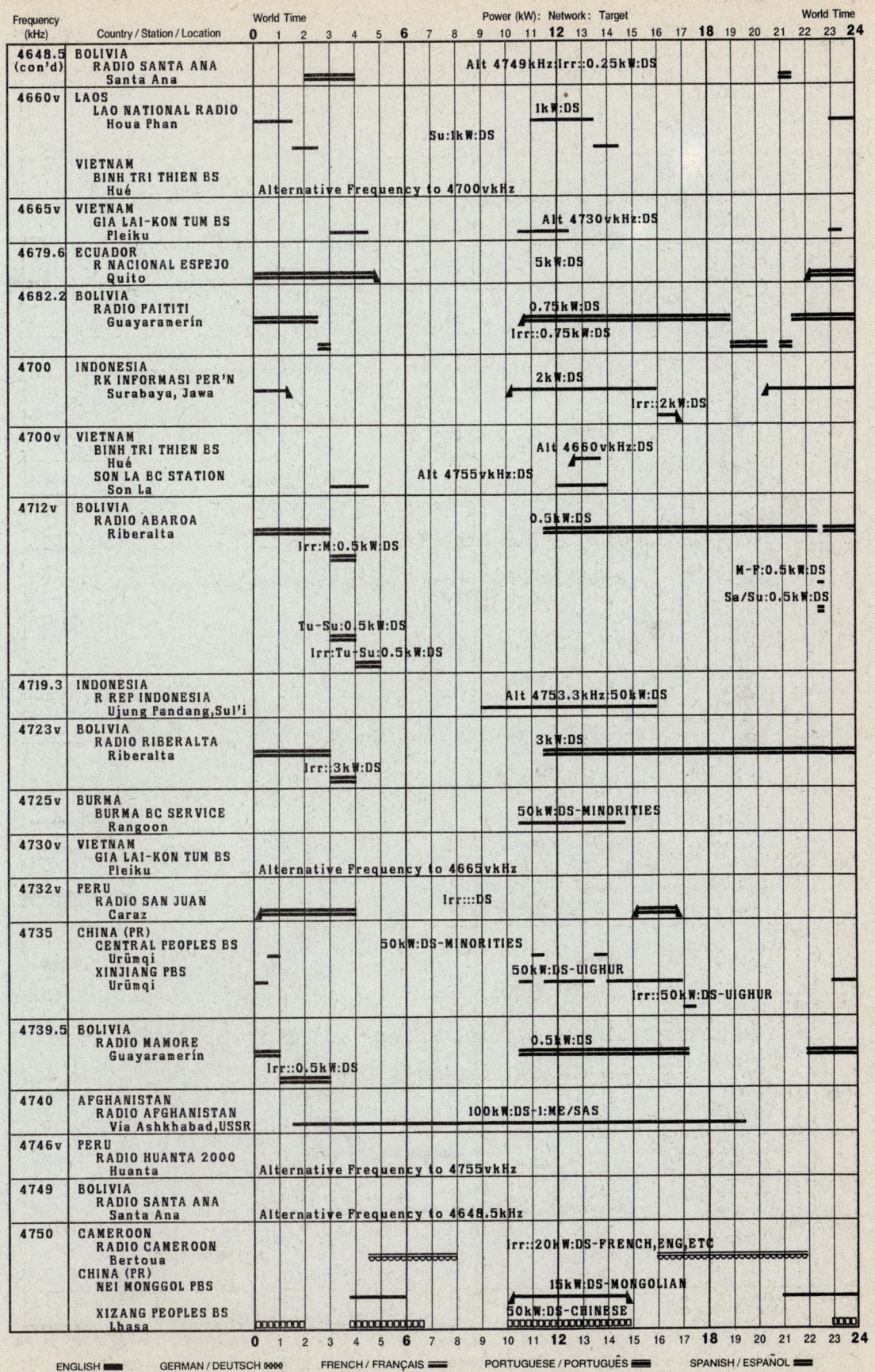

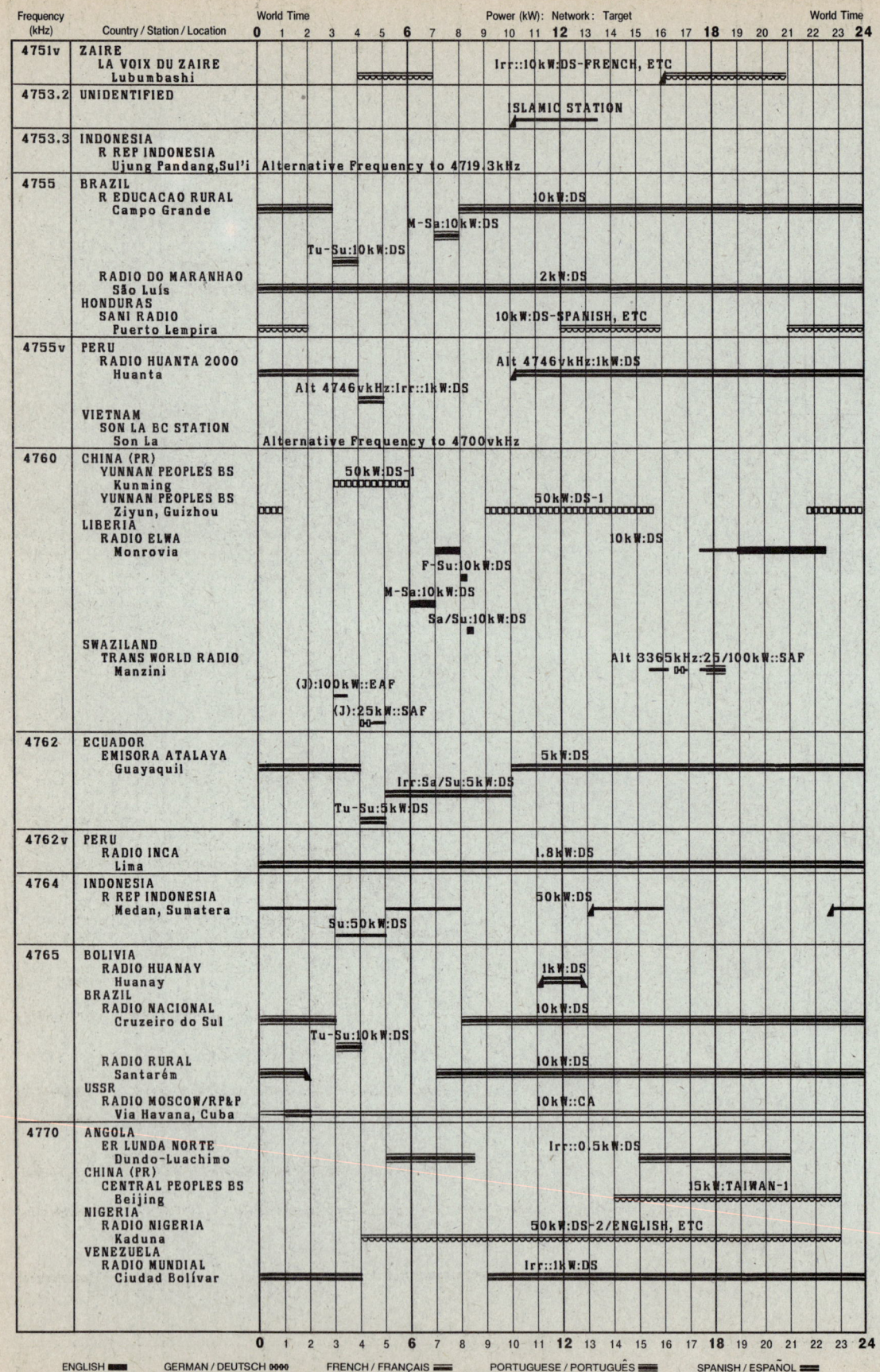

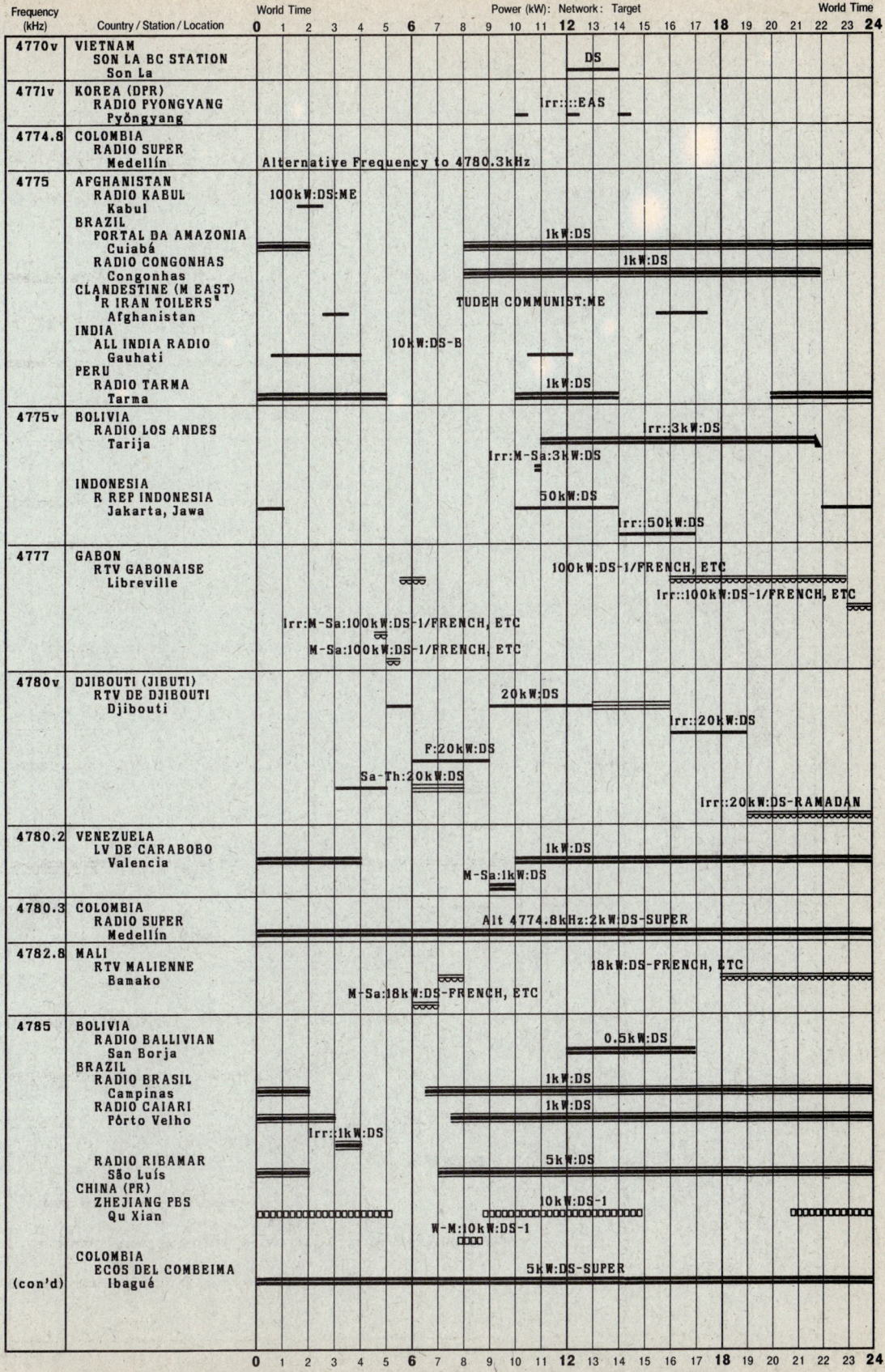

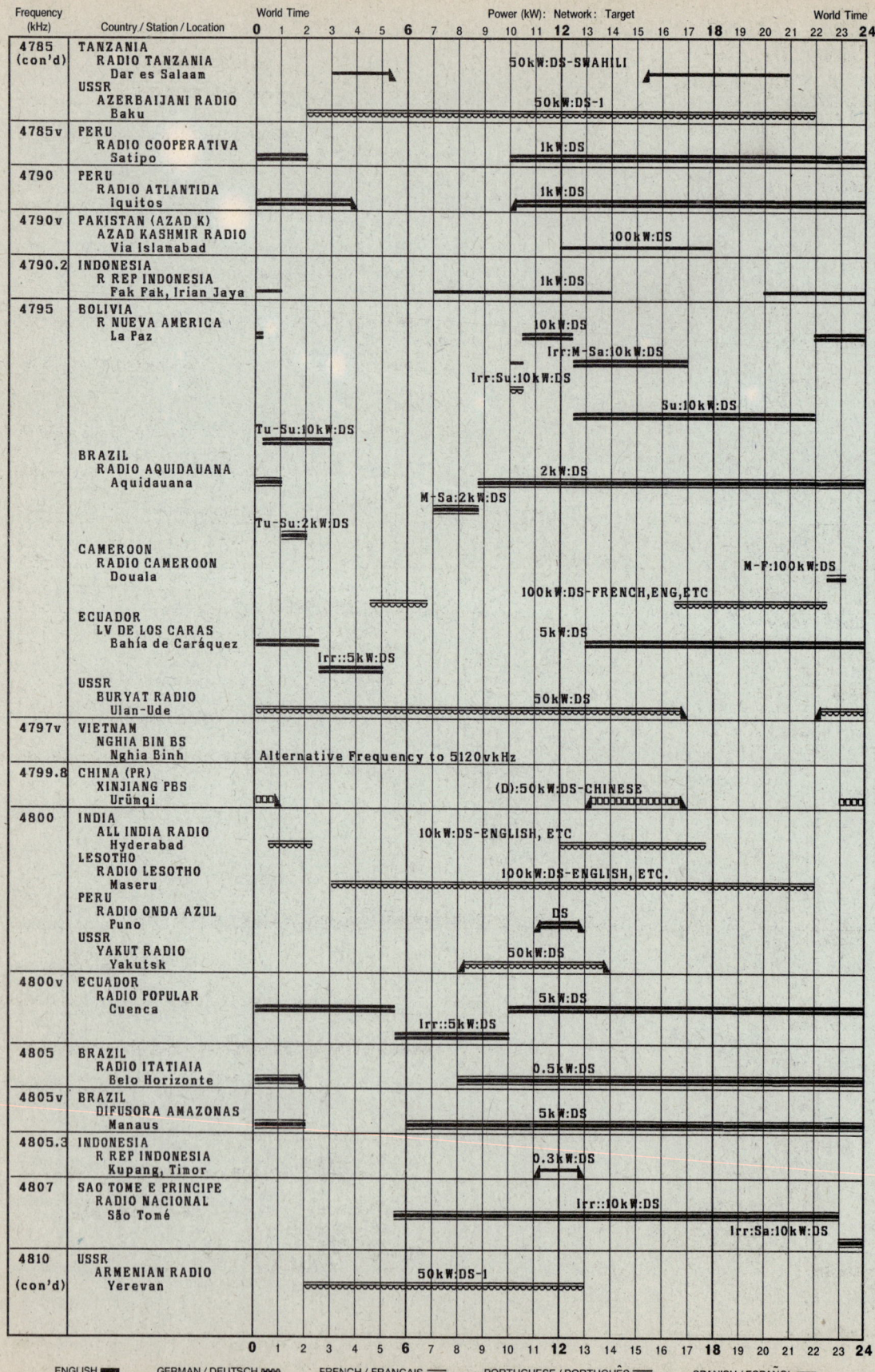

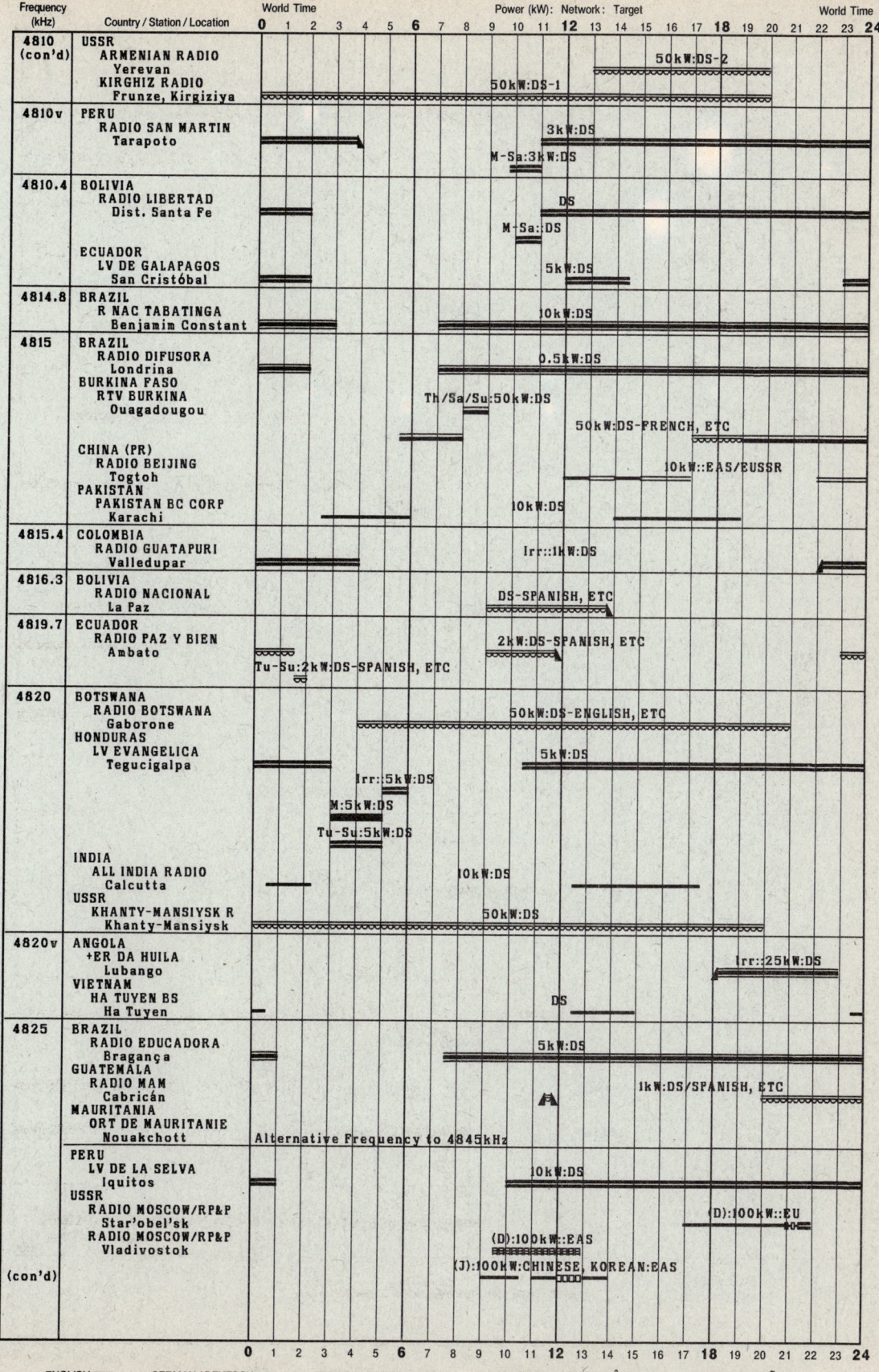

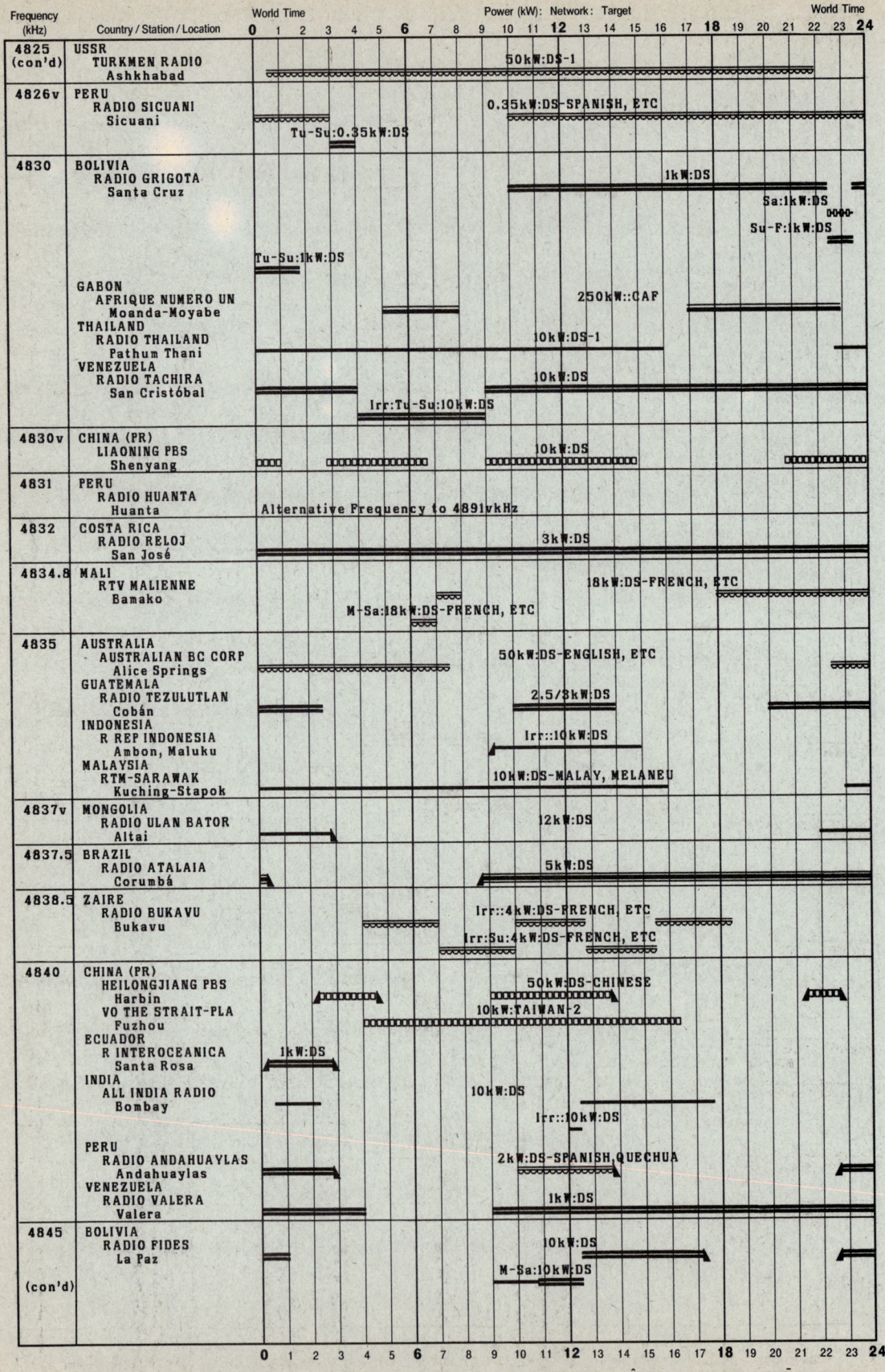

WORLDSCAN

4845–4860 kHz

Frequency (kHz)	Country / Station / Location	Broadcast Schedule
4845 (con'd)	**BOLIVIA** RADIO FIDES, La Paz	Su:10kW:DS; Tu-Su:10kW:DS
	BRAZIL RADIO NACIONAL, Manaus	250kW:DS; Tu-Sa:250kW:DS
	MALAYSIA RADIO MALAYSIA, Kajang	50kW:DS-TAMIL; Sa/Su:50kW:DS-TAMIL; Su:50kW:DS-TAMIL
	MAURITANIA ORT DE MAURITANIE, Nouakchott	Alt 4825kHz:100kW:DS-FRENCH,ARAB; Alt 4825kHz:F:100kW:DS-FRENCH,ARAB; Alt 4825kHz:Sa-Th:100kW:DS-FRENCH,ARAB,ETC
4845v	**COLOMBIA** RADIO BUCARAMANGA, Bucaramanga	Irr::1kW:DS
4850	**CAMEROON** RADIO CAMEROON, Yaoundé	100kW:DS-ENGLISH,FRENCH
	CHINA (PR) CENTRAL PEOPLES BS	TAIWAN-2
	INDIA ALL INDIA RADIO, Kohima	2kW:DS
	MONGOLIA RADIO ULAN BATOR, Ulan Bator	100kW:DS-1; Tu/F:100kW::EUSSR; Sa-M/W/Th:100kW:DS-1
	USSR RADIO MOSCOW/RP&P Via Ulan Bator	100kW::EAS
	UZBEK RADIO, Tashkent	50kW:DS-2
	VENEZUELA RADIO CAPITAL, Caracas	1kW:DS; Su/M:1kW:DS
4851v	**ECUADOR** RADIO LUZ Y VIDA, Loja	5kW:DS; Irr::5kW:DS
4852.7	**YEMEN (PDR)** "VO PALESTINE" Via Radio San'a	100kW:PLO; Irr::100kW:PLO
	RADIO SAN'A, San'a	100kW:DS; F:100kW:DS
4855	**BRAZIL** R POR MUNDO MELHOR, Gov Valadares	1kW:DS; M-Sa:1kW:DS
	RADIO ARUANA, Barra do Garças	1kW:DS
4855v	**MAURITIUS** MAURITIUS BC CORP, Curepipe	10kW:DS-FRENCH,ENG,ETC
4855.7	**INDONESIA** R REP INDONESIA, Palembang, Sumatera	Irr::10kW:DS; Irr:Su:10kW:DS
4857v	**BOLIVIA** RADIO EL CONDOR, Uyuni	DS
4860	**ANGOLA** ER DO LUNDA SUL, Saurimo	Irr::5kW:DS
	INDIA ALL INDIA RADIO, Delhi	10kW::SEA
(con'd)		

Legend: ENGLISH ▬ GERMAN/DEUTSCH ◊◊◊◊ FRENCH/FRANÇAIS ═ PORTUGUESE/PORTUGUÊS ▬ SPANISH/ESPAÑOL ▬ ARABIC ≡ RUSSIAN/РУССКИЙ ═ CHINESE ✱✱ ○○○○ JAPANESE 日本語 ▬▬ MULTILINGUAL ◊◊◊◊ OTHER ▬ SUMMER ONLY (J) WINTER ONLY (D) JAMMING ⋀⋀ or / or \ EARLIEST HEARD ◢ LATEST HEARD ◣ + TENTATIVE

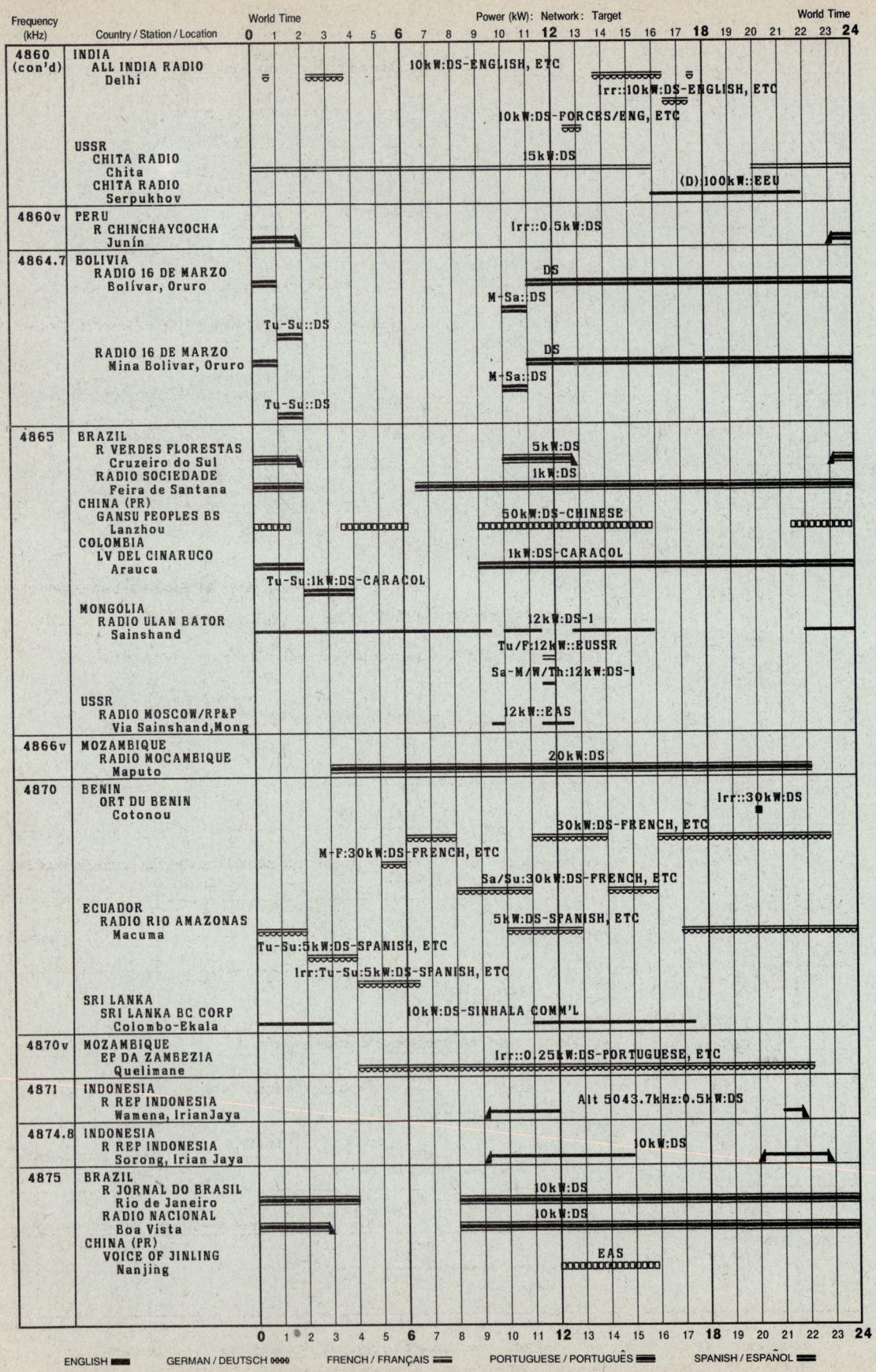

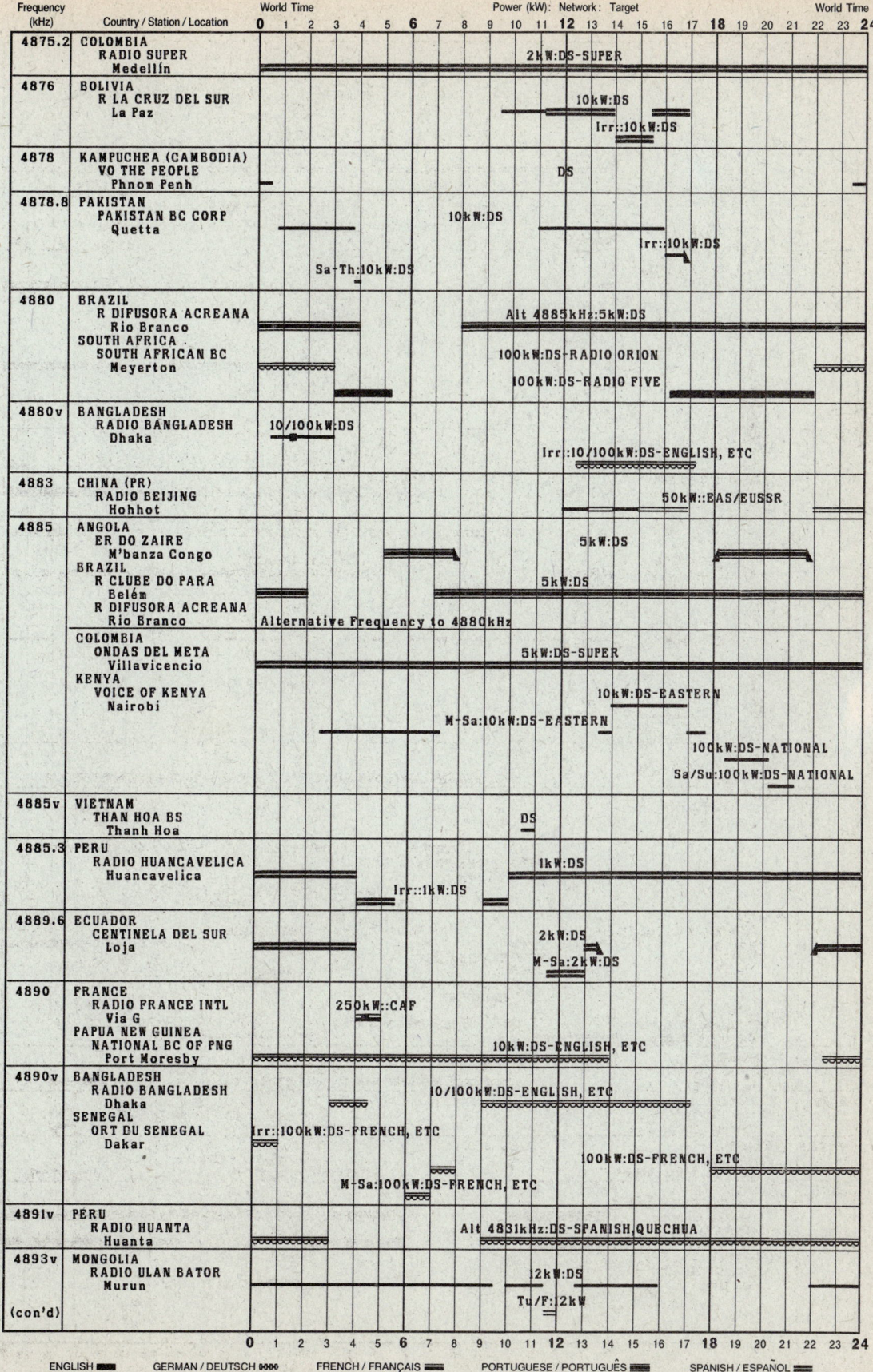

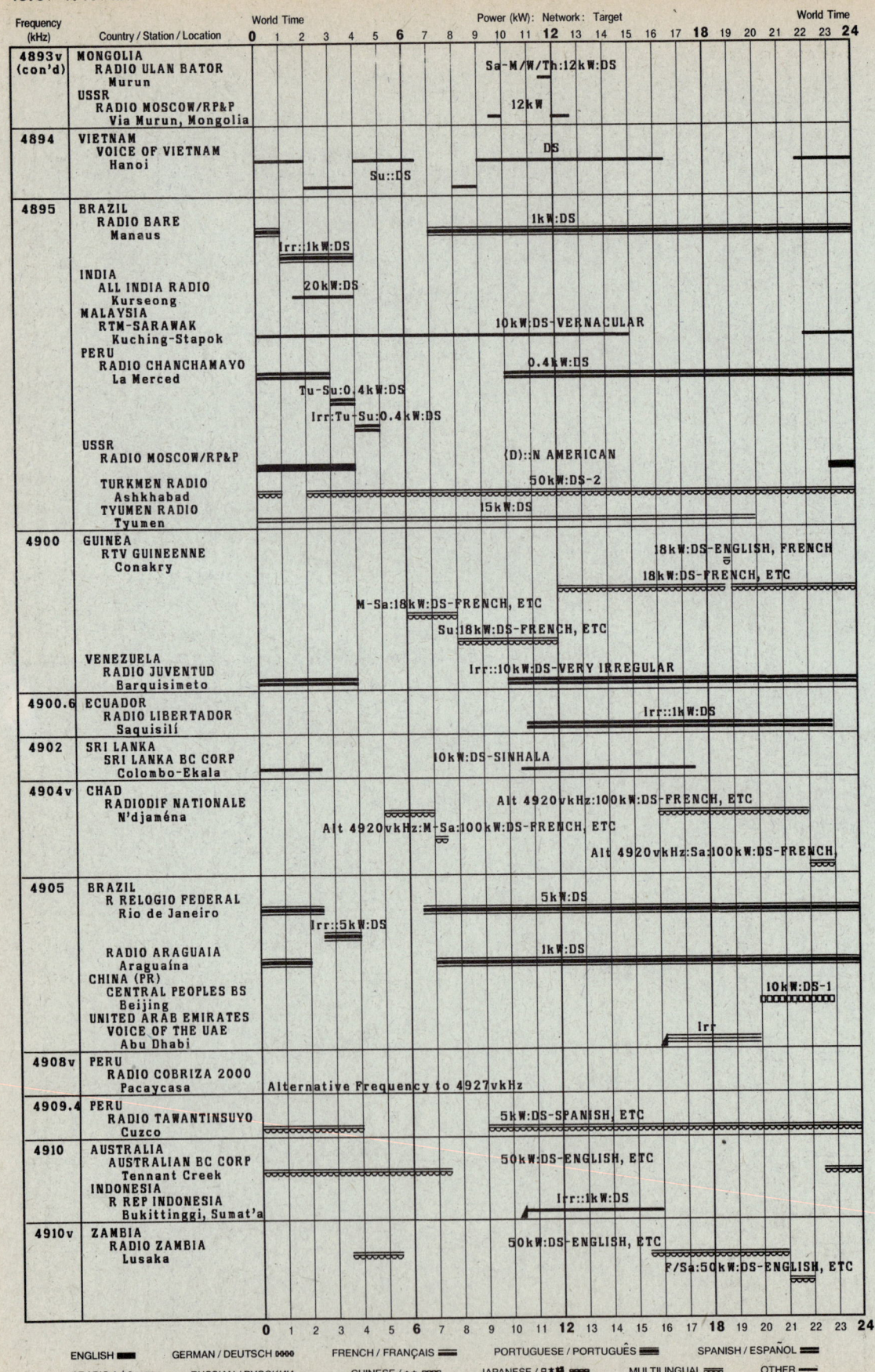

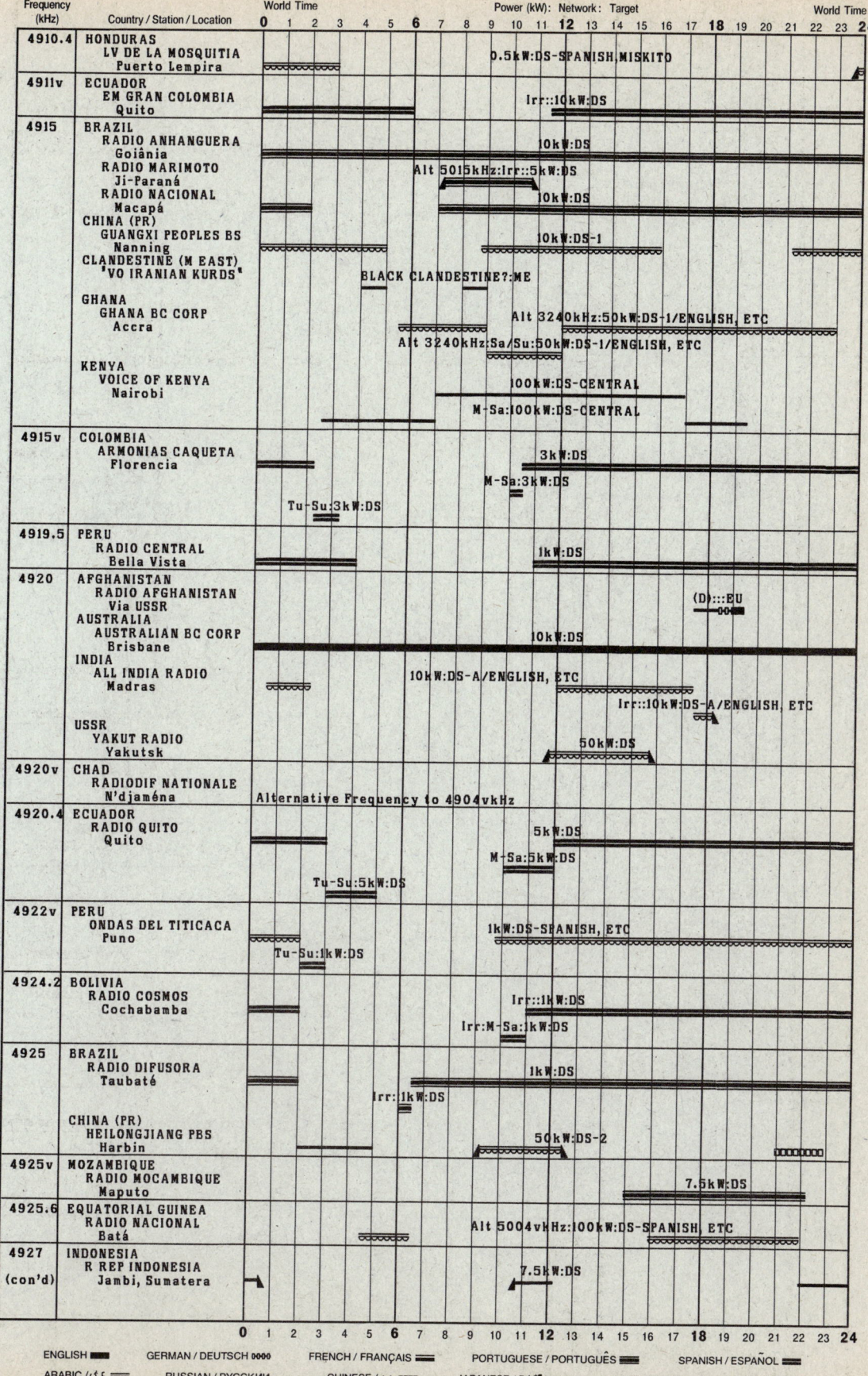

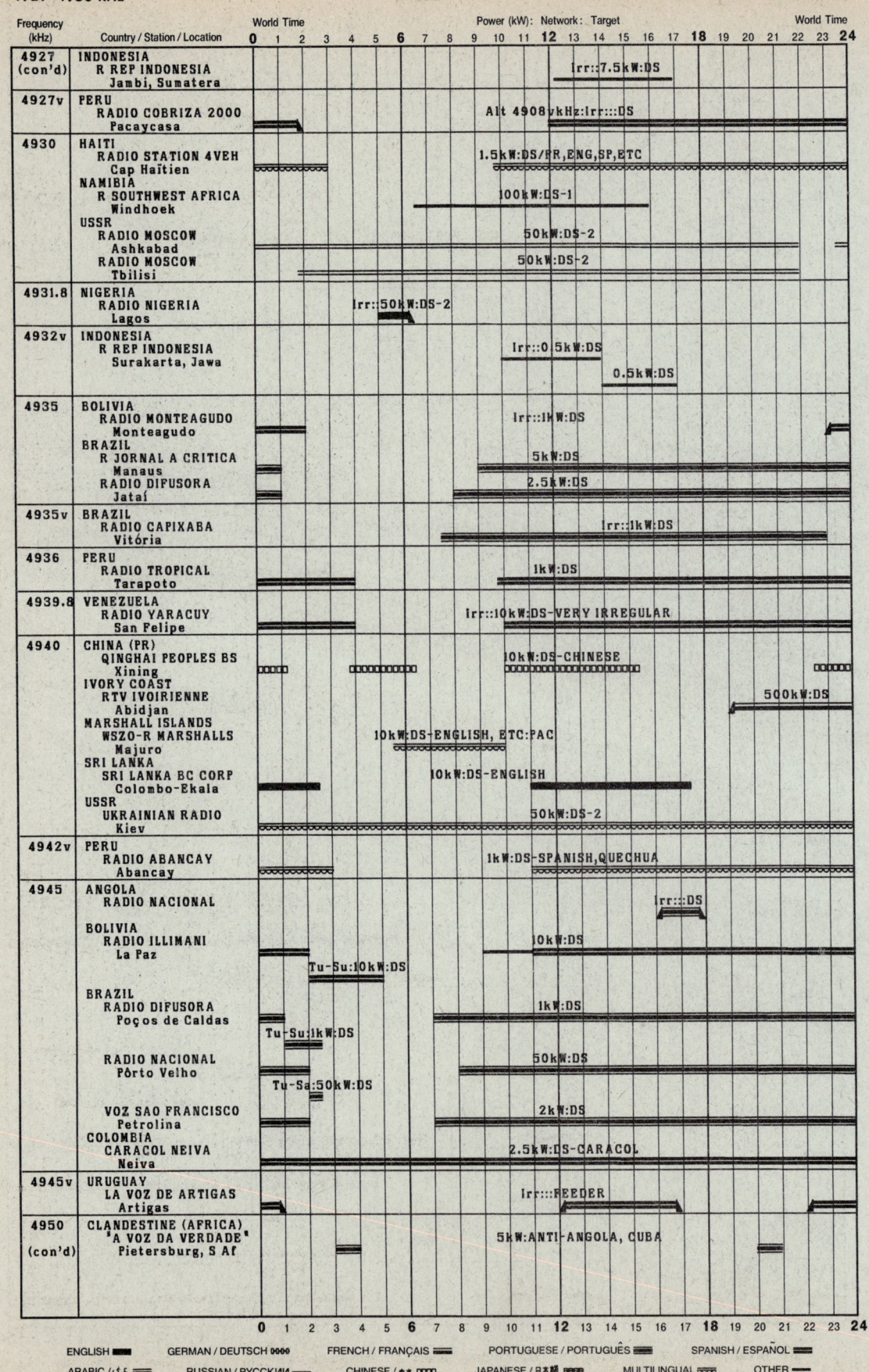

WORLDSCAN

4950–4969v kHz

Frequency (kHz)	Country / Station / Location	Schedule (World Time 0–24)
4950 (con'd)	**MALAYSIA** RTM-SARAWAK, Kuching-Stapok	10kW:DS-ENGLISH,CHINESE
	PAKISTAN PAKISTAN BC CORP, Peshawar	10kW:DS
	RADIO PAKISTAN, Peshawar	10kW::SAS
	PERU R MADRE DE DIOS, Puerto Maldonado	5kW:DS
4953v	**ANGOLA** RADIO NACIONAL	DS
4954.8	**INDONESIA** R REP INDONESIA, Banda Aceh, Sumat'a	Alternative Frequency to 3904.8 kHz
4955	**BRAZIL** RADIO CLUBE, Rondonópolis	Irr::2.5kW:DS; Irr:M-Sa:2.5kW:DS; Irr:Tu-Su:2.5kW:DS
	RADIO CULTURA, Campos	2.5kW:DS
4955v	**BRAZIL** RADIO MARAJOARA, Belém	10kW:DS
	PERU R CULTURAL AMAUTA, Huanta	1kW:DS-SPANISH, ETC
4957.5	**USSR** AZERBAIJANI RADIO, Baku	50kW:DS-2
4960	**CHINA (PR)** RADIO BEIJING, Kunming	50kW::EAS
	INDIA ALL INDIA RADIO, Delhi	10kW:DS
	ALL INDIA RADIO, Ranchi	2kW:DS
4960v	**ECUADOR** RADIO FEDERACION, Sucúa	5kW:DS-SPANISH, ETC; M-W:5kW:DS-SPANISH, ETC; Tu-Su:5kW:DS-SPANISH, ETC; Irr:Tu-Su:5kW:DS-SPANISH, ETC
	MADAGASCAR RADIO MADAGASIKARA, Antananarivo	Irr:::DS; Irr:M-Sa::DS
4960.3	**PERU** RADIO LA MERCED, La Merced	Tu-Su:0.5kW:DS; 0.5kW:DS
4965	**BRAZIL** RADIO ALVORADA, Parintins	5kW:DS
	NAMIBIA SW AFRICA BC CORP, Windhoek	100kW:DS-GERMAN, ETC
4965v	**BOLIVIA** RADIO JUAN XXIII, San Ignacio Velasco	3kW:DS; M-Sa:3kW:DS; Su:3kW:DS; Irr:Su:3kW:DS
	BRAZIL RADIO POTI, Natal	1kW:DS
4966	**PERU** RADIO SAN MIGUEL, Cuzco	5kW:DS-SPANISH, ETC; Tu-Su:5kW:DS
4968	**SRI LANKA** SRI LANKA BC CORP, Colombo-Ekala	10kW:DS-TAMIL
4969v (con'd)	**VENEZUELA** RADIO RUMBOS, Caracas	10kW:DS

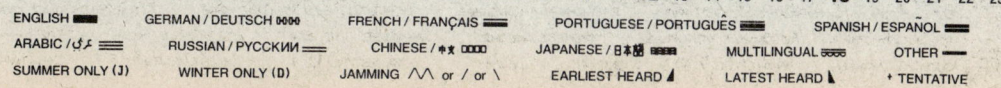

ENGLISH ▬   GERMAN / DEUTSCH ▬   FRENCH / FRANÇAIS ▬   PORTUGUESE / PORTUGUÊS ▬   SPANISH / ESPAÑOL ▬
ARABIC / ﻋﺮﺑﻲ ▬   RUSSIAN / РУССКИЙ ▬   CHINESE / 中文 ▬   JAPANESE / 日本語 ▬   MULTILINGUAL ▬   OTHER ▬
SUMMER ONLY (J)   WINTER ONLY (D)   JAMMING ∧∧ or / or \   EARLIEST HEARD ◢   LATEST HEARD ◣   ✦ TENTATIVE

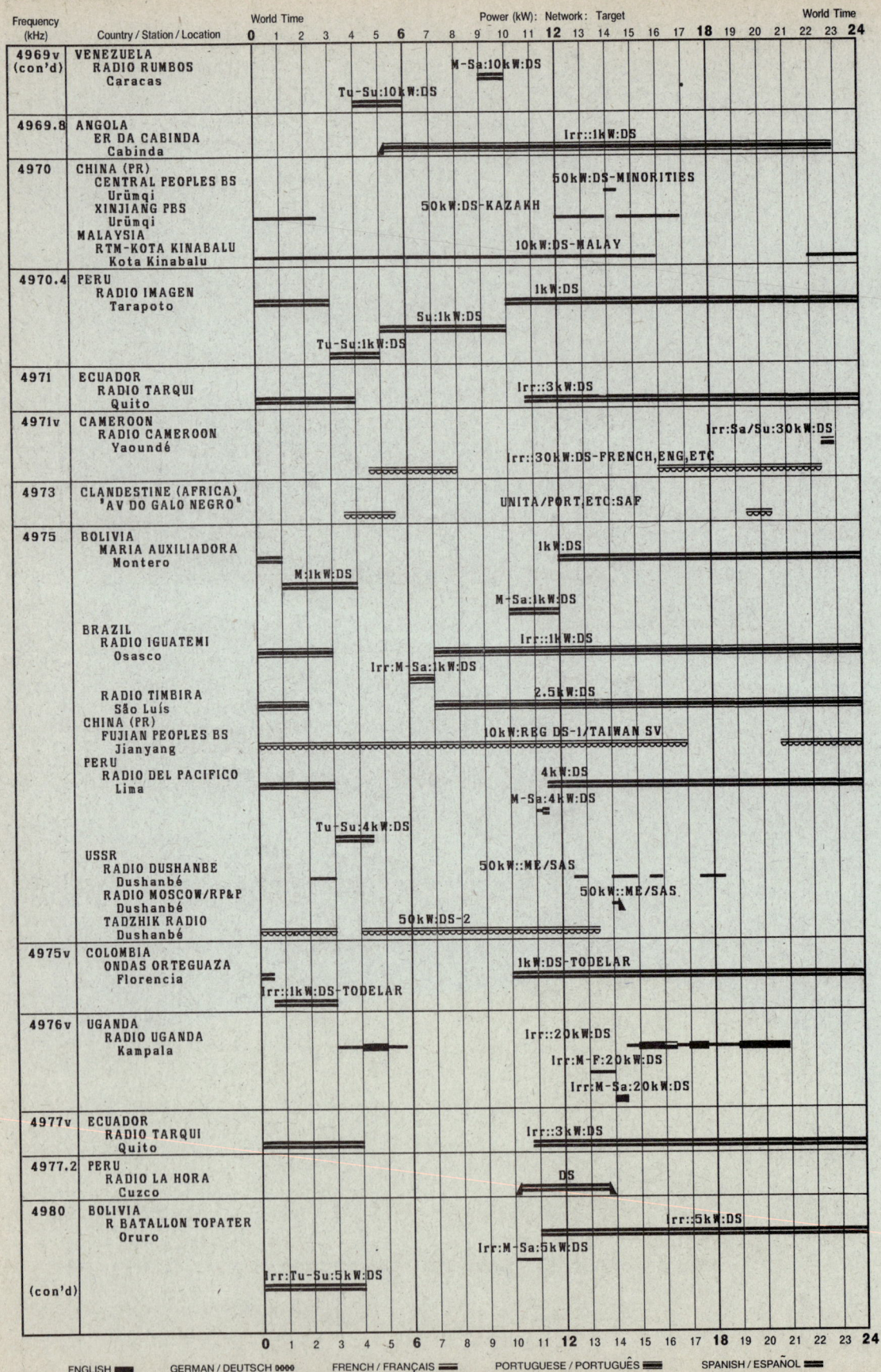

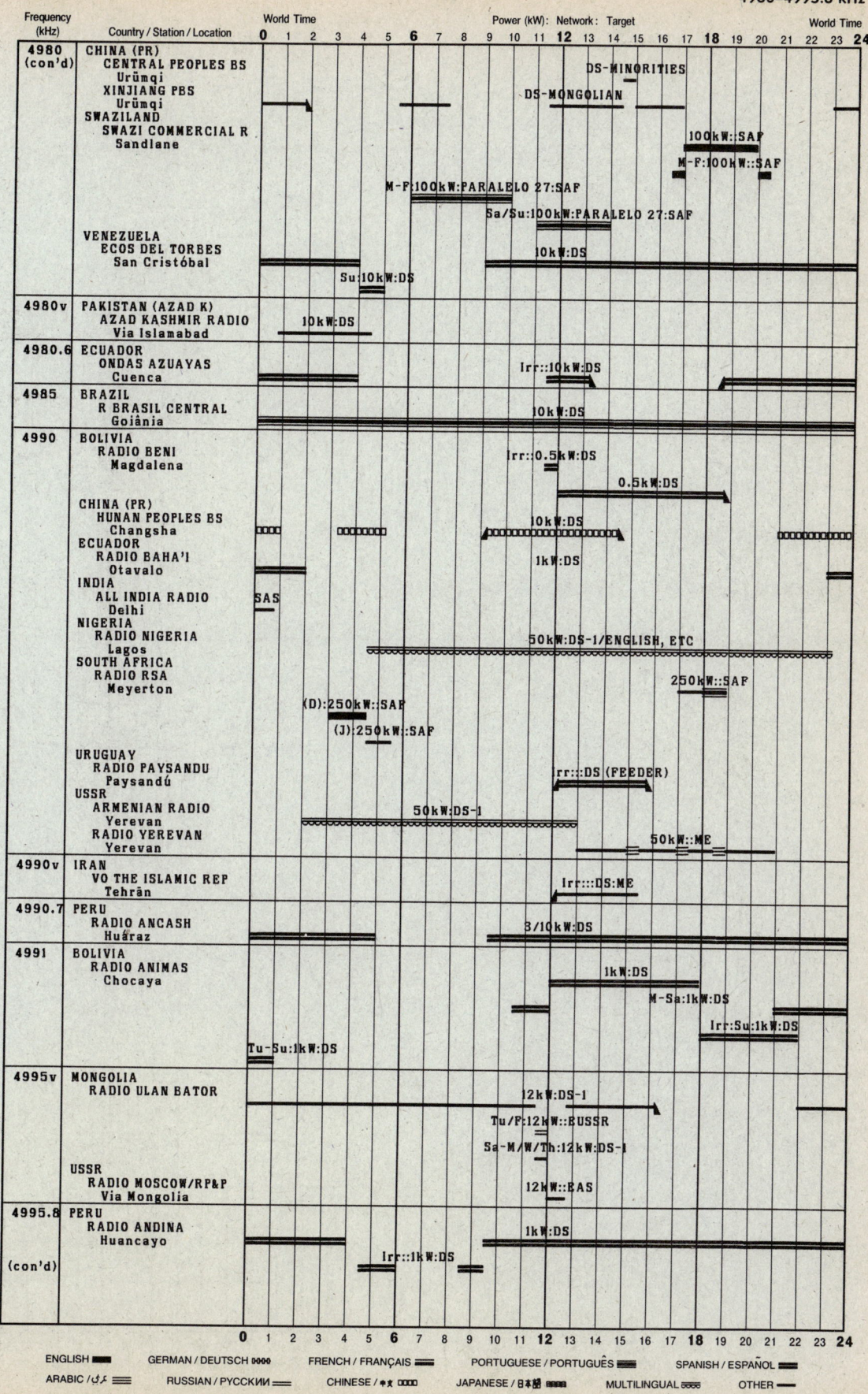

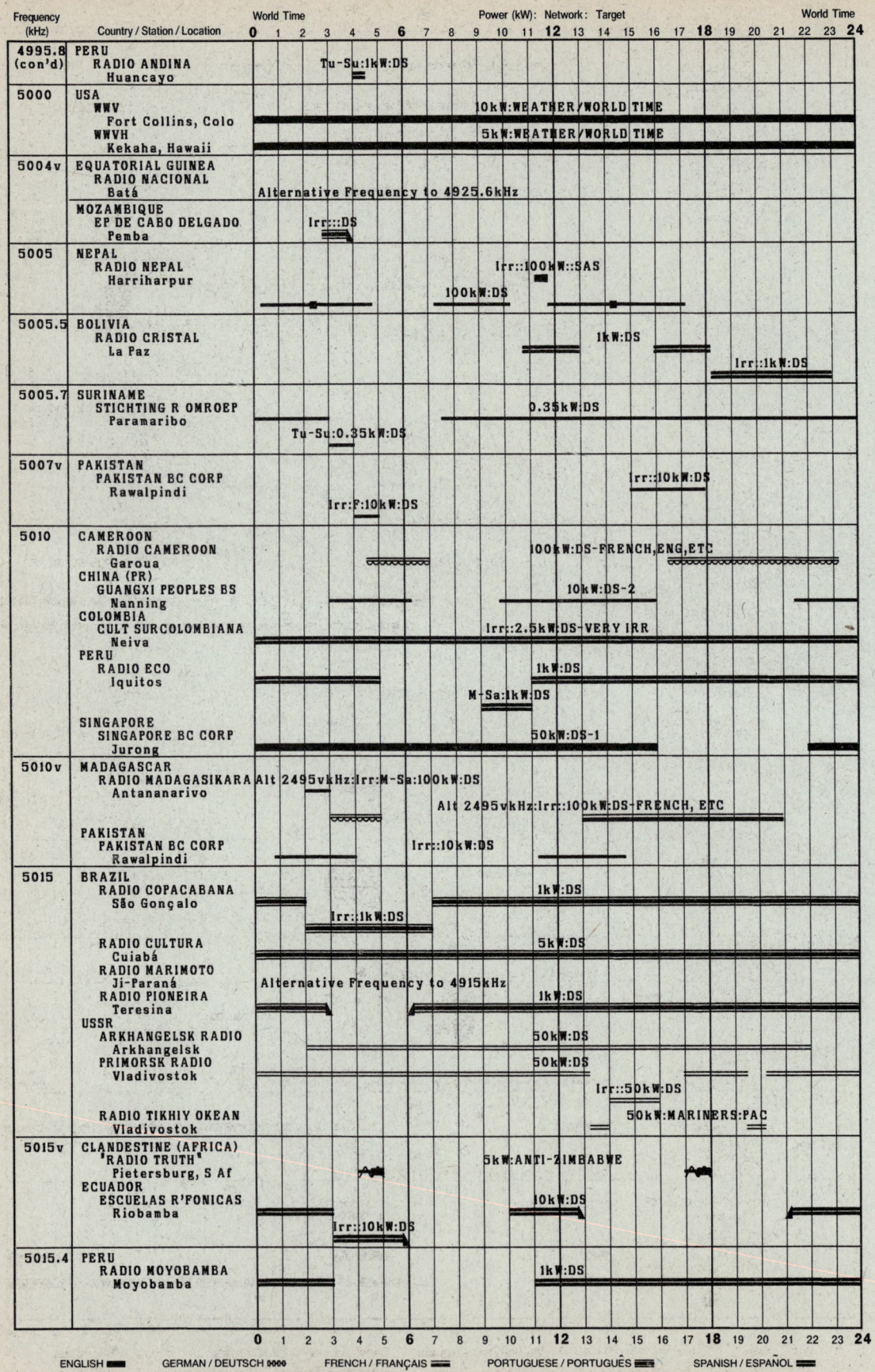

WORLDSCAN

5020–5035 kHz

Frequency (kHz)	Country / Station / Location	Schedule / Power / Network / Target
5020	**CHINA (PR)** JIANGXI PEOPLES BS — Nanchang	10kW:DS
	SOLOMON IS SOLOMON ISLANDS BC — Honiara	10kW:DS-ENGLISH, ETC
	SRI LANKA SRI LANKA BC CORP — Colombo-Ekala	10kW:DS-TAMIL
5020v	**ALBANIA** SHQIPTAR RTV — Gjirokastër	Alternative Frequency to 5057vkHz
	NIGER LA VOIX DU SAHEL — Niamey	Sa:20/100kW:DS-2 / 20/100kW:DS-2/FRENCH, ETC / Sa/Su:20/100kW:DS-2/FRENCH, ETC
	VENEZUELA RADIO NACIONAL — Caracas	Alt 5031vkHz:1kW:DS
5020.3	**BOLIVIA** CUARTO CENTENARIO — Tupiza	Alt 5030kHz:0.25kW:DS / Alt 5030kHz:Irr::0.25kW:DS / Alt 5030kHz:M-Sa:0.25kW:DS / Alt 5030kHz:Tu-Su:0.25kW:DS / Alt 5030kHz:Irr:Tu-Su:0.25kW:DS
5025	**AUSTRALIA** AUSTRALIAN BC CORP — Katherine	50kW:DS
	BENIN ORT DU BENIN — Parakou	20kW:DS-FRENCH, ETC
	BRAZIL R TRANSAMAZONICA — Altamira	5kW:DS
	RADIO BORBOREMA — Campina Grande	1kW:DS / Irr::1kW:DS
	CUBA RADIO REBELDE — Havana	Irr::10kW:DS:AM
	PERU RADIO QUILLABAMBA — Quillabamba	0.5/5kW:DS-SPANISH, ETC / Tu-Su:0.5/5kW:DS / M-Sa:0.5/5kW:DS-SPANISH, ETC
5025.6	**ECUADOR** RADIO SPLENDIT — Cuenca	Irr::3kW:DS-VERY IRR
5026v	**UGANDA** RADIO UGANDA — Kampala	20kW:DS / M-F:20kW:DS / M-Sa:20kW:DS
5030	**BOLIVIA** CUARTO CENTENARIO — Tupiza	Alternative Frequency to 5020.3kHz
	CHINA (PR) CENTRAL PEOPLES BS — Xi'an	10kW:DS-2 / 10kW:DS-MINORITIES
	MALAYSIA RTM-SARAWAK — Kuching-Stapok	10kW:DS-BIDAYUTH
5030v	**PERU** RADIO LOS ANDES — Huamachuco	1kW:DS
5031v	**VENEZUELA** RADIO NACIONAL — Caracas	Alternative Frequency to 5020vkHz
5035	**AUSTRIA** SCHULUNGSSENDER — Vienna	10kW:DS-ARMY TRAINING
	BRAZIL R EDUCACAO RURAL — Coarí	1kW:DS
(con'd)	RADIO APARECIDA — Aparecida	2.5kW:DS

Legend: ENGLISH — GERMAN/DEUTSCH — FRENCH/FRANÇAIS — PORTUGUESE/PORTUGUÊS — SPANISH/ESPAÑOL — ARABIC/عربي — RUSSIAN/РУССКИЙ — CHINESE/中文 — JAPANESE/日本語 — MULTILINGUAL — OTHER — SUMMER ONLY (J) — WINTER ONLY (D) — JAMMING — EARLIEST HEARD — LATEST HEARD — + TENTATIVE

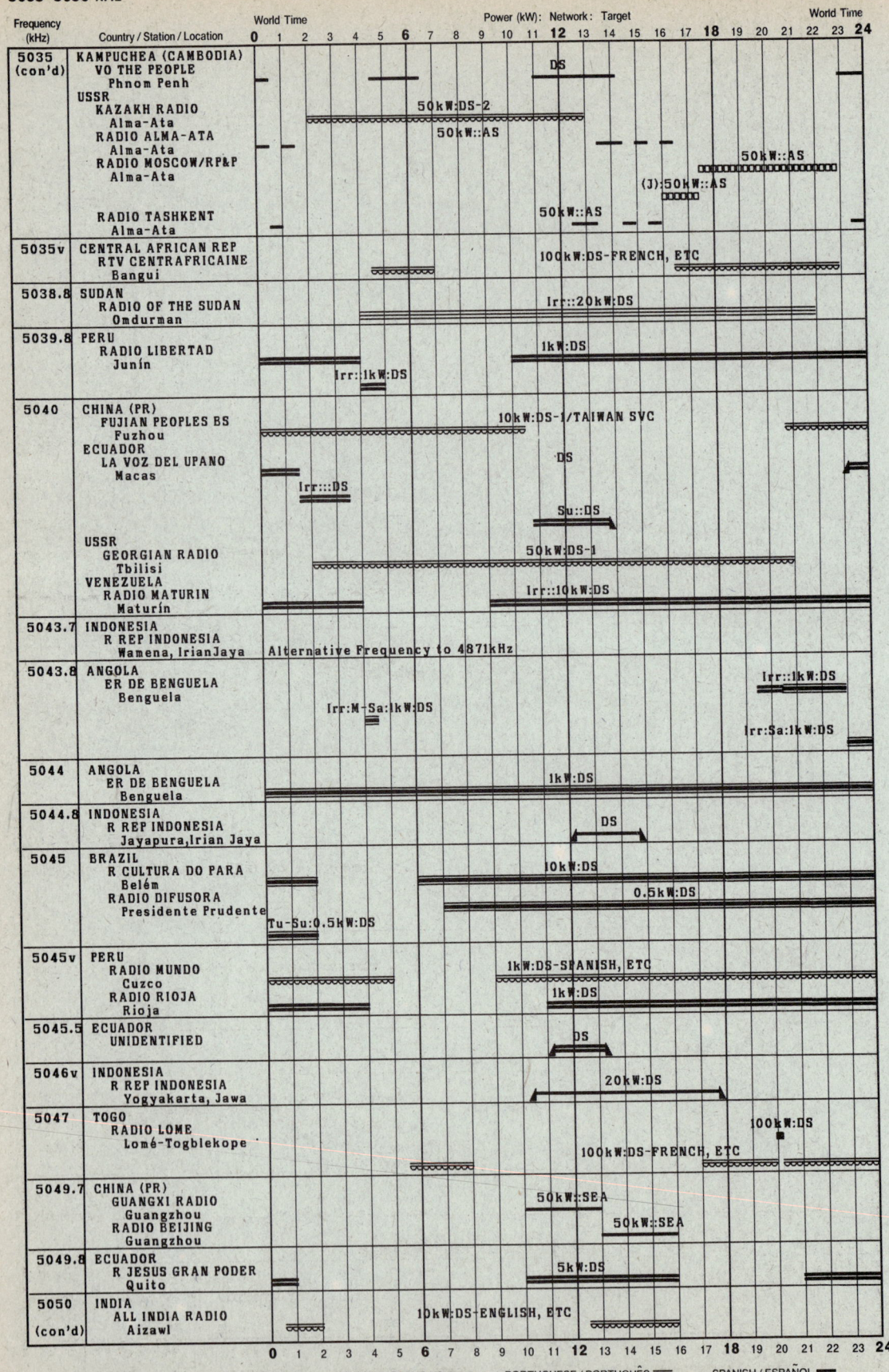

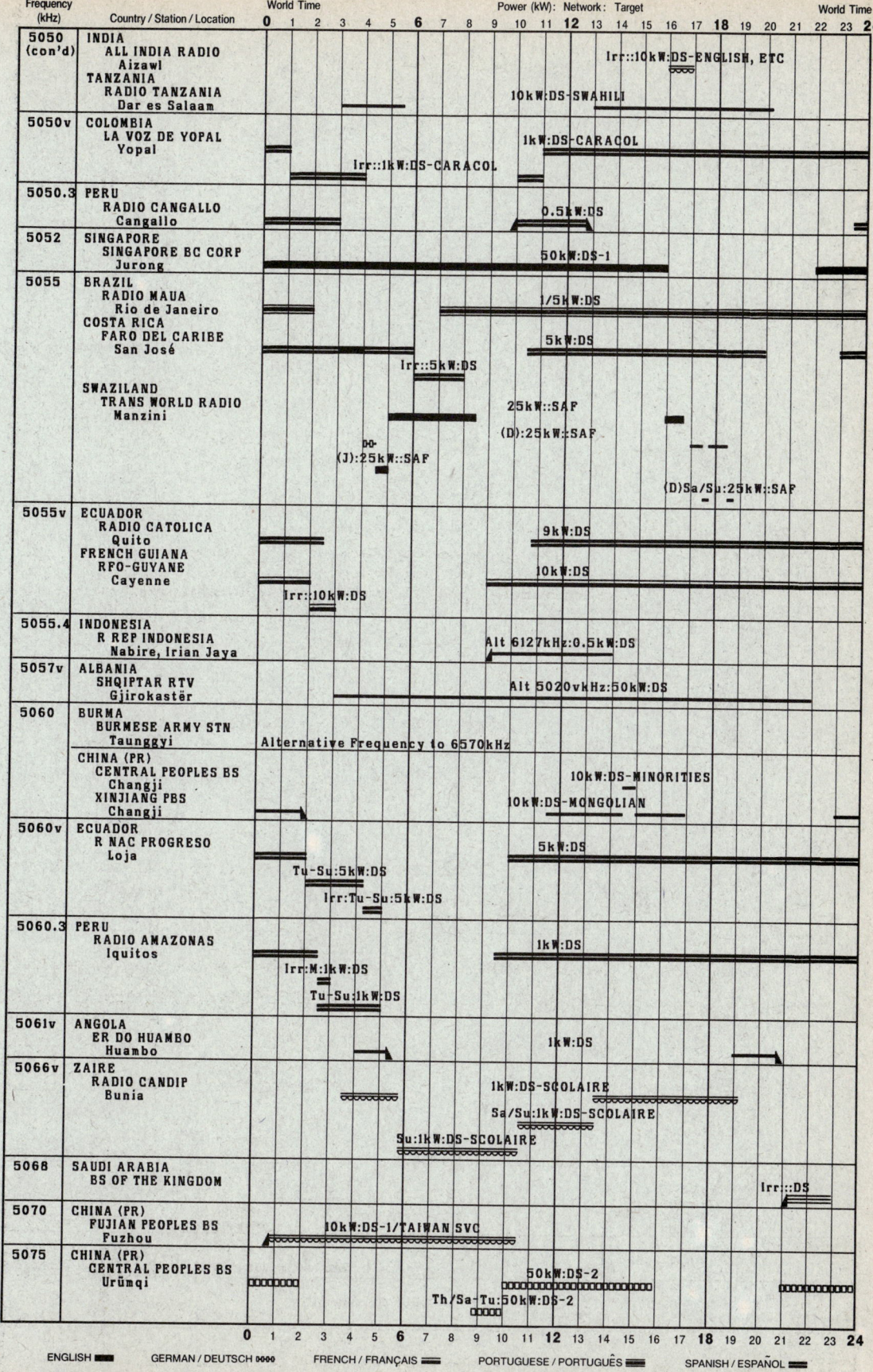

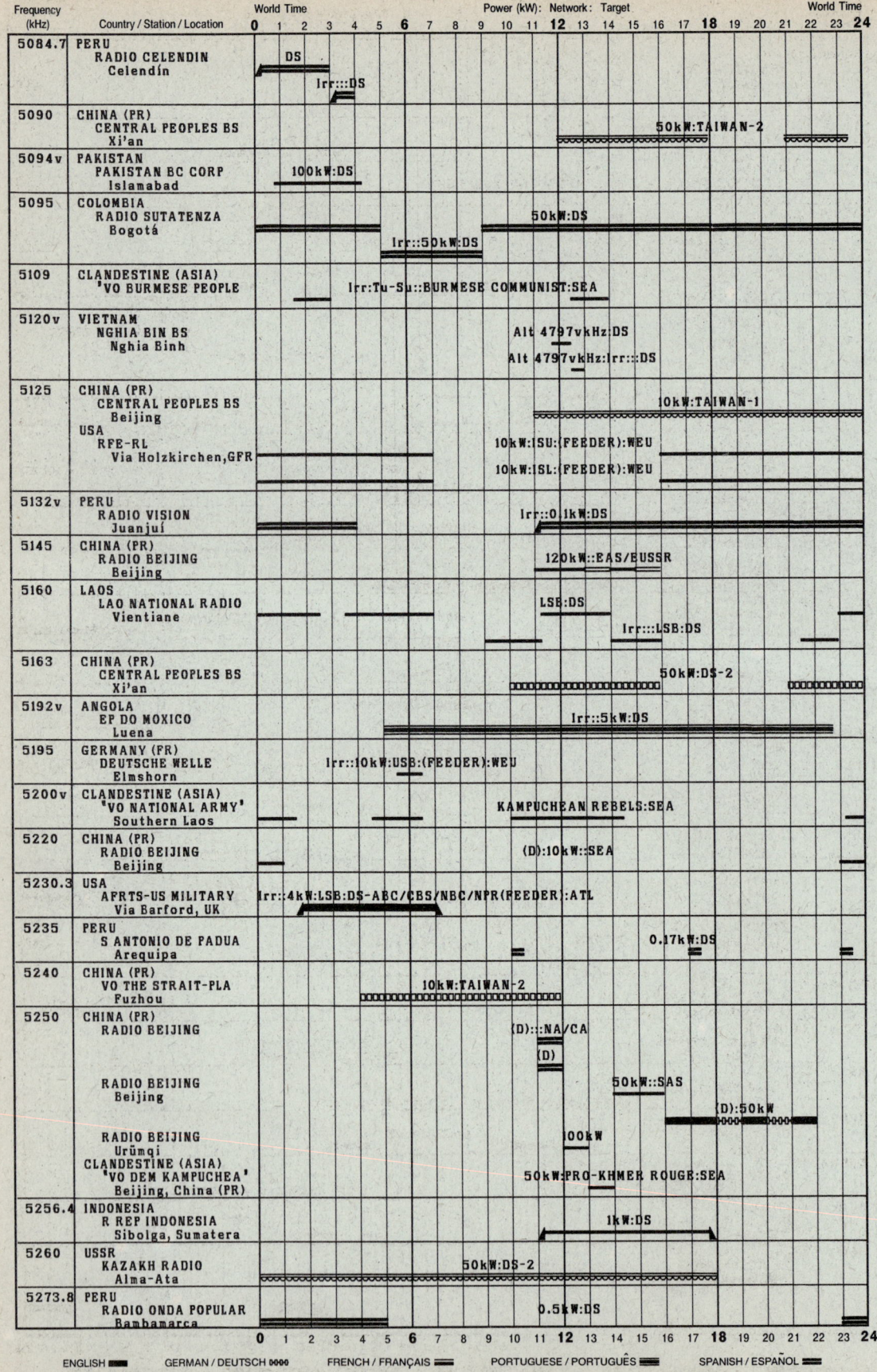

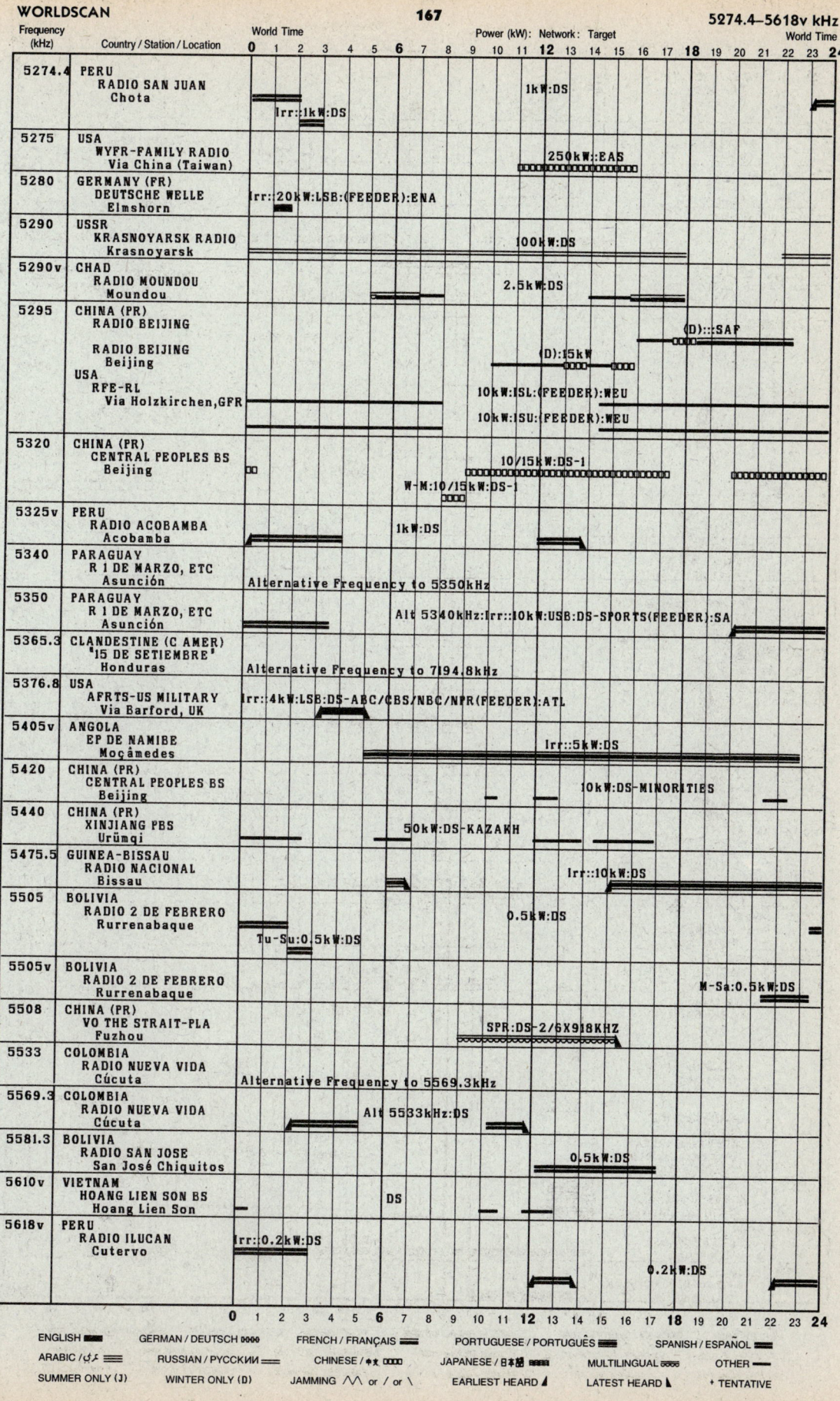

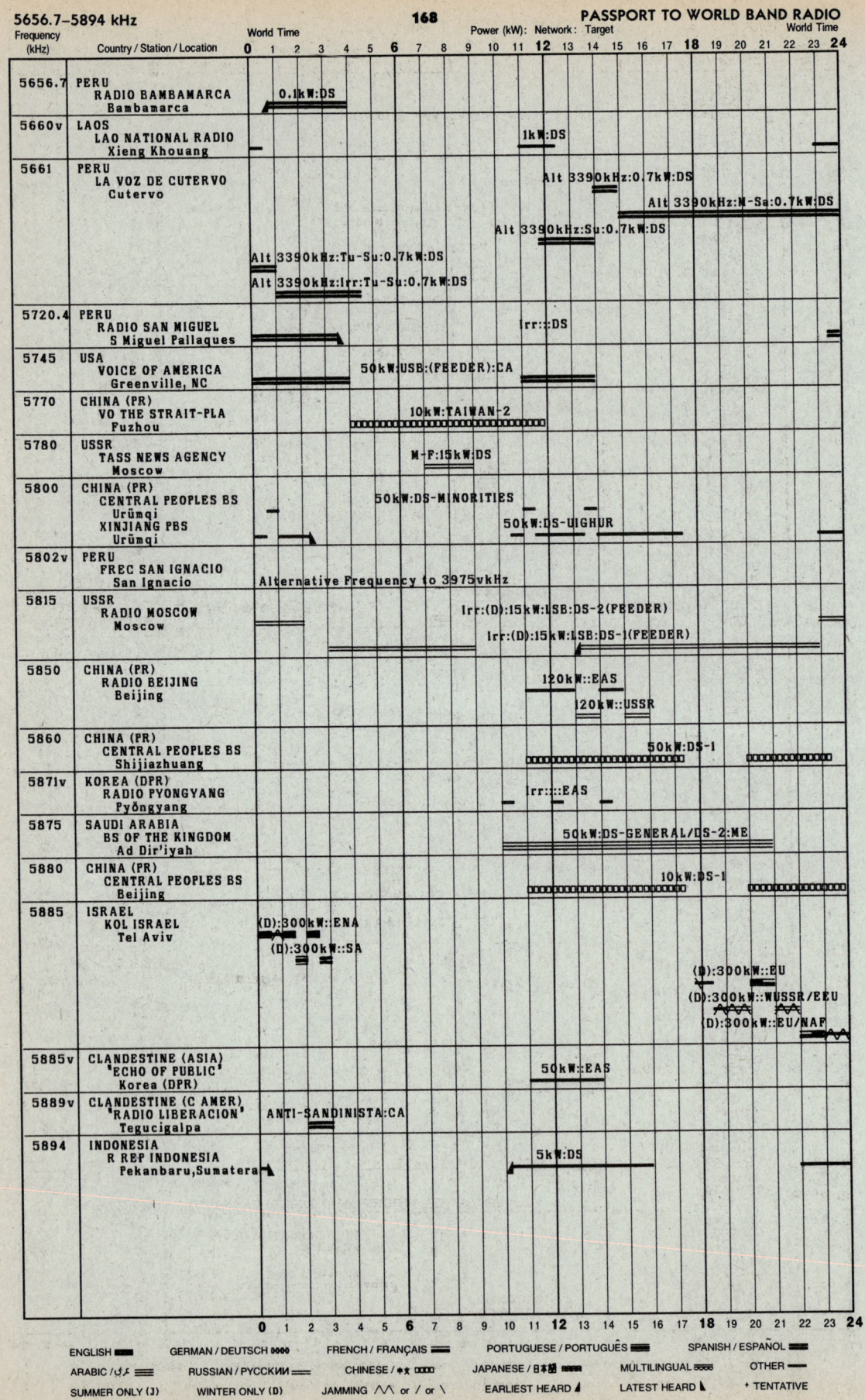

WORLDSCAN — 5900–5920 kHz

Frequency (kHz)	Country / Station / Location	Schedule (World Time 0–24) — Power (kW): Network: Target
5900	**BELGIUM** — BELGISCHE RADIO TV, Wavre	Alternative Frequency to 5910 kHz
	USSR — RADIO MOSCOW	(D)::DS-1
	RADIO MOSCOW, Kenga	100kW:DS-2; (D):100kW:DS-1
	RADIO MOSCOW, Petropavlovsk-Kam	(D):100kW:DS-1
	RADIO MOSCOW, Vologda	100kW:DS-3
	RADIO MOSCOW/RP&P, Petropavlovsk-Kam	(J):100kW:WS:NNA
	RADIO TIKHIY OKEAN, Petropavlovsk-Kam	(D):100kW:MARINERS; (D)Sa:100kW:MARINERS; (D)Su-F:100kW:MARINERS
5900v	**CHINA (PR)** — SICHUAN PEOPLES BS, Chengdu	15kW:DS-2
	UNIDENTIFIED, Via Sichuan PBS?	DS
5905	**PAKISTAN** — PAKISTAN BC CORP, Islamabad	10kW:DS/ES:SAS
	RADIO PAKISTAN, Islamabad	10kW:SAS
	+RADIO PAKISTAN, Islamabad	(D):::SAS
	RADIO PAKISTAN, Islamabad	(J):10kW::SAS
	USSR — RADIO MOSCOW/RP&P	(D):::CA
	RADIO MOSCOW/RP&P, Ryazan'	100kW:WORLD, UK SCES:EU
5910	**BELGIUM** — BELGISCHE RADIO TV, Wavre	Alt 5900kHz:100kW::EU; Alt 5900kHz:(D):100kW::ENA; Alt 5900kHz:M-Sa:100kW::EU; Alt 5900kHz:Su:100kW::EU
	USSR — RADIO MOSCOW, Chita	100kW:DS-1
	RADIO MOSCOW, Moscow	50kW:DS-1; 50kW:DS-2
	RADIO MOSCOW/RP&P	(D):::EU
	RADIO TIKHIY OKEAN	(D)::MARINERS:PAC; (D)Sa::MARINERS:PAC; (D)Su-F::MARINERS:PAC
5915	**CHINA (PR)** — CENTRAL PEOPLES BS, Beijing	50kW:DS-1
	CLANDESTINE (EUROPE) 'OUR RADIO', Cluj, Romania	15kW:TURKISH COMMUNIST:EU
	USSR — RADIO ALMA-ATA, Alma-Ata	100kW::AS
	RADIO MOSCOW/RP&P	(D)::N AMERICAN:CA; (D):::CA; (D)::WS:CA
	RADIO TASHKENT, Alma-Ata	100kW::AS
5915v	**ISRAEL** — KOL ISRAEL, Tel Aviv	20kW:DS-D:ME
5920	**CHINA (PR)** — GUANGXI PBS, Guangzhou	DS-3/EDUCATIONAL
	USSR — RADIO MOSCOW, Khabarovsk	(D):100kW:DS-2
	RADIO MOSCOW, Orcha (con'd)	(D):100kW:DS-3

Legend: ENGLISH ▬ · GERMAN / DEUTSCH ◊◊◊◊ · FRENCH / FRANÇAIS ═ · PORTUGUESE / PORTUGUÊS ▭▭▭ · SPANISH / ESPAÑOL ▬▬ · ARABIC / عربي ≡ · RUSSIAN / РУССКИЙ ─── · CHINESE / 中文 ○○○○ · JAPANESE / 日本語 ▬▬▬ · MULTILINGUAL ⌘⌘⌘ · OTHER ▬ · SUMMER ONLY (J) · WINTER ONLY (D) · JAMMING ∧∧ or / or \ · EARLIEST HEARD ◢ · LATEST HEARD ◣ · + TENTATIVE

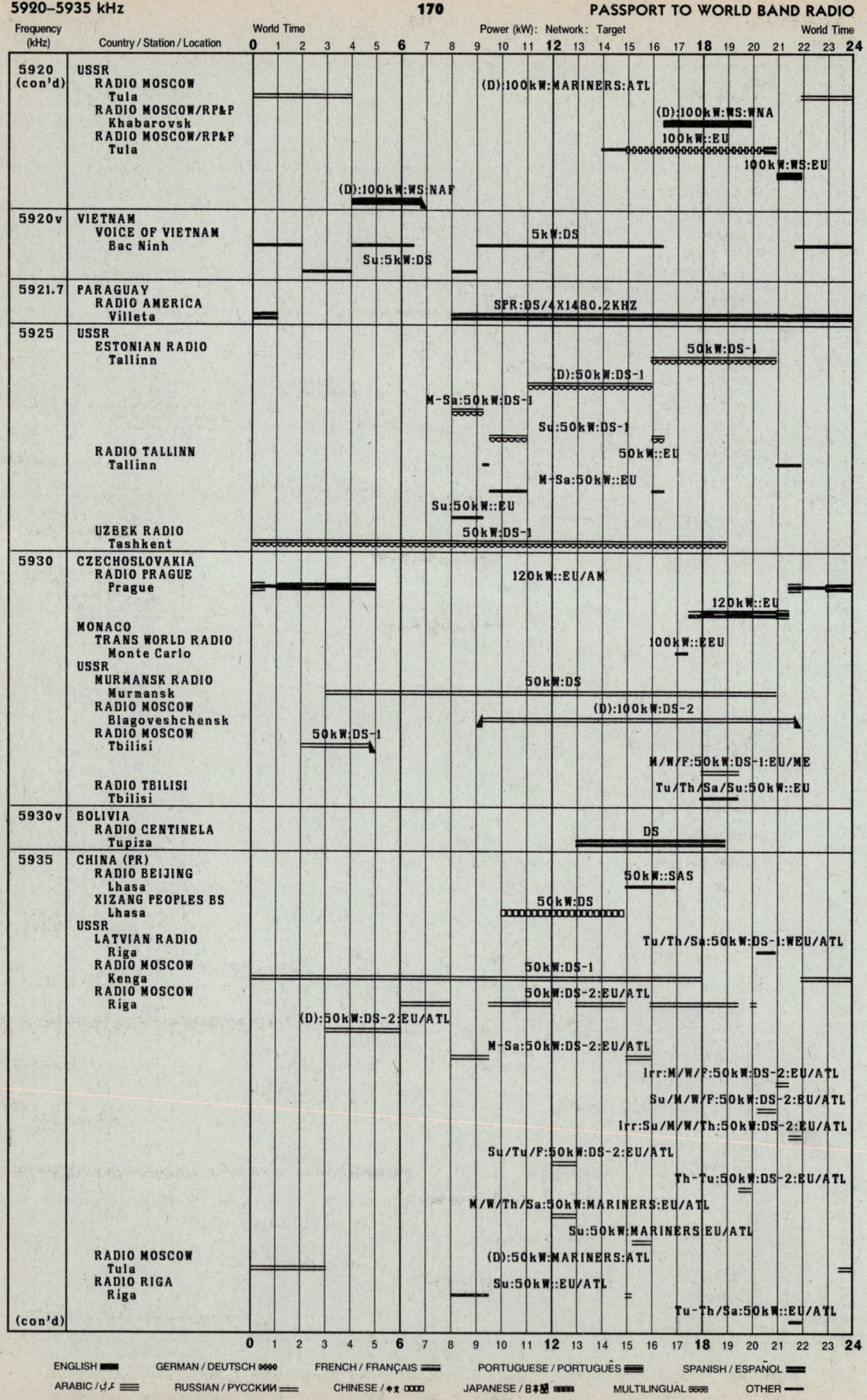

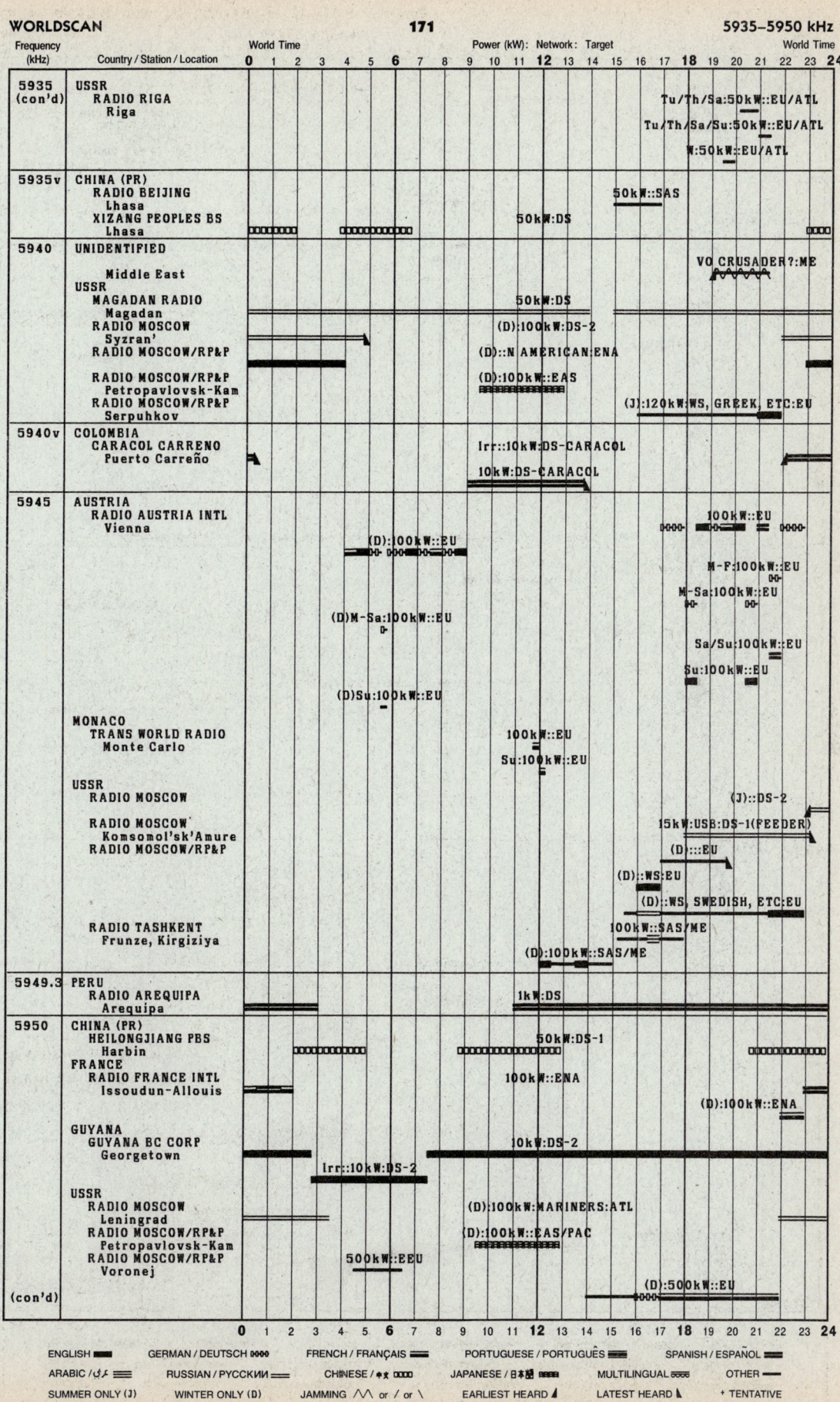

5950–5960 kHz

Frequency (kHz)	Country / Station / Location	Broadcast Schedule (World Time 0–24)
5950 (con'd)	**USSR** — RADIO TIKHIY OKEAN, Petropavlovsk-Kam	100kW:MARINERS:EAS/PAC; Sa:100kW:MARINERS:EAS/PAC; Su-F:100kW:MARINERS:EAS/PAC
5950v	**CLANDESTINE (C AMER)** — "R 15 SETIEMBRE", Honduras	FDN/ANTI-SANDINIST:CA
5954.2	**COSTA RICA** — RADIO CASINO, Limón	Irr::0.7kW:DS; Irr:M:0.7kW:DS; Irr:Tu-Su:0.7kW:DS
5954.4	**BOLIVIA** — RADIO PIO DOCE, Llallagua-Siglo XX	1kW:DS-SPANISH, ETC; M-Sa:1kW:DS-SPANISH, ETC
5954.8	**COLOMBIA** — LA VOZ CENTAUROS, Villavicencio	5kW:DS-CARACOL; Irr::5kW:DS-TOUR DE FRANCE
5955	**BOTSWANA** — RADIO BOTSWANA, Gaborone	50kW:DS-ENGLISH, ETC
	BRAZIL — RADIO GAZETA, São Paulo	7.5kW:DS
	CAMEROON — RADIO CAMEROON, Bafoussam	20kW:DS-FRENCH, ENG, ETC; Sa/Su:20kW:DS-FRENCH, ENG, ETC
	GUATEMALA — RADIO CULTURAL, Guatemala City	0.25/10kW:DS
	HOLLAND — RADIO NEDERLAND, Flevoland	500kW::EU; Su:500kW::EU
	SWAZILAND — TRANS WORLD RADIO, Manzini	(D):25kW::SAF
	TURKEY — TURKISH RTV CORP, Ankara	(J):250kW::EU
	USA — RFE-RL Via Germany (FR)	(J):20/250kW::WUSSR; (D):100kW::WUSSR
	RFE-RL Via Lisbon, Portugal	(D):50/250kW::WUSSR
	RFE-RL Via Pals, Spain	(D):250kW::WUSSR
	VOICE OF AMERICA Via Kaválla, Greece	250kW::EEU; (D):250kW::EEU
5955.3	**PERU** — RADIO HUANCAYO, Huancayo	0.5kW:DS
5960	**CANADA** — RADIO CANADA INTL, Sackville, NB	(J):250kW::ENA/CA; (D):250kW::ENA/CA; M-F:250kW::CA; (D)Su/M:250kW::ENA/CA; (J)Su/M:250kW::ENA/CA; (D)Tu-Sa:250kW::ENA/CA; (J)Tu-Sa:250kW::ENA/CA
	CHINA (PR) — YUNNAN PEOPLES BS, Kunming	50kW:DS-2
	CLANDESTINE (M EAST) — "VO THE CRUSADER", Via Radio Baghdad	MOJAHEDIN-E KHALQ:ME
(con'd)	**GERMANY (FR)** — DEUTSCHE WELLE, Jülich	100kW::EU

Legend: ENGLISH ▬ · GERMAN / DEUTSCH ◊◊◊◊ · FRENCH / FRANÇAIS ═ · PORTUGUESE / PORTUGUÊS ≡ · SPANISH / ESPAÑOL ▬ · ARABIC / عربي · RUSSIAN / РУССКИЙ ▬ · CHINESE / 中文 ▭▭ · JAPANESE / 日本語 ▬ · MULTILINGUAL ▦ · OTHER — · SUMMER ONLY (J) · WINTER ONLY (D) · JAMMING ∧∧ or / or \ · EARLIEST HEARD ◢ · LATEST HEARD ◣ · † TENTATIVE

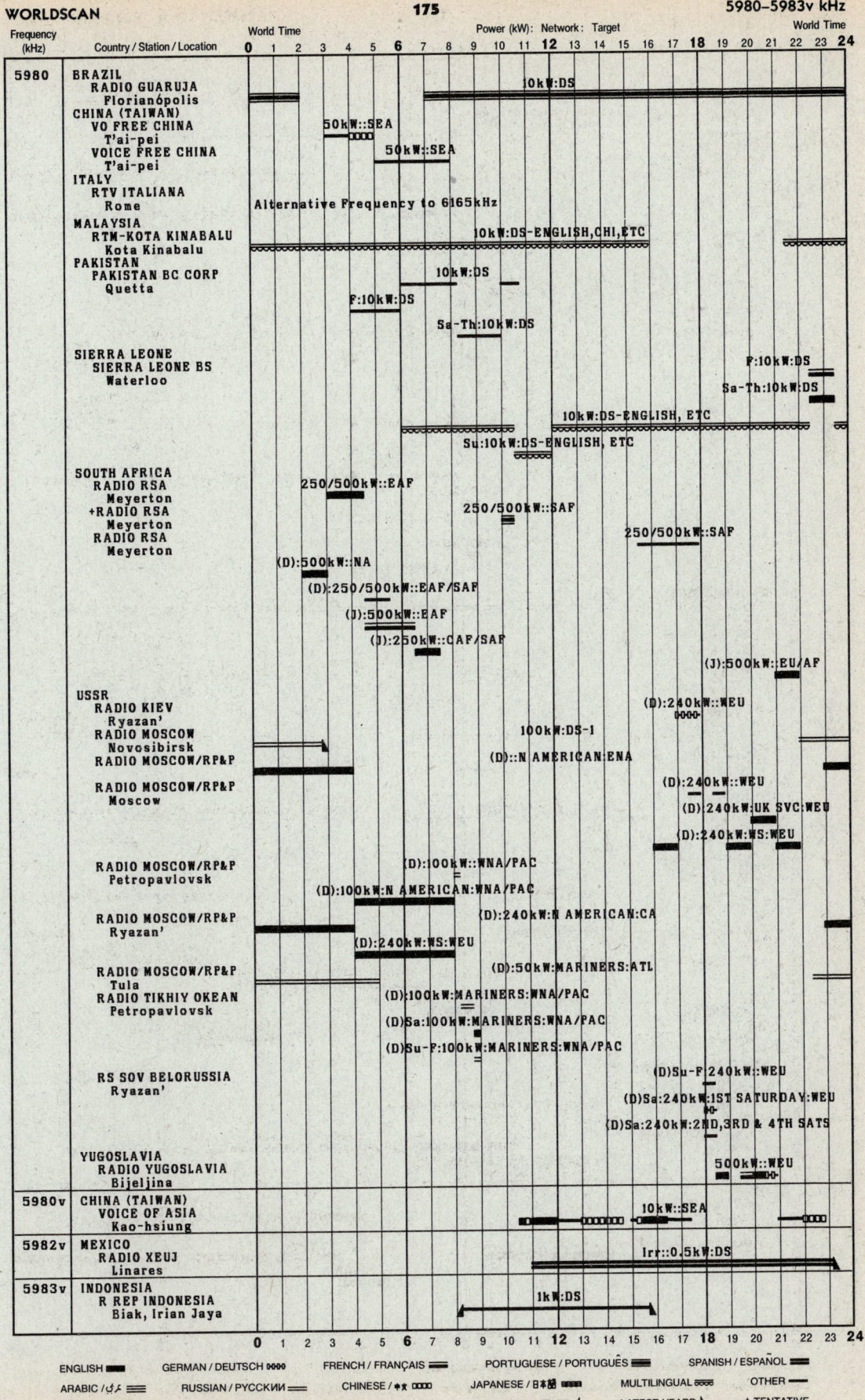

5985–5990 kHz

Frequency (kHz)	Country / Station / Location	Schedule
5985	**BOLIVIA** – RADIO SAN JOSE, Oruro	1kW:DS–TEMP INACTIVE
	BURMA – BURMA BC SERVICE, Rangoon	50kW:DS–ENGLISH, BURMESE
	CHINA (TAIWAN) – VOICE FREE CHINA, Via Okeechobee, USA	100kW::ENA; 100kW::ENA/CA; 100kW::CA
	MEXICO – RADIO MEXICO INTL, México City	10kW::CA/NA
	PAPUA NEW GUINEA – R EAST NEW BRITAIN, Rabaul	10kW:TEMP INACTIVE
	TANZANIA – RADIO TANZANIA, Dar es Salaam	AF
	TURKEY – TURKISH RTV CORP, Ankara	(D):250kW::EU
	USA – RFE-RL, Via Germany (FR)	100kW::EEU; (D):250kW::WUSSR
	RFE-RL, Via Lisbon, Portugal	(J):100kW::EEU; (D):50/250kW::EEU; (J):50/250kW::WUSSR; (J)M-F:50/250kW::EEU; (J)Sa/Su:50/250kW::WUSSR
	VOICE OF AMERICA, Via Kaválla, Greece	(D):250kW::WUSSR; (J):250kW::SAS
	WYFR-FAMILY RADIO, Okeechobee, Florida	(J):250kW::WUSSR; 100kW::ENA
	USSR – RADIO TASHKENT, Tashkent	100kW::ME/SAS; (D):100kW::ME/SAS
5990	**ARGENTINA** – RADIO NACIONAL, Viedma	1kW:DS
	BRAZIL – RADIO MEC, Rio de Janeiro	7.5kW:DS
	ETHIOPIA – VO REV ETHIOPIA, Bantu Liben	100kW:DS; M-F:100kW:DS; Sa/Su:100kW:DS; Su:100kW:DS; M-Sa:100kW:DS
	FRANCE – RADIO FRANCE INTL, Issoudun-Allouis	500kW::EEU
	RADIO FRANCE INTL, Via French Guiana	500kW::CA/NA
	GERMANY (FR) – DEUTSCHE WELLE, Jülich	(D):100kW::WUSSR
	DEUTSCHE WELLE, Via Sri Lanka	(D):250kW::SEA
	HOLLAND – RADIO NEDERLAND, Flevoland	(D):500kW::NAF/WAF
	INDIA – ALL INDIA RADIO, Bhopal	10kW:DS; Su:10kW:DS
(con'd)	**ITALY** – RTV ITALIANA, Rome	(D):100kW::ENA/CA

PASSPORT TO WORLD BAND RADIO

Legend: ENGLISH ▬ | GERMAN/DEUTSCH ○○○○ | FRENCH/FRANÇAIS ═══ | PORTUGUESE/PORTUGUÊS ≡≡≡ | SPANISH/ESPAÑOL ▬▬ | ARABIC | RUSSIAN/РУССКИЙ | CHINESE | JAPANESE/日本語 | MULTILINGUAL | OTHER ▬ | SUMMER ONLY (J) | WINTER ONLY (D) | JAMMING ∧∧ or / or \ | EARLIEST HEARD ◢ | LATEST HEARD ◣ | ✦ TENTATIVE

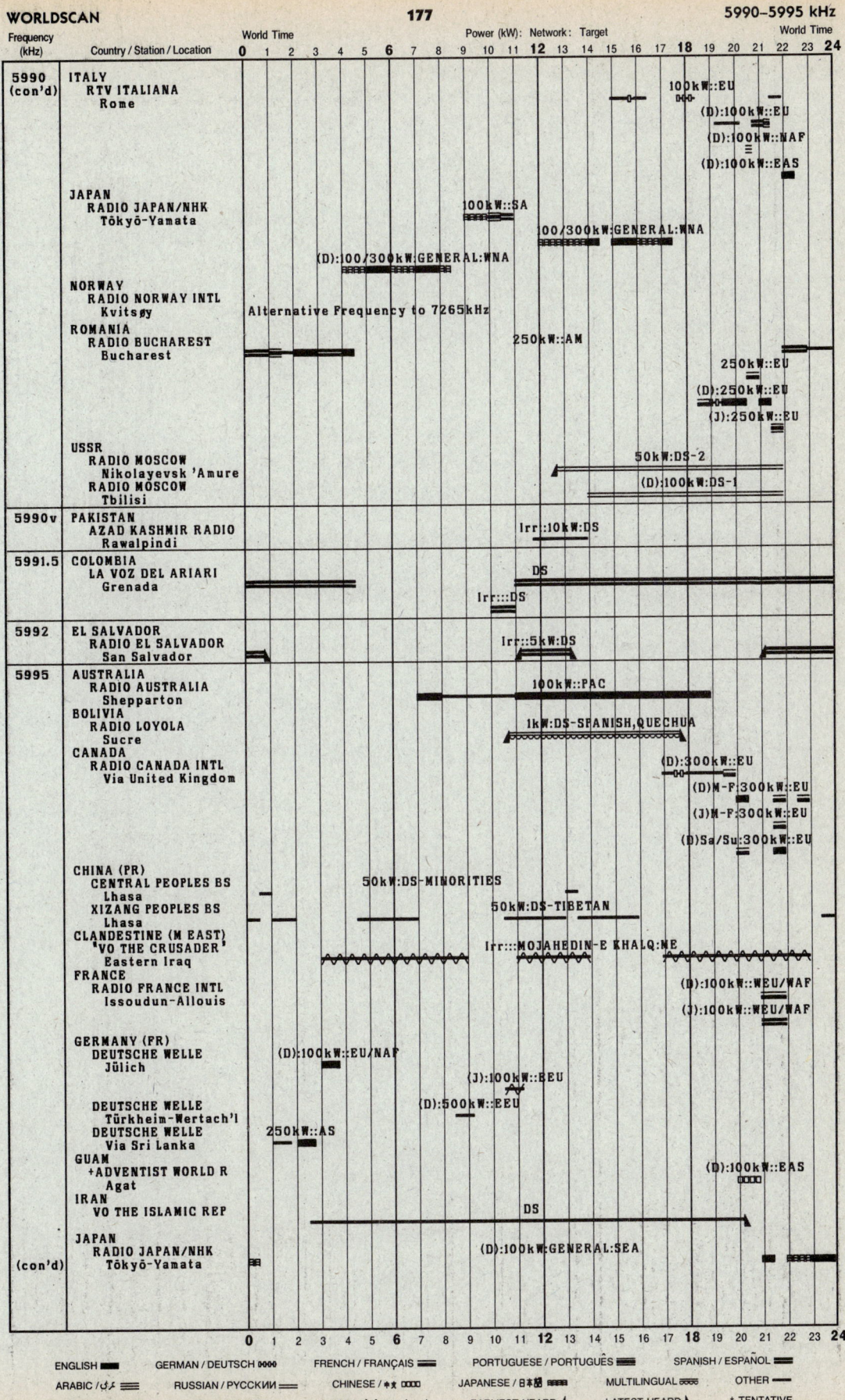

5995–6005 kHz

Frequency (kHz)	Country / Station / Location	Transmission Schedule (World Time, Power, Target)
5995 (con'd)	**MALAWI** — MALAWI BC CORP, Limbe	20/100kW:DS-ENGLISH, ETC; (D):20/100kW:DS-ENGLISH, ETC
	MALI — RTV MALIENNE, Bamako	M-Sa:50kW:DS-FRENCH, ETC; 50kW:DS-FRENCH, ETC
	POLAND — RADIO POLONIA, Warsaw	8kW::EU
	USA — RFE-RL, Via Germany (FR)	(J):100kW::EEU
	VOICE OF AMERICA, Greenville, NC	250/500kW::CA/SA; (J):250kW::EU
	VOICE OF AMERICA, Via Tangier, Morocco	(D):100kW::EU
	WORLD HARVEST R, Noblesville, Indiana	100kW::ENA
5995.3	**ECUADOR** — LA VOZ DE UPANO, Macas	DS
	PERU — RADIO MELODIA, Arequipa	5kW:DS
6000	**AUSTRIA** — RADIO AUSTRIA INTL, Vienna	(J):100kW::SA; 10/50/100kW::EU/NAF/ME; M-F:10/50/100kW::EU/NAF/ME; M-Sa:10/50/100kW::EU/NAF/ME; Sa/Su:10/50/100kW::EU/NAF/ME; Su:10/50/100kW::EU/NAF/ME; (J)Su/M:100kW::SA; (J)Tu-Sa:100kW::SA
	BRAZIL — RADIO GUAIBA, Pôrto Alegre	7.5kW:DS
	CHINA (PR) — VO THE STRAIT-PLA, Fuzhou	50kW:TAIWAN-2
	GERMANY (FR) — DEUTSCHE WELLE, Jülich	100kW:EU/NAF
	DEUTSCHE WELLE, Türkheim-Wertach'l	(D):100kW::SEA; (D):500kW::WUSSR
	DEUTSCHE WELLE, Via Cyclops, Malta	(J):250kW::ME; (J):250kW::NAF/ME
	SINGAPORE — SINGAPORE BC CORP, Jurong	50kW:DS-3
	USA — WORLD HARVEST R, Noblesville, Indiana	100kW::ENA
	USSR — RADIO MOSCOW, Ul'yanovsk	(D):240kW:DS-1
	RADIO MOSCOW/RP&P, Ul'yanovsk	(D):240kW::EEU
6005	**CAMEROON** — RADIO CAMEROON, Buea	Irr::4kW:DS-FRENCH,ENG,ETC; Irr:Su:4kW:DS-FRENCH,ENG,ETC
	CANADA — CFCX-CFCF, Montréal, Québec	0.5kW:DS
	CHINA (PR) — GANSU PEOPLES BS, Lanzhou	Irr::15kW:DS-CHINESE
	GERMANY (FR) — R AMERICAN SECTOR, Berlin	100kW:RIAS-1:EU
(con'd)	**JAPAN** — NIPPON HOSO KYOKAI, Fukuoka-Kasuga	Irr::0.6kW:USB:DS-1(FEEDER)

Legend: ENGLISH; GERMAN/DEUTSCH; FRENCH/FRANÇAIS; PORTUGUESE/PORTUGUÊS; SPANISH/ESPAÑOL; ARABIC; RUSSIAN/РУССКИЙ; CHINESE/中文; JAPANESE/日本語; MULTILINGUAL; OTHER; SUMMER ONLY (J); WINTER ONLY (D); JAMMING; EARLIEST HEARD; LATEST HEARD; † TENTATIVE

WORLDSCAN 6005–6010 kHz

Frequency (kHz)	Country / Station / Location	Schedule
6005 (con'd)	**JAPAN** NIPPON HOSO KYOKAI Matsuyama-Harita	Irr::0.6kW:DS-1
	NIPPON HOSO KYOKAI Nagoya-Nabeta	Irr::0.3kW:USB:DS-1(FEEDER)
	PHILIPPINES FAR EAST BC CO Bocaue	(J):10kW::EAS
	ROMANIA RADIO BUCHAREST Bucharest	(D):250kW:WUSSR
	SRI LANKA SRI LANKA BC CORP Colombo-Ekala	10kW
	UNITED KINGDOM BBC Via Ascension	250kW:WAF; 125kW:AFRICAN SERVICE:AF; Su:125kW:AFRICAN SERVICE:AF; 125kW:WORLD SERVICE:AF; M-Sa:125kW:WORLD SERVICE:AF
	USA KUSW Salt Lake City, Utah	100kW:STARTS 11/87:AM; Su:100kW:STARTS 11/87:AM
	USSR RADIO MOSCOW/RP&P Krasnodar	(D):120kW::ME
6006	**COSTA RICA** RADIO RELOJ San José	3kW:DS
6008v	**MEXICO** RADIO MIL México City	Irr::5kW:DS
6009	**CLANDESTINE (AFRICA)** "RADIO BARDAI" Sabrātah, Libya	500kW:ANTI-CHAD:CAF; Irr:500kW:ANTI-CHAD:CAF; Sa/Su:500kW:ANTI-CHAD:CAF
6010	**BRAZIL** R INCONFIDENCIA Belo Horizonte	25kW:DS
	CANADA RADIO CANADA INTL Via United Kingdom	(D):300kW::EEU/WUSSR
	GERMANY (DR) RADIO BERLIN INTL Königswusterhausen	(D):100kW::NA/CA; (D):100kW::SA
	GERMANY (FR) DEUTSCHE WELLE Jülich	(D):100kW::EU/NAF; (J):100kW::EU/NA/CA; (D):100kW::WUSSR; (J):100kW::SEA; (D):250kW::EEU
	DEUTSCHE WELLE Via Sines, Portugal	
	INDIA ALL INDIA RADIO Calcutta	10kW:DS
	SOUTH AFRICA RADIO RSA Meyerton	250kW::NA
	UNITED KINGDOM BBC Multiple Locations	250/500kW:WORLD SERVICE:EU
	BBC Skelton, Cumbria	250kW::NEU/NAF; (D):250kW::EU
	BBC Via Maşirah, Oman	100kW::SAS
	USA KGEI-VO FRIENDSHIP Redwood City, Ca	(D):250kW::USSR
	WORLD HARVEST R Noblesville, Indiana	100kW::ENA
(con'd)	**USSR** RADIO MOSCOW Novosibirsk	100kW:DS-1

Legend: ENGLISH ▬ GERMAN / DEUTSCH ◊◊◊◊ FRENCH / FRANÇAIS ≡ PORTUGUESE / PORTUGUÊS ▬ SPANISH / ESPAÑOL ▬
ARABIC / عربي ≡ RUSSIAN / РУССКИЙ ▬ CHINESE / 中文 ◊◊◊◊ JAPANESE / 日本語 ▬ MULTILINGUAL ▬ OTHER ▬
SUMMER ONLY (J)   WINTER ONLY (D)   JAMMING ∧∧ or / or \   EARLIEST HEARD ◄   LATEST HEARD ►   + TENTATIVE

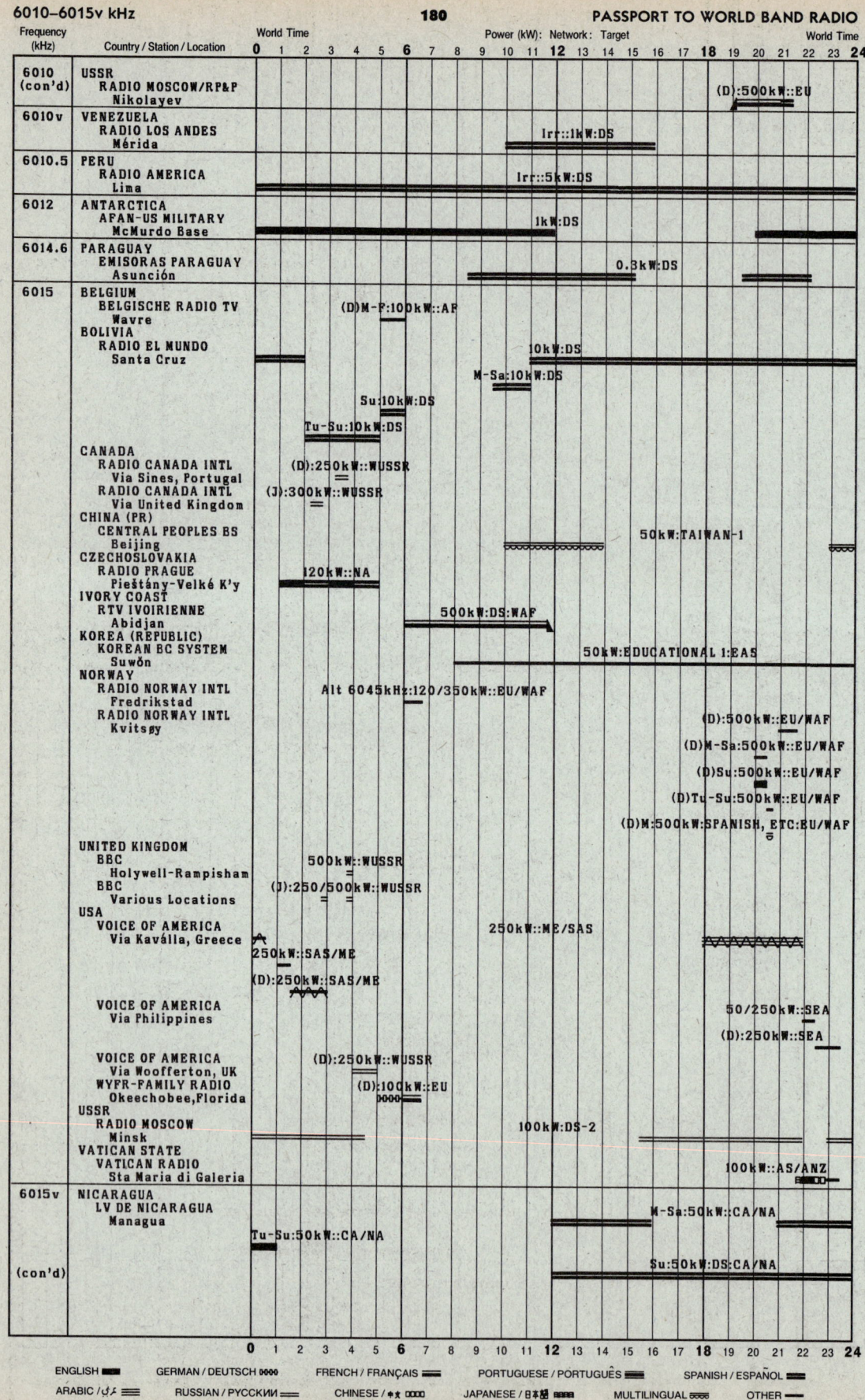

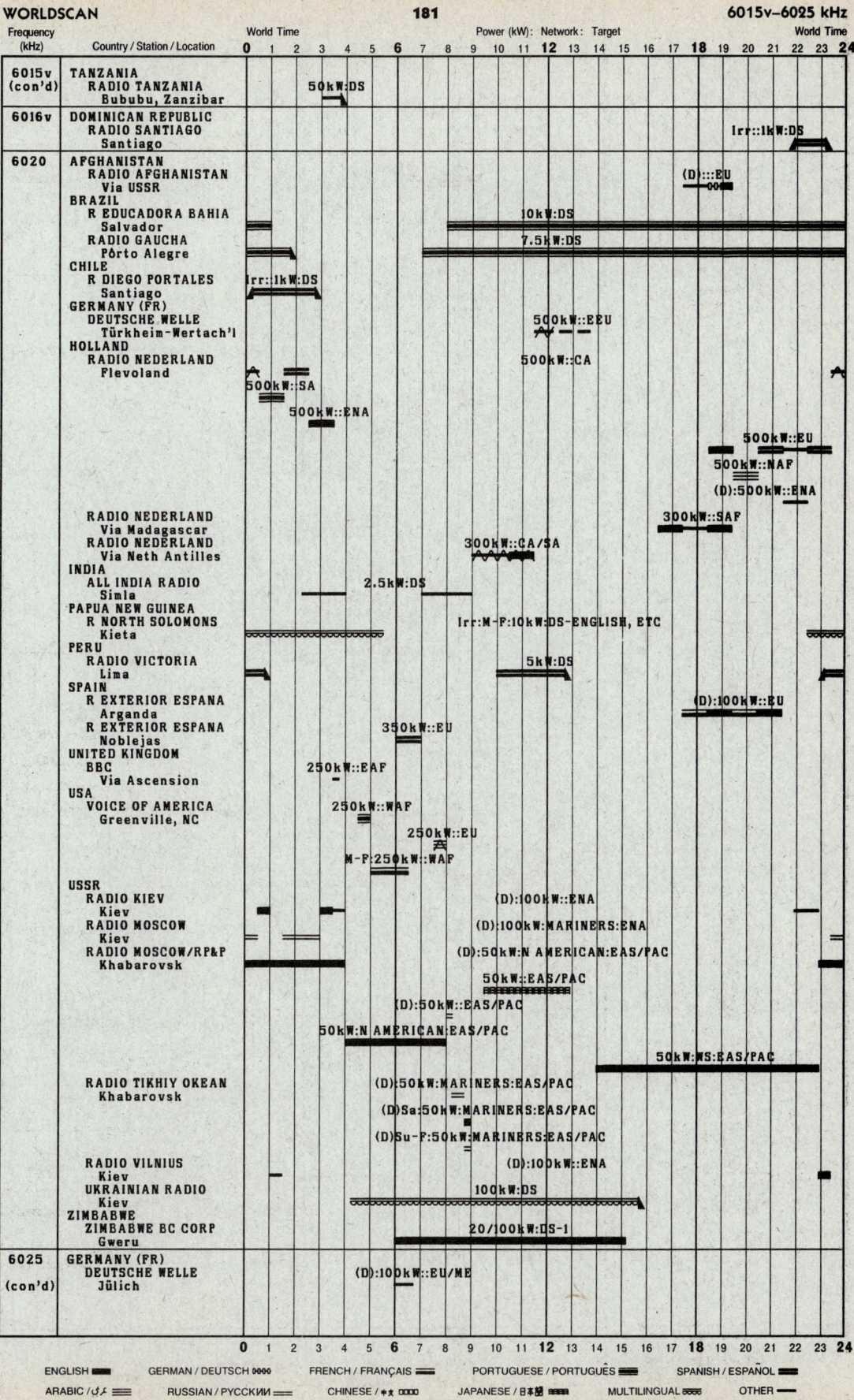

6025–6030 kHz

Frequency (kHz)	Country / Station / Location	Schedule
6025 (con'd)	**GERMANY (FR)** DEUTSCHE WELLE — Jülich	(D):100kW::EEU (19–22)
	DEUTSCHE WELLE — Via Cyclops, Malta	250kW::EU/NAF/ME (3–6)
	HUNGARY RADIO BUDAPEST — Jászberény	250kW::SA (11–13); 250kW::NA; M:250kW::SA; M:250kW::NA; M/W/Sa:250kW::NA; Th/F/Su-Tu:250kW::NA; Tu-Su:250kW::SA; Tu-Su:250kW::NA; Tu/F:250kW::SA; W/Sa:250kW::NA
	RADIO KOSSUTH — Jászberény	250kW:DS:EU
	MALAYSIA RADIO MALAYSIA — Kajang	100kW:DS-CHINESE; Sa/Su:100kW:DS-CHINESE
	NIGERIA FEDERAL RADIO CORP — Enugu	10kW:DS-ENGLISH, ETC
	PARAGUAY RADIO NACIONAL — Asunción	0.6/2kW:DS-SPANISH,GUARANI; Irr:Su:0.6/2kW:DS
	USA VOICE OF AMERICA — Via Ismaning, GFR	(J):100kW::WUSSR
	VOICE OF AMERICA — Via Woofferton, UK	(D):300kW::WUSSR
	USSR RADIO TASHKENT — Tashkent	(J):100kW::ME
6025v	**MOZAMBIQUE** EP DE SOFALA — Beira	10kW:DS
6026	**COLOMBIA** ECOS DE COMBEIMA — Ibagué	5kW:DS
6029	**ANTARCTICA** R NACIONAL-LRA36 — Base Esperanza	1.5kW:DS
6029.4	**CHILE** RADIO SANTA MARIA — Coihaique	1kW:DS
6030	**CANADA** CFVP-CFCN — Calgary, Alberta	0.1kW:DS
	RADIO CANADA INTL — Via United Kingdom	(D):300kW::EEU/WUSSR
	CLANDESTINE (AFRICA) "AV DO GALO NEGRO"	USB UNITA/PORT,ETC:SAF
	CYPRUS CYPRUS BC CORP — Zyyi	F-Su:250kW::WEU
	+CYPRUS BC CORP — Zyyi	(D)F-Su:250kW::WEU
	GERMANY (FR) SUDDEUTSCHER RFUNK — Mühlacker	20kW:DS:EU; M-Sa:20kW:DS:EU
	PHILIPPINES FAR EAST BC CO — Bocaue	50kW::SEA; (D):10kW::EAS
	UNITED KINGDOM BBC — Holywell-Rampisham	250kW::WEU
	BBC — Via Maṣirah, Oman	100kW:TURKISH:ME
(con'd)	BBC — Via Zyyi, Cyprus	100kW::EU

ENGLISH ▬▬ GERMAN / DEUTSCH ◊◊◊◊ FRENCH / FRANÇAIS ═══ PORTUGUESE / PORTUGUÊS ▬▬ SPANISH / ESPAÑOL ▬▬

ARABIC / ﻋﺮﺑﻲ ▬▬ RUSSIAN / РУССКИЙ ═══ CHINESE / 中文 ◊◊◊◊ JAPANESE / 日本語 ▬▬ MULTILINGUAL ◊◊◊◊ OTHER ▬▬

SUMMER ONLY (J)   WINTER ONLY (D)   JAMMING /\/\ or / or \   EARLIEST HEARD ◢   LATEST HEARD ◣   + TENTATIVE

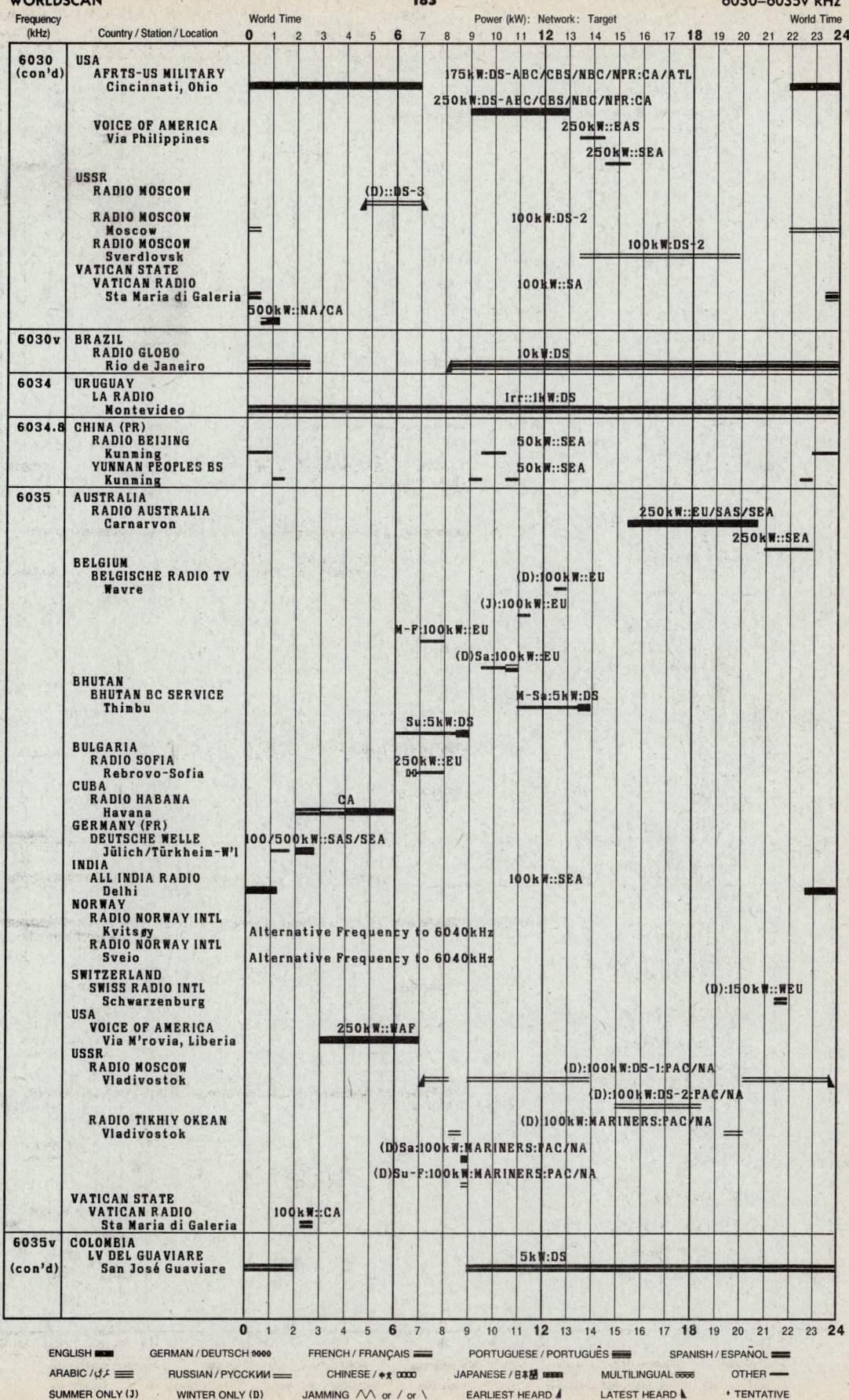

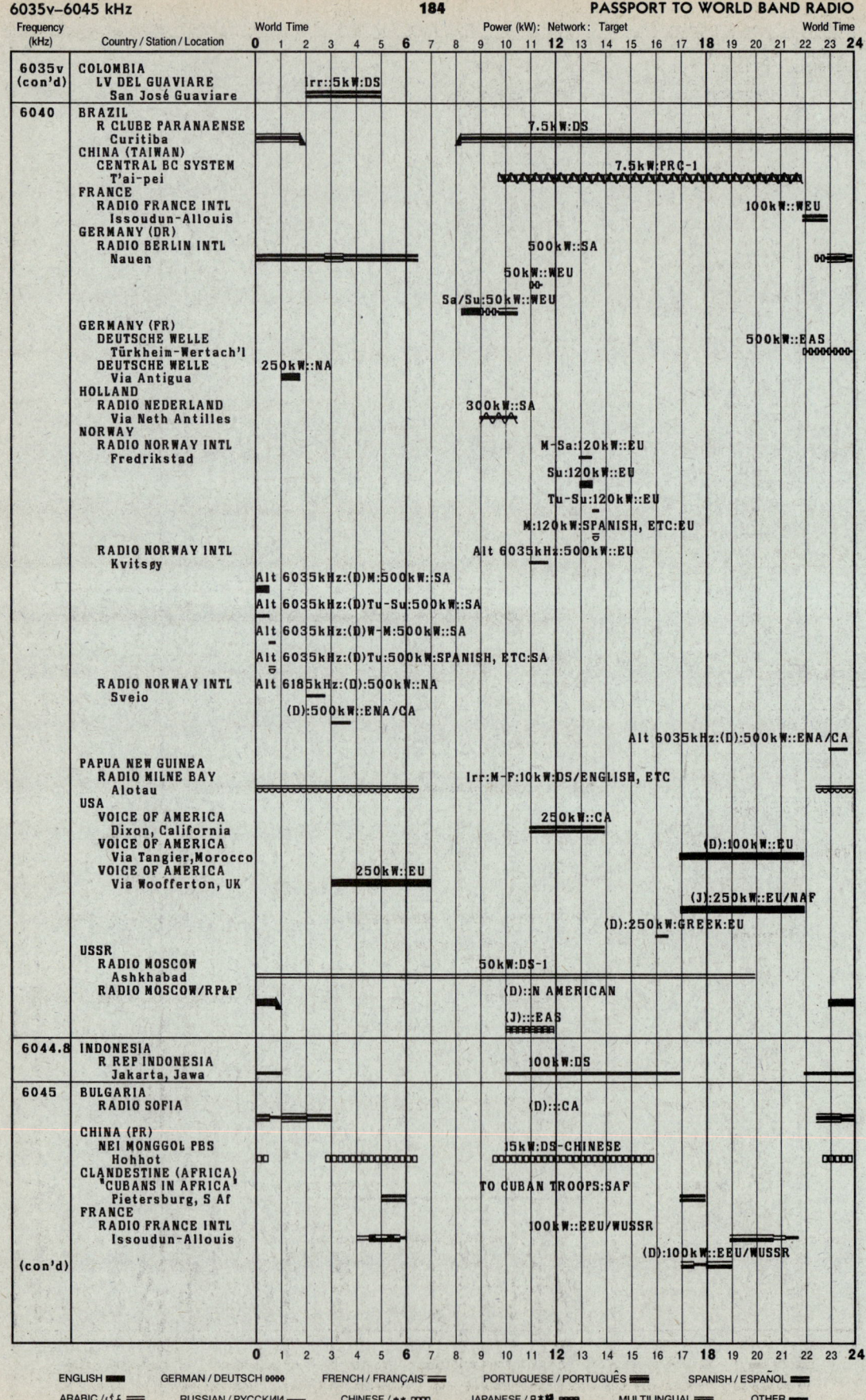

Frequency (kHz)	Country / Station / Location	Schedule (World Time 0–24)	Power (kW) : Network : Target
6045 (con'd)	**GERMANY (FR)**		
	DEUTSCHE WELLE Jülich/Türkheim-W'l	2–3	100/500kW::CA/SA
	DEUTSCHE WELLE Via Sackville, Can	4–6	250kW::NA
	INDIA		
	ALL INDIA RADIO Delhi	0–1	100kW:DS
		2–5 (Irr)	Irr::100kW:DS-ENGLISH, ETC
	KENYA		
	VOICE OF KENYA Nairobi	3–6	250kW:DS-GENERAL
		2–5 M-Sa	M-Sa:250kW:DS-GENERAL
	NORWAY		
	RADIO NORWAY INTL Fredrikstad		Alternative Frequency to 6015 kHz
	ROMANIA		
	RADIO BUCHAREST Bucharest	17–18	(D):250kW::WUSSR
	SENEGAL		
	ORT DU SENEGAL Tambacounda	8–14	4kW:DS-FRENCH, ETC
	SWEDEN		
	RADIO SWEDEN INTL Hörby	21–22	(D):350kW::ENA/CA
	UNITED KINGDOM		
	BBC Skelton, Cumbria	7–17	100/250kW:WORLD SERVICE:EU
	BBC Via Maṣirah, Oman	0–1	100kW::SAS
	URUGUAY		
	RADIO SPORT Montevideo	0–24	1kW:DS
	USA		
	VOICE OF AMERICA Via M'rovia, Liberia	18–22	250kW::WAF/SAF
	USSR		
	RADIO MOSCOW/RP&P Moscow	4–10	(D):240kW::EEU
	RADIO MOSCOW/RP&P Simferopol'	3–14	(D):240kW:WS, N AMERICAN:ENA
	ZIMBABWE		
	ZIMBABWE BC CORP Gweru	9–17 M-F	M-F:10kW:DS-EDUCATIONAL
6045v	**MEXICO**		
	RADIO UNIVERSIDAD San Luis Potosí	(Irr)	Irr::0.5kW:DS
6045.5	**PERU**		
	RADIO SANTA ROSA Lima	0–5, 9–?	3/10kW:DS
6045.6	**COLOMBIA**		
	RADIO MELODIA Bogotá	(Irr)	Irr::5kW:DS
6049.2	**GUATEMALA**		
	UNION RADIO-AWR Guatemala City		Alternative Frequency to 6090.5 kHz
6050	**BELGIUM**		
	RT BELGE FRANCAISE Wavre	12–13	100kW::EU
	BRAZIL		
	RADIO GUARANI Belo Horizonte	0–24	10kW:DS
	CANADA		
	RADIO CANADA INTL Via United Kingdom	5–6 M-F (J)	(J)M-F:100/300kW::EU
	ECUADOR		
	HCJB-VO THE ANDES Quito	9–16	100kW:DS:AM
	GERMANY (FR)		
	DEUTSCHE WELLE Türkheim-Wertach'l	5–6 (J)	(J):500kW::EEU
	HOLLAND		
	RADIO NEDERLAND Flevoland	13–14	500kW::EU
	INDIA		
	ALL INDIA RADIO Delhi	8–10	50kW:DS-ENGLISH, ETC
	ITALY		
	RTV ITALIANA Rome	18–19	(D):100kW::EEU
	MALAYSIA		
	RTM-SARAWAK Sibu	9–23	10kW:DS-VERNACULARS
	NIGERIA		
(con'd)	FEDERAL RADIO CORP Ibadan	6–24	50kW:DS-ENGLISH, ETC

Legend: ENGLISH, GERMAN/DEUTSCH, FRENCH/FRANÇAIS, PORTUGUESE/PORTUGUÊS, SPANISH/ESPAÑOL, ARABIC, RUSSIAN/РУССКИЙ, CHINESE, JAPANESE/日本語, MULTILINGUAL, OTHER. SUMMER ONLY (J), WINTER ONLY (D), JAMMING ∧∧ or / or \, EARLIEST HEARD ◢, LATEST HEARD ◣, + TENTATIVE

6050–6060 kHz

Frequency (kHz)	Country / Station / Location	Schedule
6050 (con'd)	**UNITED KINGDOM** BBC, Holywell-Rampisham	500kW::EEU/ME (approx 15–22)
	BBC, Multiple Locations	100/500kW:WORLD SERVICE:EU (approx 02–08)
	BBC Via Zyyi, Cyprus	100kW:WORLD SERVICE:ME (approx 00–05)
	USA RFE-RL Via Germany (FR)	100kW::WUSSR (approx 21–24); (J):100/250kW::WUSSR (approx 10–17)
	RFE-RL Via Lisbon, Portugal	(D):50/250kW::EEU/WUSSR (approx 00–06)
	USSR RADIO MOSCOW/RP&P, Khabarovsk	(D):100kW::EUSSR/WNA (approx 12–20); (J):100kW::EUSSR/WNA (approx 11–15); 100kW:WS, JAPANESE:EUSSR/WNA (approx 11–17)
6055	**CANADA** RADIO CANADA INTL Via Tōkyō, Japan	50kW::EAS/PAC (approx 09–12)
	CZECHOSLOVAKIA RADIO PRAGUE, Pieštány-Velké K'y	250kW:ENG,FR,GERMAN,ETC:WEU/ENA/CA (approx 10–15); 250kW::WEU/ENA/CA (03–06, 17–23)
	RADIO PRAGUE, Prague	400kW:EU (approx 10–13); Sa/Su:400kW::EU (approx 13–15); 400kW:ENG,FR,GERMAN,ETC:EU (approx 05–10)
	FRANCE RADIO FRANCE INTL, Issoudun-Allouis	500kW:CA/SA (approx 01–03); (D):500kW::CA/SA (approx 16–21)
	RADIO FRANCE INTL Via French Guiana	500kW:NA/CA (approx 03–05); (J):500kW:SA (approx 01–04)
	JAPAN NIHON SHORTWAVE BC, Tōkyō-Nagara	50kW:DS-1:EAS/PAC (approx 08–18); M-F:50kW:DS-1:EAS/PAC (approx 18–22); Su-F:50kW:DS-1:EAS/PAC (approx 15–18)
	KUWAIT RADIO KUWAIT, Jadâdiyah	Irr::250kW:DS-RAMADAN:ME (approx 02–05, 22–24); 250kW:DS-MAIN PROGRAM:ME (approx 05–22)
	ROMANIA RADIO BUCHAREST, Bucharest	(D):250kW::EU (approx 16–18)
	RWANDA R REP RWANDAISE, Kigali	50kW:DS-FRENCH, ETC (approx 08–12); Sa/Su:50kW:DS-FRENCH, ETC (approx 10–13); Su:50kW:DS-FRENCH, ETC (approx 06–09)
	UNITED KINGDOM BBC Via Delano, USA	250kW::CA/SA (approx 01–04)
6055.3	**PERU** RADIO CONTINENTAL, Arequipa	2kW:DS (all hours)
6058v	**CHINA (PR)** SICHUAN PEOPLES BS, Xichang	50kW:DS-CHINESE (approx 10–14, 22–24, 00–05); (D):50kW:DS-CHINESE (approx 11–23)
6060	**ARGENTINA** R ARGENTINA-RAE, Buenos Aires	50kW:TEMP INACTIVE:SA (approx 00–01, 23–24)
	RADIO NACIONAL, Buenos Aires	50kW:DS-TEMP INACTIVE:SA (approx 02–11); Sa/Su:50kW:DS-TEMP INACTIVE:SA (approx 11–14)
	AUSTRALIA RADIO AUSTRALIA, Shepparton	50/100kW::PAC/NA (approx 13–21)
6060 (con'd)	**BRAZIL** RADIO UNIVERSO, Curitiba	10kW:DS (all hours)

Legend:
ENGLISH, GERMAN / DEUTSCH, FRENCH / FRANÇAIS, PORTUGUESE / PORTUGUÊS, SPANISH / ESPAÑOL
ARABIC, RUSSIAN / РУССКИЙ, CHINESE / 中文, JAPANESE / 日本語, MULTILINGUAL, OTHER
SUMMER ONLY (J) • WINTER ONLY (D) • JAMMING • EARLIEST HEARD • LATEST HEARD • † TENTATIVE

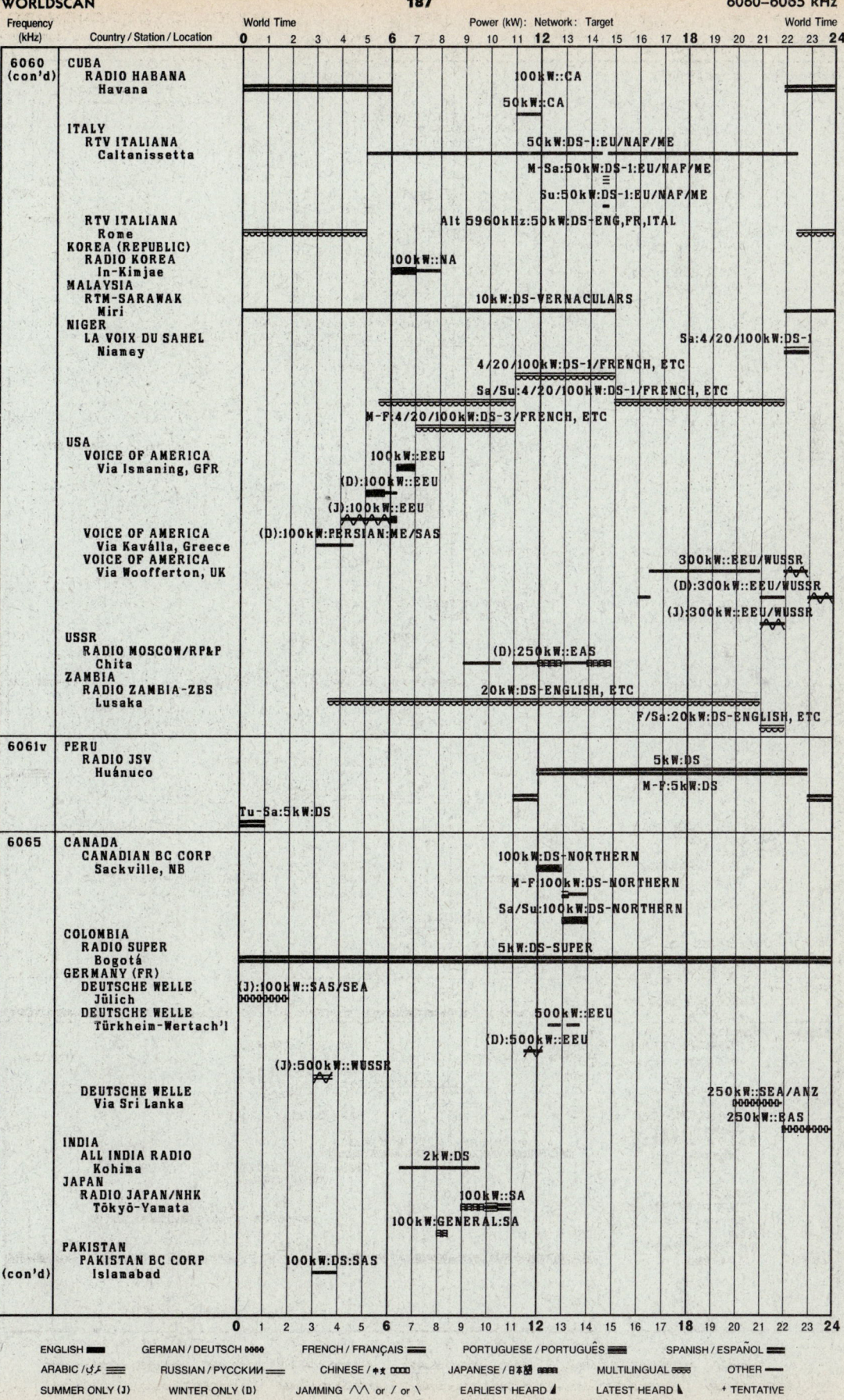

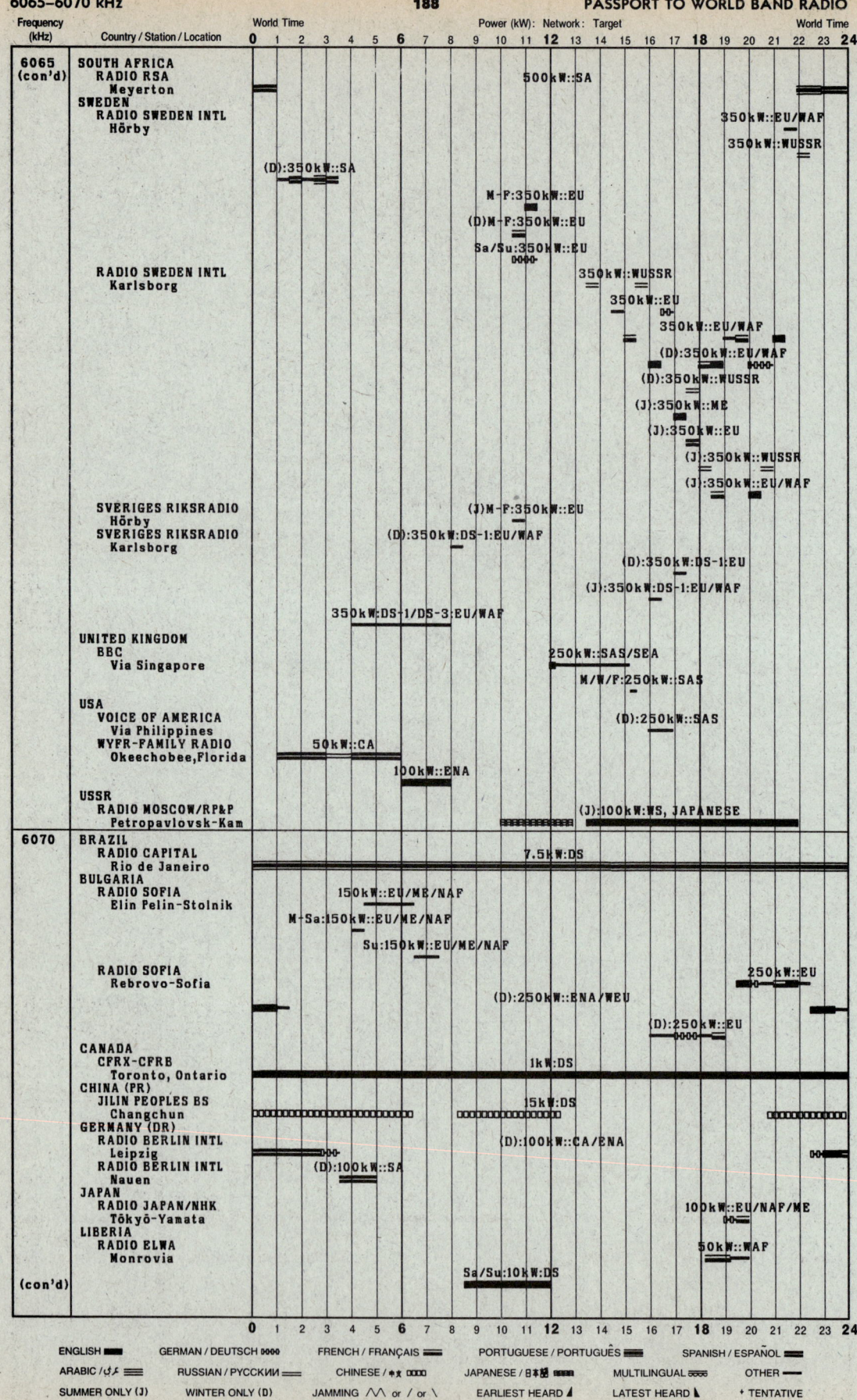

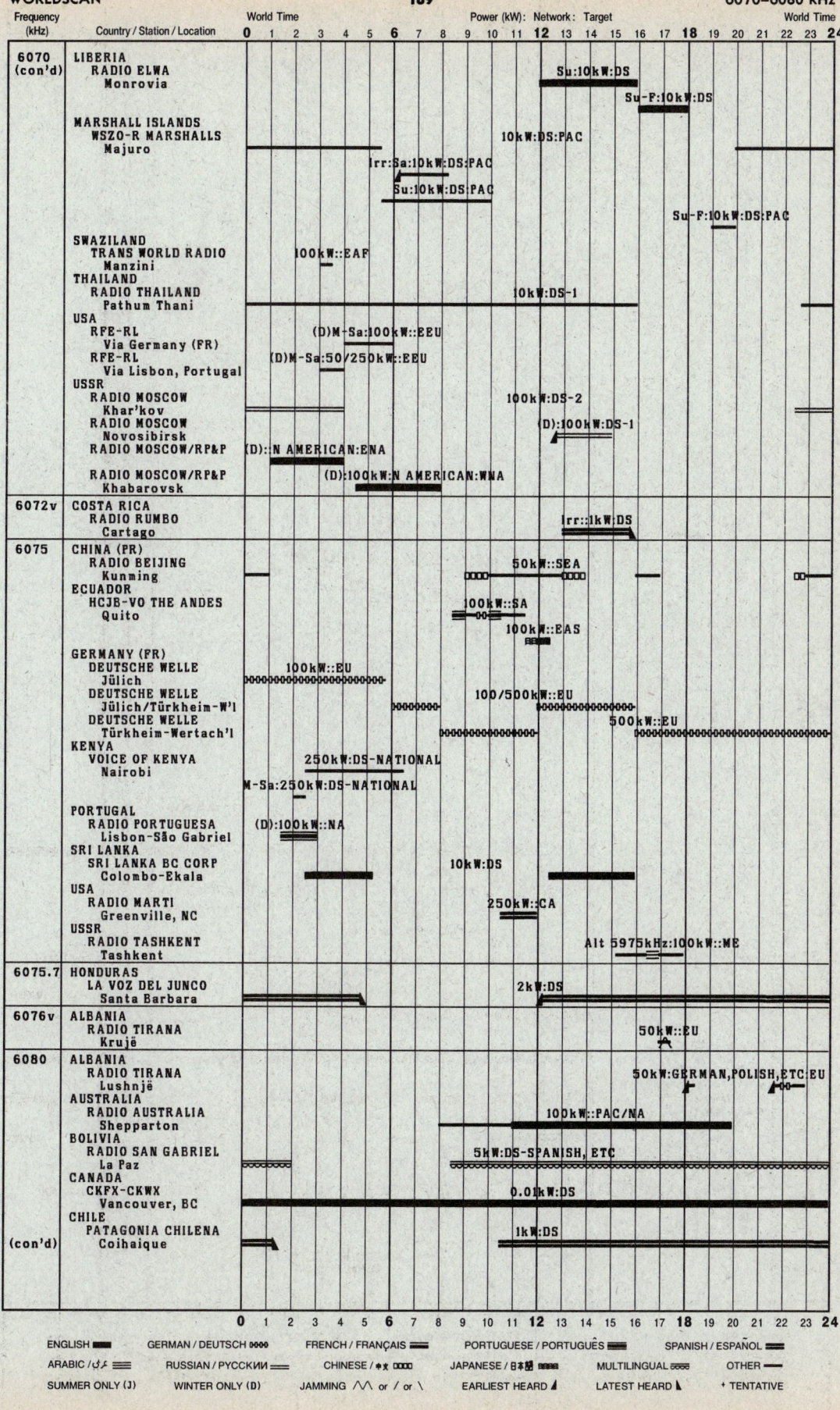

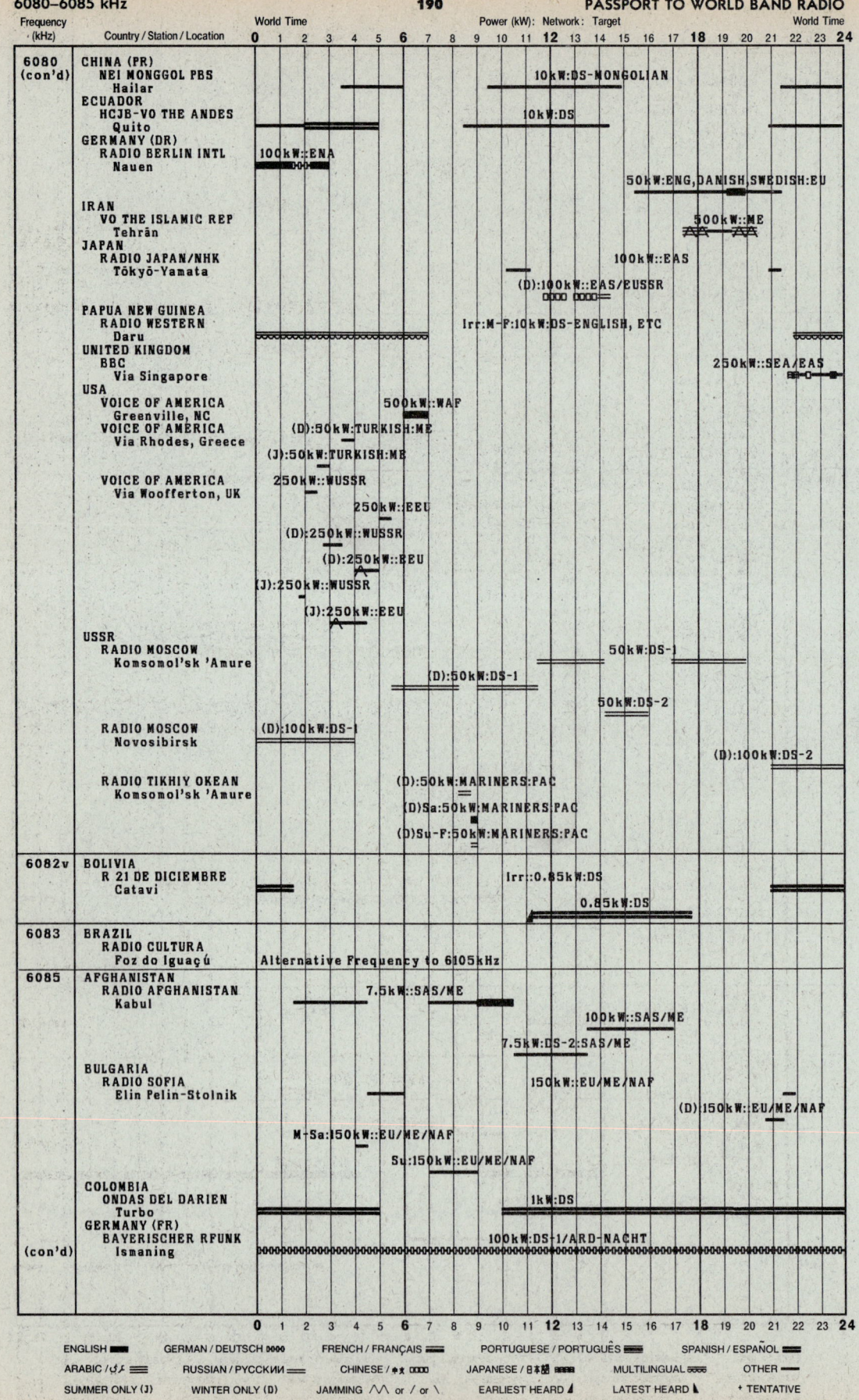

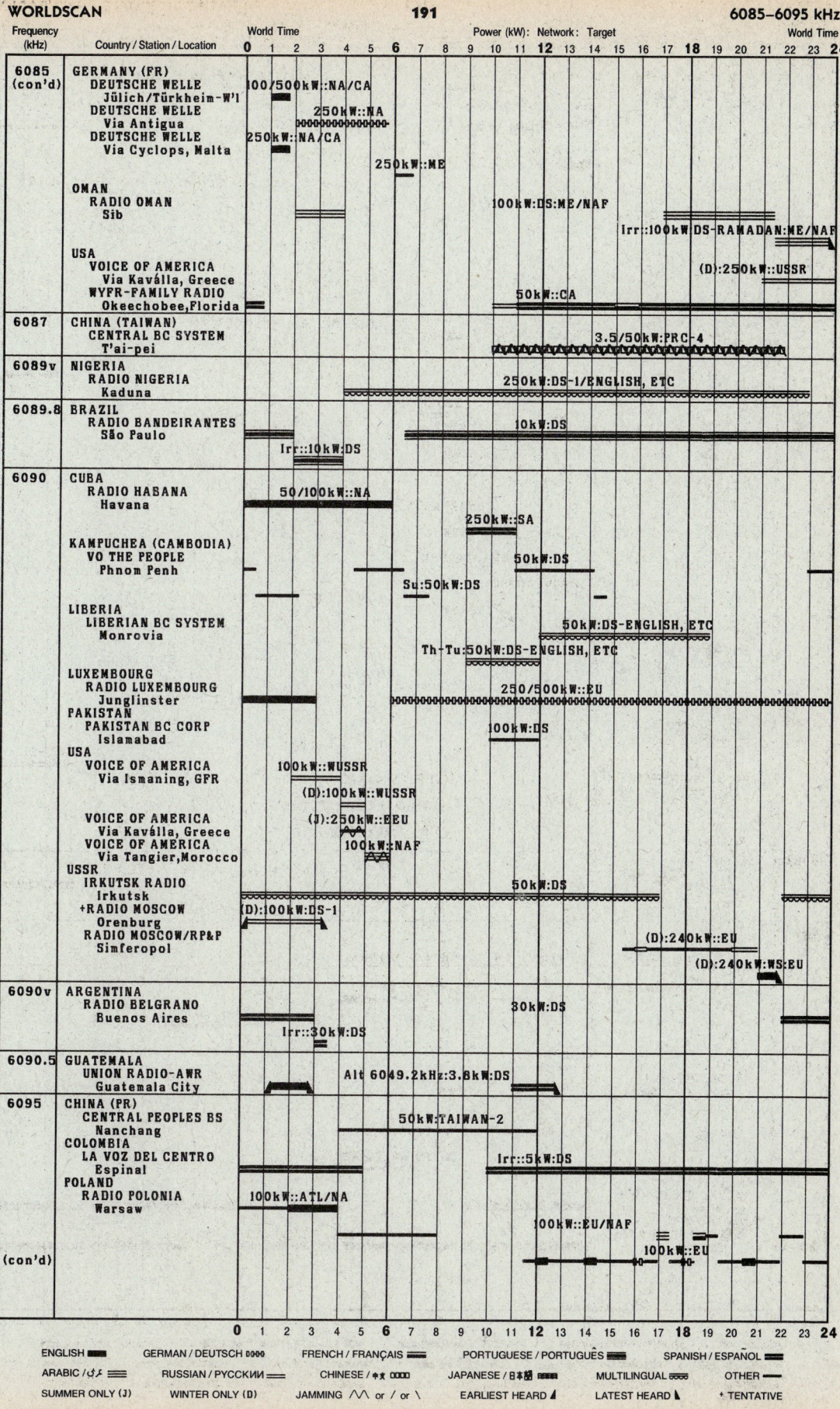

6095–6100 kHz

Frequency (kHz)	Country / Station / Location	Schedule (World Time 0–24)
6095 (con'd)	**PORTUGAL** — RADIO PORTUGUESA, Lisbon–São Gabriel	100 kW::ENA
	SOMALIA — "VO SOMALI ABO LF" via R Mogadishu	50 kW:ANTI-ETHIOPIAN GOV:EAF
	RADIO MOGADISHU, Mogadishu	50 kW::EAF
	USA — KGEI-VO FRIENDSHIP, Redwood City, Ca	(D):250 kW::USSR
	RFE-RL, Via Germany (FR)	(D):100 kW::EEU; (D):100 kW::WUSSR; 100 kW::WUSSR; (D):100 kW::WUSSR
	VOICE OF AMERICA, Via Ismaning, GFR	(J):100 kW::EEU; (J):100 kW::WUSSR
	VOICE OF AMERICA, Via Tangier, Morocco	50/100 kW::NAF
	USSR — RADIO MOSCOW, Kalinin	(D):100 kW:DS-1
	+RADIO MOSCOW, Serpuhkov	(D):100 kW:DS
6099.3	**CHILE** — RADIO CALAMA, Calama	Irr::1 kW:DS
6100	**AFGHANISTAN** — RADIO KABUL, Kabul	100 kW:DS:ME
	ARGENTINA — R PROV SANTA CRUZ, Rio Gallegos	Irr::0.5 kW:DS
	CHINA (PR) — XINJIANG PBS, Urümqi	50 kW:DS-CHINESE
	+XINJIANG PBS, Urümqi	50 kW:DS-CHINESE
	GERMANY (FR) — DEUTSCHE WELLE, Jülich/Türkheim-W'l	100/500 kW::NA/CA
	DEUTSCHE WELLE, Türkheim-Wertach'l	500 kW::NA/CA
	KENYA — VOICE OF KENYA, Nairobi	250 kW:DS-GENERAL; Sa/Su:250 kW:DS-GENERAL
	KOREA (DPR) — KOREAN CENTRAL BS, Pyŏngyang	Alternative Frequency to 6599vkHz
	MALAYSIA — VOICE OF MALAYSIA, Kajang	100 kW:SEA
	NEW ZEALAND — RADIO NEW ZEALAND, Wellington	(J):7.5 kW:PAC
	PORTUGAL — RADIO PORTUGUESA, Lisbon–São Gabriel	(D):100 kW::EU; (D)M-F:100 kW::EU; (D)Sa/Su:100 kW::EU
	UNITED KINGDOM — BBC, Holywell-Rampisham	(D):500 kW::EAF
	USA — VOICE OF AMERICA, Via Philippines	50 kW::SEA
	WORLD HARVEST R, Noblesville, Indiana	100 kW::ENA/EU
	WYFR-FAMILY RADIO, Okeechobee, Florida	(D):100 kW::EU
	USSR — RADIO MOSCOW, Kaunas	50 kW:DS-1:EU
	RADIO MOSCOW/RP&P, Kenga	(D):100 kW::EAS
	RADIO VILNIUS, Kaunas	50 kW::EU
	YUGOSLAVIA — RADIO KOPER, Belgrade	100 kW:DS:EU/NAF/ME
(con'd)		

Legend: ENGLISH ▬▬ GERMAN/DEUTSCH ○○○○ FRENCH/FRANÇAIS ≡≡ PORTUGUESE/PORTUGUÊS ≡≡ SPANISH/ESPAÑOL ≡≡ ARABIC/عربى ≡≡ RUSSIAN/РУССКИЙ ≡≡ CHINESE/中文 ○○○○ JAPANESE/日本語 ▬▬ MULTILINGUAL ○○○○ OTHER ▬▬

SUMMER ONLY (J) · WINTER ONLY (D) · JAMMING ∧∧ or / or \ · EARLIEST HEARD ◢ · LATEST HEARD ◣ · + TENTATIVE

WORLDSCAN   6100–6110 kHz

Frequency (kHz)	Country / Station / Location	Schedule (World Time 0–24)
6100 (con'd)	**YUGOSLAVIA** RADIO LJUBLJANA, Belgrade	M:100kW:DS-ENGLISH,ETC:EU/NAF/ME
	RADIO YUGOSLAVIA, Belgrade	100kW::EU/NAF/ME; (D):100kW::EU/NAF/ME
	RADIO ZAGREB, Belgrade	Th:100kW:DS-MARINERS:EU/NAF
	VARIOUS LOCAL STNS, Belgrade	Tu-Su:100kW:DS:EU/NAF/ME
6100v	**NICARAGUA** LV DE NICARAGUA, Managua	Irr::50kW:DS:CA
	VENEZUELA OBSERVATORIO NAVAL, Caracas	Irr::1kW:DS
6105	**BRAZIL** RADIO CULTURA, Foz do Iguaçú	Alt 6083kHz:5kW:DS
	GERMANY (DR) RADIO BERLIN INTL, Nauen	100kW::WEU
	INDIA ALL INDIA RADIO, Delhi	100kW::SAS
	MEXICO SU PANTERA, Mérida	Irr::0.25kW:DS
	ROMANIA RADIO BUCHAREST, Bucharest	(D):250kW::EU
	SOUTH AFRICA SOUTH AFRICAN BC, Meyerton	Alt 7205kHz:100kW:DS-RADIO ORANJE
	TANZANIA RADIO TANZANIA, Dar es Salaam	AF / DS
	USA RFE-RL Via Germany (FR)	(D):20/250kW::WUSSR; 20/100/250kW::WUSSR; (J):100kW::EEU
		(J)M-Sa:100kW::EEU; (D)Sa/Su:100kW::WUSSR; (J)Su:100kW::WUSSR
	RFE-RL Via Lisbon, Portugal	(D):50/250kW::WUSSR; (J):50/250kW::EEU
		(D)M-F:50/250kW::EEU
	VOICE OF AMERICA Via Rhodes, Greece	50kW::WUSSR
	WYFR-FAMILY RADIO Okeechobee, Florida	100kW::SA
	USSR RADIO MOSCOW, Kalinin	120kW:DS-1; (D):120kW:DS-1
	RADIO MOSCOW/RP&P, L'vov	(D):500kW::ENA/CA
6105.5	**BOLIVIA** RADIO PANAMERICANA, La Paz	10kW:DS; M-Sa:10kW:DS; Tu-Sa:10kW:DS
6105.6	**COSTA RICA** RADIO UNIVERSIDAD, San José	Irr::2kW:DS
6110 (con'd)	**HOLLAND** RADIO NEDERLAND, Flevoland	500kW::WEU/ATL
	HUNGARY RADIO BUDAPEST, Erd-Diósd	100kW::SA; 100kW::NA; 100kW::WEU; M:100kW::SA

ENGLISH ▬ GERMAN / DEUTSCH ◊◊◊◊ FRENCH / FRANÇAIS ▬ PORTUGUESE / PORTUGUÊS ≡ SPANISH / ESPAÑOL ▬
ARABIC / ﻋﺮﺑﻲ ≡ RUSSIAN / РУССКИЙ ▬ CHINESE / 中文 ○○○○ JAPANESE / 日本語 ▬ MULTILINGUAL ≋ OTHER ▬
SUMMER ONLY (J)   WINTER ONLY (D)   JAMMING ∧∧ or / or \   EARLIEST HEARD ◢   LATEST HEARD ◣   + TENTATIVE

6110–6115 kHz

Frequency (kHz)	Country / Station / Location	Schedule
6110 (con'd)	**HUNGARY** RADIO BUDAPEST Érd-Diósd	M:100kW::NA M/W/Sa:100kW::NA Su-Tu/Th/F:100kW::NA Tu-Su:100kW::SA Tu-Su:100kW::NA Tu/F:100kW::SA W/Sa:100kW::NA
	RADIO BUDAPEST Székésfehérvár	20kW::EU M-F:20kW::EU M-Sa:20kW::EU Sa:20kW::EU Sa/Su:20kW::EU Su:20kW::EU Su/M/Th:20kW::EU Tu/F:20kW::EU W/Th/Sa-M:20kW::EU
	INDIA RADIO KASHMIR Srinagar	7.5kW:DS-ENGLISH, ETC
	RADIO KASHMIR Srinigar	Su:7.5kW:DS-ENGLISH, ETC
	MALTA R MEDITERRANEAN Marsaxlokk Cyclops	250kW::EU
	SWEDEN IBRA RADIO Via Cyclops, Malta	(D):250kW::EU (D)M-F:250kW::EU (D)M-W:250kW::EU (D)Sa/Su:250kW::EU (D)Th-Su:250kW::EU
	UNITED KINGDOM BBC Multiple Locations	250kW::AM
	BBC Skelton, Cumbria	(D):250kW::NAF
	USA VOICE OF AMERICA Via Philippines	250kW::SAS/SEA 250kW::EAS
	USSR RADIO BAKU Baku	50kW:ARABIC,PERSIAN,ETC:ME
	RADIO MOSCOW Baku	50kW:DS-2:ME
	RADIO MOSCOW/RP&P Baku	50kW:PERSIAN, ETC:ME
	RADIO MOSCOW/RP&P Novosibirsk	(D):100kW::EAS
6115	**BULGARIA** RADIO SOFIA Plovdiv	500kW::EU (D):500kW::EU Su:500kW::EU
	CANADA RADIO CANADA INTL Via Sines, Portugal	(J):250kW::WUSSR
	FRANCE RADIO FRANCE INTL Issoudun-Allouis	100/500kW::CA
	GERMANY (DR) RADIO BERLIN INTL Königswusterhausen	50kW::EU
	STIMME DER DDR Königswusterhausen	50kW:DS
	JAPAN NIHON SHORTWAVE BC Tōkyō-Nagara	50kW:DS-2:EAS/PAC
(con'd)		

ENGLISH ▬▬ GERMAN / DEUTSCH ○○○○ FRENCH / FRANÇAIS ═══ PORTUGUESE / PORTUGUÊS ▬▬ SPANISH / ESPAÑOL ≡≡≡

ARABIC / عربي ≡ RUSSIAN / РУССКИЙ ═ CHINESE / 中文 □□□□ JAPANESE / 日本語 ▬▬ MULTILINGUAL ∞∞∞ OTHER ━━

SUMMER ONLY (J) WINTER ONLY (D) JAMMING ∧∧ or / or \ EARLIEST HEARD ◂ LATEST HEARD ▸ † TENTATIVE

PASSPORT TO WORLD BAND RADIO

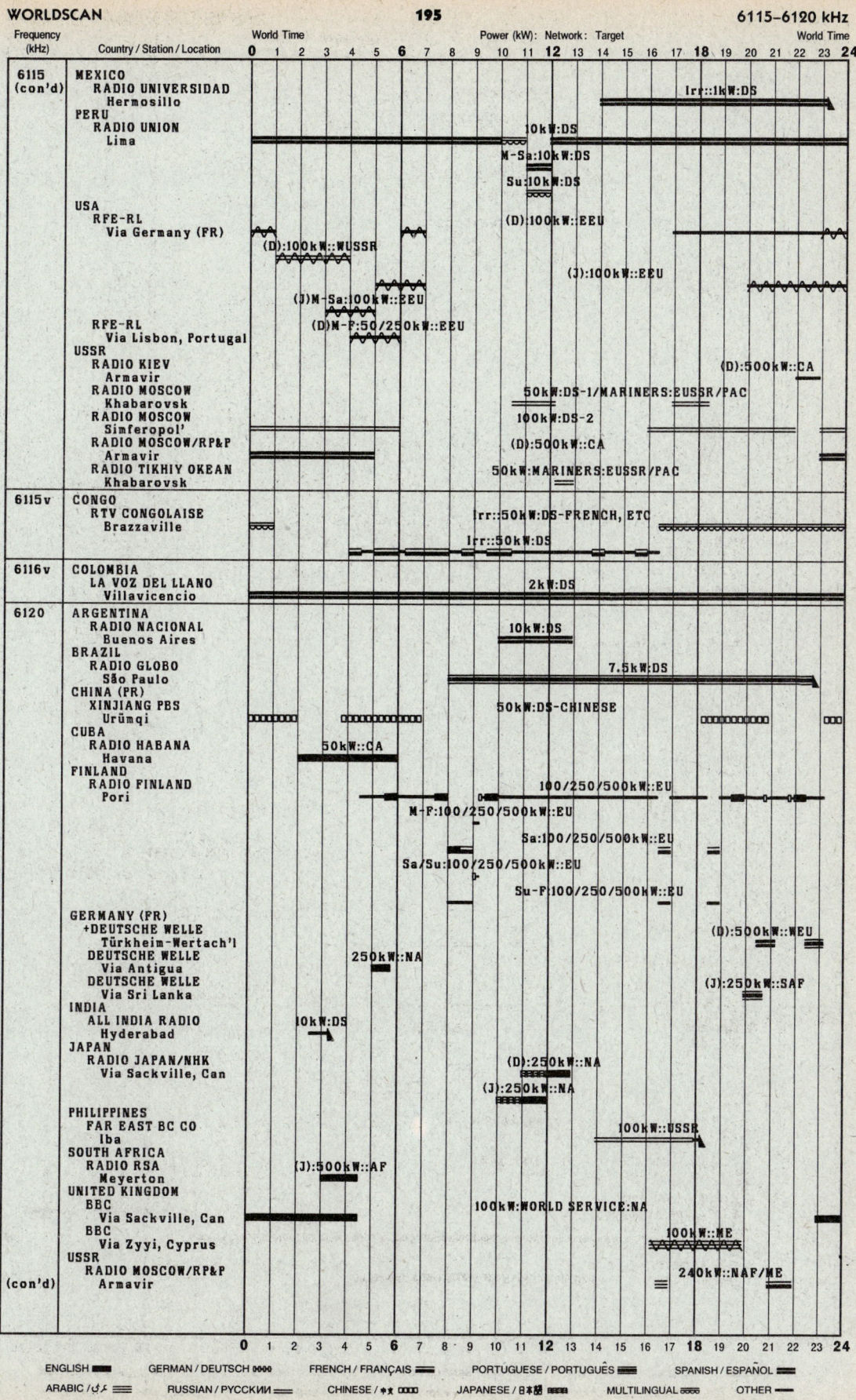

6120–6130 kHz

Frequency (kHz)	Country / Station / Location	Transmission Schedule (World Time 0–24)
6120 (con'd)	**USSR** RADIO MOSCOW/RP&P, Armavir	240kW:WS:NAF/ME (22–24)
	RADIO YEREVAN, Armavir	240kW::NAF/ME (15–19)
6120v	**NICARAGUA** RADIO ZINICA, Bluefields	2kW:DS-SPANISH,ENGLISH (0–5, 11–15)
6125	**AFGHANISTAN** 'R IRAN TOILERS' Via R Afghanistan	100kW:TUDEH COMMUNIST:ME (11–15); (3–4)
	RADIO AFGHANISTAN, Kabul	100kW::ME/SAS (13–15)
	ALBANIA RADIO TIRANA, Lushnjë	50kW:GREEK:EU (3–5)
	CHINA (PR) CENTRAL PEOPLES BS, Shijiazhuang	50kW:DS-1 (0–1, 10–17, 21–24)
	CYPRUS CYPRUS BC CORP, Zyyi	F-Su:250kW::EU (16–17)
	GERMANY (DR) RADIO BERLIN INTL, Nauen	500kW::CA (1–5); (22–24)
	IRAQ RADIO BAGHDAD, Baghdad-Abu Ghraib	(D):250kW::ME (6–11)
	SPAIN R EXTERIOR ESPAÑA, Noblejas	350kW::NA/CA (3–6); (D):350kW::NA/CA (9–15)
	UNITED ARAB EMIRATES VOICE OF THE UAE, Abu Dhabi	Irr (14–18)
	UNITED KINGDOM BBC, Holywell-Rampisham	500kW::WEU (11–13); 500kW::EEU/WUSSR (18–24); Su:500kW::WEU (13); Su:500kW::EEU/WUSSR (14)
	USA AFRTS-US MILITARY, Delano, California	(D):250kW:DS-ABC/CBS/NBC/NPR:EAS/SEA (10–17)
	VOICE OF AMERICA, Cincinnati, Ohio	250kW::WAF (5–7)
	VOICE OF AMERICA, Via Woofferton, UK	300kW::NAF (4–6); (D):250kW::WUSSR (3–5); (J):250/300kW::WUSSR (3–5); (J):250kW:TURKISH,ETC:EEU/ME (4–6)
	USSR +RADIO MOSCOW, Ashkhabad	(D):100kW:DS-1 (11–24)
	+RADIO MOSCOW, Ul'yanovsk	(D):100kW:DS-1 (3–6)
6127	**INDONESIA** R REP INDONESIA, Nabire, Irian Jaya	Alternative Frequency to 5055.4 kHz
6127.5	**INDONESIA** R REP INDONESIA, Nabire, Irian Jaya	0.5kW:DS (5–11)
6130	**CANADA** CHNX-CHNS, Halifax, NS	0.5kW:DS (0–24)
	DENMARK DANMARKS RADIO, Copenhagen	50kW::EU/WAF (9–10)
	ECUADOR HCJB-VO THE ANDES, Quito	100kW::PAC (7–11)
	GERMANY (FR) DEUTSCHE WELLE, Jülich/Türkheim-W'l	100/500kW::NA (4–6)
	DEUTSCHE WELLE, Türkheim-Wertach'l	500kW::EEU (9–12); (D):500kW::WUSSR (5–7)
(con'd)	DEUTSCHE WELLE, Via Cyclops, Malta	(J):500kW::WEU (19–21); (D):250kW::NAF/WAF (18–22)

Legend: ENGLISH ▬ GERMAN/DEUTSCH ○○○○ FRENCH/FRANÇAIS ▬▬ PORTUGUESE/PORTUGUÊS ≡ SPANISH/ESPAÑOL ═ ARABIC ≡ RUSSIAN/РУССКИЙ ══ CHINESE ◊◊◊◊ JAPANESE/日本語 ▬▬ MULTILINGUAL ▭▭ OTHER ▬
SUMMER ONLY (J) WINTER ONLY (D) JAMMING ∧∧ or / or \ EARLIEST HEARD ◢ LATEST HEARD ◣ + TENTATIVE

Frequency (kHz)	Country / Station / Location	Schedule / Power / Network / Target
6130 (con'd)	**GHANA** RADIO GHANA, Accra	100kW::WAF
	JAPAN NIPPON HOSO KYOKAI, Kumamoto-Shimizu	Irr::1kW:DS-1
	PAKISTAN PAKISTAN BC CORP, Islamabad	100kW:DS
	RADIO PAKISTAN, Karachi	(D):50kW::SAS
	PORTUGAL RADIO PORTUGUESA, Lisbon-São Gabriel	M-F:100kW::EU
	SPAIN R EXTERIOR ESPANA, Arganda	100kW::NAF
	SRI LANKA SRI LANKA BC CORP, Colombo-Ekala	10kW:DS
		M-F:10kW:DS
		Sa/Su:10kW:DS
	USA VOICE OF AMERICA, Greenville, NC	500kW:CA/SA
	VOICE OF AMERICA, Via Philippines	250kW::EAS
	VOICE OF AMERICA, Via Woofferton, UK	(D):250kW::EEU
	USSR RADIO MOSCOW, Novosibirsk	(D):100kW:DS-2
	RADIO MOSCOW/RP&P, Moscow	240kW::EU
	RADIO MOSCOW/RP&P, Star'obel'sk	(J):100kW::EU
	RADIO MOSCOW/RP&P, Vladivostok	(D):240kW:N AMERICAN, WS:WNA
6130v	**LAOS** LAO NATIONAL RADIO, Vientiane	10kW:DS
	PAKISTAN RADIO PAKISTAN, Karachi	50kW::SAS
	VENEZUELA R VALLES DEL TUY, Ocumare Del Tuy	Irr::1kW:DS/VERY IRR
6134.3	**BOLIVIA** RADIO SANTA CRUZ, Santa Cruz	1kW:DS
		M-Sa:1kW:DS
6134.4	**BOLIVIA** RADIO SANTA CRUZ, Santa Cruz	M-Sa:1kW:DS
6135	**BRAZIL** RADIO APARECIDA, Aparecida	7.5kW:DS
	BULGARIA RADIO SOFIA, Plovdiv	500kW::ME
		(D):500kW::ME
	KOREA (REPUBLIC) KOREAN BC SYSTEM, Suwŏn	10kW:EDUCATIONAL-2:EAS
	RADIO KOREA, Suwŏn	10kW::EAS
	MADAGASCAR R MADAGASIKARA, Antananarivo	100kW:DS-FRENCH, ETC
	POLAND RADIO POLONIA, Warsaw	100kW::WAF/ATL/SA
		100kW::WEU/NAF
		100kW::EU
	SWITZERLAND RED CROSS BC SVC, Schwarzenburg	Irr:Tu/F:150kW::NA/CA
	SWISS RADIO INTL, Schwarzenburg	150kW::NA/CA
		M:150kW::NA/CA
(con'd)		Tu-Su:150kW::NA/CA

ENGLISH ▬▬ GERMAN / DEUTSCH ○○○○ FRENCH / FRANÇAIS ═══ PORTUGUESE / PORTUGUÊS ≡≡≡ SPANISH / ESPAÑOL ▬▬
ARABIC / عربي RUSSIAN / РУССКИЙ CHINESE / 中文 ○○○○ JAPANESE / 日本語 MULTILINGUAL OTHER
SUMMER ONLY (J) WINTER ONLY (D) JAMMING /\/\ or / or \ EARLIEST HEARD ◢ LATEST HEARD ◣ + TENTATIVE

Frequency (kHz)	Country / Station / Location	Power (kW) : Network : Target
6135 (con'd)	USA RFE-RL Via Germany (FR)	(D):100kW::EEU
	RFE-RL Via Lisbon, Portugal	(D):50/250kW::EEU
	RFE-RL Via Pals, Spain	(D):250kW::WUSSR
	USSR AZERBAIJANI RADIO Baku	100kW:DS-2
	RADIO BAKU Baku	100kW:ARABIC,PERSIAN,ETC:ME
	RADIO MOSCOW Baku	100kW:DS-2
	YEMEN (REPUBLIC) "VO PALESTINE" Via Radio San'ā	50kW:PLO:EAF / Irr::50kW:PLO:EAF
	RADIO SAN'A San'ā	50kW:DS:EAF / F:50kW:DS:EAF
6135.3	SOCIETY ISLANDS RFO-TAHITI Papeete	4kW:DS-FRENCH,TAHITIAN
6140	AUSTRALIA AUSTRALIAN BC CORP Perth	10kW:DS
	CANADA RADIO CANADA INTL Sackville, NB	M-F 250kW::EU
	CHINA (PR) RADIO BEIJING Kunming	50kW::SEA
	COSTA RICA RADIO IMPACTO San José	Alternative Frequency to 6150 kHz
	CUBA RADIO HABANA Havana	100/250kW::NA
	GERMANY (FR) DEUTSCHE WELLE Jülich	100kW::EEU/ME / 100kW::EU/NAF
	DEUTSCHE WELLE Türkheim-Wertach'l	500kW::EEU/ME
	INDIA ALL INDIA RADIO Ranchi	10kW:DS
	PAPUA NEW GUINEA RADIO EAST SEPIK Wewak	Irr:M-F:10kW:DS-ENGLISH, ETC
	PERU RADIO HUAYLLAY Huayllay	1kW:DS / M-Sa:1kW:DS
	SOUTH AFRICA RADIO RSA Meyerton	(J):250kW::SA
	UNITED KINGDOM BBC Holywell-Rampisham	500kW::EEU/WUSSR
	BBC Multiple Locations	250/500kW::EEU/ME
	URUGUAY RADIO MONTE CARLO Montevideo	.5kW:DS
	USA VOICE OF AMERICA Greenville, NC	500kW::CA
	VOICE OF AMERICA Via Kavála, Greece	(D):250kW::USSR
	VOICE OF AMERICA Via Philippines	(J):50kW::SEA
	VOICE OF AMERICA Via Woofferton, UK	300kW::USSR / (D):300kW::USSR / (J):300kW::USSR
	USSR RADIO MOSCOW Voronej	100kW:DS-1
	RADIO MOSCOW/RP&P Zhigulevsk	(D):100kW:WS, SWEDISH, ETC:EU
6140v (con'd)	BURUNDI LA VOIX DE LA REV Bujumbura	25kW:DS

ENGLISH ▬▬ GERMAN / DEUTSCH ◊◊◊◊ FRENCH / FRANÇAIS ═══ PORTUGUESE / PORTUGUÊS ▬▬ SPANISH / ESPAÑOL ═══

ARABIC / عربي ≡≡≡ RUSSIAN / РУССКИЙ ═══ CHINESE / 中文 ◊◊◊◊ JAPANESE / 日本語 ▬▬ MULTILINGUAL ≈≈≈ OTHER ───

SUMMER ONLY (J)   WINTER ONLY (D)   JAMMING ∧∧ or / or \   EARLIEST HEARD ◢   LATEST HEARD ◣   + TENTATIVE

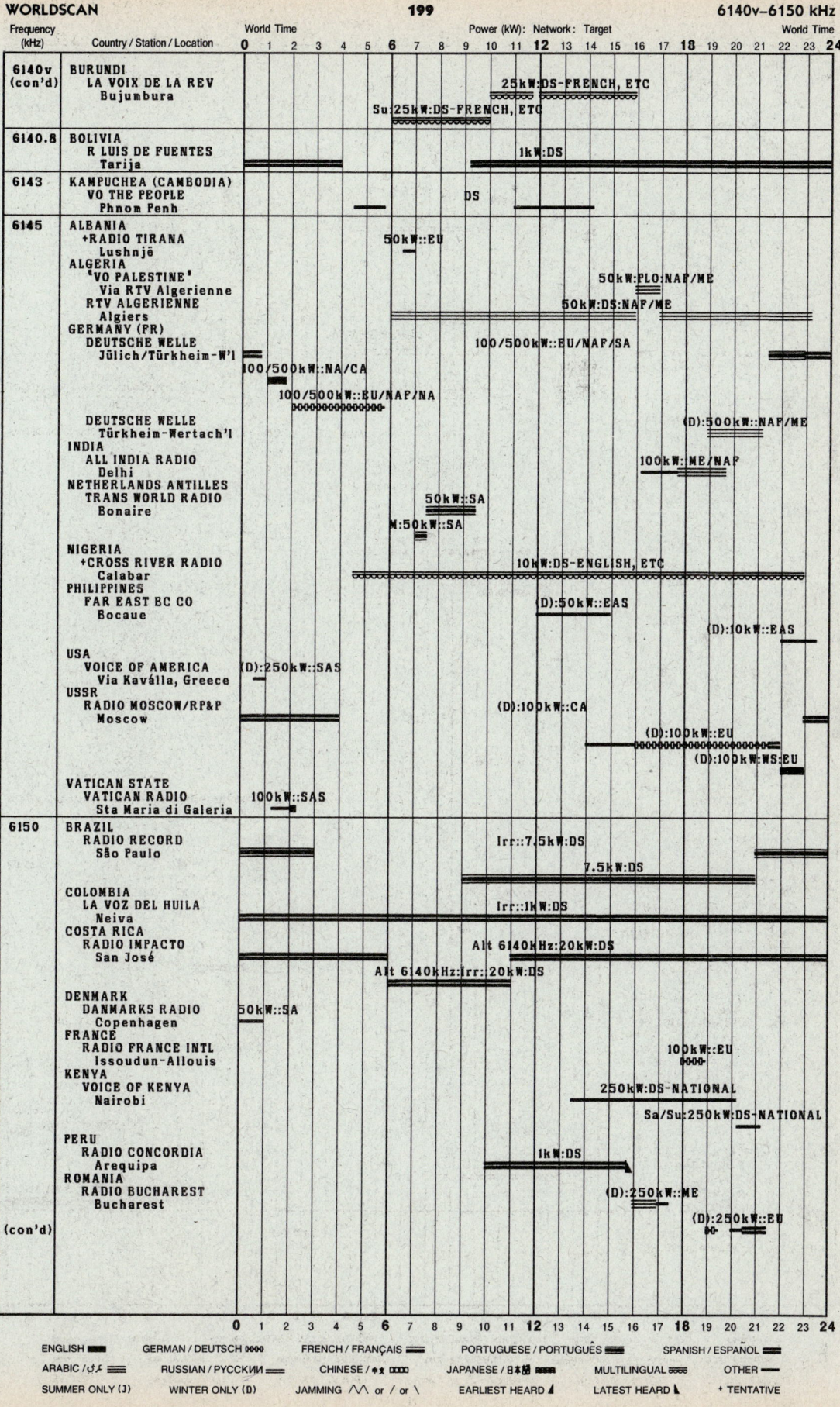

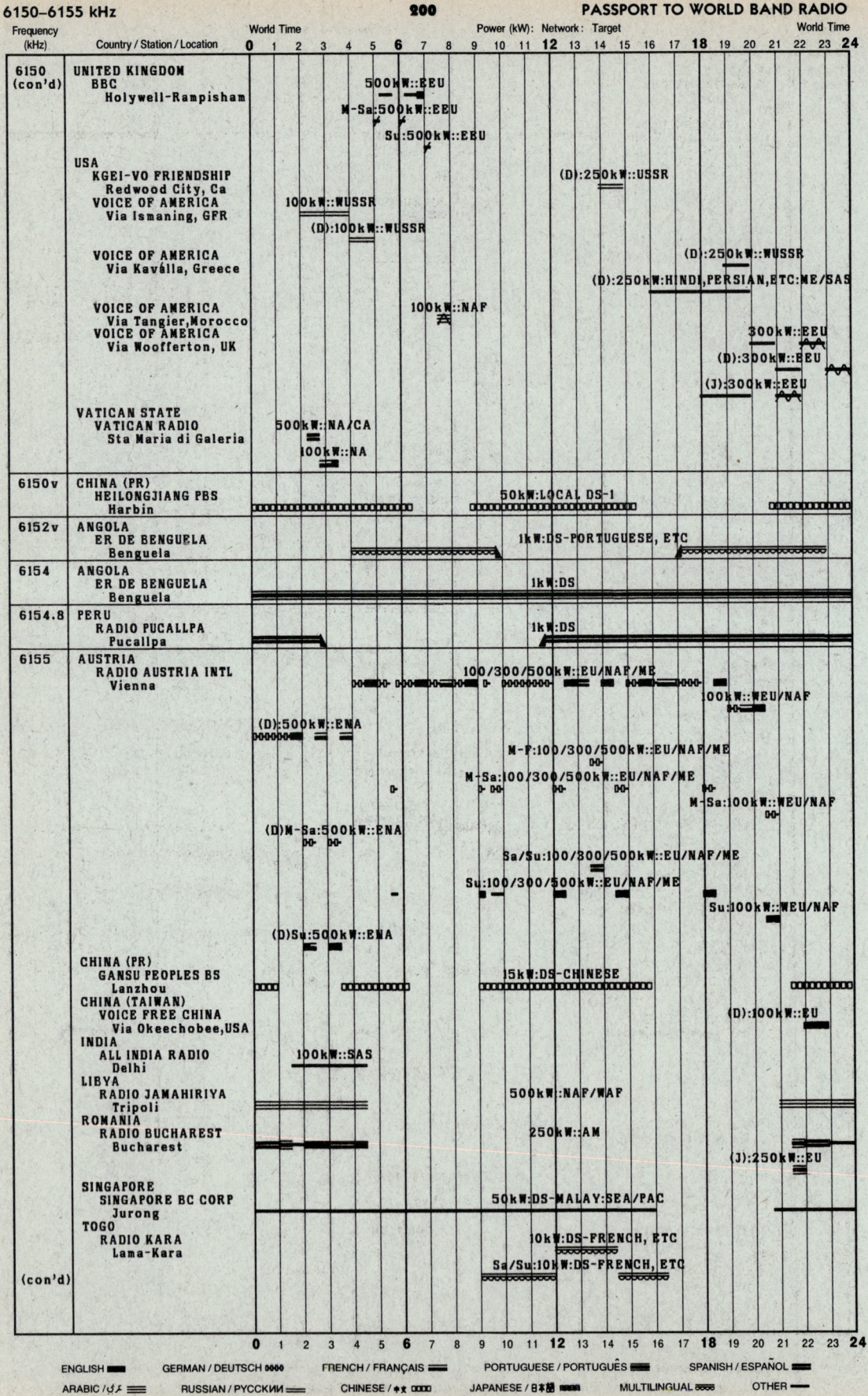

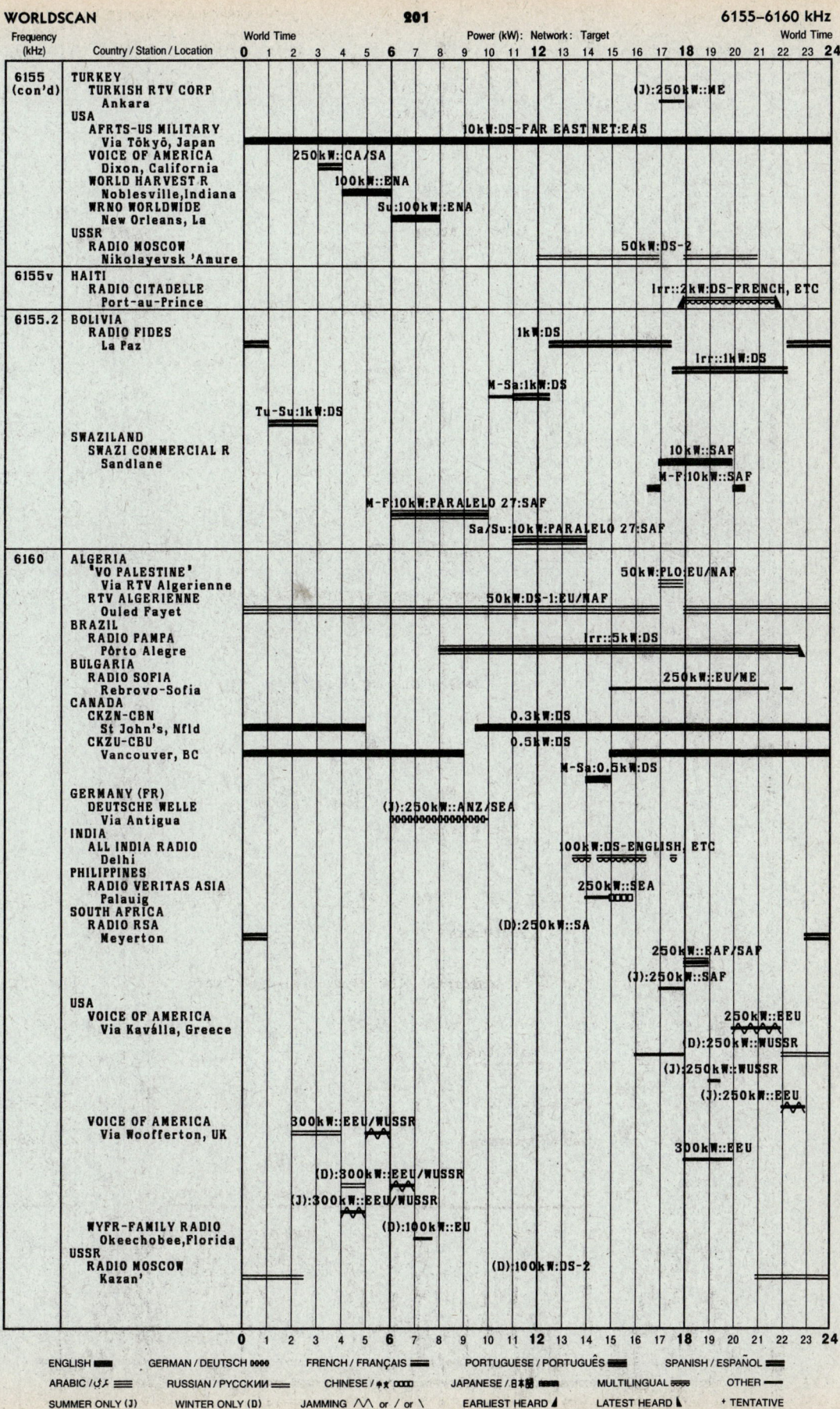

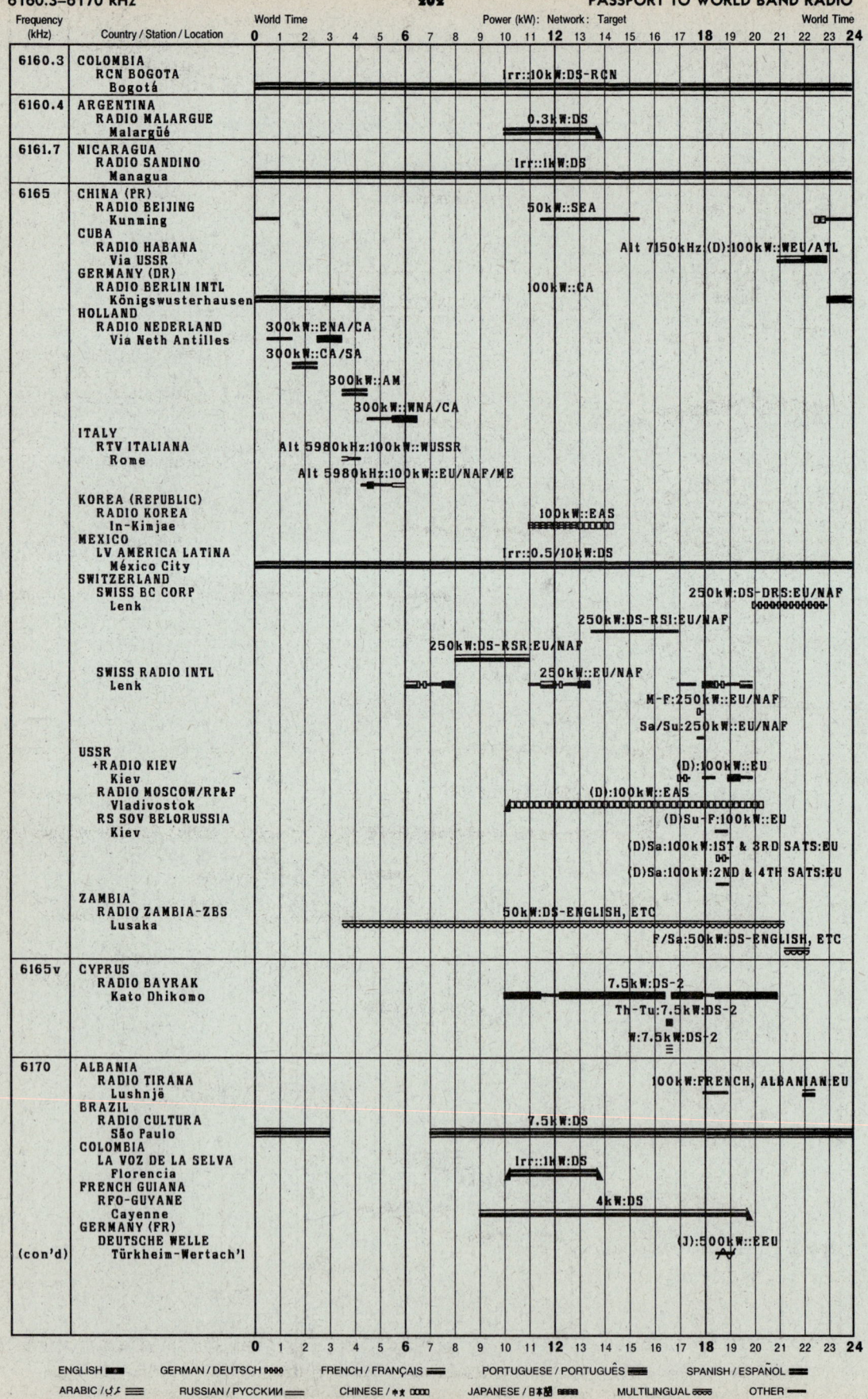

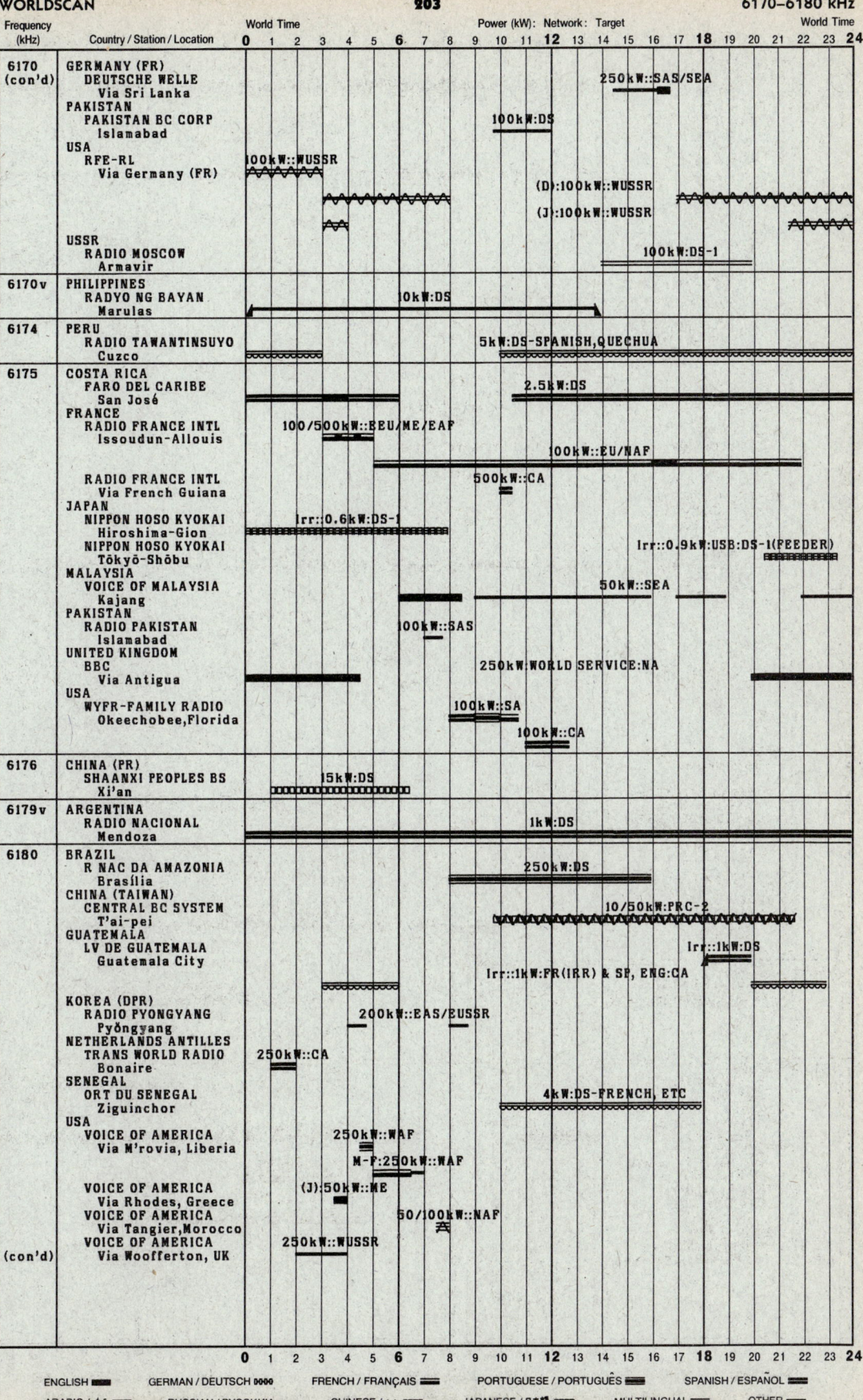

6180–6190 kHz

Broadcast schedule chart showing frequencies 6180–6190 kHz with time (World Time, 0–24) vs. Country/Station/Location, indicating Power (kW), Network, and Target.

Frequency (kHz)	Country / Station / Location	Schedule details
6180 (con'd)	**USSR** — KAZAKH RADIO, Alma-Ata	100kW:DS-1; Sa:100kW:DS-1; Sa-M:100kW:DS-1; Su-F:100kW:DS-1; Tu-F:100kW:DS-1
	+RADIO MOSCOW/RP&P, Tula	(D):100kW::ME
	+RADIO MOSCOW/RP&P, Tula	(D):100kW::NAF/ME
	VENEZUELA — RADIO TURISMO, Valera	1kW:DS
6185	**ALBANIA** — RADIO TIRANA, Lushnjë	50kW::EU
	CHINA (PR) — +RADIO BEIJING, Kunming	(D):50kW::SEA
	GERMANY (FR) — +DEUTSCHE WELLE, Jülich	(D):100kW::NA/CA
	+DEUTSCHE WELLE, Via Cyclops, Malta	(D):250kW::NA/CA
	DEUTSCHE WELLE, Via Sri Lanka	250kW::AS
	LIBYA — RADIO JAMAHIRIYA, Tripoli	Irr::100kW:DS-RAMADAN; 100kW:DS
	MEXICO — RADIO EDUCACION, México City	Irr::1kW:DS
	NORWAY — RADIO NORWAY INTL, Sveio	Alternative Frequency to 6040 kHz
	SRI LANKA — SRI LANKA BC CORP, Colombo-Ekala	10kW:DS-SINHALA
	USA — VOICE OF AMERICA, Via Philippines	250kW::EAS; 250kW::EUSSR
	WRNO WORLDWIDE, New Orleans, La	100kW::ENA; Su:100kW::ENG, RUSSIAN, ETC:ENA
	USSR — RS SOV BELORUSSIA, Ryazan'	Su-F:240kW::EU; Sa:240kW:1ST & 3RD SATS:EU; Sa:240kW:2ND & 4TH SATS:EU
	VATICAN STATE — VATICAN RADIO, Sta Maria di Galeria	100kW::EU; 100kW::EU/WUSSR
6189v	**PERU** — RADIO ORIENTE, Yurimaguas	1kW:DS
6190	**ALBANIA** — RADIO TIRANA, Lushnjë	50kW::WUSSR
	CHINA (PR) — CENTRAL PEOPLES BS, Hohhot	50kW:DS-2; 50kW:DS-MINORITIES
	GERMANY (FR) — DEUTSCHE WELLE, Türkheim-Wertach'l	(D):500kW::WUSSR
	DEUTSCHE WELLE, Via Sines, Portugal	(D):250kW::EEU
	FREIES BERLIN-SFB, Bremen	10kW:DS:EU; Sa:10kW:DS:EU; Su-F:10kW:DS:EU
	RADIO BREMEN, Bremen	Sa:10kW:DS:EU; Su-F:10kW:DS:EU
(con'd)		

Legend: ENGLISH ▬ · GERMAN/DEUTSCH ○○○○ · FRENCH/FRANÇAIS ═ · PORTUGUESE/PORTUGUÊS ≡ · SPANISH/ESPAÑOL ═ · ARABIC ΄϶΄ ≡ · RUSSIAN/РУССКИЙ ═ · CHINESE/中文 □□□□ · JAPANESE/日本語 ▬▬ · MULTILINGUAL ▭▭▭ · OTHER ▬ · SUMMER ONLY (J) · WINTER ONLY (D) · JAMMING ∧∧ or / or \ · EARLIEST HEARD ◢ · LATEST HEARD ◣ · +TENTATIVE

Frequency (kHz)	Country / Station / Location	Schedule
6230 (con'd)	**MONACO** – TRANS WORLD RADIO, Monte Carlo	Su-Th:100kW::WUSSR; W:500kW:EU
	PIRATE (EUROPE) – "EAST COAST COMML", United Kingdom	Irr:Su:::WEU
6230v	**CLANDESTINE (M EAST)** – "R IRAN TOILERS", Afghanistan	TUDEH COMMUNIST:ME
6232	**PIRATE (EUROPE)** – "BRITAIN R INTL", Germany (FR)	Irr:Su::ENGLISH,GERMAN:WEU
	"SUDWEST RADIO", Germany (FR)	Irr:Su::ENGLISH,GERMAN:WEU
6240	**BANGLADESH** – RADIO BANGLADESH, Dhaka	250kW::EU
	PIRATE (EUROPE) – "R RAINBOW INTL", Ireland	Su:0.1kW::WEU
6243v	**PERU** – RADIO MUNICIPAL, Calca	0.15kW:DS
6248v	**VATICAN STATE** – VATICAN RADIO, Vatican City	80kW::EU; Su:80kW:EU; Th:80kW:EU; M-Sa:80kW:ENG,FR,SPANISH,ETC:EU; M-Sa:80kW:ENG,FRENCH,SPANISH
6249.6	**EQUATORIAL GUINEA** – RADIO NACIONAL, Malabo	10kW:DS-SPANISH, ETC
6250v	**KOREA (DPR)** – RADIO PYONGYANG, Pyŏngyang	50/100kW::EAS
6260	**CHINA (PR)** – QINGHAI PEOPLES BS, Xining	10kW:DS-CHINESE
6266	**PIRATE (EUROPE)** – "RADIO ORION", England	Alternative Frequency to 6290kHz
6275	**PIRATE (EUROPE)** – "RADIO WAVES INTL", France	Irr:Su:0.015kW:ENGLISH,FRENCH:WEU
6280	**LEBANON** – VOICE OF HOPE, Marjayoûn	12kW:DS
	PIRATE (EUROPE) – "WESTSIDE R INTL", Ireland	Su:0.08kW::WEU
6280.7	**PERU** – RADIO HUANCABAMBA, Huancabamba	1.5kW:DS; Tu-Su:1.5kW:DS
6285	**PIRATE (EUROPE)** – "W.L.R.", Scotland	Alt 7383.5kHz:Irr:Su:0.06kW::WEU
6290	**CHINA (PR)** – RADIO BEIJING	SEA; SAS; (D)::EU
	PIRATE (EUROPE) – "RADIO ORION", England	Alt 6266kHz:Sa/Su:0.018kW::WEU
6290v	**PIRATE (EUROPE)** – "R KRYSTAL INTL", England	Irr:Su:::WEU
6293	**PIRATE (EUROPE)** – "PIRATE FREAKS BS", Germany (FR)	Su::ENGLISH,GERMAN:WEU
6295v	**CLANDESTINE (AFRICA)** – "VO BROAD MASSES", Argadom, Ethiopia	PLF OF ERITREA:EAF
	PIRATE (EUROPE) – "WEEKEND MUSIC R", Scotland	Alt 6200kHz:Irr:Su:0.05kW::WEU
6297v (con'd)	**PERU** – RADIO CHOTA, Chota	DS

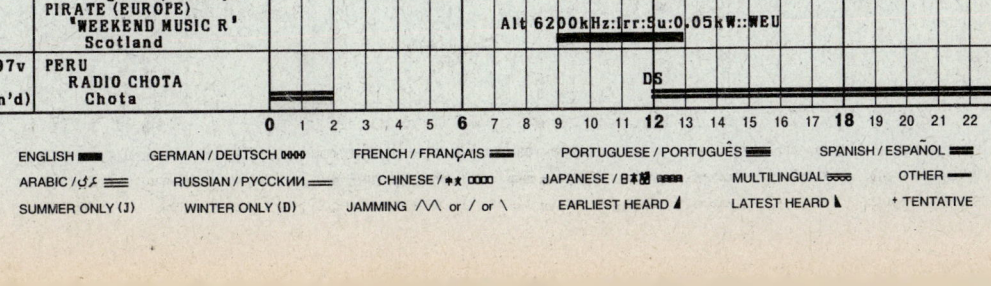

6297v–6540 kHz — PASSPORT TO WORLD BAND RADIO

Frequency (kHz)	Country / Station / Location	Schedule (World Time 0–24)
6297v (con'd)	PERU — RADIO CHOTA, Chota	Irr:::DS (≈2–3)
6300	USA — WYFR-FAMILY RADIO, Via China (Taiwan)	250kW::EAS (≈21–23)
6300v	PIRATE (EUROPE) — "WONDERFUL FREE R", London, England	Irr:Su:0.1kW::WEU (≈9–12)
6304v	PERU — UNIDENTIFIED	DS (≈2–3)
6305	CLANDESTINE (C AMER) — "LA VOZ DEL CID", Guatemala?	ANTI-CASTRO:CA (≈10–13)
6305v	PIRATE (EUROPE) — "RADIO BRIGITTE", Belgium	Irr:Su::ENGLISH, DUTCH:WEU (≈9–13)
6312	PIRATE (EUROPE) — "R IRELAND INTL", Ireland	Su:0.08kW::WEU (≈10–12)
6324v	PERU — ESTACION C, Moyobamba	0.3kW:DS (≈10–13); Tu-Su:0.3kW:DS (≈2–5)
6325	CLANDESTINE (ASIA) — "VO THE KHMER", Kampuchea/Thailand	PRO-REBEL STATION:SEA (≈11–14, 22–24)
6330v	VIETNAM — SON LA BC STATION, Son La	DS (≈8–9)
6339.2	TURKEY — TURKISH POLICE R, Ankara	1kW:DS (≈9–11)
6348v	CLANDESTINE (ASIA) — "ECHO OF HOPE", Suwŏn	50kW:KOREANS IN JAPAN:EAS (≈9–13)
6355	CLANDESTINE (M EAST) — "IRAQI KURDISTAN", Middle East	ANTI-IRAQI GOVT:ME (≈5–11)
6363v	PERU — RADIO HUALLAGA, Saposoa	0.3kW:DS (≈10–12)
6383	MONGOLIA — RADIO ULAN BATOR, Ulan Bator	50kW:DS-2 (≈7–8)
6400	CHINA (PR) — VO THE STRAIT-PLA, Fuzhou	10kW:TAIWAN-2 (≈5–9)
6400v	KOREA (DPR) — RADIO PYONGYANG, Pyŏngyang	50kW::EAS (≈10–12)
6429v	PERU — RADIO ESPINAR, Yauri	Irr:::DS-SPANISH,QUECHUA (≈10–13)
6430	CHINA (PR) — CENTRAL PEOPLES BS / RADIO BEIJING	DS-MINORITIES (≈4–6); (D):::NA (≈10–12); (D):::EU/WUSSR (≈18–20)
6441v	PERU — RADIO HUAYNO, Bodega	Irr:::DS (≈10–11)
6446v	VIETNAM — VOICE OF VIETNAM, Hanoi	DS (≈9–12); Su::DS (≈4–6)
6475v	VIETNAM — LAI CHAU BS, Lai Chau	Alternative Frequency to 6622vkHz
6480	KOREA (REPUBLIC) — RADIO KOREA, In-Kimjae	250kW:EU (≈18–23)
6500	CHINA (PR) — QINGHAI PEOPLES BS, Xining	10kW:DS-TIBETAN (≈9–13)
6523v	VIETNAM — CAO BANG BS, Cao Bang	Alt 6564vkHz:DS (≈10–12); Alt 6564vkHz:Irr:::DS (≈12–15)
6540 (con'd)	KOREA (DPR) — RADIO PYONGYANG, Pyŏngyang	100kW::EAS (≈9–14)

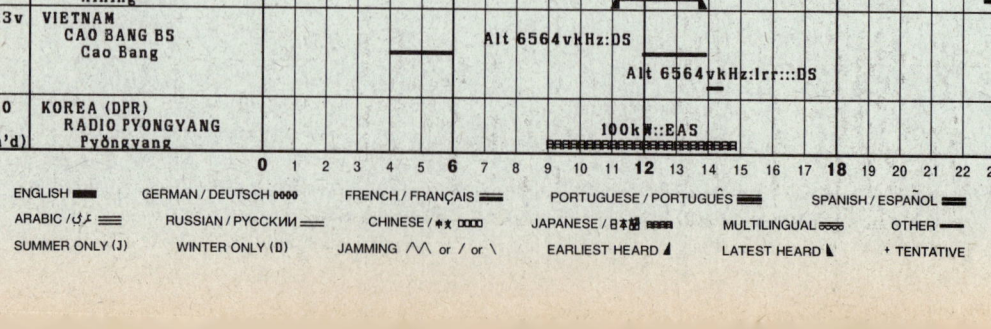

ENGLISH ▬ GERMAN / DEUTSCH ◊◊◊◊ FRENCH / FRANÇAIS ═ PORTUGUESE / PORTUGUÊS ≡ SPANISH / ESPAÑOL ≣
ARABIC /عربي RUSSIAN / РУССКИЙ ═ CHINESE / 中文 ◊◊◊◊ JAPANESE / 日本語 ▬▬ MULTILINGUAL ◊◊◊◊ OTHER ▬
SUMMER ONLY (J)   WINTER ONLY (D)   JAMMING /\/\/ or / or \   EARLIEST HEARD ◢   LATEST HEARD ◣   + TENTATIVE

Frequency (kHz)	Country / Station / Location	Schedule / Notes
6540 (con'd)	KOREA (DPR) RADIO PYONGYANG Pyŏngyang	Alt 11335kHz 200kW::ME/NAF
6548.6	LEBANON VOICE OF LEBANON Beirut-Ashrafiyah	8kW:DS-PHALANGE
6550	CHINA (PR) RADIO BEIJING	ANZ; SEA; SAS; (D):::EEU
6555v	CLANDESTINE (C AMER) "RADIO VENCEREMOS" Morazán, Salvador	Alt 6650vkHz:FMLN/ANTI-SALVADOR:CA; Alt 6650vkHz:M-Sa::FMLN/ANTI-SALVADOR:CA; Alt 6650vkHz:Su::FMLN/ANTI-SALVADOR
6560	CHINA (PR) RADIO BEIJING	USSR
6564v	VIETNAM CAO BANG BS Cao Bang	Alternative Frequency to 6523vkHz
6565	USA WYFR-FAMILY RADIO Via China (Taiwan)	250kW::EAS
6570	BURMA BURMESE ARMY STN Taunggyi	Alt 5060kHz:0.05kW:DS
6571v	PERU RADIO TACNA Tacna	0.15/0.18kW:DS; M-Sa:0.15/0.18kW:DS; Tu-Su:0.15/0.18kW:DS
6576	KOREA (DPR) RADIO PYONGYANG Pyŏngyang	400kW:CA; 200kW::EAS/EUSSR; 200kW::WUSSR/EU
6578v	ECUADOR R OMEGA/V JUVENTUD Catacocha	DS; Irr:::DS
6584.6	LEBANON VOICE OF LEBANON Beirut-Ashrafiyah	Irr::8kW:DS-PHALANGE; 8kW:DS-ARAB,FRENCH,ENG
6590	CHINA (PR) RADIO BEIJING +RADIO BEIJING	(D):::SEA; (D):::EAF/SAF; (D):::EU; (D)
6599v	KOREA (DPR) KOREAN CENTRAL BS Pyŏngyang	Alt 6100kHz:50kW:DS
6622v	VIETNAM LAI CHAU BS Lai Chau	Alt 6475vkHz:DS
6623	ARGENTINA +RADIO EL MUNDO Buenos Aires	Irr:::USB:DS(FEEDER)
6647v	LAOS LAO NATIONAL RADIO Paksé	Alternative Frequency to 6660vkHz
6650v	CLANDESTINE (C AMER) "RADIO VENCEREMOS" Morazán, Salvador	Alternative Frequency to 6555vkHz
6660v	LAOS LAO NATIONAL RADIO Paksé	Alt 6647vkHz:1kW:DS
6665	CHINA (PR) CENTRAL PEOPLES BS	DS-1; (D)W-M::DS-1

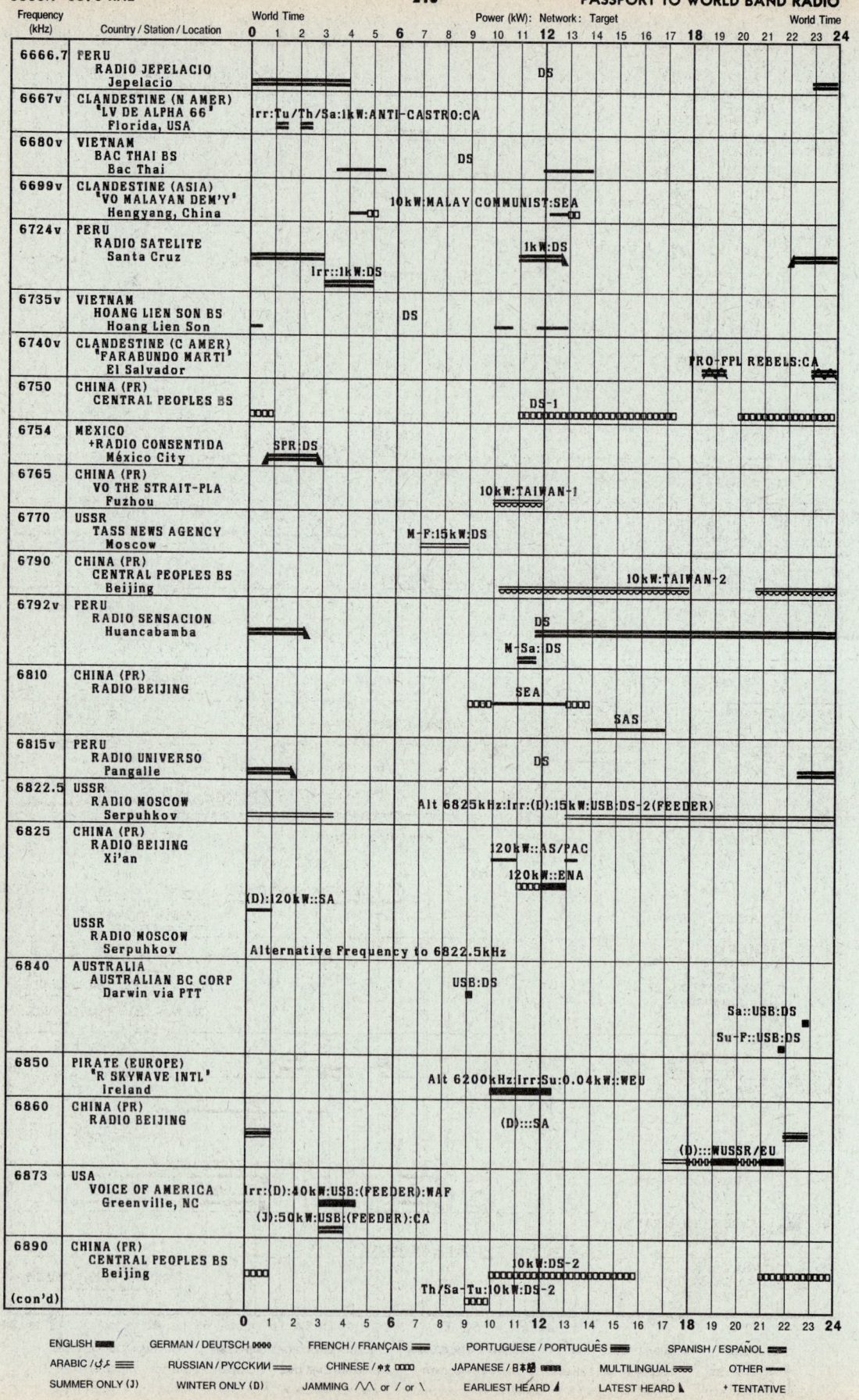

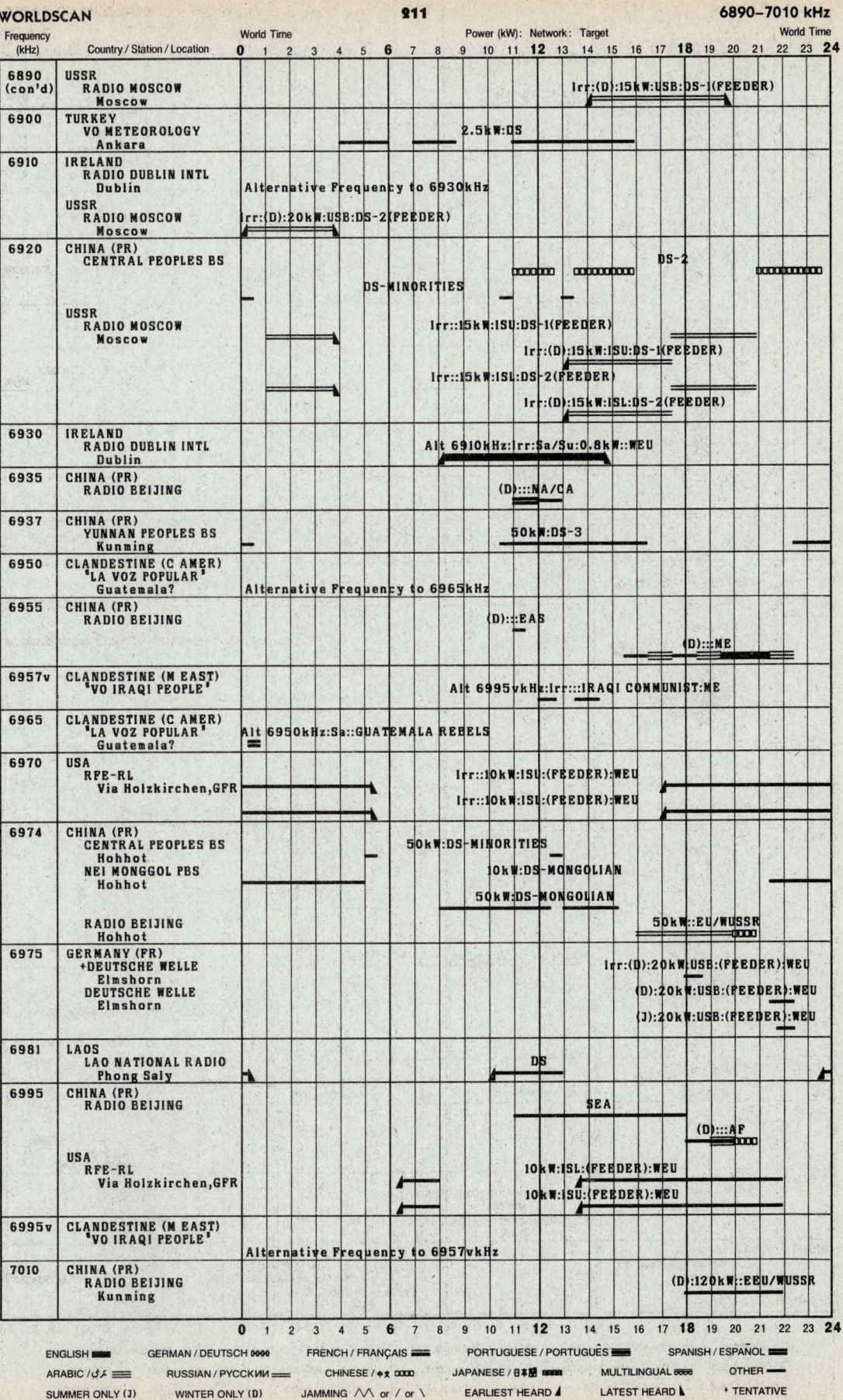

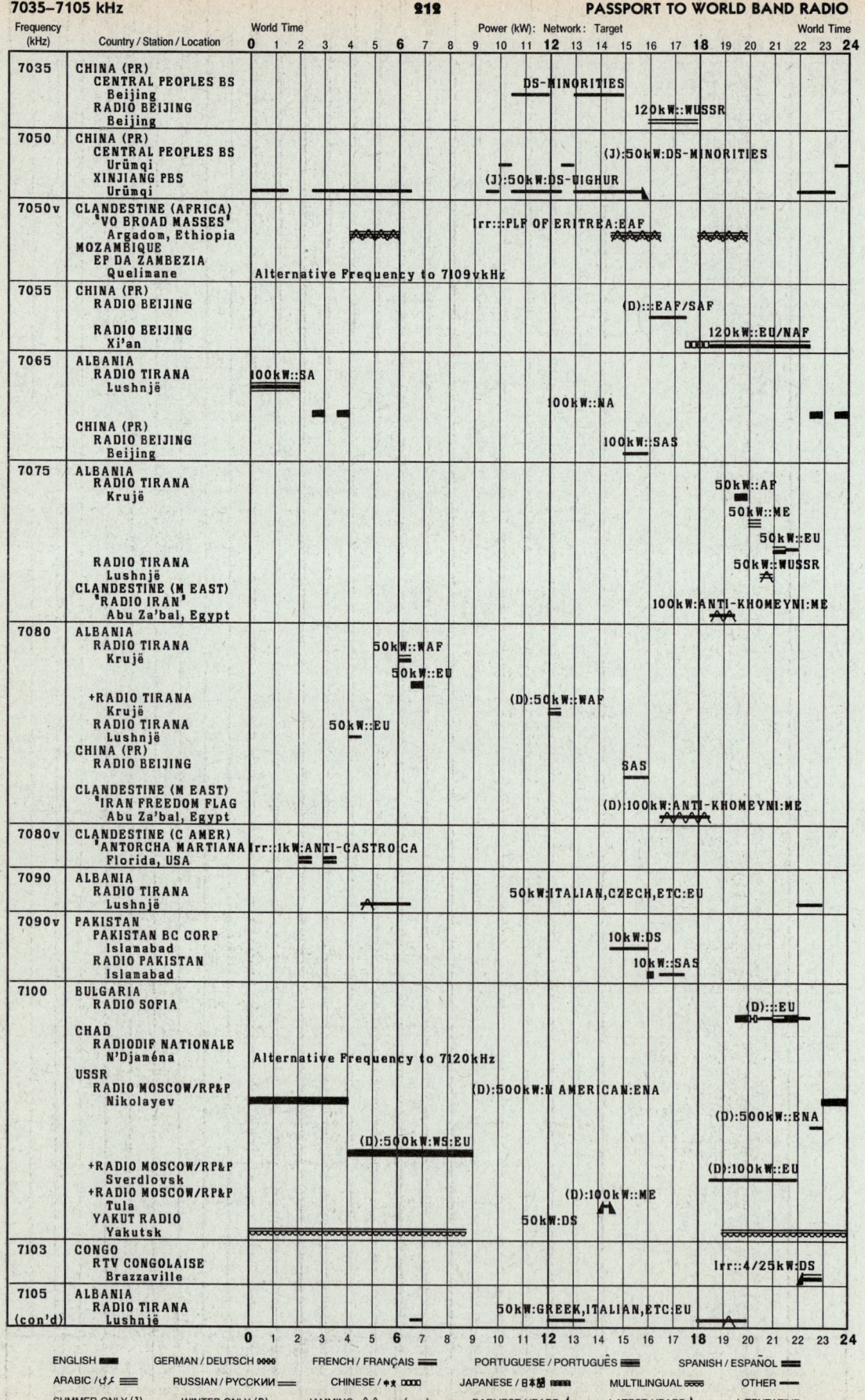

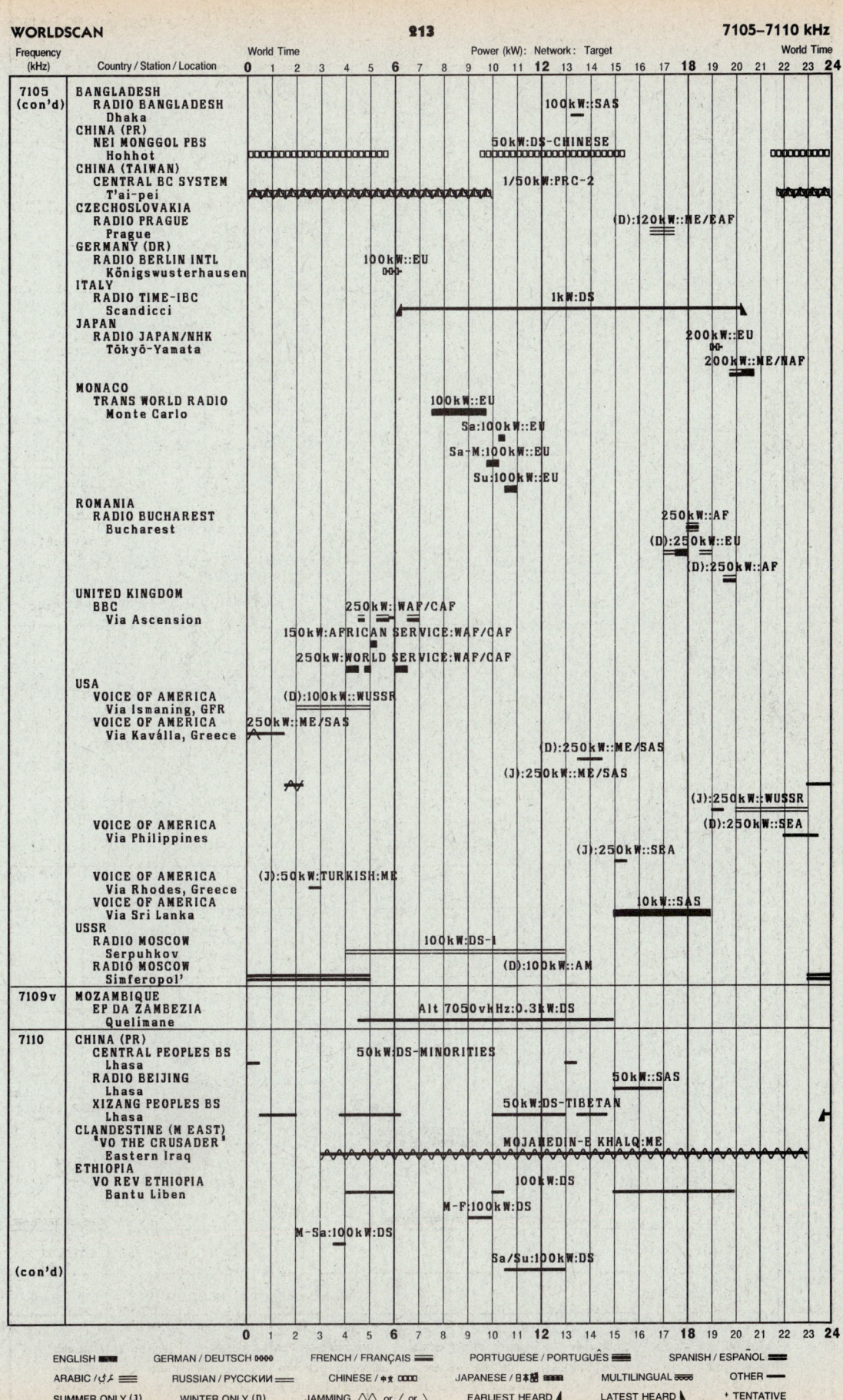

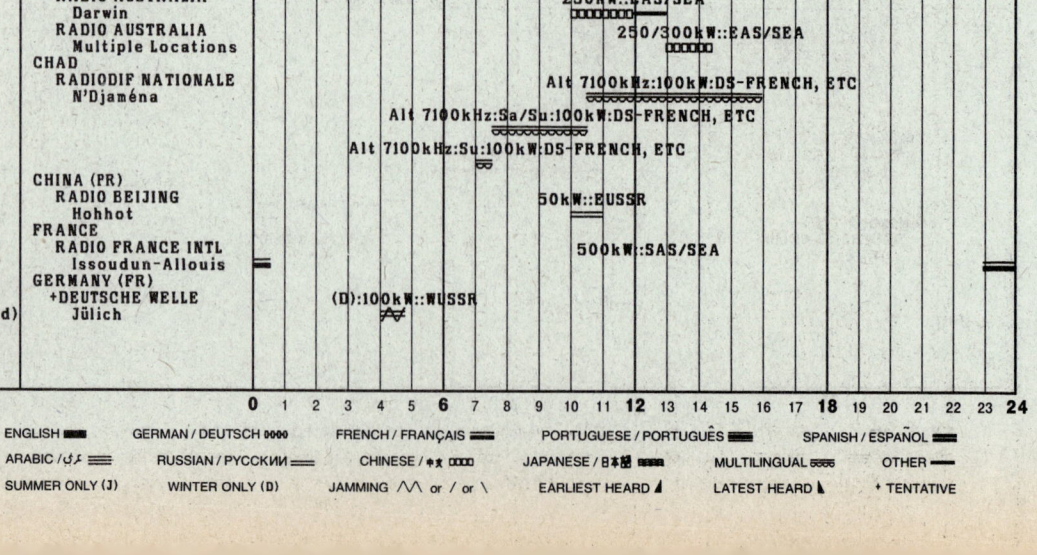

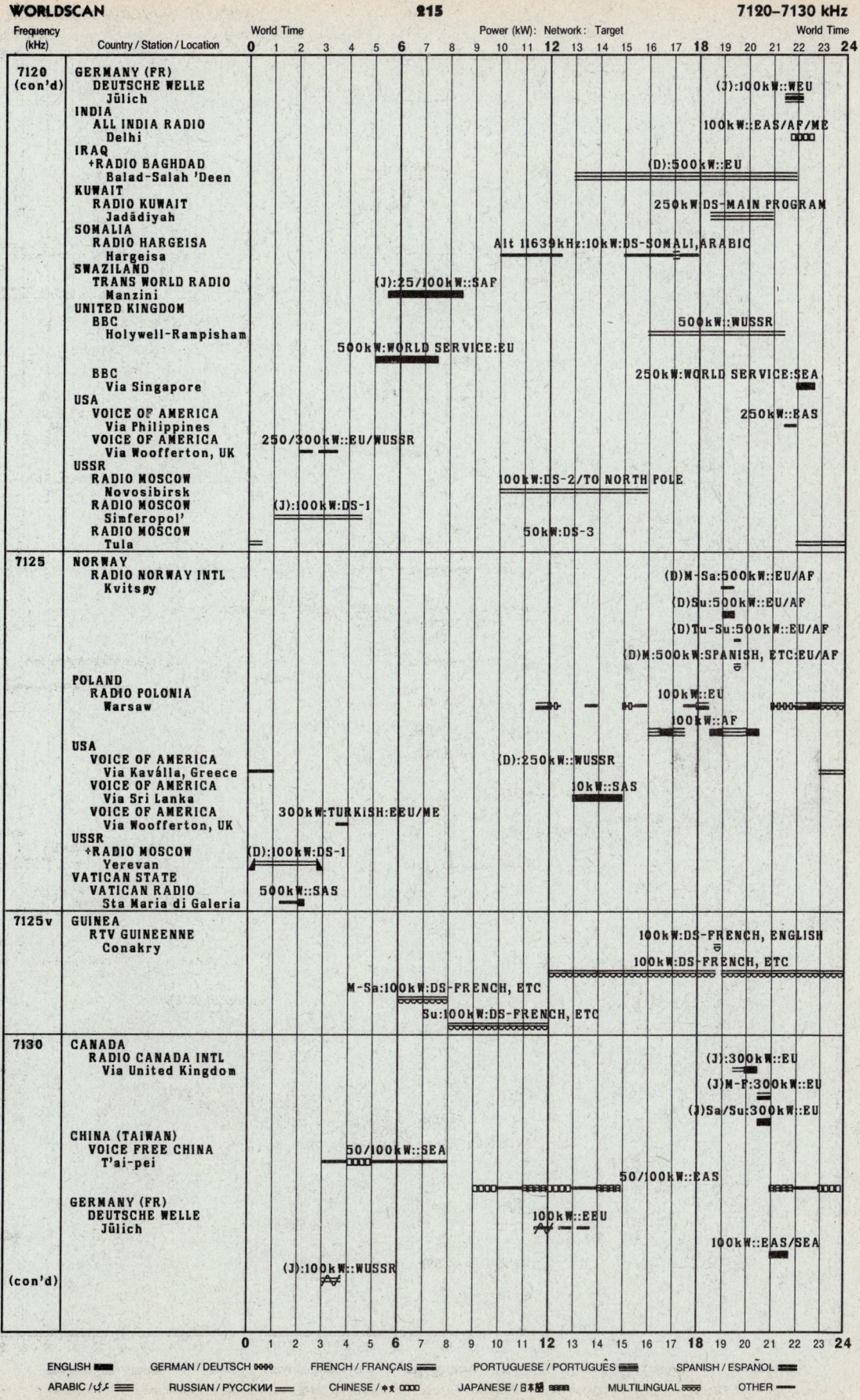

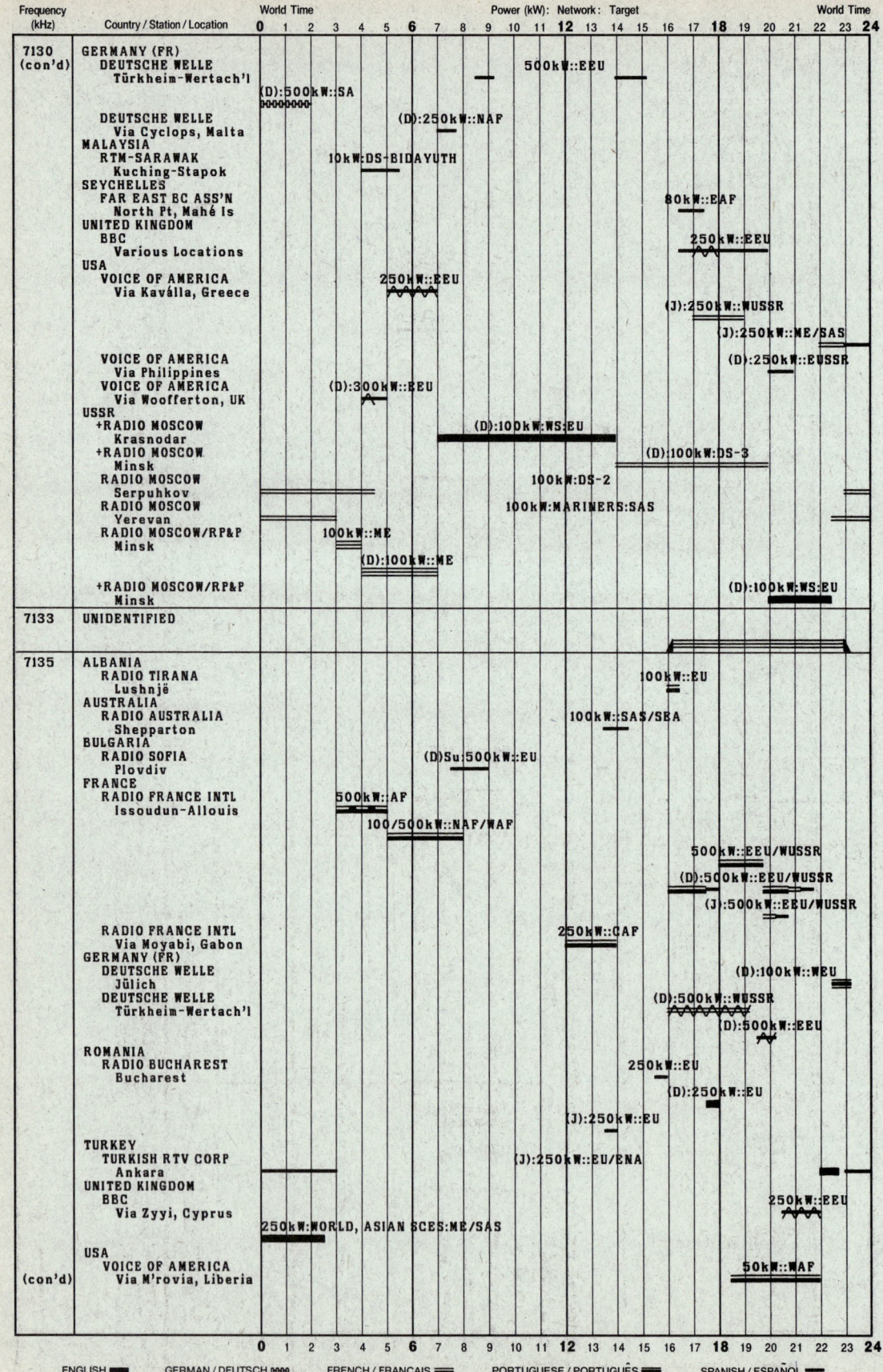

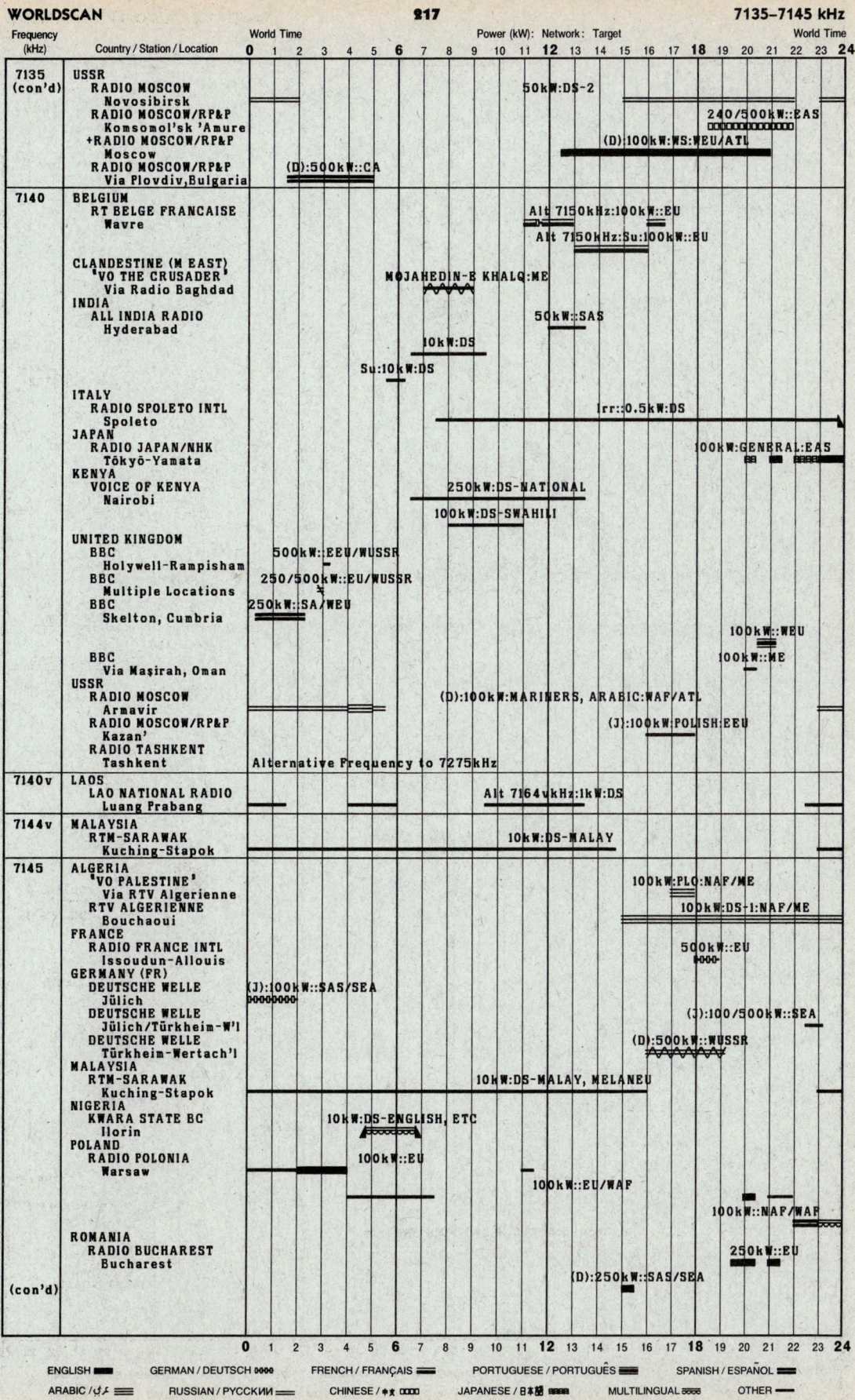

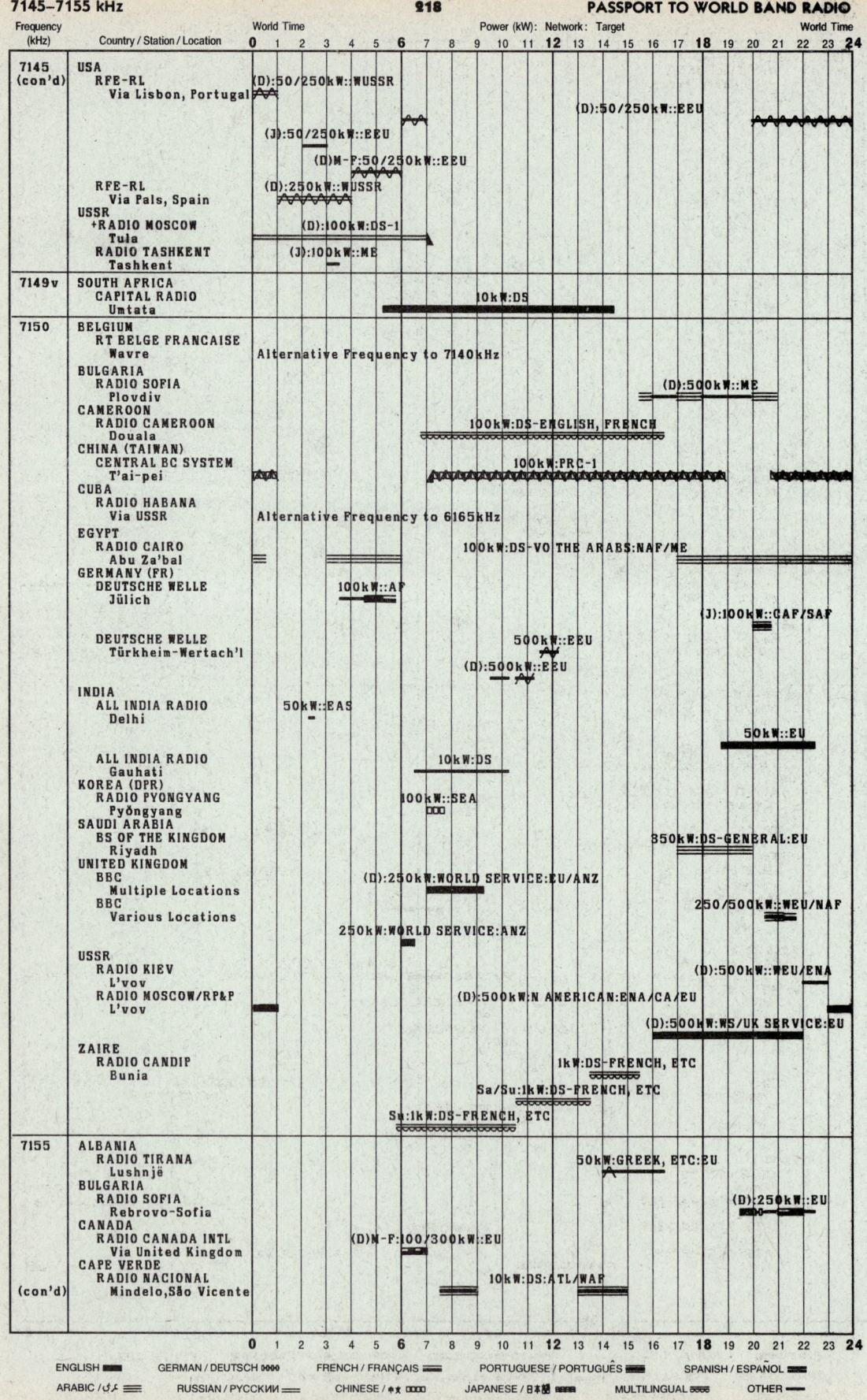

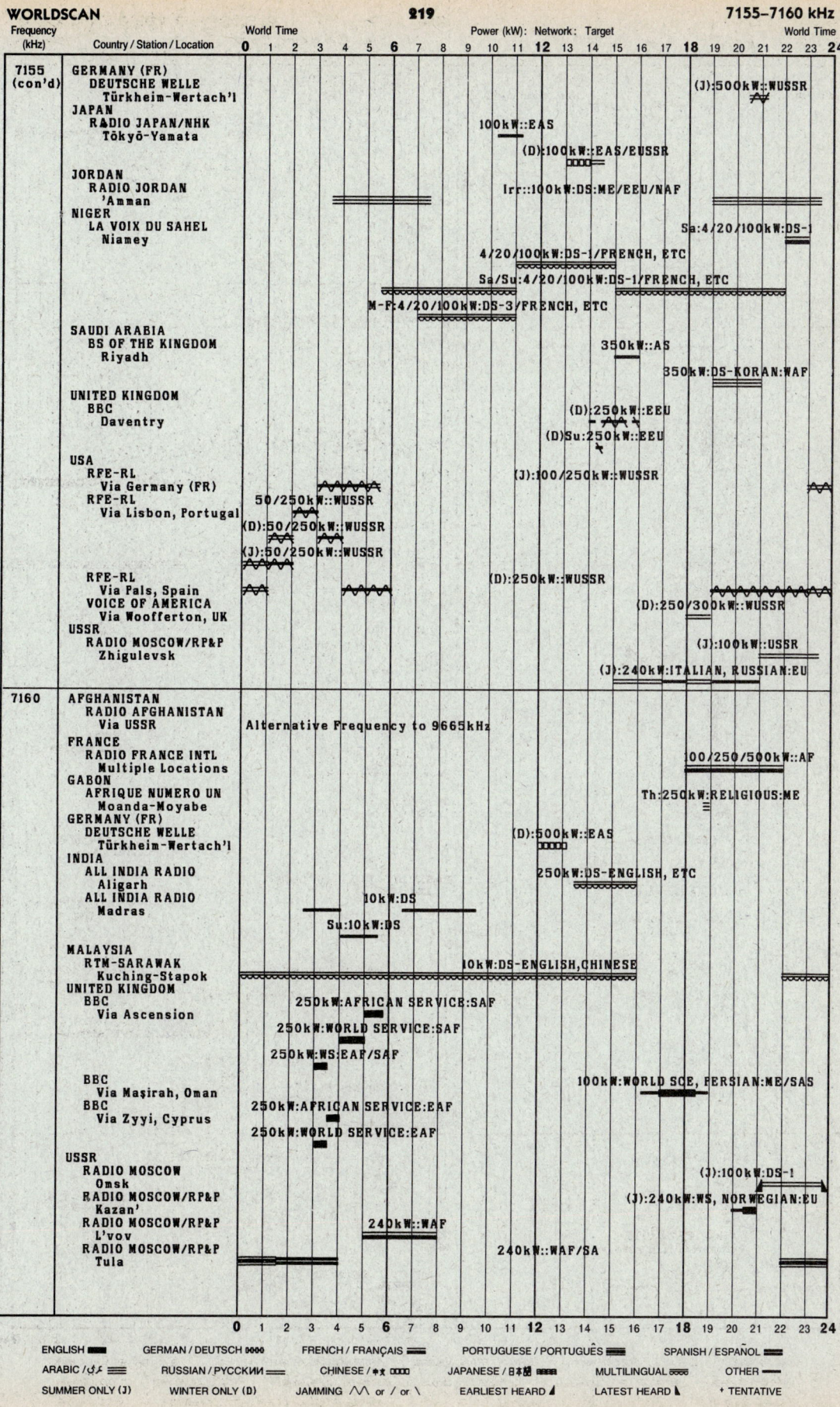

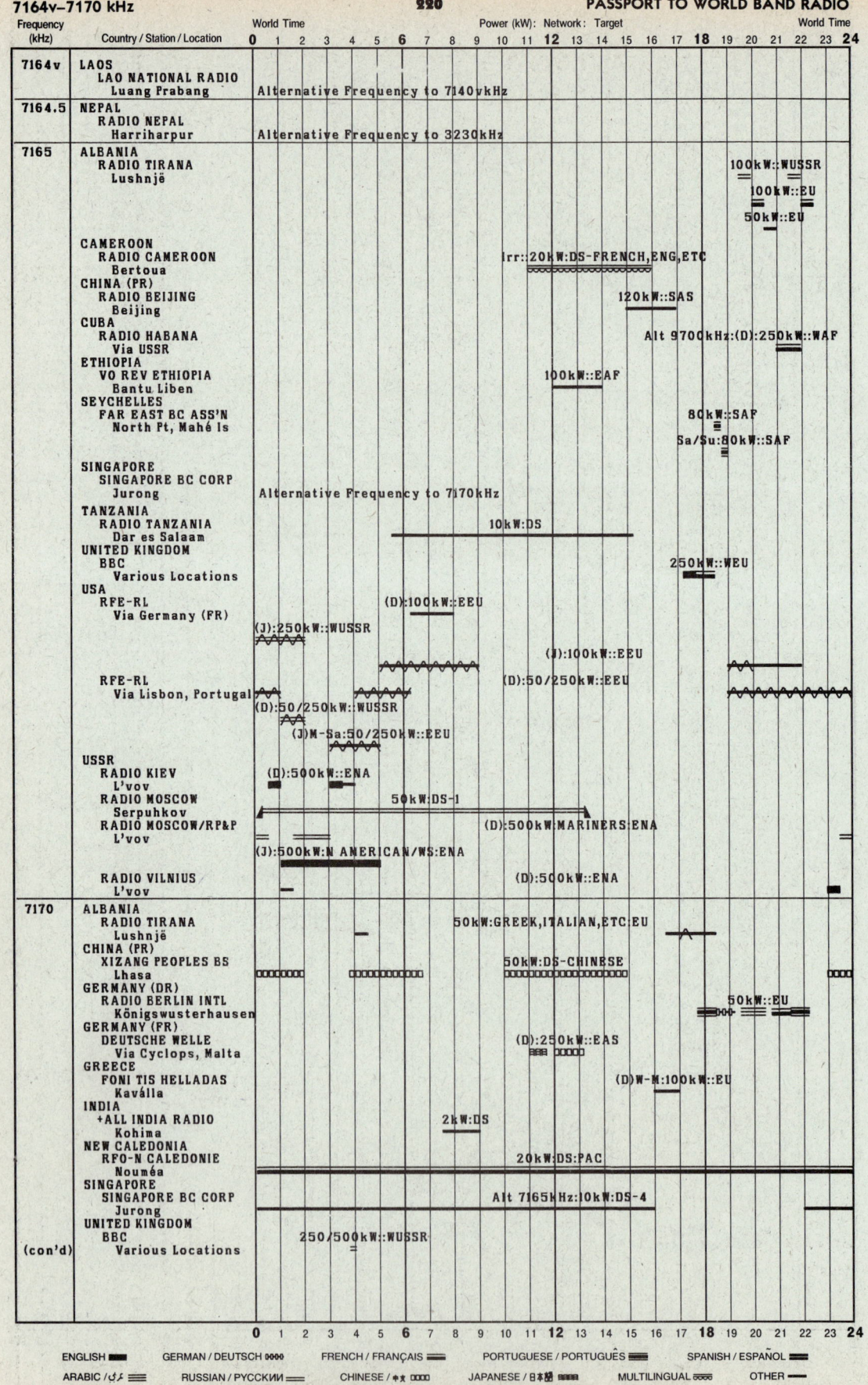

WORLDSCAN

7170–7180 kHz

Frequency (kHz)	Country / Station / Location	Schedule
7170 (con'd)	**USA**	
	VOICE OF AMERICA Via Kaválla, Greece	(D):250kW::WUSSR; (J):250kW::WUSSR
	VOICE OF AMERICA Via Woofferton, UK	300kW::NAF; (J):300kW::NAF
	USSR	
	RADIO MOSCOW Komsomol'sk 'Amure	(D):100kW::PAC
	RADIO MOSCOW/RP&P Krasnoyarsk	(D):500kW::EAS
	+RADIO MOSCOW/RP&P Krasnoyarsk	(D):500kW::EAS
7172v	**ANGOLA**	
	ER DO LOBITO Lobito	Irr:1kW:DS; 1kW:DS
7173v	**SENEGAL**	
	ORT DU SENEGAL Dakar	100kW:DS-FRENCH, ETC
7175	**AUSTRALIA**	
	RADIO AUSTRALIA Darwin	(J):250kW::EAS/SEA
	GERMANY (FR)	
	DEUTSCHE WELLE Türkheim-Wertach'l	500kW::AF
	HOLLAND	
	RADIO NEDERLAND Flevoland	500kW::ME
	ITALY	
	RTV ITALIANA Caltanissetta	50kW:DS-2:EU/NAF
	ROMANIA	
	RADIO BUCHAREST Bucharest	(D):250kW::ME; (D):250kW::WUSSR
	SWEDEN	
	RADIO SWEDEN INTL Hörby	(D):350kW::EAS
	USA	
	VOICE OF AMERICA Via M'rovia, Liberia	M-F:250kW:ENGLISH,FRENCH,ETC:WAF/CAF
	USSR	
	RADIO MOSCOW Star'obel'sk	(J):240kW:MARINERS:EU/ATL
	RADIO MOSCOW Zhigulevsk	(J):100kW::DS-1
	RADIO MOSCOW/RP&P Khabarovsk	(J):100kW::EUSSR
	RADIO MOSCOW/RP&P Novosibirsk	(D):500kW:WS, HAUSA, ETC:WAF
	RS SOV BELORUSSIA Orcha	(J)Su-F:240kW:EU/ATL; (J)Sa:240kW:1ST & 3RD SATS:EU/ATL; (J)Sa:240kW:2ND & 4TH SATS:EU/ATL
7180	**CLANDESTINE (M EAST)**	
	"RADIO IRAN" Via Radio Baghdad	ANTI-KHOMEYNI:ME; Irr::ANTI-KHOMEYNI:ME
	JAPAN	
	RADIO JAPAN/NHK Tōkyō-Yamata	100kW::EU/NAF/ME
	TURKEY	
	TURKISH RTV CORP Ankara	(D):250kW::ME
	UNITED KINGDOM	
	BBC Via Singapore	100/250kW::SEA/EAS; 100/250kW::SAS
	USA	
	RFE-RL Via Germany (FR)	100kW::WUSSR; (D):100kW::WUSSR; (D):100kW::EEU; (J):100kW::WUSSR
	VOICE OF AMERICA Via Woofferton, UK	300kW::EEU; (D):300kW::EEU
(con'd)		

Legend: ENGLISH ▬ GERMAN / DEUTSCH ◊◊◊◊ FRENCH / FRANÇAIS ═ PORTUGUESE / PORTUGUÊS ▬ SPANISH / ESPAÑOL ▬
ARABIC / ألف ═ RUSSIAN / РУССКИЙ ═ CHINESE / 中文 ◊◊◊◊ JAPANESE / 日本語 ▬ MULTILINGUAL ▬ OTHER ▬
SUMMER ONLY (J) WINTER ONLY (D) JAMMING ∧∧ or / or \ EARLIEST HEARD ◢ LATEST HEARD ◣ + TENTATIVE

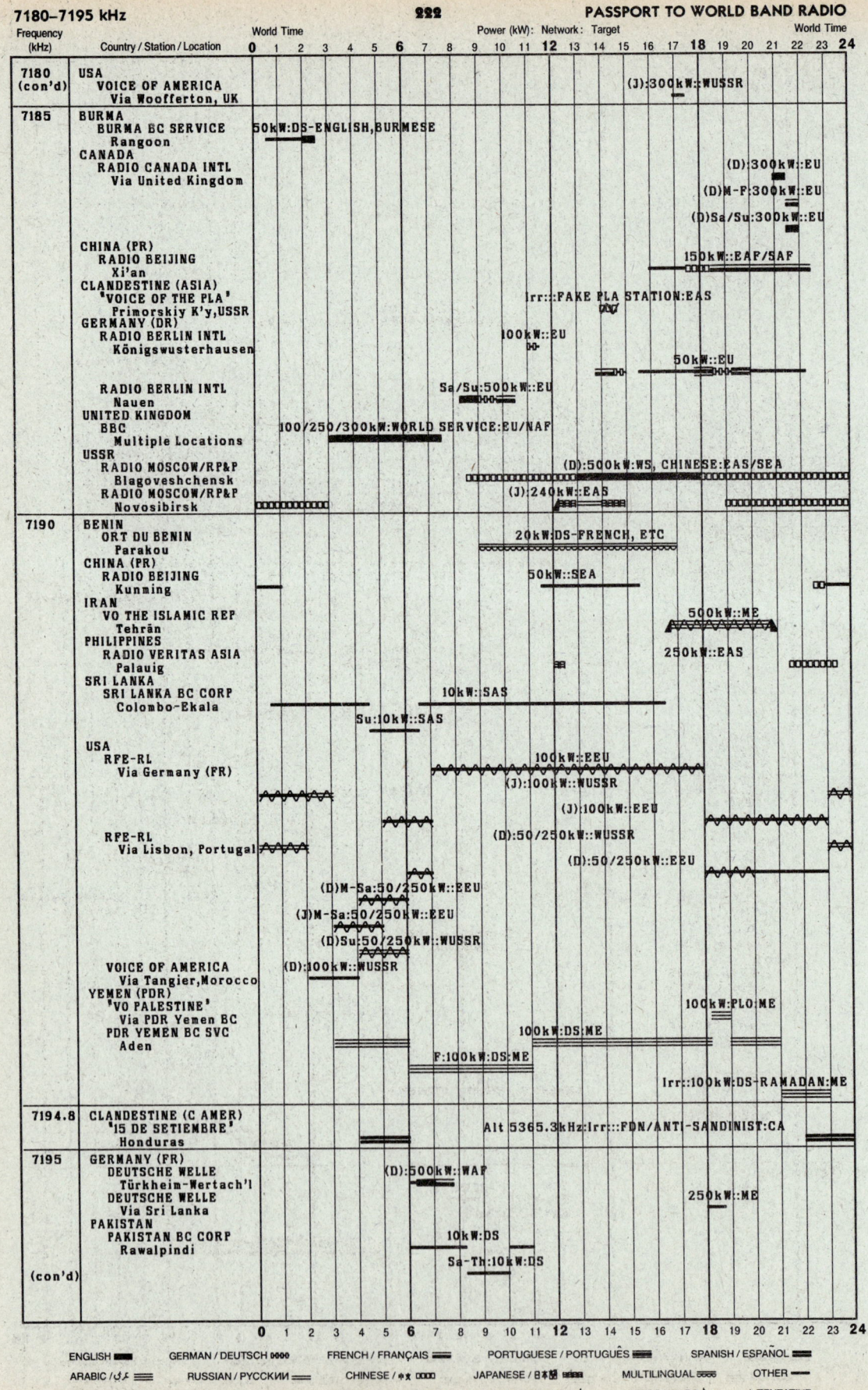

WORLDSCAN

7195–7200v kHz

Frequency (kHz)	Country / Station / Location	Schedule
7195 (con'd)	**ROMANIA** RADIO BUCHAREST — Bucharest	250kW::EU; (D):250kW::ME; (D):250kW::EU; (D):250kW::ME/EU
	SAUDI ARABIA BS OF THE KINGDOM — Riyadh	350kW:PERSIAN:ME
	SWAZILAND TRANS WORLD RADIO — Manzini	(J):100kW::SAF/CAF
	UGANDA RADIO UGANDA — Kampala	Irr::20kW:DS; 20kW:DS
	USA VOICE OF AMERICA — Via M'rovia, Liberia	50kW::WAF/CAF
	USSR RADIO KIEV — Tula	(D):240kW::EU
	RADIO MOSCOW/RP&P — Komsomol'sk 'Amure	(J):100kW::EAS/SEA
	RADIO MOSCOW/RP&P — Tula	(D):240kW:WS,FRENCH,RUSSIAN:EU/ENA
7195v	**PAKISTAN** PAKISTAN BC CORP — Rawalpindi	10kW:DS; Sa-Th:10kW:DS
7199v	**AFGHANISTAN** RADIO AFGHANISTAN — Kabul	50kW:DS-1:ME
7200	**CLANDESTINE (AFRICA)** VO ETHIOPIA UNITY — Sudan	ANTI-ETHIOPIA GOVT:EAF
	GABON AFRIQUE NUMERO UN — Moanda-Moyabe	250kW::CAF; Sa/Su:250kW::CAF; 250kW:ENGLISH,FRENCH:CAF; M-F:250kW:ENGLISH,FRENCH:CAF
	GERMANY (FR) DEUTSCHE WELLE — Via Sri Lanka	250kW::ME/SAS
	KOREA (DPR) RADIO PYONGYANG — Pyŏngyang	200kW:EAS/EUSSR; 200kW::EAS/SEA
	USA RFE-RL — Via Germany (FR)	(D):100kW::EEU
	RFE-RL — Via Lisbon, Portugal	(D):50/250kW::EEU
	VOICE OF AMERICA — Via Philippines	(J):100kW::EAS
	VOICE OF AMERICA — Via Woofferton, UK	250/300kW::EU/ME
	USSR RADIO MOSCOW — Vladivostok	50kW:DS-2; (D):50kW:DS-2
	RADIO MOSCOW — Zhigulevsk	100kW:DS-1; (D):100kW:DS-1
	YAKUT RADIO — Yakutsk	50kW:DS
	YUGOSLAVIA RADIO BEOGRAD — Belgrade	100kW:DS-1:EU/NAF/ME; Su:100kW:DS-1
7200v	**SOMALIA** RADIO MOGADISHU — Mogadishu	F:100kW:DS; Sa-Th:100kW:DS; 100kW:DS-SOMALI,ARABIC

ENGLISH ▬▬ GERMAN / DEUTSCH ◊◊◊◊ FRENCH / FRANÇAIS ═══ PORTUGUESE / PORTUGUÊS ≡≡≡ SPANISH / ESPAÑOL ▬▬
ARABIC / ﻋﺮﺑﻲ ≡≡ RUSSIAN / РУССКИЙ ═══ CHINESE / 中文 ○○○○ JAPANESE / 日本語 ▬▬▬ MULTILINGUAL ◊◊◊◊ OTHER ───
SUMMER ONLY (J)   WINTER ONLY (D)   JAMMING ∧∧ or / or \   EARLIEST HEARD ◢   LATEST HEARD ◣   † TENTATIVE

7205–7210 kHz

Frequency (kHz)	Country / Station / Location	Broadcast Schedule
7205	**ALBANIA** RADIO TIRANA — Lushnjë	100kW::W USSR; 100kW::EU; 100kW::ME
	AUSTRALIA RADIO AUSTRALIA — Carnarvon	300kW::SAS/SEA
	RADIO AUSTRALIA — Shepparton	100kW::SEA
	CAMEROON RADIO CAMEROON — Yaoundé	Irr::30kW:DS-FRENCH,ENG,ETC; Irr:Su:30kW:DS-FRENCH,ENG,ETC
	MONACO TRANS WORLD RADIO — Monte Carlo	500kW::EU; 100kW::EU; M-Sa:500kW::EU; Sa/Su:100kW::EU; Su:100kW::EU; 100kW:SPANISH,CATALAN:EU
	SOUTH AFRICA SOUTH AFRICAN BC — Meyerton	Alternative Frequency to 6105 kHz
	USA VOICE OF AMERICA — Via Kaválla, Greece	250kW::SAS; (D):250kW::SAS
	VOICE OF AMERICA — Via Rhodes, Greece	50kW::ME; (J):50kW::ME; (D):50kW::TURKISH:ME
	USSR +RADIO KIEV — L'vov	(D):500kW::ENA/ATL
	RADIO MOSCOW — Armavir	100kW:MARINERS:EU/ATL
	+RADIO MOSCOW — L'vov	(D):500kW::ENA/ATL
	+RADIO MOSCOW/RP&P — Krasnoyarsk	(D):500kW::SAS
	RS SOV BELORUSSIA — Kiev	(J)Su-F:100kW::EU/ATL; (J)Sa:100kW:1ST & 3RD SATS:EU/ATL; (J)Sa:100kW:2ND & 4TH SATS:EU/ATL
7205v	**ZAIRE** LA VOIX DU ZAIRE — Lubumbashi	Irr::10kW:DS-FRENCH,ETC; Irr:Sa/Su:10kW:DS-FRENCH,ETC; Irr:Su:10kW:DS-FRENCH,ETC
7210	**AUSTRIA** +RADIO AUSTRIA INTL — Vienna	(D):100kW::EU
	+RADIO AUSTRIA INTL — Vienna	(D)M-Sa:100kW::EU
	+RADIO AUSTRIA INTL — Vienna	(D)Su:100kW::EU
	GERMANY (FR) DEUTSCHE WELLE — Türkheim-Wertach'l	(J):500kW::SEA
	INDIA +ALL INDIA RADIO — Calcutta	10kW:DS
	JAPAN RADIO JAPAN/NHK — Tōkyō-Yamata	(D):100kW::EAS/EUSSR; (D):100kW::SEA
	SAUDI ARABIA BS OF THE KINGDOM — Riyadh	350kW:DS-GENERAL:EU; Irr::350kW:DS-RAMADAN:EU
(con'd)	**SWEDEN** RADIO SWEDEN INTL — Hörby	Alternative Frequency to 9645 kHz

Legend: ENGLISH; GERMAN / DEUTSCH; FRENCH / FRANÇAIS; PORTUGUESE / PORTUGUÊS; SPANISH / ESPAÑOL; ARABIC; RUSSIAN / РУССКИЙ; CHINESE / 中文; JAPANESE / 日本語; MULTILINGUAL; OTHER; SUMMER ONLY (J); WINTER ONLY (D); JAMMING; EARLIEST HEARD; LATEST HEARD; + TENTATIVE

WORLDSCAN   7210–7215v kHz

Frequency (kHz)	Country / Station / Location	Schedule (World Time 0–24, Power:Network:Target)
7210 (con'd)	**SWITZERLAND** – RED CROSS BC SVC, Schwarzenburg	Irr:M:150kW:1ST OR LAST MONDAY:EU; Irr:Su:150kW:1ST OR LAST SUNDAY:EU/NAF
	UNITED KINGDOM – BBC, Various Locations	250/500kW::EEU/WUSSR; 250kW::NEU/NAF; 250/500kW::EEU/ME
	BBC Via Zyyi, Cyprus	250kW:WORLD SERVICE, ETC:EEU
	USA – VOICE OF AMERICA, Via Kavála, Greece	(J):250kW::SAS
	VOICE OF AMERICA, Via Philippines	(D):250kW::EAS
	VOICE OF AMERICA, Via Woofferton, UK	(D):250/300kW::EEU/WUSSR
	USSR – BELORUSSIAN RADIO, Minsk	Alt 12005kHz:15kW:DS-1
	KHABAROVSK RADIO, Khabarovsk	50kW:DS-1
	RADIO MOSCOW, Moscow	100kW:DS-1; 100kW:DS-2
	RADIO TIKHIY OKEAN, Khabarovsk	50kW:MARINERS:EUSSR/PAC; Sa:50kW:MARINERS:EUSSR/PAC; Su-F:50kW:MARINERS:EUSSR/PAC
7210v	**CHINA (PR)** – YUNNAN PEOPLES BS, Kunming	50kW:DS-1
	SENEGAL – ORT DU SENEGAL, Dakar	4kW:DS-FRENCH, ETC
7215	**AUSTRALIA** – RADIO AUSTRALIA, Shepparton	100kW::PAC
	+RADIO AUSTRALIA, Shepparton	(D):100kW::PAC
	CHINA (PR) – VOICE OF JINLING, Nanjing	PROJECTED:EAS
	CLANDESTINE (M EAST) – "FREE VOICE IRAN" Via Radio Baghdad	ANTI-KHOMEYNI:ME
	"VO THE CRUSADER" Via Radio Baghdad	MOJAHEDIN-E KHALQ:ME
	INDIA – ALL INDIA RADIO, Delhi	100kW::SEA
	IRAN – VO THE ISLAMIC REP, Tehrān	Irr::500kW:RAMADAN:ME; 500kW::ME/NAF; 100kW::ME/SAS; 500kW::ME
	IVORY COAST – RTV IVOIRIENNE, Abidjan	20kW:DS:WAF
	NORWAY – RADIO NORWAY INTL, Kvitsøy	(D):500kW::SAS
	TURKEY – TURKISH RTV CORP, Ankara	250kW::EU; (D):250kW::EU/ENA
	UNITED ARAB EMIRATES – +VOICE OF THE UAE, Abu Dhabi	Irr::500kW::ME/SAS
	USA – RFE-RL, Via Lisbon, Portugal	(D)M-Sa:50/250kW::EEU
	USSR – +RADIO MOSCOW/RP&P, Via Plovdiv, Bulgaria	(D):250kW:N AMERICAN:ENA
7215v	**ANGOLA** – RADIO NACIONAL, Luanda	10kW:DS

Key: ENGLISH ▬ GERMAN/DEUTSCH ◊◊◊◊ FRENCH/FRANÇAIS ≡ PORTUGUESE/PORTUGUÊS ≣ SPANISH/ESPAÑOL ▬ ARABIC/عربي ▬ RUSSIAN/РУССКИЙ ═ CHINESE/中文 □□□□ JAPANESE/日本語 ▬ MULTILINGUAL ∞∞∞ OTHER ▬

SUMMER ONLY (J)   WINTER ONLY (D)   JAMMING ∧∧∧ or / or \   EARLIEST HEARD ▲   LATEST HEARD ▼   + TENTATIVE

7220–7230 kHz

Frequency (kHz)	Country / Station / Location	Schedule (World Time 0–24)
7220	**CLANDESTINE (M EAST)** "RADIO IRAN" Via Radio Baghdad	500kW:ANTI-KHOMEYNI:ME
	HUNGARY RADIO BUDAPEST Jászberény	250kW::EU
	RADIO BUDAPEST Szekésfehérvár	20kW::EU
	+RADIO BUDAPEST Szekésfehérvár	(D):20kW::EU
	+RADIO BUDAPEST Szekésfehérvár	(D)M-F:20kW::EU
	+RADIO BUDAPEST Szekésfehérvár	(D)Sa-M/W:20kW::EU
	+RADIO BUDAPEST Szekésfehérvár	(D)Su:20kW::EU
	RADIO BUDAPEST Szekésfehérvár	(D)Su:20kW::EU
	+RADIO BUDAPEST Szekésfehérvár	(D)Tu/F:20kW::EU
	USA RFE-RL Via Germany (FR)	100/250kW:WUSSR
		(J):250kW:WUSSR
	VOICE OF AMERICA Via Woofferton, UK	300kW:EEU
		(D):300kW::EEU
		(J):300kW::EEU
	USSR RADIO MOSCOW/RP&P Chita	500kW::EAS/SEA
7220v	**ZAMBIA** RADIO ZAMBIA-ZBS Lusaka	50kW:DS-ENGLISH, ETC
7225	**CHINA (PR)** SICHUAN PEOPLES BS Chengdu	15kW:DS-CHINESE
	GERMANY (FR) DEUTSCHE WELLE Jülich	(D):100kW::EAF/SAF
	DEUTSCHE WELLE Türkheim-Wertach'l	(D):500kW::AF
	DEUTSCHE WELLE Via Kigali, Rwanda	250kW:CAF/SAF
		(J):250kW:CAF/SAF
	HUNGARY RADIO BUDAPEST Erd-Diósd	100kW::EU
		M-Sa:100kW::EU
		M/Th:100kW::EU
		Sa:100kW::EU
		Tu/F:100kW::EU
		W:100kW::EU
	INDIA ALL INDIA RADIO Aligarh	250kW::SAS
	PORTUGAL RADIO PORTUGUESA Lisbon-São Gabriel	(D)Sa/Su:100kW::EU
	ROMANIA RADIO BUCHAREST Bucharest	(D):250kW::WEU
	SAUDI ARABIA BS OF THE KINGDOM Riyadh	350kW:DS-GENERAL:ME
	TUNISIA RTV TUNISIENNE Sfax	Alternative Frequency to 7310 kHz
	TURKEY TURKISH RTV CORP Ankara	(D):250kW::ME
	USA VOICE OF AMERICA Via Philippines	(D):250kW::SEA
		(J):250kW::SEA
7230 (con'd)	**BURKINA FASO** RTV BURKINA Ouagadougou	50kW:DS
		M-Sa:50kW:DS
		Su:50kW:DS

ENGLISH ▬ GERMAN / DEUTSCH ◊◊◊◊ FRENCH / FRANÇAIS ═ PORTUGUESE / PORTUGUÊS ≡ SPANISH / ESPAÑOL ≣
ARABIC / عربى ≡ RUSSIAN / РУССКИЙ ═ CHINESE / 中文 ◊◊◊◊ JAPANESE / 日本語 ▬ MULTILINGUAL ◊◊◊◊ OTHER ▬
SUMMER ONLY (J)   WINTER ONLY (D)   JAMMING ∧∧ or / or \   EARLIEST HEARD ◢   LATEST HEARD ◣   + TENTATIVE

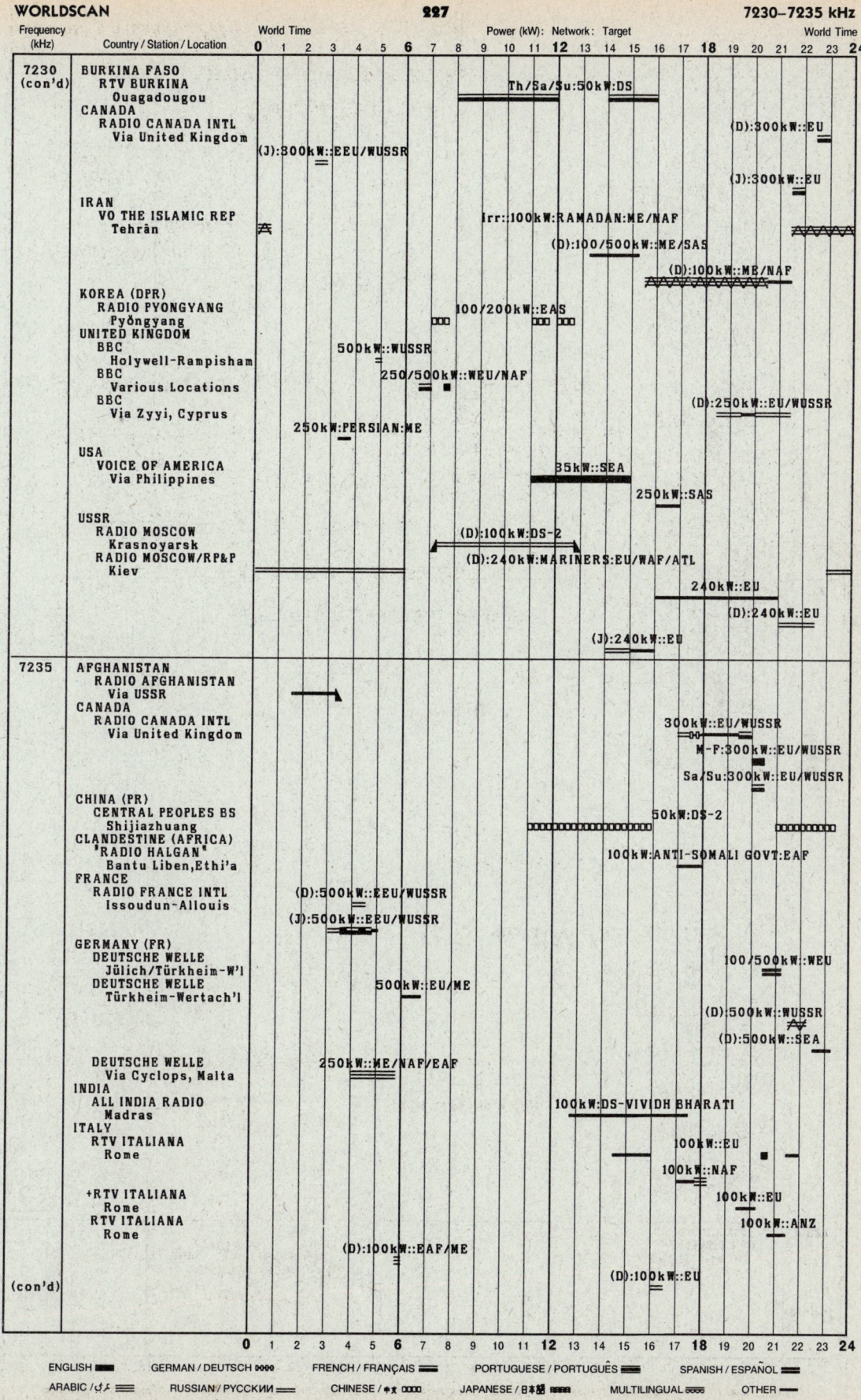

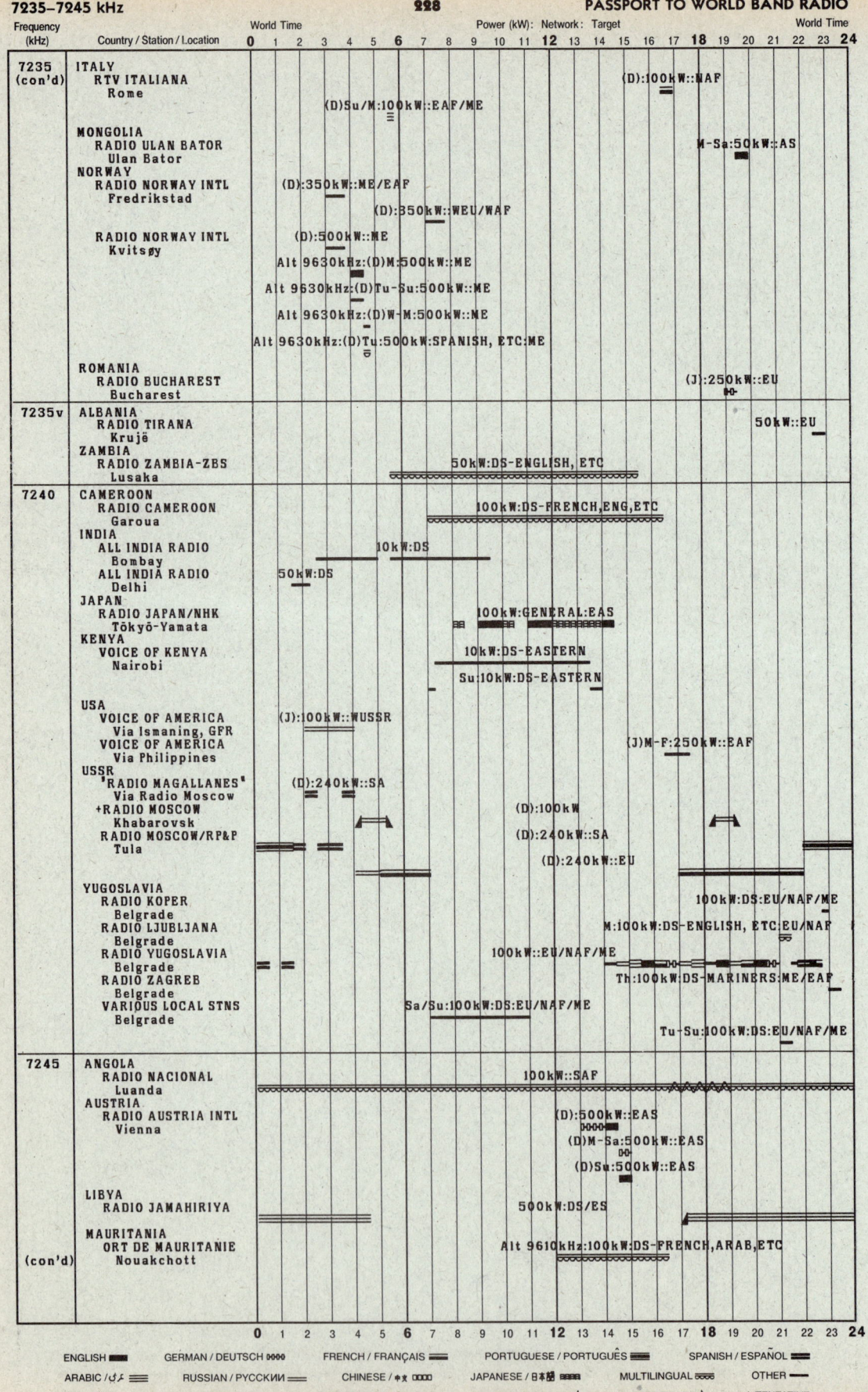

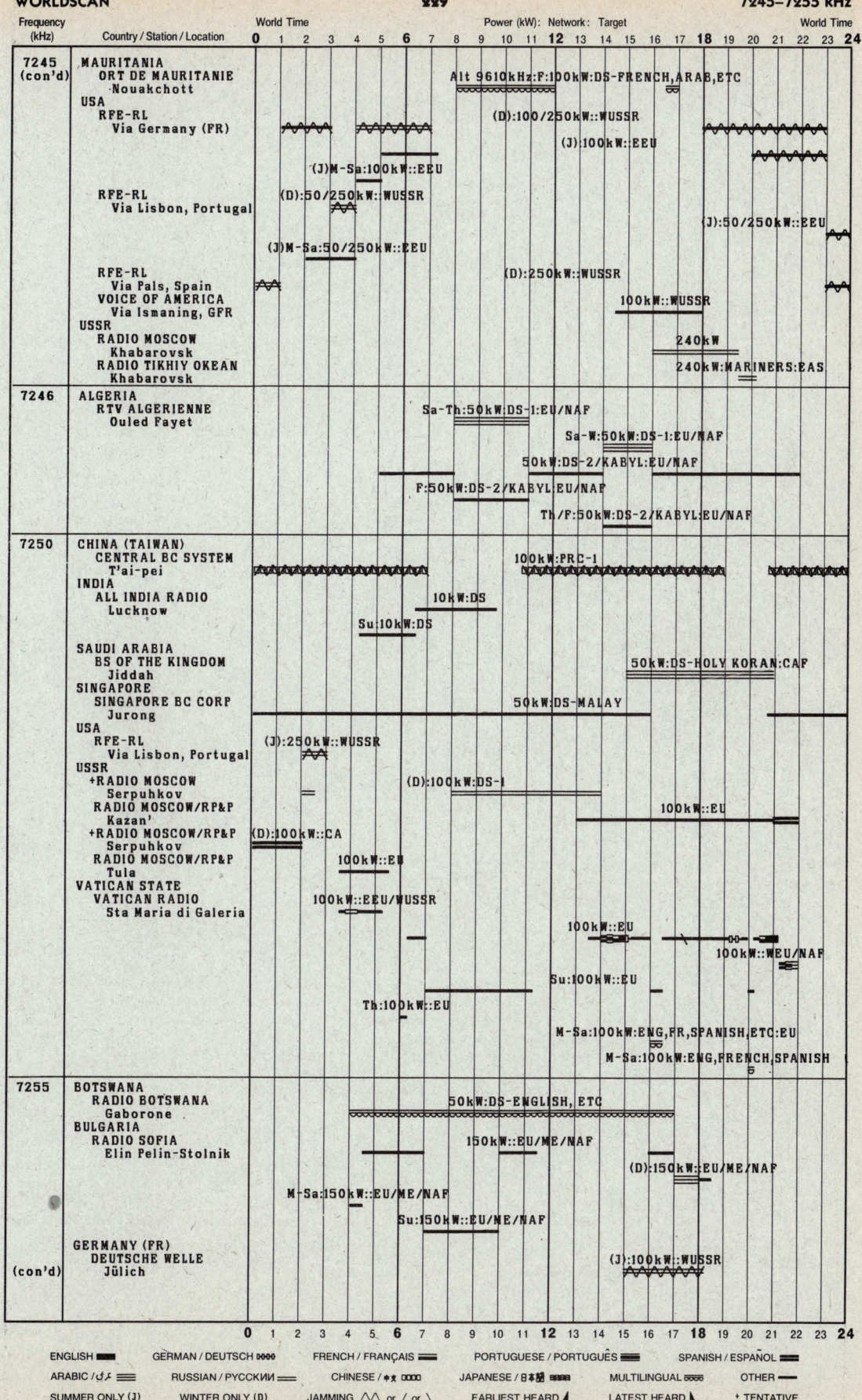

7255–7260 kHz

7255 kHz (con'd)

- **GERMANY (FR)** — DEUTSCHE WELLE, Türkheim-Wertach'l — (J):500kW::EEU (around 18)
- **INDIA** — ALL INDIA RADIO, Aligarh — 250kW::SAS (around 12–13)
- **IRAQ** — RADIO BAGHDAD — ME (around 14–16, jammed)
- **NIGERIA** — VOICE OF NIGERIA, Ikorodu — 50kW::WAF (approx 5–22)
- **NORWAY** — RADIO NORWAY INTL, Kvitsøy:
  - (D)M-Sa:500kW::ANZ (around 13)
  - (D)Su:500kW::ANZ (around 14)
  - (D)Tu-Su:500kW::ANZ (around 15)
  - (D)M:500kW:SPANISH, ETC:ANZ (around 15)
- **UNITED KINGDOM** — BBC, Holywell-Rampisham:
  - (D):500kW::EU (around 15)
  - (D)Sa-M/W/Th:500kW::EU (around 15–16)
  - (D)Su:500kW::EU (around 16)
- **BBC** Via Zyyi, Cyprus — 250kW::WUSSR (around 16–17)
- **USA** — RFE-RL Via Germany (FR):
  - 100kW::WUSSR (around 18–22)
  - (D):100kW::WUSSR (around 9–12)
  - (D):100kW::EEU (around 6–12, jammed)
- **RFE-RL** Via Lisbon, Portugal:
  - (J):50/250kW::WUSSR (around 0–3, jammed)
  - (D)M-Sa:50/250kW::EEU (around 3–6)
- **VOICE OF AMERICA** Via Philippines — (D):250kW::SEA (around 13–15)
- **USSR** — RADIO ALMA-ATA, Alma-Ata — (J):100kW::ME/SAS (around 13–15)
- **+RADIO MOSCOW**, Minsk — (D):100kW:DS-3 (around 17–19)
- **RADIO TASHKENT**, Alma-Ata — (J):100kW::ME/SAS (around 13–15)

7259.8 kHz

- **VANUATU** — RADIO VANUATU, Vila, Efate Island — 2kW:DS-ENGLISH,FR,ETC (approx 0–10 and 20–24)

7260 kHz

- **AUSTRIA** — RADIO AUSTRIA INTL, Vienna:
  - (J):300kW::SAF (around 18)
  - (J)M-Sa:300kW::SAF (around 19)
  - (J)Su:300kW::SAF (around 19)
- **CANADA** — RADIO CANADA INTL Via Sines, Portugal — (J):250kW::EU/WUSSR (around 2–3)
- **RADIO CANADA INTL** Via United Kingdom — (D):300kW::EU/WUSSR (around 2–3)
- **CHINA (PR)** — RADIO BEIJING, Kunming — 50kW::SEA (around 15–17)
- **GERMANY (DR)** — RADIO BERLIN INTL, Nauen — 500kW::EU (around 18–20)
- **INDIA** — ALL INDIA RADIO, Bombay — 100kW:DS (around 2–3)
- **UNITED KINGDOM** — BBC, Holywell-Rampisham:
  - 500kW::EEU (around 5)
  - M-F:500kW::EEU (around 5–6)
  - M-Sa:500kW::EEU (around 5–6)
  - Su:500kW::EEU (around 6)
- **BBC** Via Zyyi, Cyprus — 250kW:WORLD SCE, RUSSIAN:WUSSR (around 4–6)
- **USA** — VOICE OF AMERICA Via Philippines:
  - 250kW::EAF (around 19–20)
  - (J):50kW::SEA (around 13–14)

(con'd)

Legend: ENGLISH ▬▬ | GERMAN/DEUTSCH ○○○ | FRENCH/FRANÇAIS ═══ | PORTUGUESE/PORTUGUÊS ≡≡≡ | SPANISH/ESPAÑOL ≣≣≣ | ARABIC/عربي ▬ | RUSSIAN/РУССКИЙ ═══ | CHINESE/中文 ○○○○ | JAPANESE/日本語 ▬▬ | MULTILINGUAL ○○○ | OTHER ▬ | SUMMER ONLY (J) | WINTER ONLY (D) | JAMMING ∧∧ or / or \ | EARLIEST HEARD ◢ | LATEST HEARD ◣ | + TENTATIVE

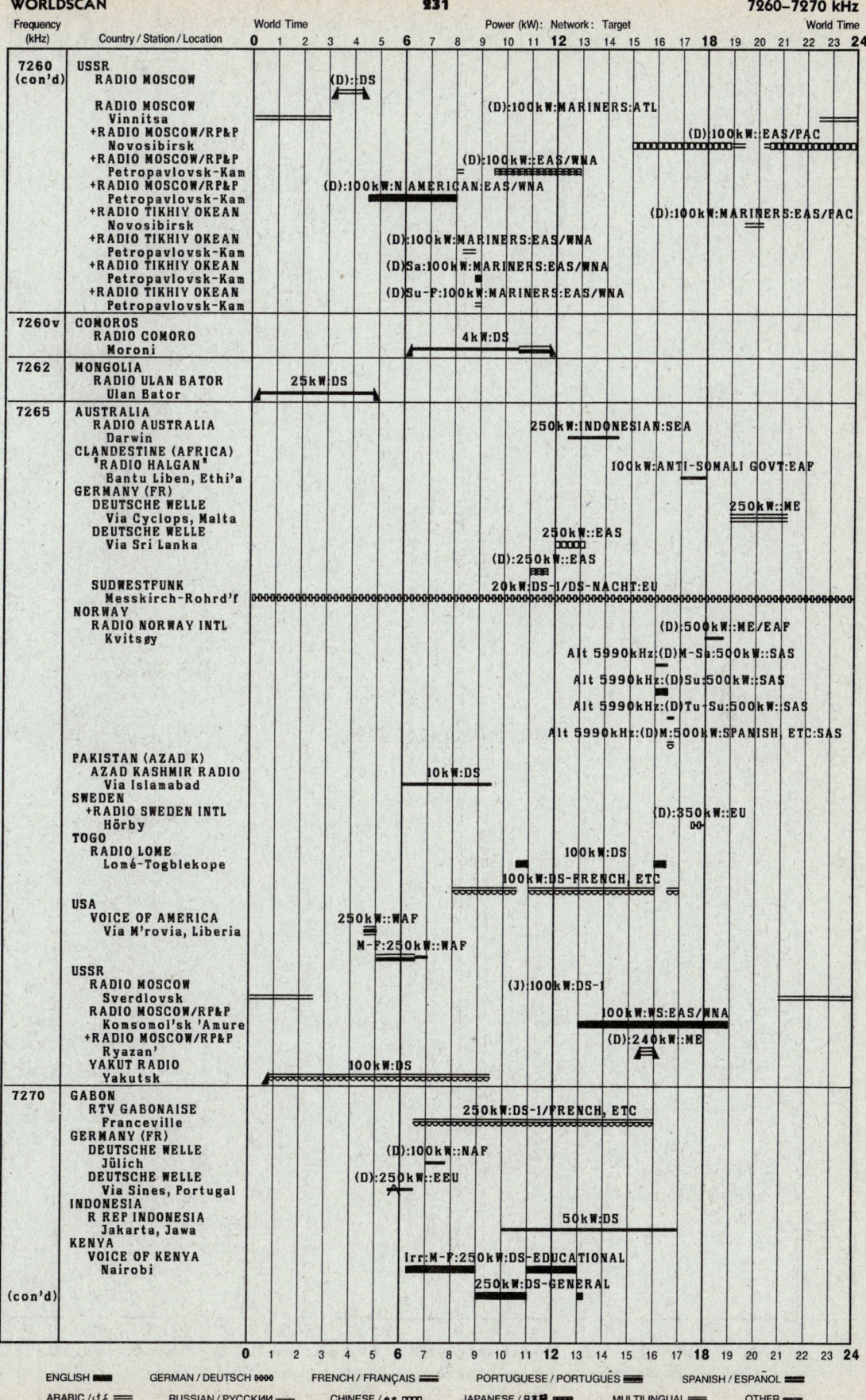

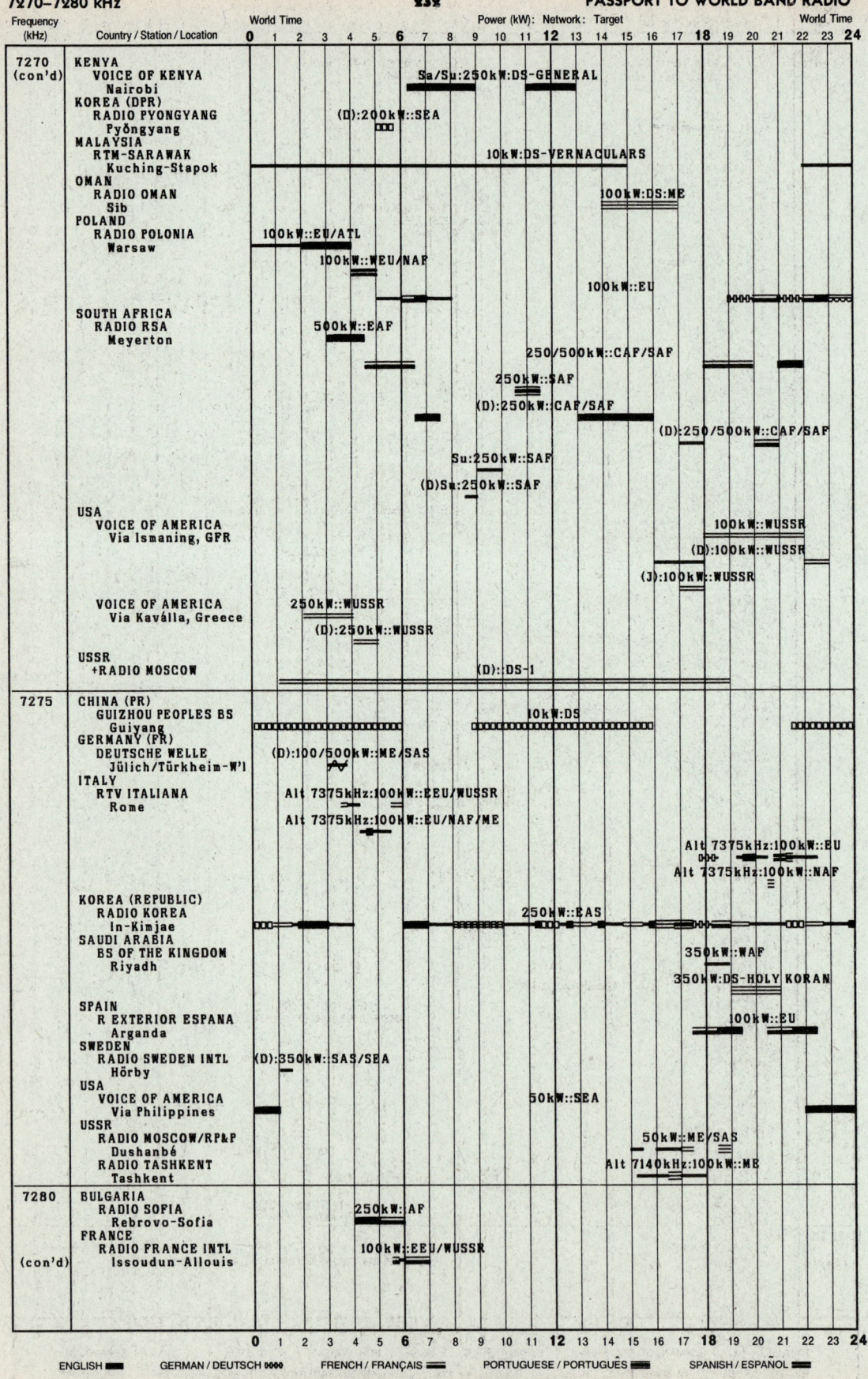

7280–7285v kHz

Frequency (kHz)	Country / Station / Location	Schedule
7280 (con'd)	**FRANCE** RADIO FRANCE INTL — Issoudun-Allouis	(D):100kW::EEU/WUSSR 06–09; (J):500kW::AF 03–06
	INDIA ALL INDIA RADIO — Delhi	50kW:DS-ENGLISH, ETC 13–14
	JAPAN RADIO JAPAN/NHK — Tōkyō-Yamata	100kW:GENERAL:SEA 12–14
	SAUDI ARABIA BS OF THE KINGDOM — Jiddah	50kW:DS-GENERAL:ME/EAF 03–08
	USA VOICE OF AMERICA — Via Ismaning, GFR	(D):100kW::WUSSR 19–21
	VOICE OF AMERICA — Via Kaválla, Greece	250kW::WUSSR 19–21; (D):250kW::WUSSR; (J):250kW::WUSSR; 250kW:HINDI,PERSIAN,ETC:ME/SAS 15–18
	VOICE OF AMERICA — Via M'rovia, Liberia	250kW::WAF/CAF 03–07
	USSR +RADIO KIEV — Moscow	(D):200kW::EU 18–20
	+RADIO KIEV — Moscow	(D):240kW:G:EU 18–20
	RADIO MOSCOW — Moscow	(J):200kW:MARINERS:ATL 08–12; 22–24
	+RADIO MOSCOW — Moscow	(D):200kW:DS-1 03–09
	RADIO MOSCOW/RP&P — Komsomol'sk 'Amure	(D):240kW:EAS 12–14
	+RADIO MOSCOW/RP&P — Moscow	(D):200kW:CA 09–12
	+RADIO MOSCOW/RP&P — Moscow	(D):200kW::EU 18–21
7285	**CANADA** RADIO CANADA INTL — Via Sines, Portugal	(D):250kW::EU/WUSSR 03–04
	CHINA (PR) GANSU PEOPLES BS — Lanzhou	15kW:DS-VERNACULARS 11–16
	CHINA (TAIWAN) VOICE FREE CHINA — T'ai-pei	SEA 08–09; EAS 10–11
	VOICE OF ASIA — Kao-hsiung	100kW::SEA 04–07
	GERMANY (FR) DEUTSCHE WELLE — Jülich	100kW::SEA/ANZ 05–07; 100kW::ME 17–18
	DEUTSCHE WELLE — Türkheim-Wertach'l	500kW:ME/SAS/SEA 03–05; 500kW::WUSSR 04–05
	DEUTSCHE WELLE — Via Sines, Portugal	(D):250kW::EEU 19–21; (J):500kW::AF 17–19
	HOLLAND RADIO NEDERLAND — Via Madagascar	300kW::SEA 09–12
	NIGERIA RADIO NIGERIA — Lagos	50kW:DS-1/ENGLISH, ETC 10–16
	POLAND RADIO POLONIA — Warsaw	100kW::EU 13–15, 18–22; 100kW::NAF 16–19; 100kW::EU/NAF 19–22
	USA VOICE OF AMERICA — Via Philippines	100kW::EAS 10–16
	USSR +RADIO MOSCOW — Moscow	(D):240kW:DS-2 08–15
	+RADIO MOSCOW/RP&P — Chita	(D):500kW::SEA 09–12
7285v (con'd)	**MALI** RTV MALIENNE — Bamako	50kW:DS-FRENCH, ETC 11–19

Legend: ENGLISH ▬ · GERMAN/DEUTSCH ◊◊◊◊ · FRENCH/FRANÇAIS ═ · PORTUGUESE/PORTUGUÊS ≡ · SPANISH/ESPAÑOL ▬ · ARABIC ﺽﺽ · RUSSIAN/РУССКИЙ ▬ · CHINESE/中文 □□□□ · JAPANESE/日本語 ▬ · MULTILINGUAL ◊◊◊◊ · OTHER ▬ · SUMMER ONLY (J) · WINTER ONLY (D) · JAMMING ∧∧ or / or \ · EARLIEST HEARD ◢ · LATEST HEARD ◣ · + TENTATIVE

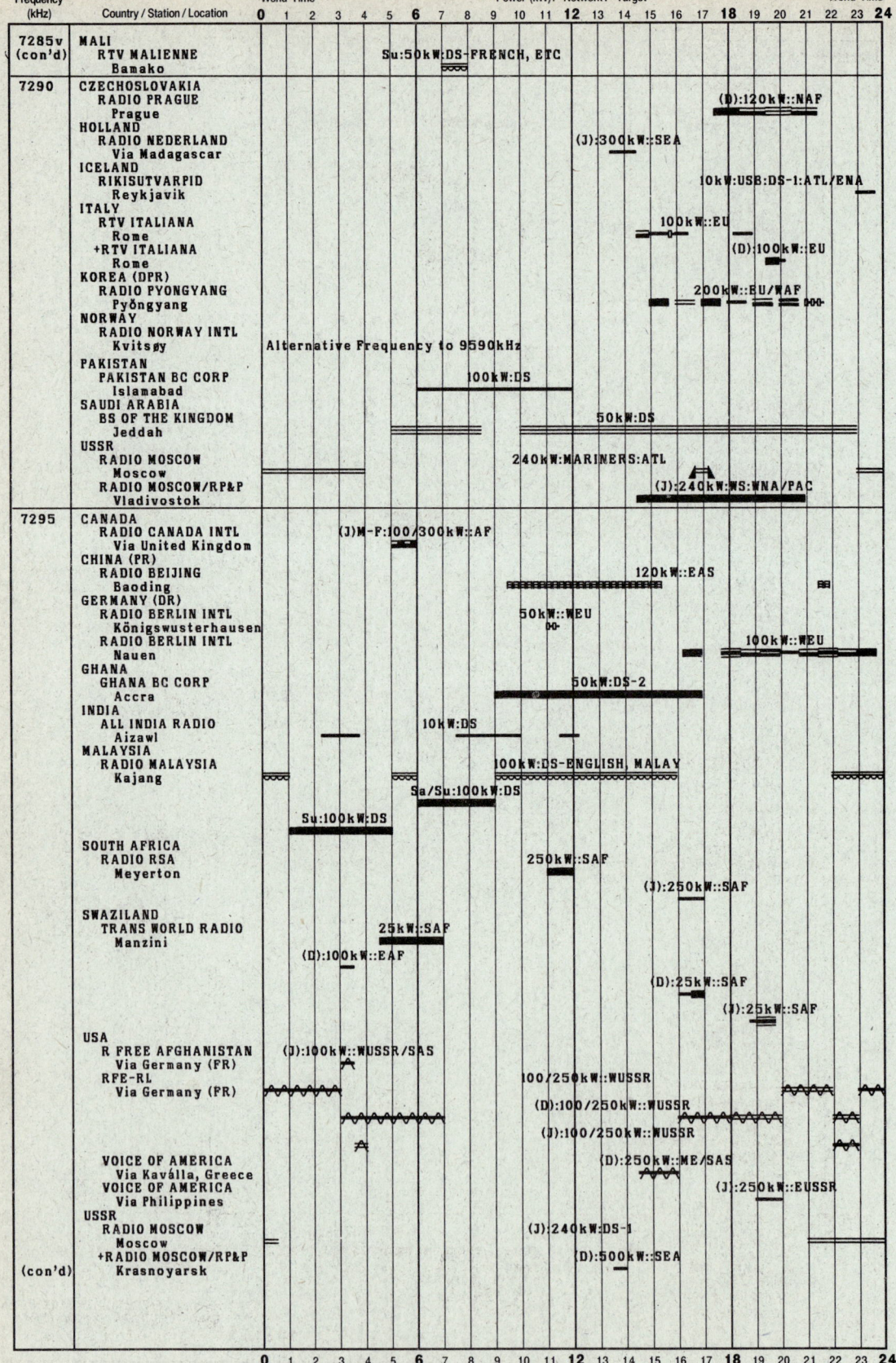

WORLDSCAN 7295–7320 kHz

Frequency (kHz)	Country / Station / Location	Schedule (World Time 0–24) — Power (kW) : Network : Target
7295 (con'd)	USSR — +RADIO MOSCOW/RP&P, Krasnoyarsk	(D):500kW:WS:SEA (13–17)
	ZAIRE — LA VOIX DU ZAIRE, Mbuji-Mayi	Irr::10kW:DS (05–13); Irr:Sa/Su:10kW:DS (12–17); Irr:Su:10kW:DS (04–07)
7300	ALBANIA — RADIO TIRANA, Lushnjë	100kW::NA/CA (0–5); 100kW::EU (13–15); 100kW::WUSSR (14–16); 50kW:ITALIAN:EU (15–17); 100kW:PERSIAN,TURKISH:ME (10–13)
	USSR — RADIO MOSCOW	(D):::CA (12–14)
7305	USSR — +RADIO MOSCOW, Krasnodar	100kW:DS-2 (09–13)
	+RADIO MOSCOW, Novosibirsk	(D):100kW:DS-1 (10–16)
	RADIO MOSCOW, Star'obel'sk	100kW:DS-2 (10–13)
	RADIO MOSCOW/RP&P, Star'obel'sk	100kW::ME (15–19); 100kW:WS:ME (05–13)
7310	ALBANIA — RADIO TIRANA, Lushnjë	50kW:GERMAN,POLISH,ETC:EU (13–22)
	MONACO — TRANS WORLD RADIO, Monte Carlo	100kW::WUSSR (14–16); M/F/Sa:100kW::WUSSR (15–17)
	TUNISIA — RTV TUNISIENNE, Sfax	Alt 7225kHz:Irr::100kW:RAMADAN:EU (09–12); Alt 7225kHz:100kW::EU (11–14)
	USSR — +RADIO MOSCOW, Khabarovsk	(D):100kW:DS-2 (06–08)
	RADIO MOSCOW, Moscow	100kW::EU/ATL (09–14)
	+RADIO MOSCOW, Moscow	(D):100kW::EU/ATL (0–6)
	RADIO MOSCOW, Nikolayevsk 'Amure	100kW:DS-2 (15–20)
	RADIO MOSCOW/RP&P, Moscow	(D):100kW:N AMERICAN:ENA (10–16); (D):100kW::ENA (0–6); (D):100kW:WS:EU/ATL (16–22)
7315	CHINA (PR) — RADIO BEIJING	WAF/CAF (15–17)
	RADIO BEIJING, Xi'an	120kW::SAS (13–15)
	PAKISTAN — RADIO PAKISTAN, Islamabad	100kW::ME (14–16)
	RADIO PAKISTAN, Karachi	50kW::SAS/SEA (1–5)
	USSR — RADIO MOSCOW	(D)::MARINERS (10–14)
	RADIO MOSCOW, Kalinin	(J):100kW:DS-1 (0–4)
	RADIO MOSCOW/RP&P, Irkutsk	(J):500kW:WS:EAS (10–14); (D):500kW::EAS (19–22); (D):500kW:WS:SAS (12–16)
7320	MOZAMBIQUE — EP DE NAMPULA, Nampula	0.25kW:DS-PORTUGUESE,ETC (04–22)
	UNITED KINGDOM — BBC, Multiple Locations	250/300kW::ME/NAF (3–6); 250/300kW:WORLD SERVICE:EU (13–22)

ENGLISH ▬▬ GERMAN / DEUTSCH ◦◦◦◦ FRENCH / FRANÇAIS ═══ PORTUGUESE / PORTUGUÊS ≡≡≡ SPANISH / ESPAÑOL ▬▬

ARABIC / عربي ═ RUSSIAN / РУССКИЙ ══ CHINESE / 中文 ▫▫▫▫ JAPANESE / 日本語 ▬▬ MULTILINGUAL ▒▒▒▒ OTHER ───

SUMMER ONLY (J) WINTER ONLY (D) JAMMING ∧∧ or / or \ EARLIEST HEARD ◢ LATEST HEARD ◣ + TENTATIVE

7320–7340 kHz

Frequency (kHz)	Country / Station / Location	Schedule (World Time 0–24)
7320 (con'd)	**USSR** MAGADAN RADIO, Yakutsk	100kW:DS
	+RADIO MOSCOW/RP&P, Krasnodar	(D):500kW:N AMERICAN:ENA
	RADIO MOSCOW/RP&P, Minsk	(J):100kW:MARINERS:ENA/CA / (J):100kW::EU / (J):100kW:WS:EU
7320v	**CLANDESTINE (ASIA)** "VIET RESISTANCE" Eastern Thailand	ANTI-HANOI:SEA
7325	**PAKISTAN** +PAKISTAN BC CORP, Rawalpindi	(D):10kW:DS
	+PAKISTAN BC CORP, Rawalpindi	(D)Sa-Th:10kW:DS
	UNITED KINGDOM BBC, Various Locations	WORLD SERVICE:ANZ / (J):::EEU
	USA VOICE OF AMERICA, Via Woofferton, UK	300kW::EU
	USSR RADIO MOSCOW, Novosibirsk	50kW:DS-1 / (J):50kW:DS-2
	+RADIO MOSCOW/RP&P, Novosibirsk	(D):500kW::EU
	RADIO TASHKENT, Frunze, Kirgiziya	(J):100kW::SAS
7330	**USSR** RADIO KIEV, Kiev	(J):100kW::EU/ATL
	RADIO MOSCOW/RP&P	(D):::WAF/SA
	RADIO MOSCOW/RP&P, Chita	100kW::EAS
	+RADIO MOSCOW/RP&P, Chita	(D):100kW::SAS/ME
	+RADIO MOSCOW/RP&P, Khabarovsk	(D):240kW::EAS
	RADIO MOSCOW/RP&P, Kiev	(J):100kW::EU/ATL
	RS SOV BELORUSSIA, Kiev	(J)Su-F:100kW::EU/ATL / (J)Sa:100kW:1ST & 3RD SATS:EU/ATL / (J)Sa:100kW:2ND & 4TH SATS:EU/ATL
7335	**CHINA (PR)** CENTRAL PEOPLES BS, Xi'an	50kW:DS-1
	CLANDESTINE (EUROPE) "OUR RADIO" Schwerin, E Germany	TURKISH COMMUNIST:EU/ME
	"VO TURKISH CP" Schwerin, E Germany	TURKISH COMMUNIST:EU/ME
	USSR RADIO MOSCOW, Khabarovsk	100kW:DS-1 / 100kW:DS-2
	+RADIO MOSCOW/RP&P, Simferopol'	(D):500kW:N AMERICAN:NA/CA
	+RADIO MOSCOW/RP&P, Simferopol'	(D):500kW::NA/CA
	+RADIO MOSCOW/RP&P, Simferopol'	(D):500kW:UK SVC:EU
	+RADIO MOSCOW/RP&P, Simferopol'	(D):500kW:WS:NA/CA
	+RADIO MOSCOW/RP&P, Simferopol'	(D):500kW:WS:EU
	RADIO TIKHIY OKEAN, Khabarovsk	100kW:MARINERS
7340	**USSR** RADIO MOSCOW, Krasnoyarsk	(J):100kW:DS-2
	+RADIO MOSCOW, Leningrad	(D):100kW:DS-1
	+RADIO MOSCOW, Yerevan	(D):120kW::ATL
	+RADIO MOSCOW/RP&P	(D)::WS
	+RADIO MOSCOW/RP&P, Irkutsk	(D):500kW::EAS
(con'd)	+RADIO MOSCOW/RP&P, Irkutsk	(D):500kW::SAS

ENGLISH ▬ GERMAN / DEUTSCH ▭ FRENCH / FRANÇAIS ═ PORTUGUESE / PORTUGUÊS ≡ SPANISH / ESPAÑOL ≋
ARABIC /ﻋﺮﺑﻲ ═ RUSSIAN / РУССКИЙ ═ CHINESE / 中文 ▭ JAPANESE / 日本語 ▭ MULTILINGUAL ▭ OTHER ▬
SUMMER ONLY (J)   WINTER ONLY (D)   JAMMING ∧∧ or / or \   EARLIEST HEARD ▲   LATEST HEARD ▼   + TENTATIVE

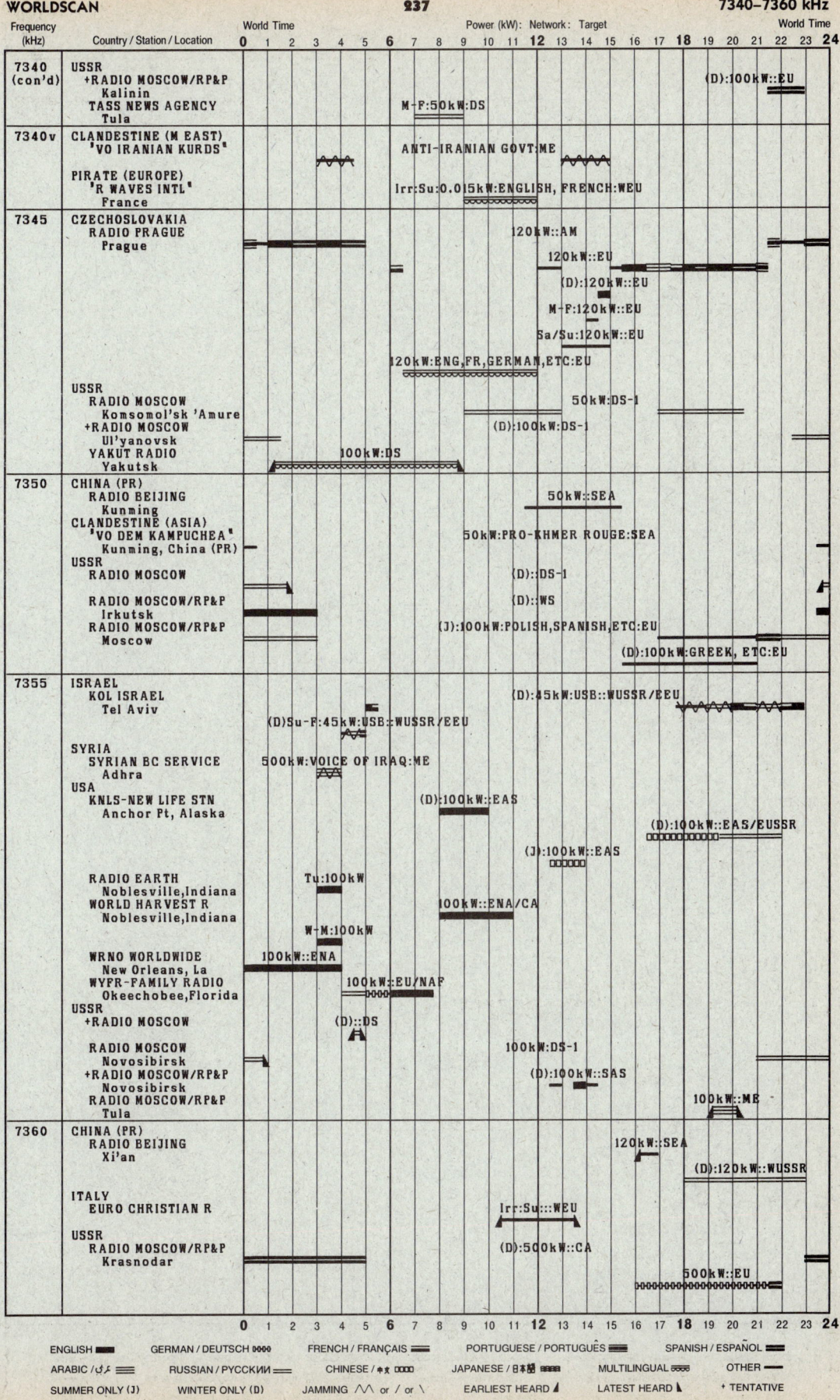

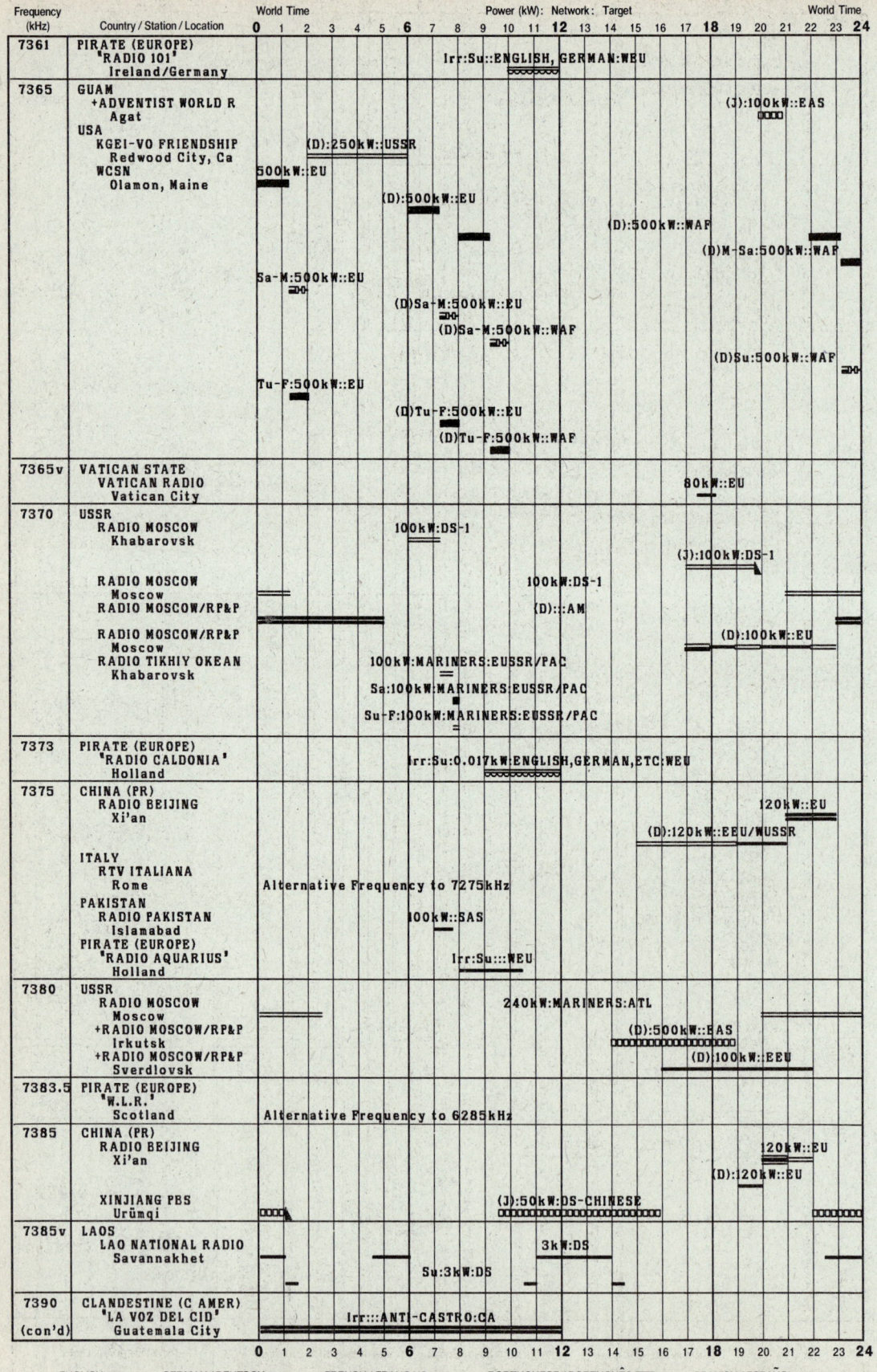

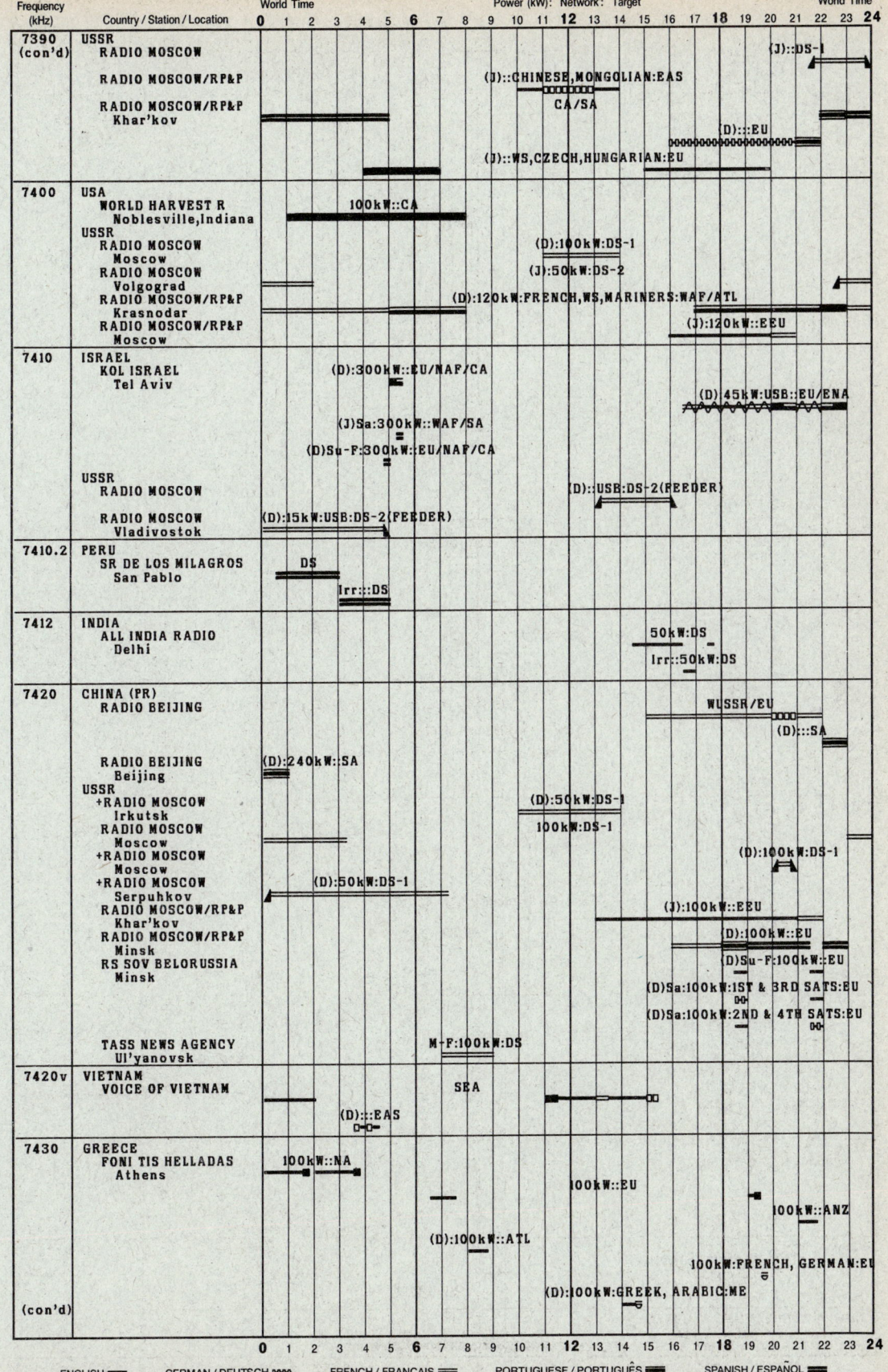

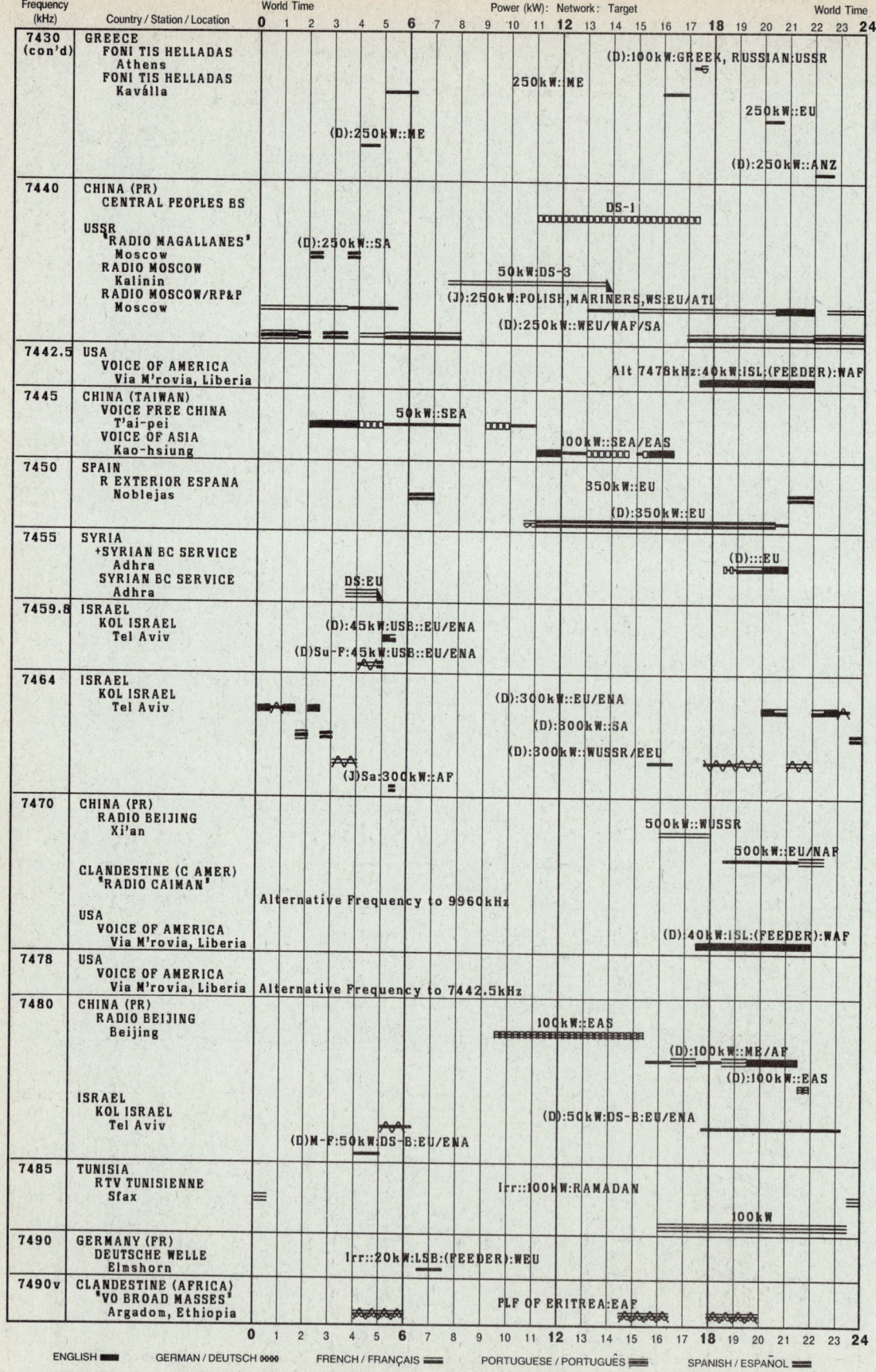

Frequency (kHz)	Country / Station / Location	Schedule (World Time 0–24)
7492.8	USSR — RADIO MOSCOW, Birobidzhan	15kW:LSB:(FEEDER):EUSSR
	RADIO MOSCOW/RP&P, Birobidzhan	15kW:LSB:N AMERICAN(FEEDER):EUSSR; 15kW:LSB:(FEEDER):EUSSR; 15kW:LSB:WS(FEEDER):EUSSR
	RADIO TIKHIY OKEAN, Birobidzhan	15kW:LSB:MARINERS(FEEDER):EUSSR; Sa:15kW:LSB:MARINERS(FEEDER):EUSSR; Su-F:15kW:LSB:MARINERS(FEEDER):EUSSR
7504	CHINA (PR) — CENTRAL PEOPLES BS, Xi'an	120kW:DS-1; W-M:120kW:DS-1
7505	BANGLADESH — RADIO BANGLADESH, Dhaka	250kW::EU; (D):250kW::ME
7516	CHINA (PR) — CENTRAL PEOPLES BS, Beijing	50kW:DS-1
7517.5	CHINA (PR) — RADIO BEIJING	Irr::::WUSSR
7525	CHINA (PR) — CENTRAL PEOPLES BS	DS-1
	CLANDESTINE (ASIA) "CENTRAL PBS", Primorskiy K'y, USSR	Irr:::FAKE CPBS STATION:EAS
	TUNISIA — RTV TUNISIENNE, Sfax	100kW:DS
7550	KOREA (REPUBLIC) — RADIO KOREA, In-Kimjae	100/250kW::ME/AF; 250kW::EU
7565	USSR — KAZAKH TELEGRAPH, Alma-Ata	M/W/F:15kW:DS-NEWSCAST
7565.5	USA — AFRTS-US MILITARY, Via Barford, UK	Irr::4kW:LSB:DS-ABC/CBS/NBC/NPR(FEEDER):ATL
7590	CHINA (PR) — RADIO BEIJING, Kunming	120kW::SEA; 120kW::SEA/SAS
7615	USSR — RADIO MOSCOW/RP&P, Moscow	Irr::20kW:LSB:(FEEDER)
7620	CHINA (PR) — CENTRAL PEOPLES BS	TAIWAN-1
7651	USA — VOICE OF AMERICA, Greenville, NC	50kW:ISL:(FEEDER):EU/ME; (D):50kW:ISU:(FEEDER):EU/ME; (J):50kW:ISU:(FEEDER):EU/ME
7670	BULGARIA — BULGARIAN RADIO, Elin Pelin-Stolnik	15kW:DS-1
7700	CHINA (PR) — RADIO BEIJING, Kunming	50kW::EEU/WUSSR
7725	USA — VOICE OF AMERICA, Via Ismaning, GFR	40kW:USB:(FEEDER):EEU/ME; (D):40kW:USB:(FEEDER):EEU/ME
7770	CHINA (PR) — CENTRAL PEOPLES BS, Kunming	50kW:DS-2; Th/Sa-M:50kW:DS-2
	ITALY — R CALABRIA INTL, Gioiosa Iónica	Sa/Su:0.2/1kW::EU
7775	CHINA (PR) — RADIO BEIJING, Beijing	240kW::SEA

ENGLISH ▬ GERMAN / DEUTSCH ▭ FRENCH / FRANÇAIS ═ PORTUGUESE / PORTUGUÊS ≡ SPANISH / ESPAÑOL ═
ARABIC / عربى RUSSIAN / РУССКИЙ CHINESE / 中文 JAPANESE / 日本語 MULTILINGUAL OTHER
SUMMER ONLY (J) WINTER ONLY (D) JAMMING ∧∧ or / or \ EARLIEST HEARD ◢ LATEST HEARD ◣ + TENTATIVE

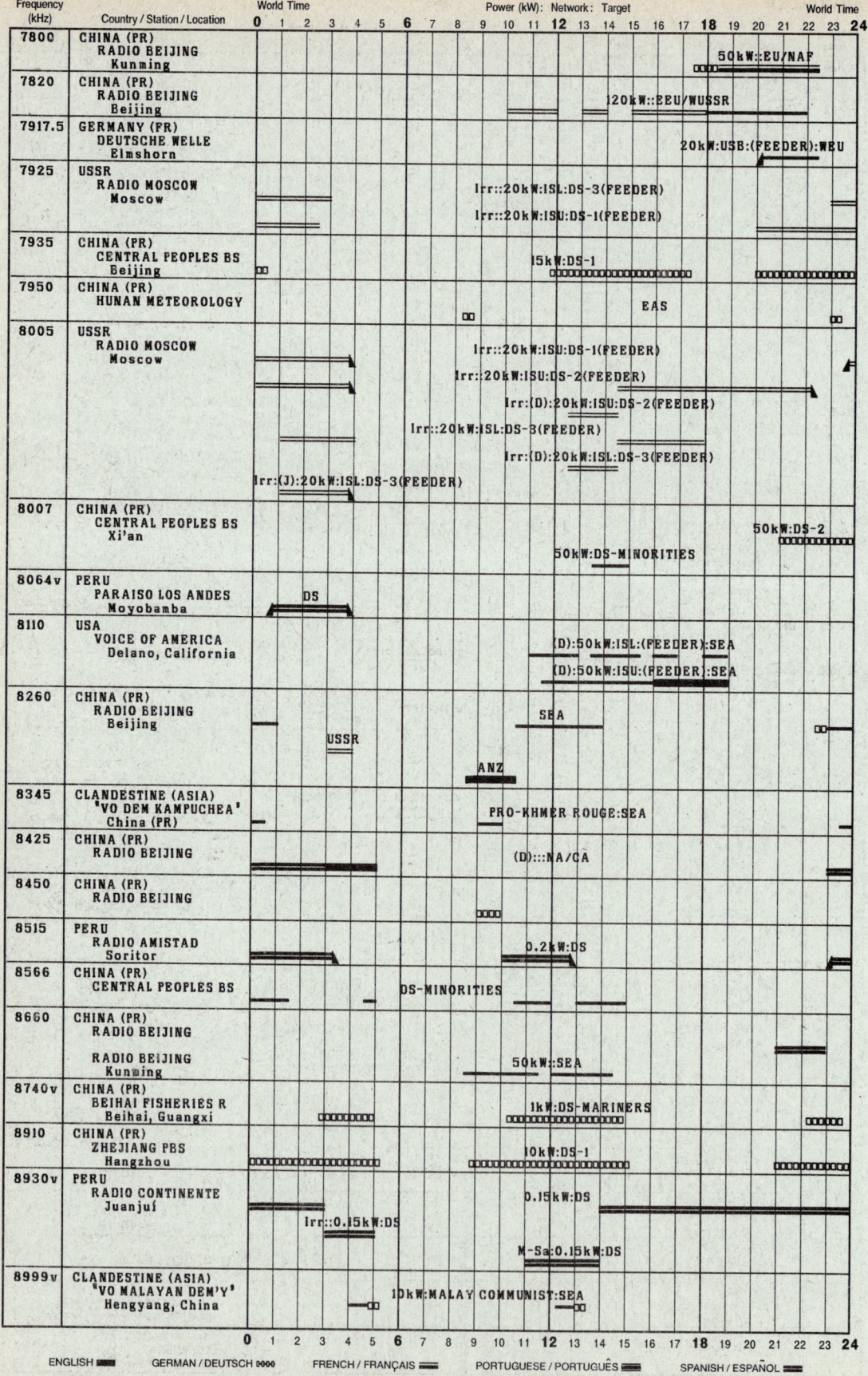

WORLDSCAN 9010.5–9325 kHz

Frequency (kHz)	Country / Station / Location	Schedule (World Time 0–24, Power kW, Network, Target)
9010.5	ISRAEL — KOL ISRAEL — Tel Aviv	Alternative Frequency to 9012 kHz
9012	ISRAEL — KOL ISRAEL — Tel Aviv	Alt 9010.5kHz:300kW::ME; Alt 9010.5kHz:300kW::AF/SA; Alt 9010.5kHz:300kW:WUSSR/EEU; (D):100kW::EU; Alt 9010.5kHz:(D):300kW:WUSSR/EEU; Alt 9010.5kHz:(D):300kW::EU; (D)Sa:300kW::AF; (D)Su-F:100kW::EU
9020	CHINA (PR) — CENTRAL PEOPLES BS	DS-2; Th/Sa-M::DS-2
9022	IRAN — VO THE ISLAMIC REP — Tehrān	350kW:PERSIAN:EU/NAF/CA; 350kW::EU/NAF/CA; 350kW::EU/NAF/AF; 350kW:PERSIAN:EU/NAF
9027	CLANDESTINE (M EAST) — "IRAN FREEDOM FLAG" — Abu Za'bal, Egypt	Alt 9040kHz:100kW:ANTI-KHOMEYNI:ME
9030	CHINA (PR) — CENTRAL PEOPLES BS	DS-2
9040	CLANDESTINE (M EAST) — "IRAN FREEDOM FLAG" — Abu Za'bal, Egypt	Alternative Frequency to 9027 kHz
9064	CHINA (PR) — CENTRAL PEOPLES BS — Kunming	15kW:DS-1; W-M:15kW:DS-1
9080	CHINA (PR) — CENTRAL PEOPLES BS — Kunming	50kW:DS-1
9090	USSR — KAZAKH TELEGRAPH — Alma-Ata	M/W/F:15kW:DS-NEWSCAST
9115	ARGENTINA — RADIO RIVADAVIA — Buenos Aires	Irr:M:10kW:USB:DS(FEEDER); Irr:Su:10kW:USB:DS(FEEDER); Su:10kW:USB:DS(FEEDER)
9140	USSR — RADIO MOSCOW — Moscow	(J):20kW:LSB:DS-1(FEEDER)
9170	CHINA (PR) — CENTRAL PEOPLES BS — Beijing	10kW:TAIWAN-2
9200	USSR — RADIO MOSCOW — Khabarovsk	Irr::15kW:LSB:DS-2(FEEDER); Irr:(D):15kW:LSB:DS-2(FEEDER)
9210	USSR — RADIO MOSCOW; RADIO MOSCOW/RP&P	(J)::ISU:DS-1(FEEDER); (J)::ISL:(FEEDER)
9215	CLANDESTINE (ASIA) — "VOICE OF UNITY" — Abu Za'bal, Egypt	(D):100kW:PRO-AFGHAN REBELS:ME/SAS
9220	KOREA (DPR) — RADIO PYONGYANG — Pyŏngyang	200kW::AF
9270	CLANDESTINE (ASIA) — "OCTOBER STORM" — Primorskiy K'y, USSR	Irr:::ANTI-DENG XIAOPING:EAS
9290	CHINA (PR) — RADIO BEIJING — Kunming	50kW::EEU/WUSSR
9325	KOREA (DPR) — RADIO PYONGYANG — Pyŏngyang	200kW::EU/WUSSR

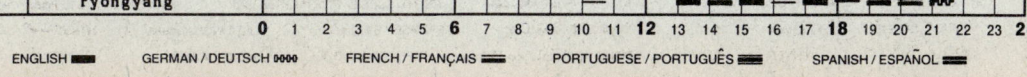

ENGLISH ▬ GERMAN / DEUTSCH ○○○○ FRENCH / FRANÇAIS ═ PORTUGUESE / PORTUGUÊS ≡ SPANISH / ESPAÑOL ═
ARABIC / عربي ═ RUSSIAN / РУССКИЙ ═ CHINESE / 中文 ▫▫▫▫ JAPANESE / 日本語 ═ MULTILINGUAL ⊗⊗⊗ OTHER ─
SUMMER ONLY (J)   WINTER ONLY (D)   JAMMING ∧∧ or / or \   EARLIEST HEARD ◢   LATEST HEARD ◣   + TENTATIVE

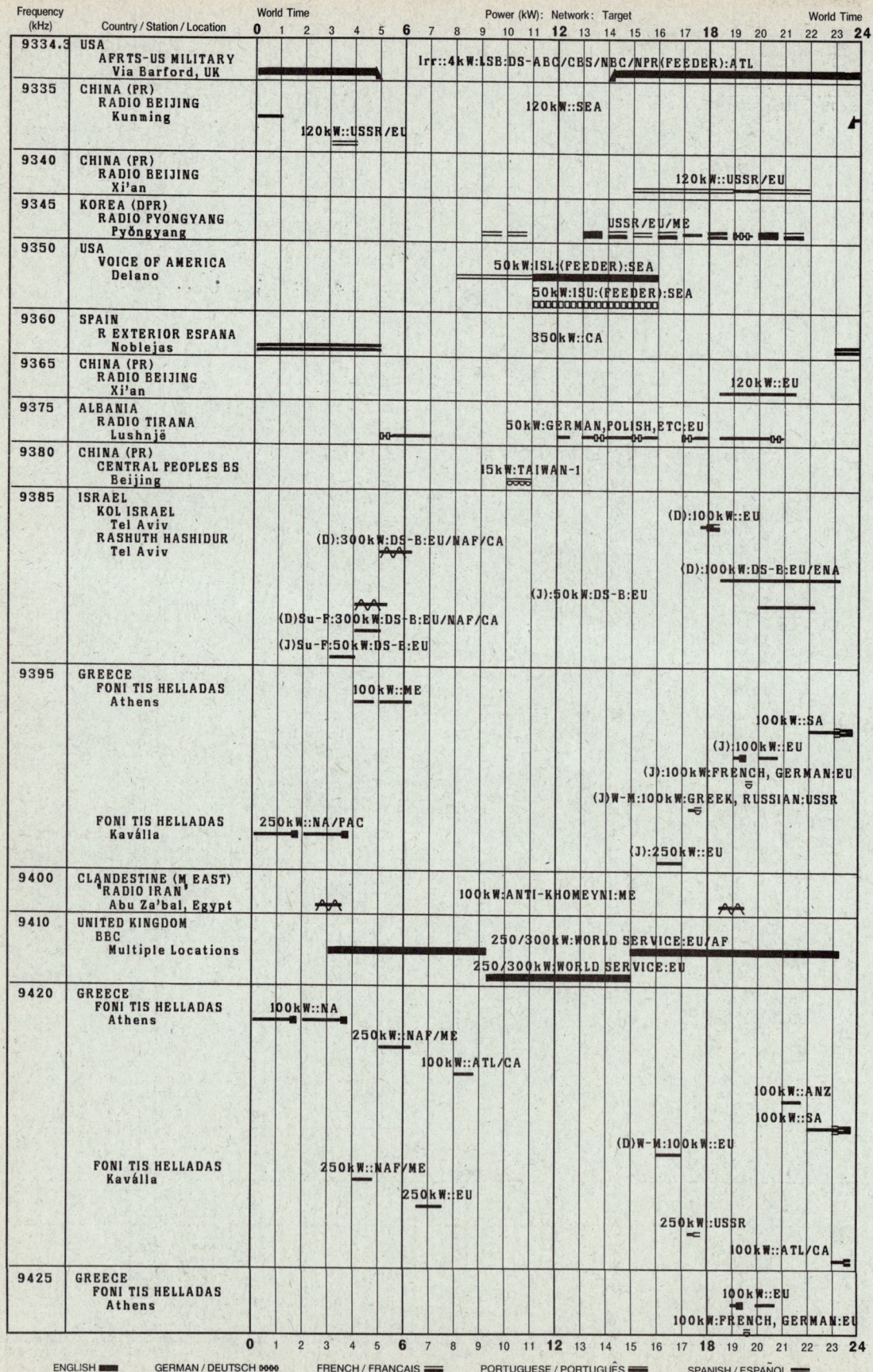

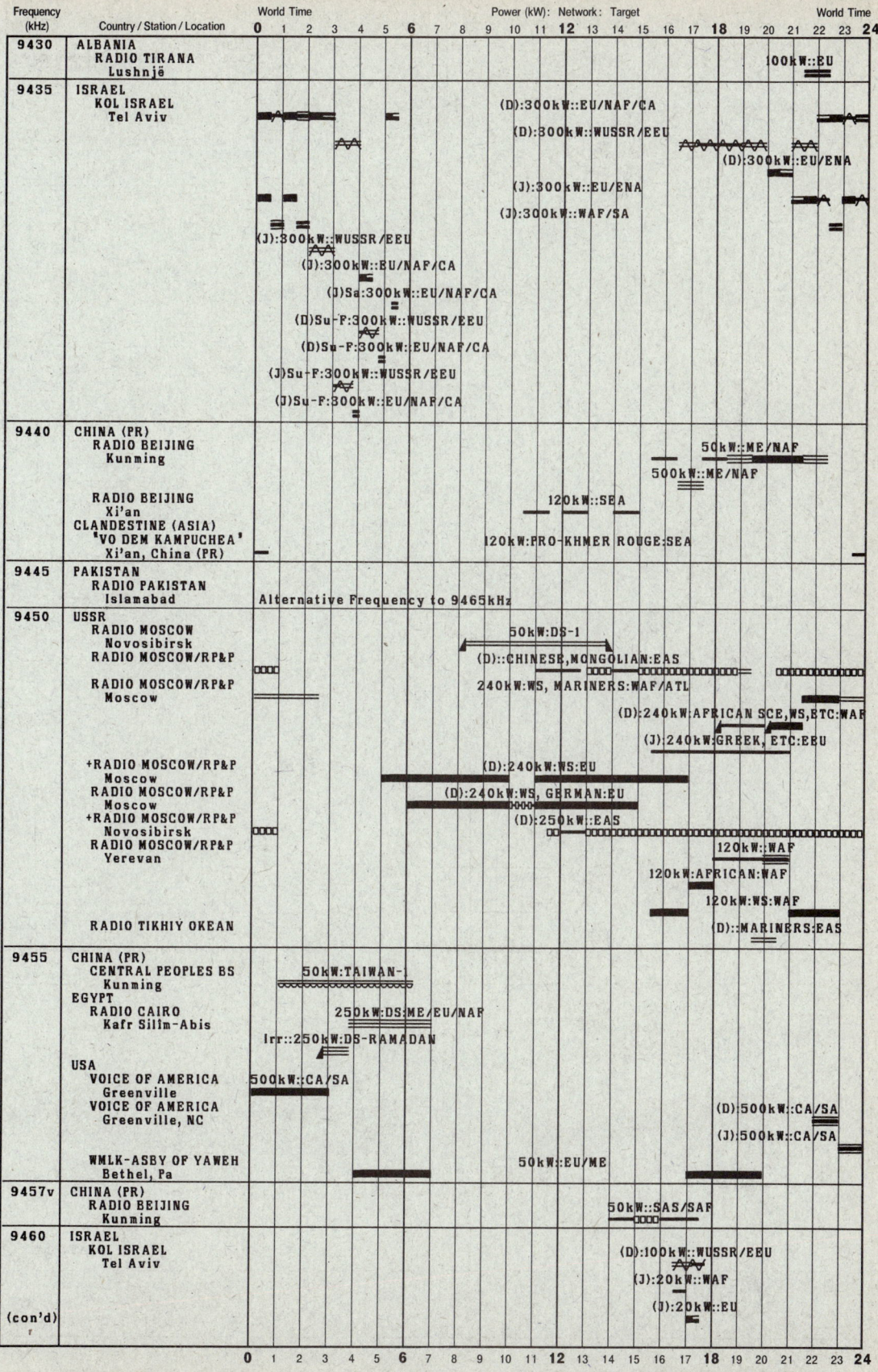

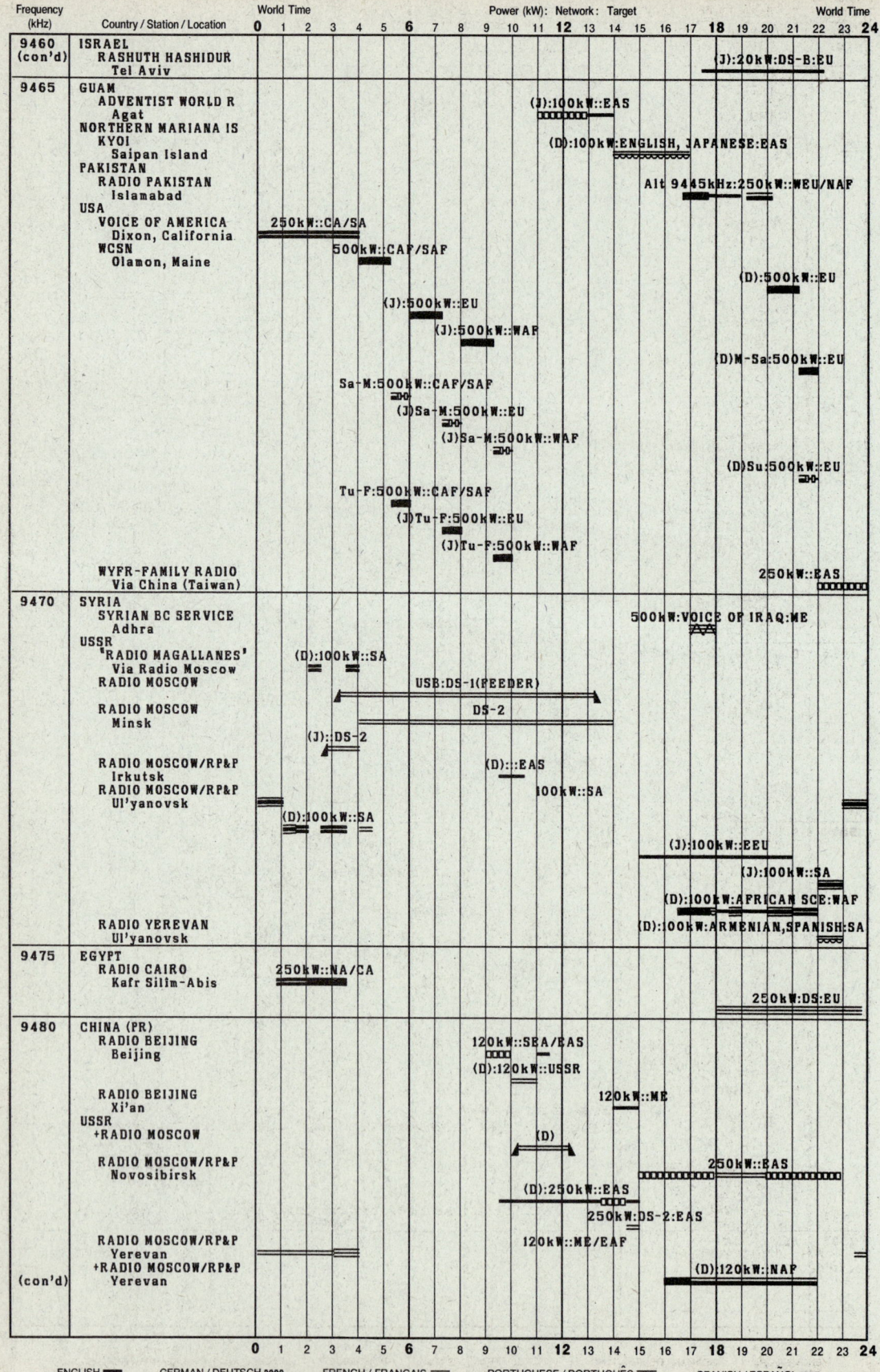

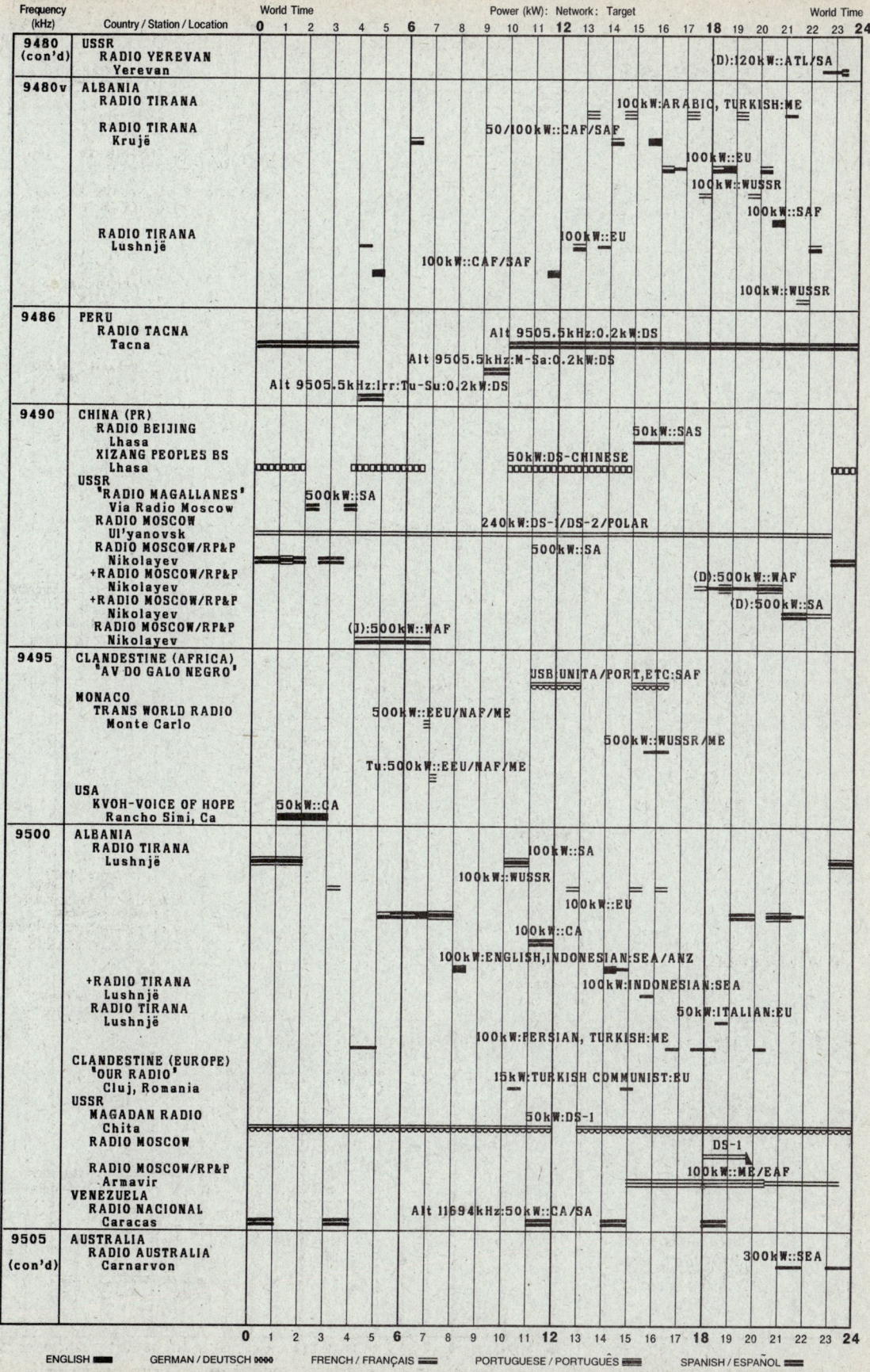

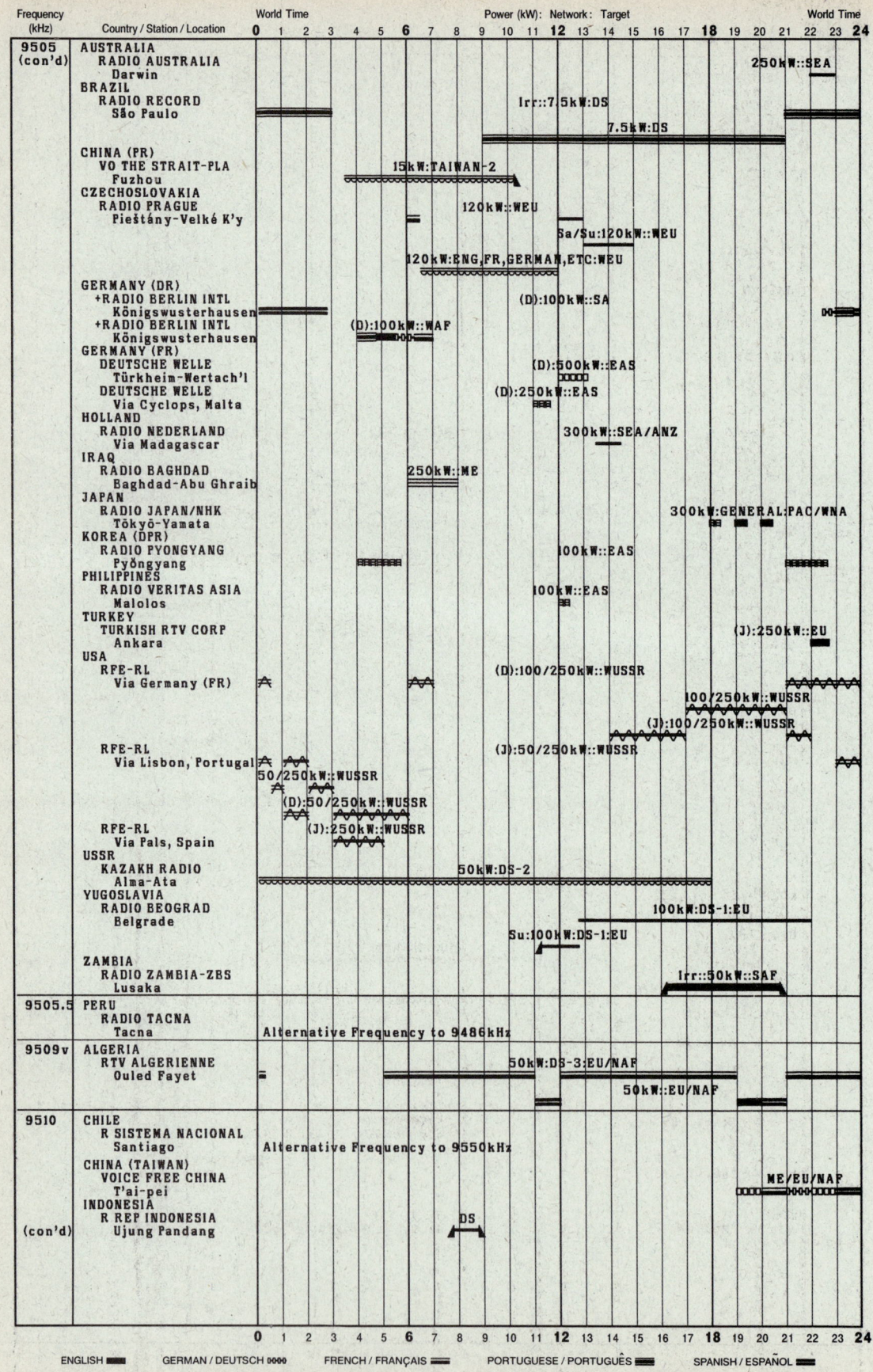

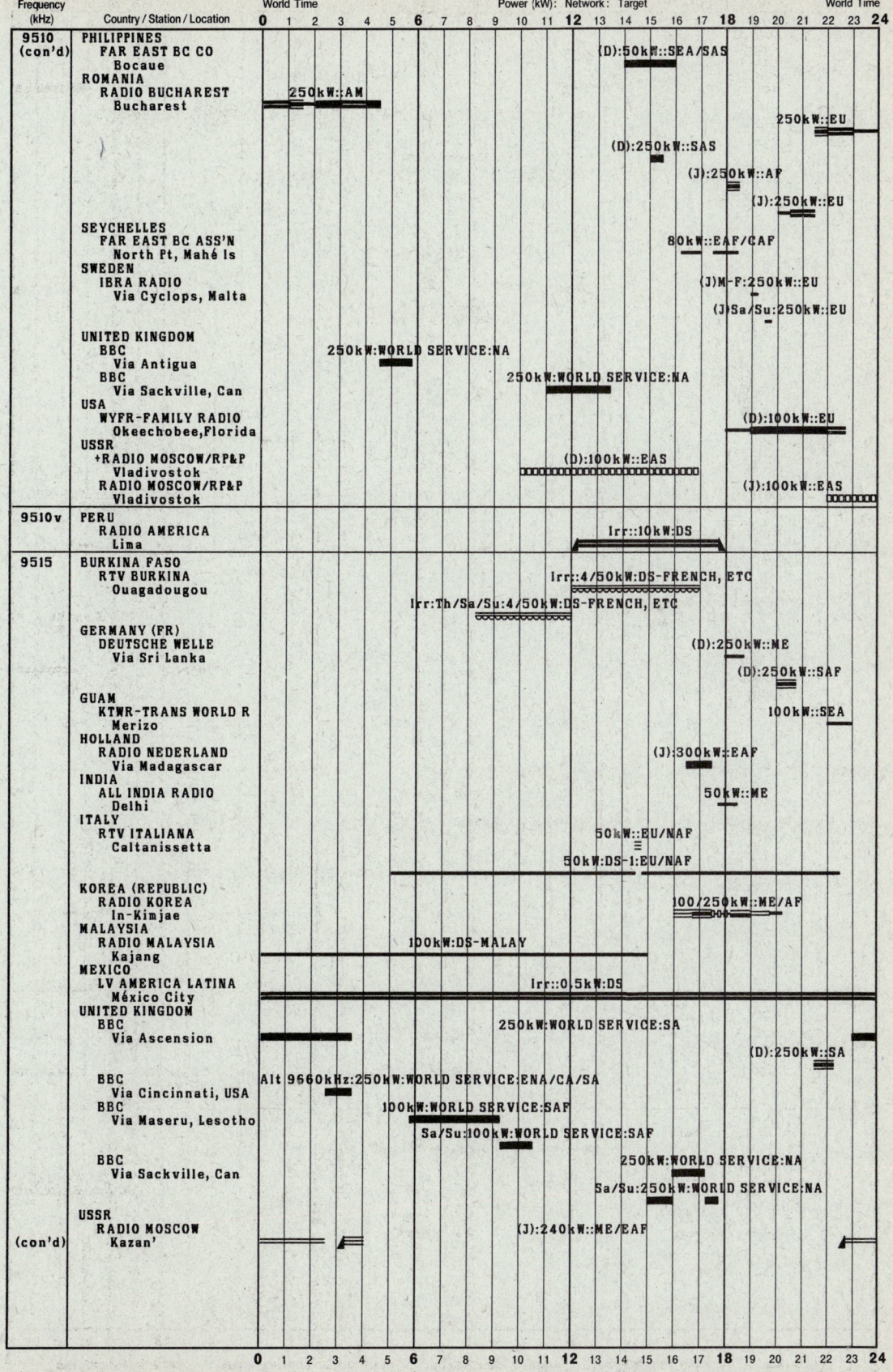

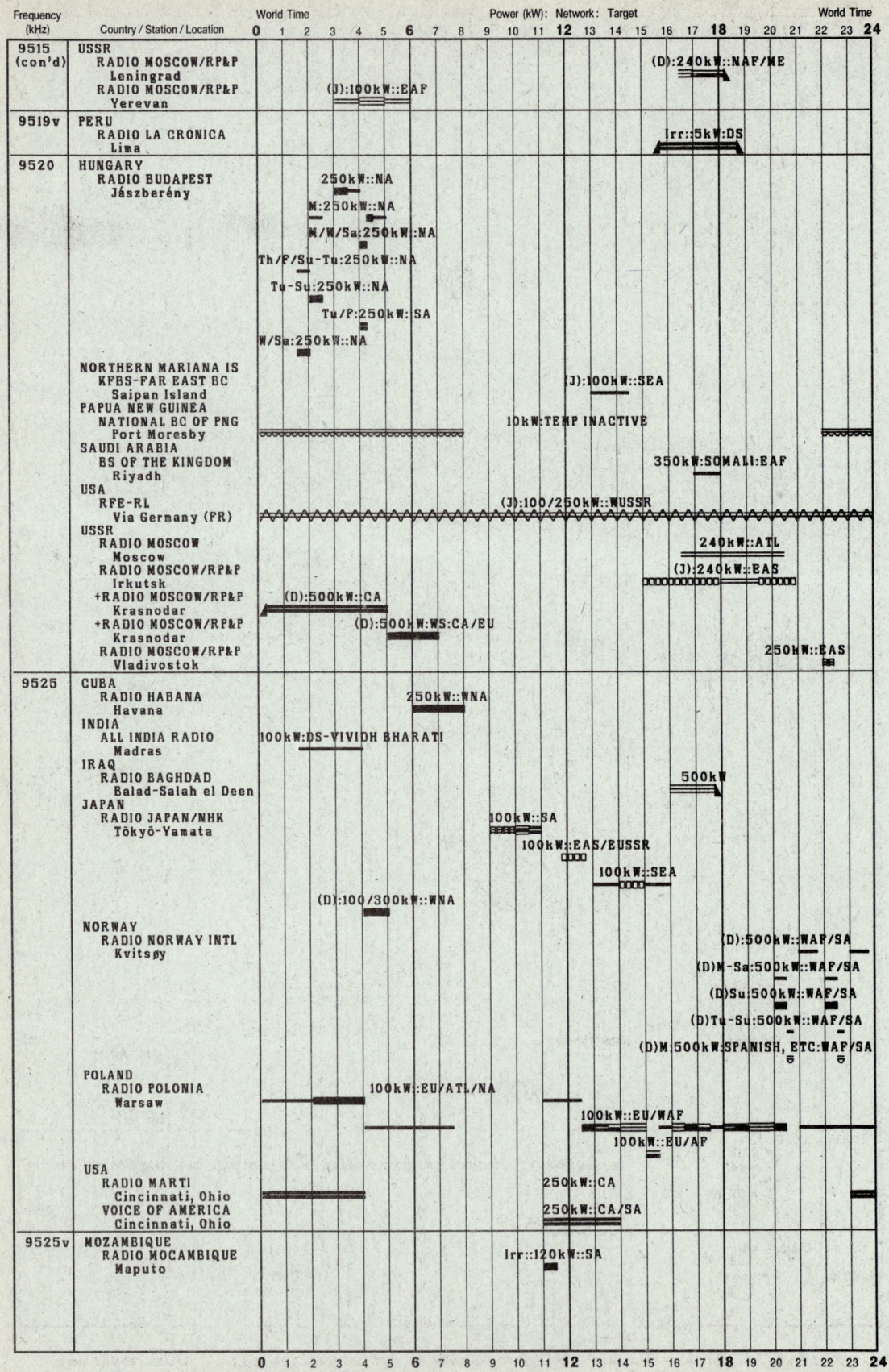

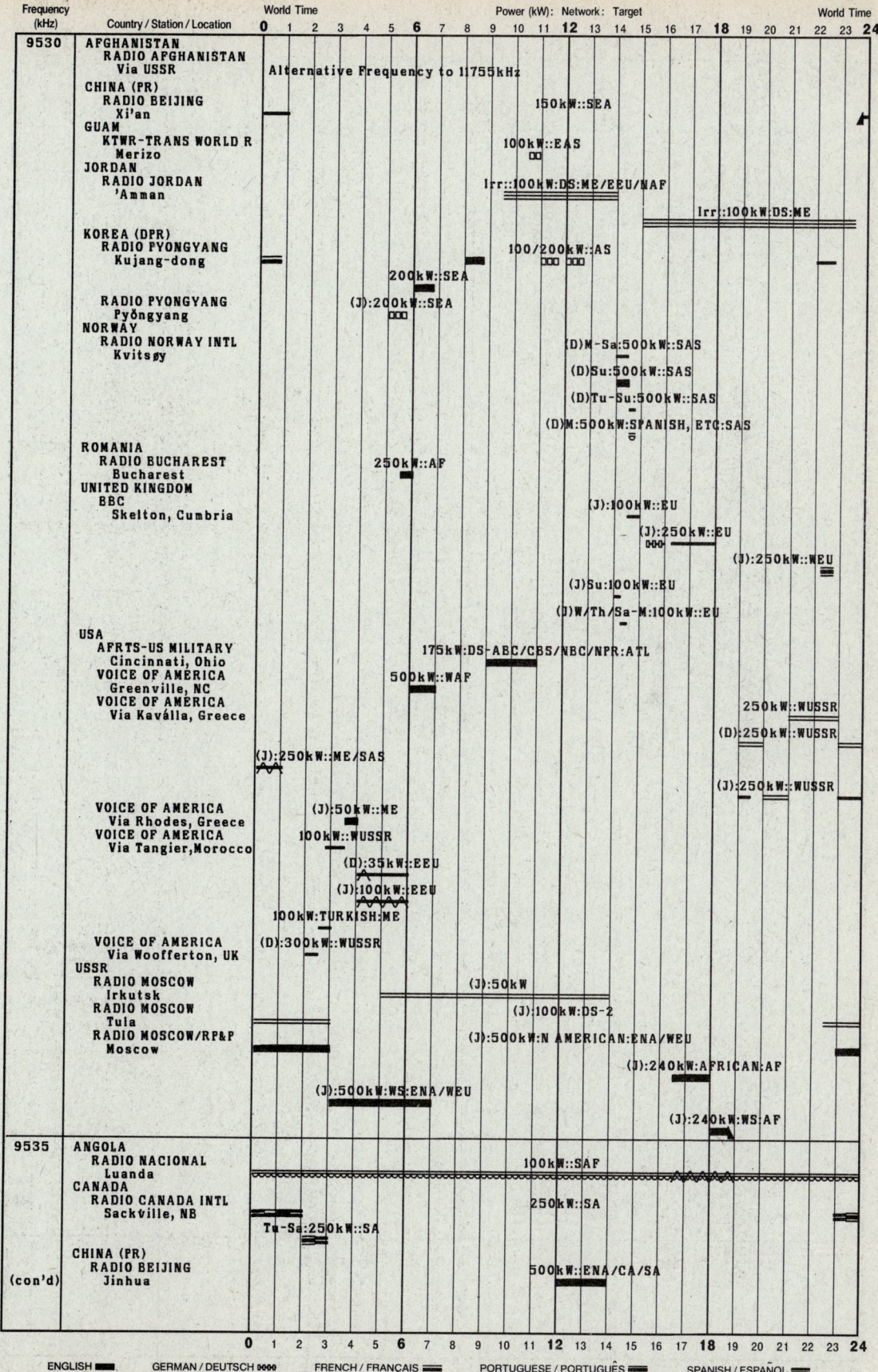

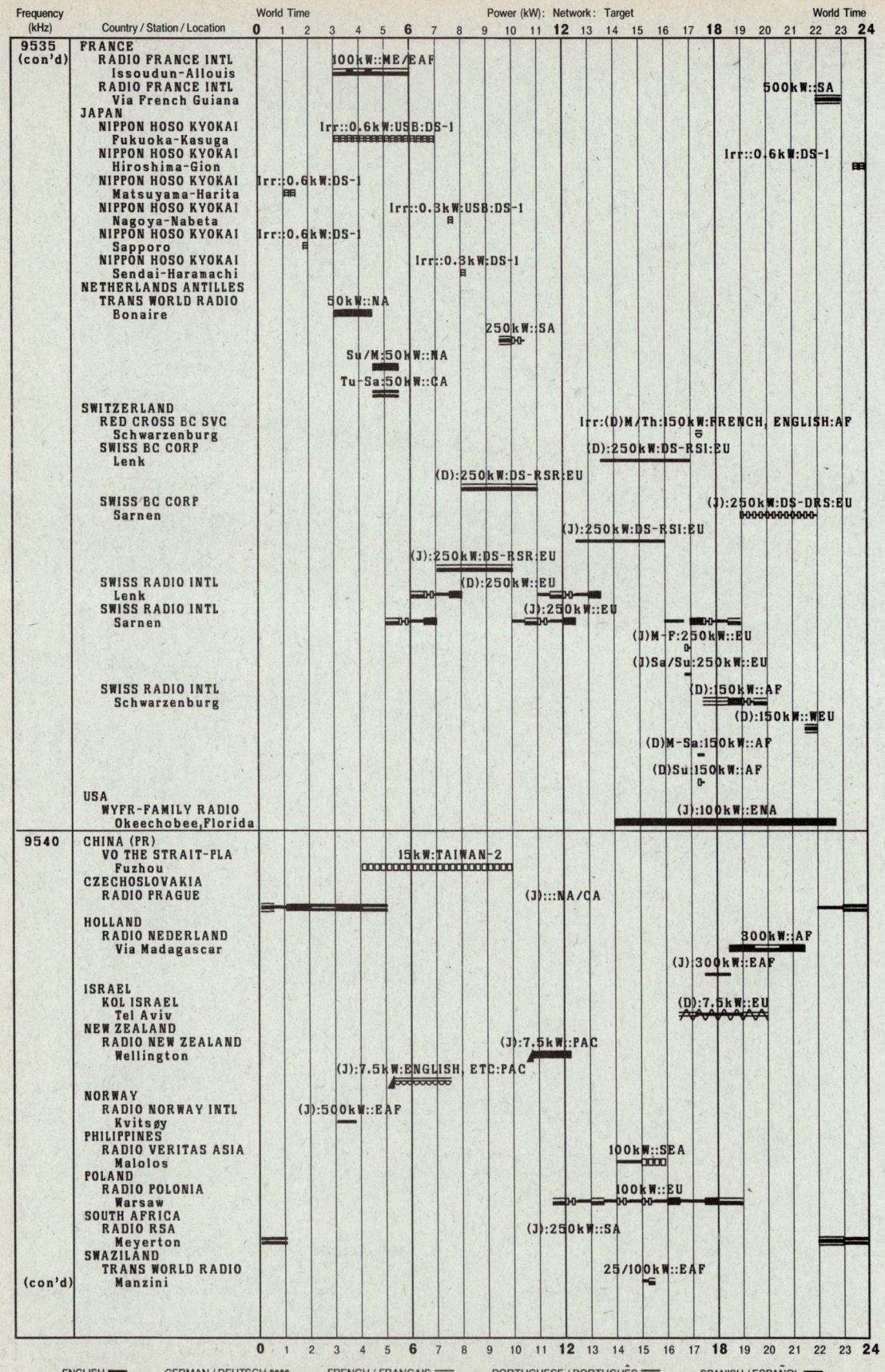

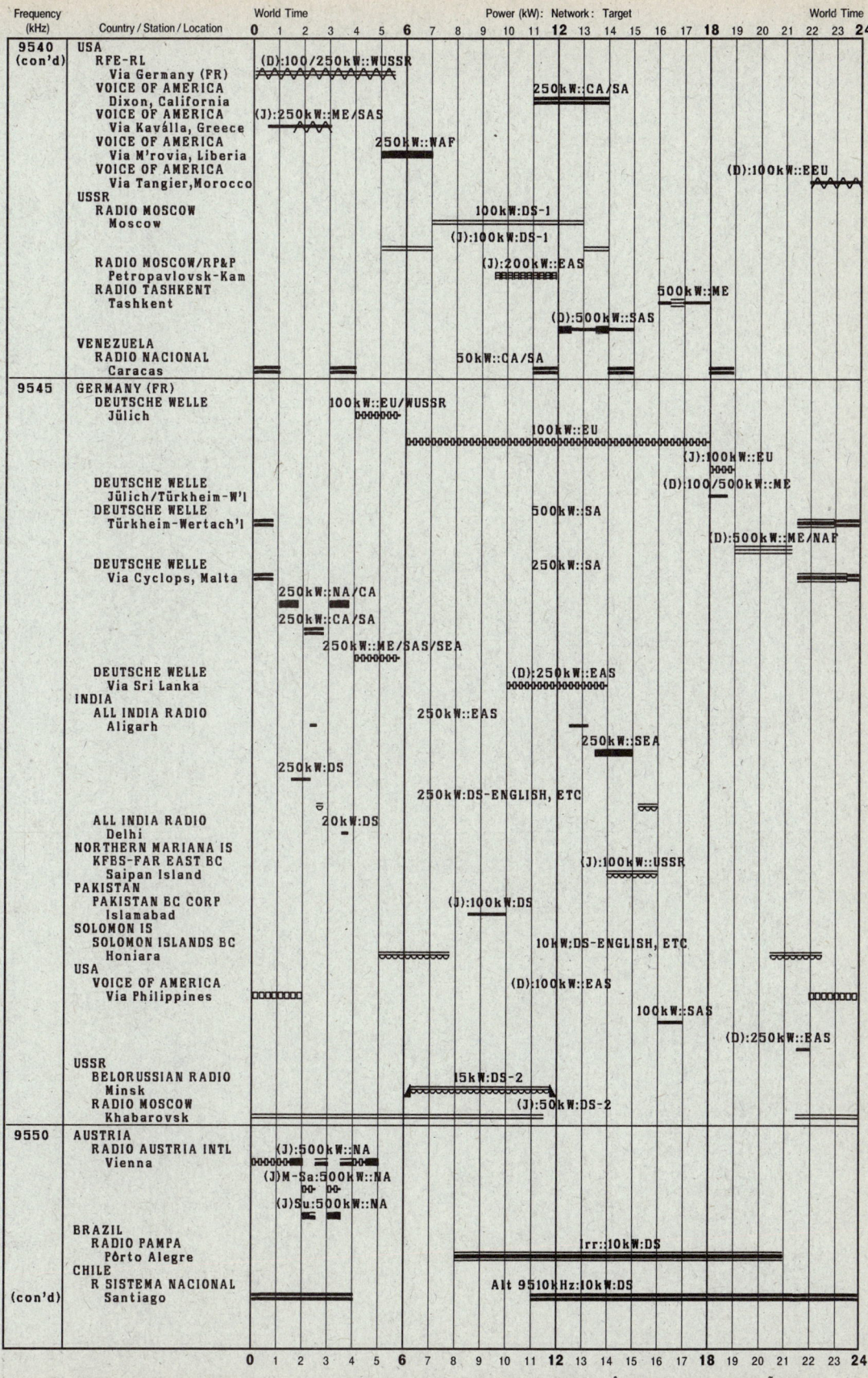

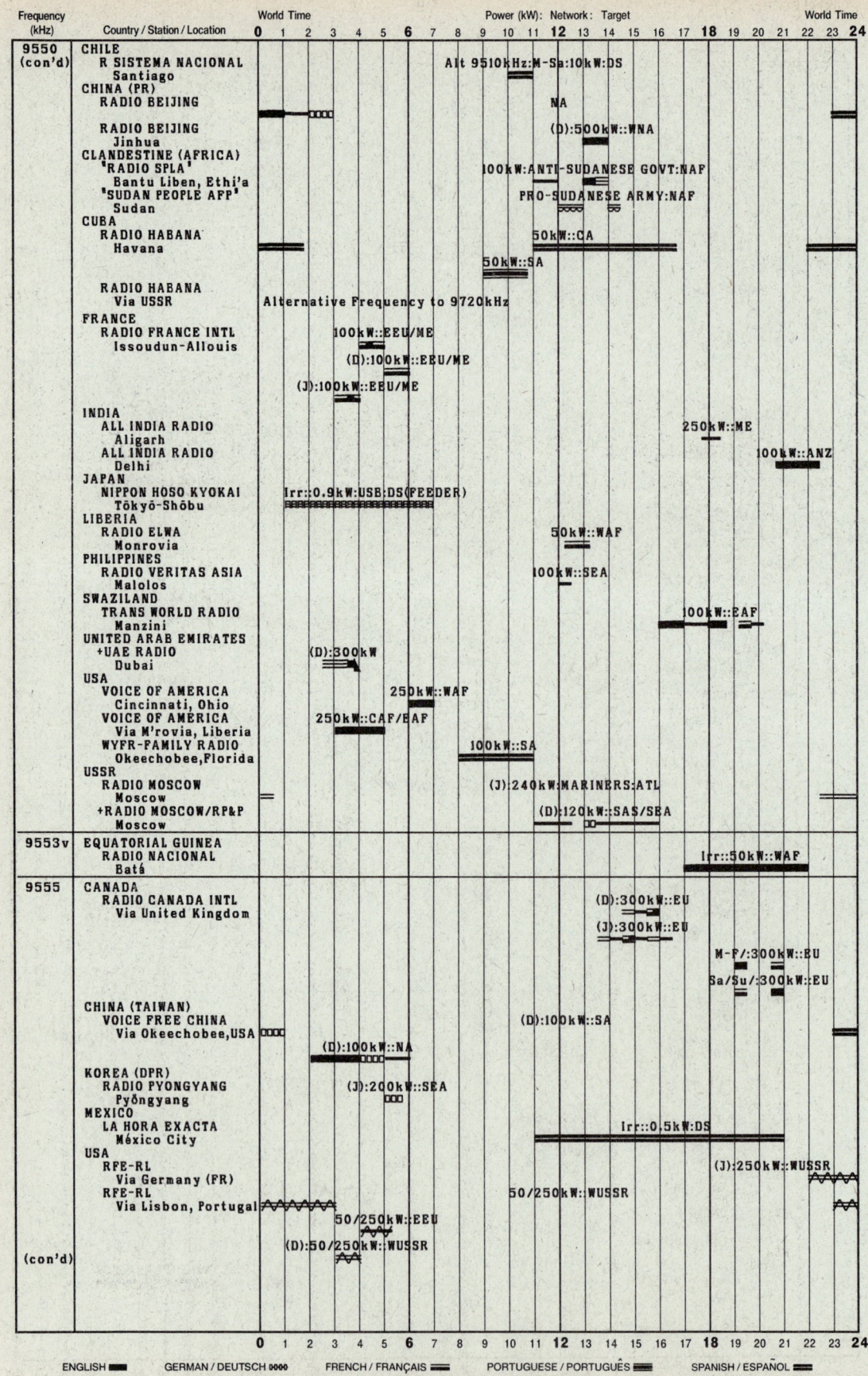

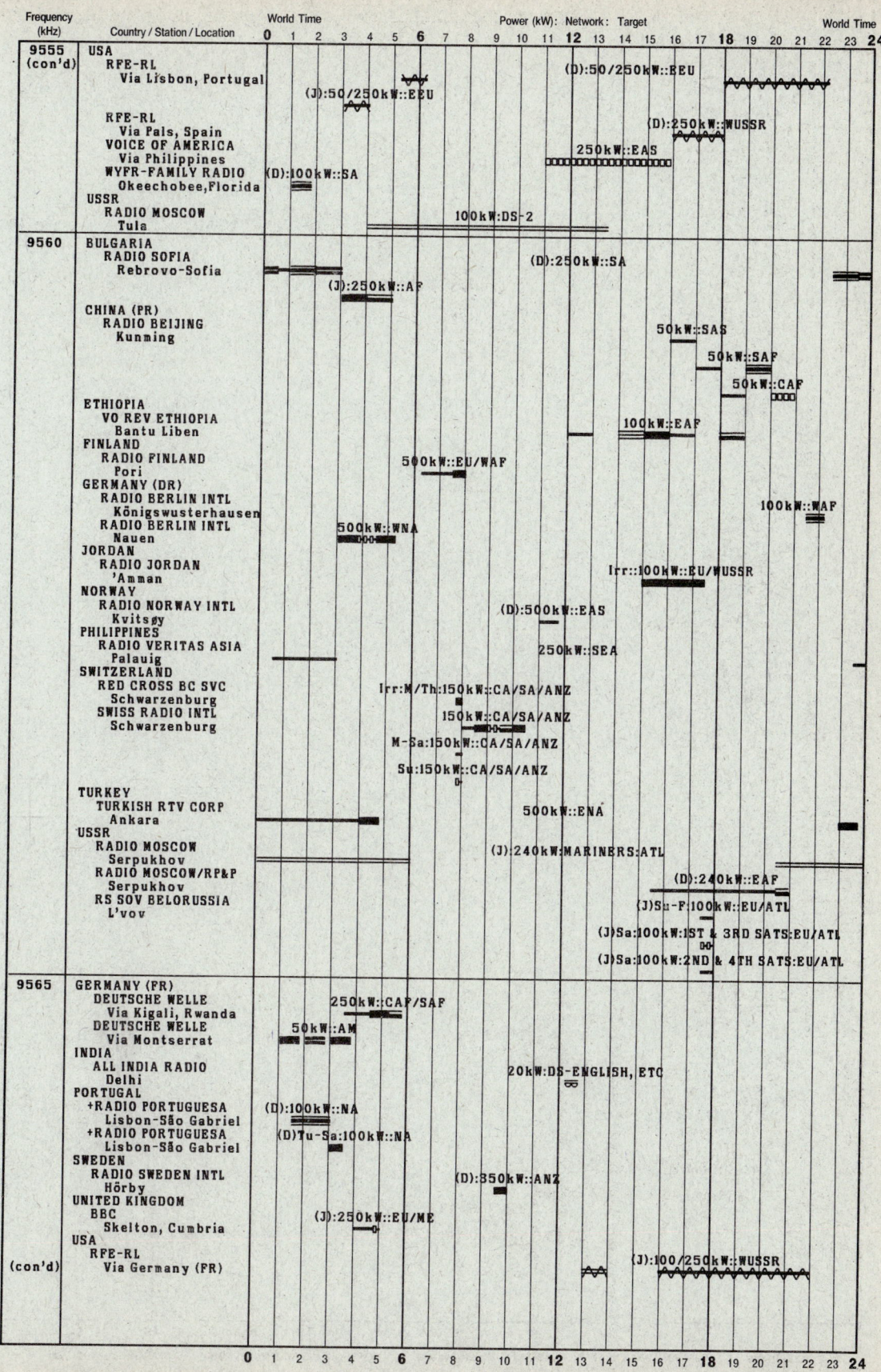

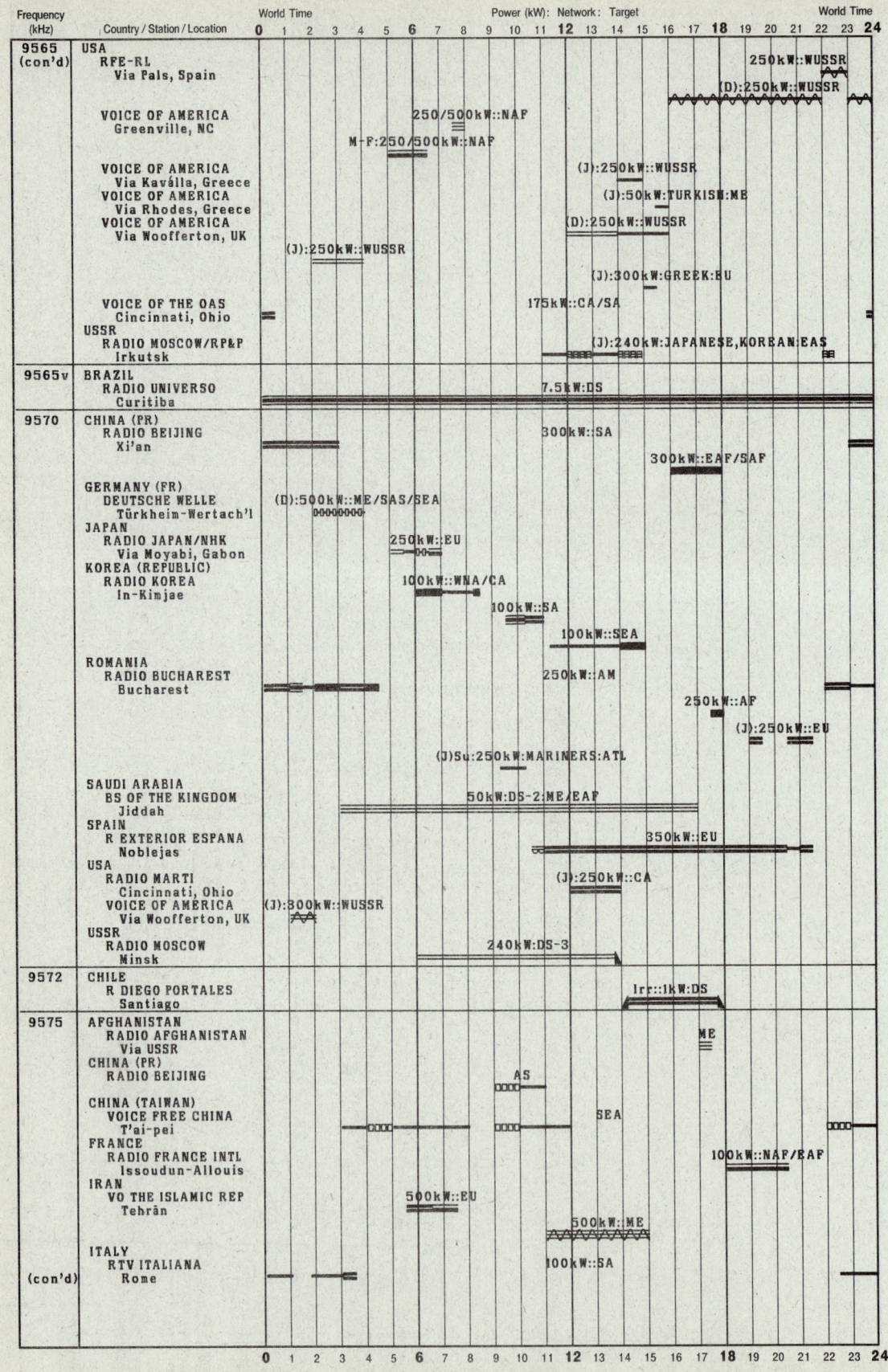

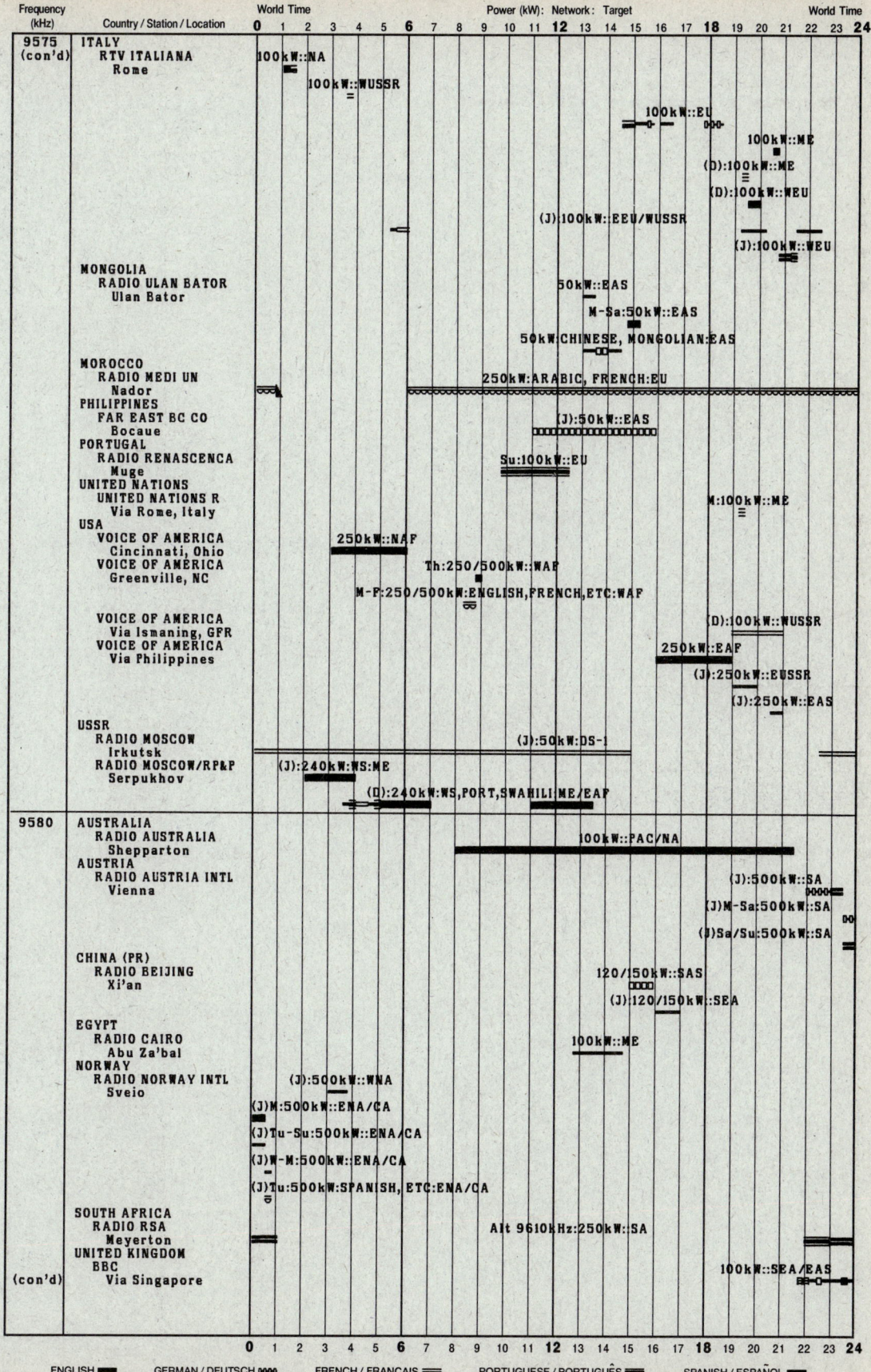

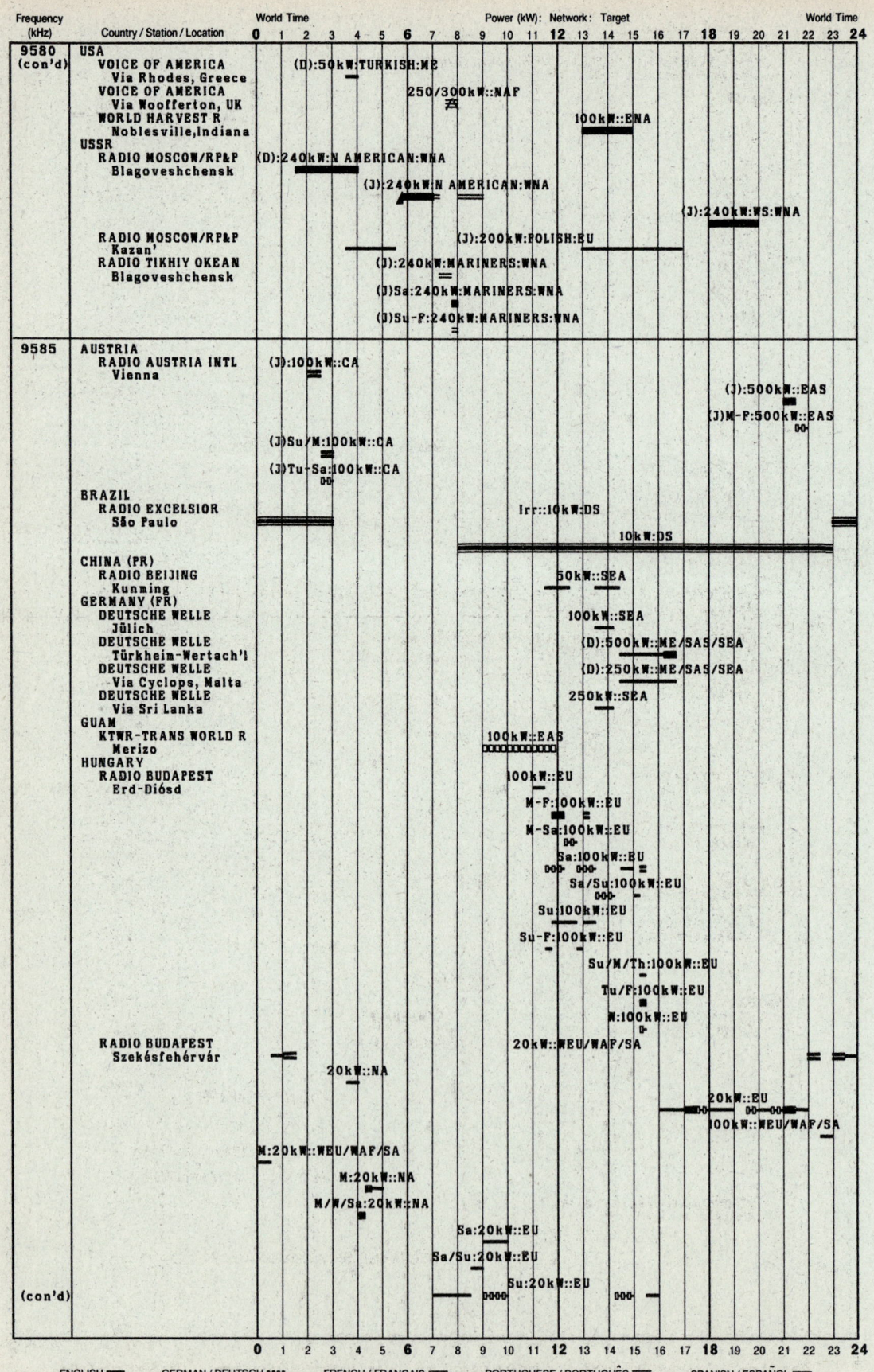

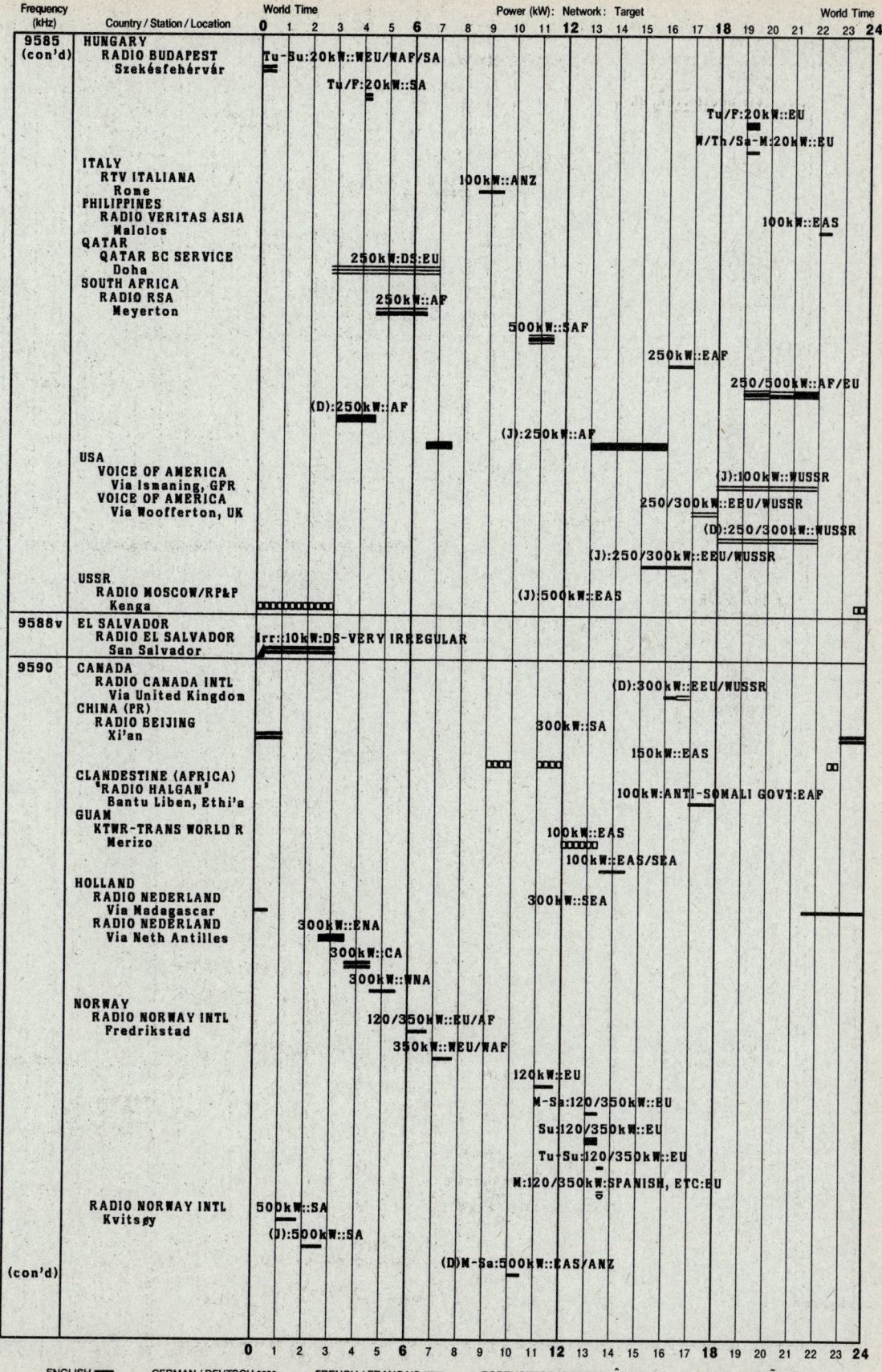

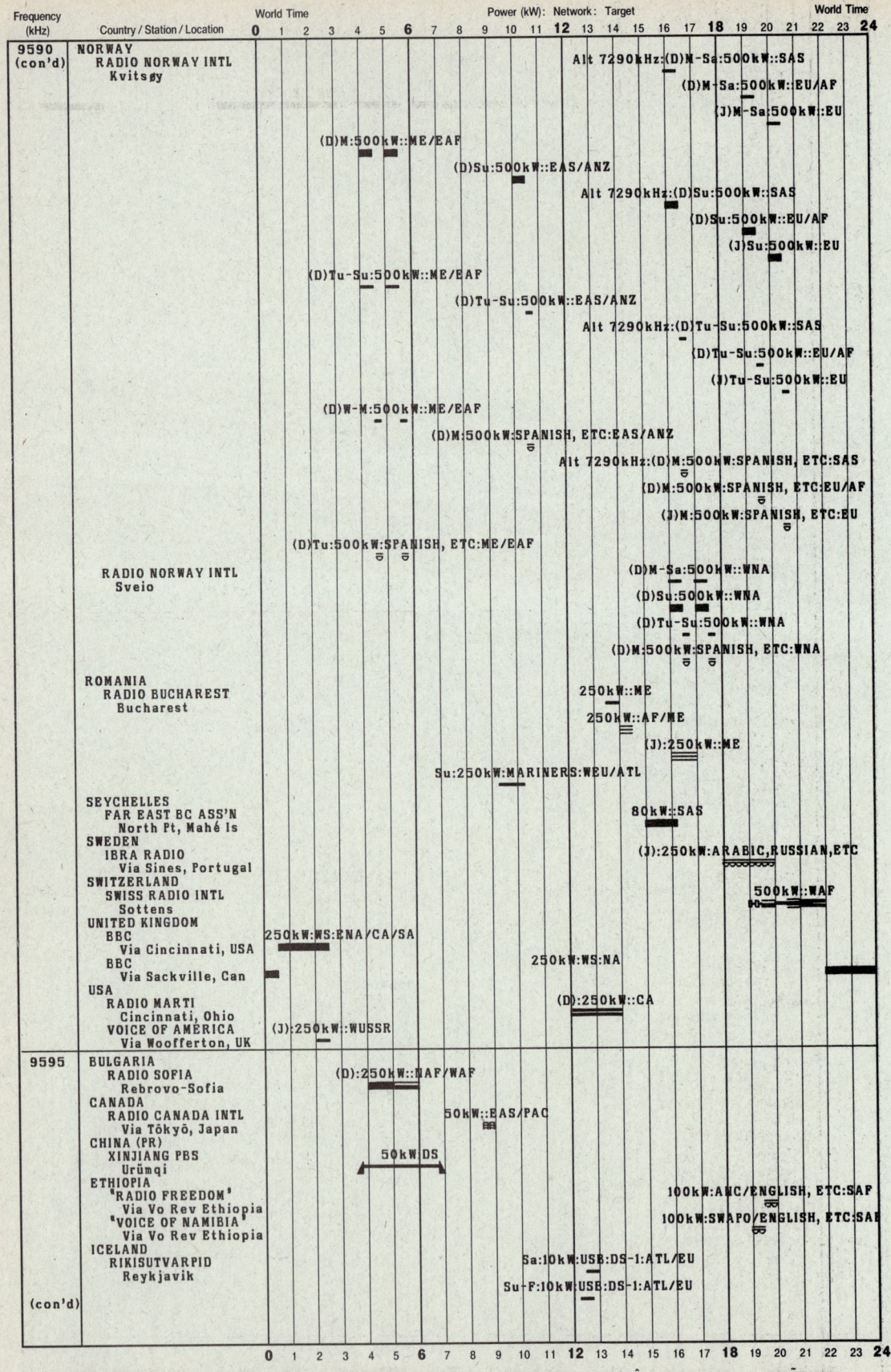

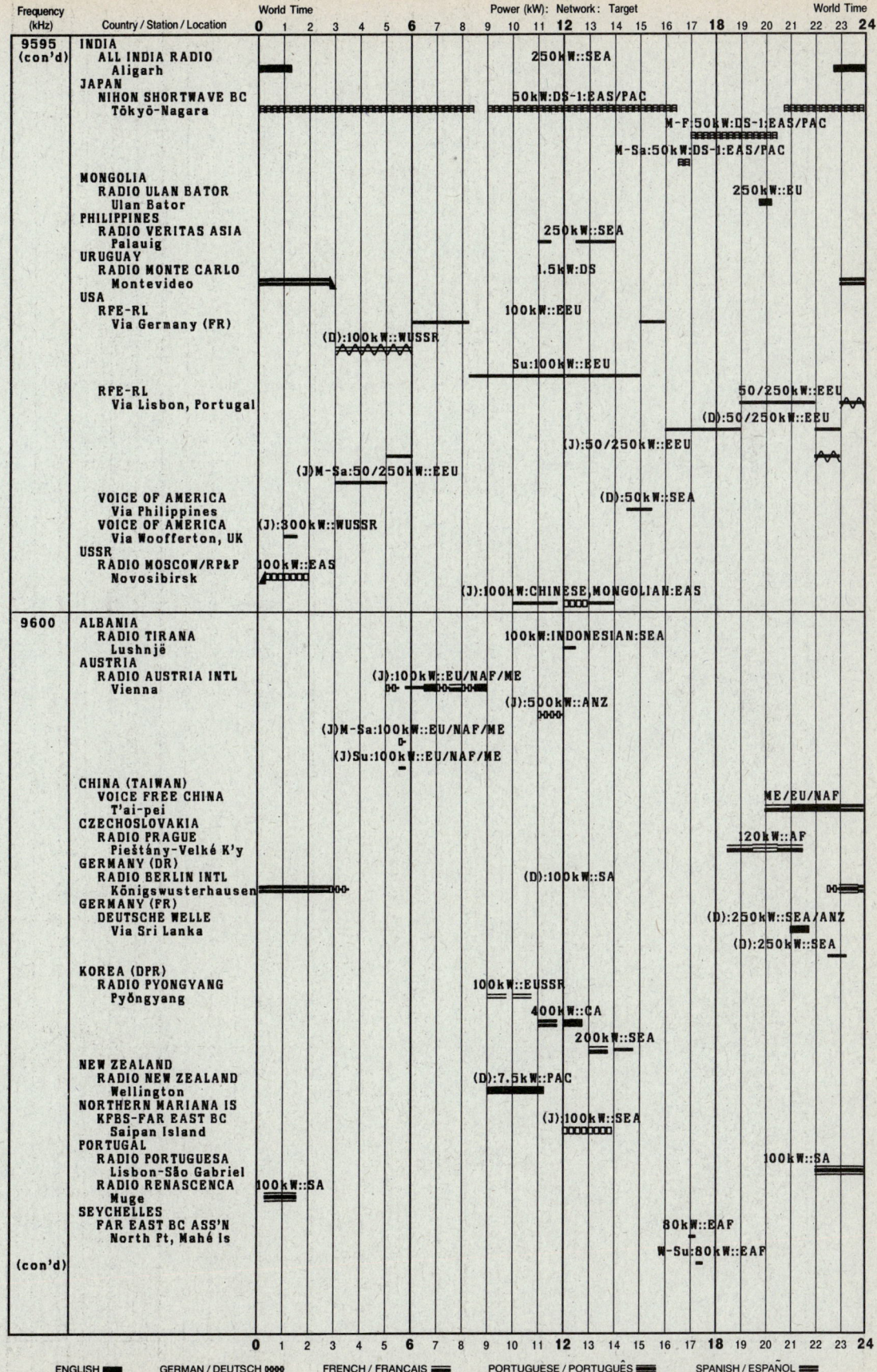

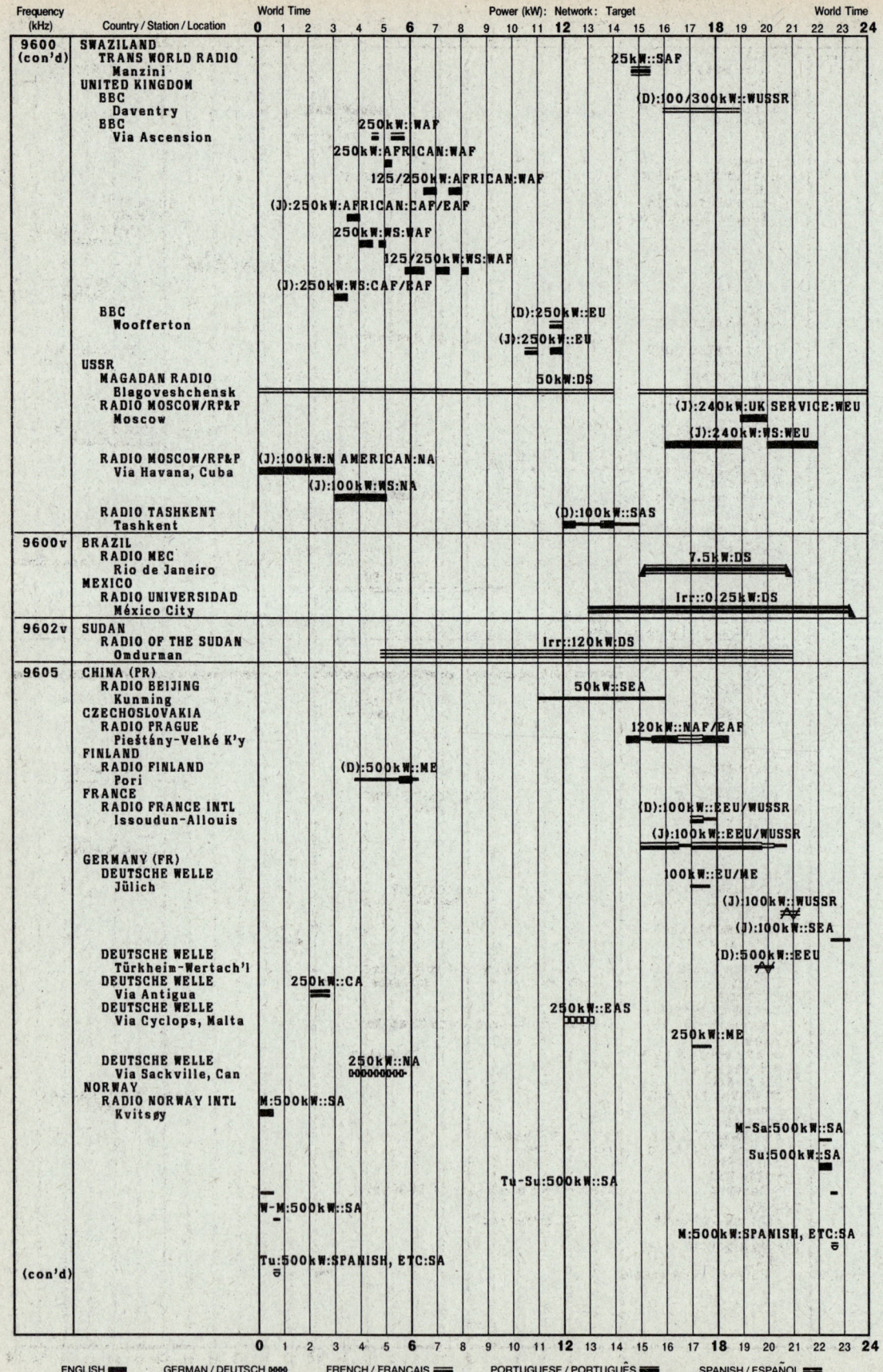

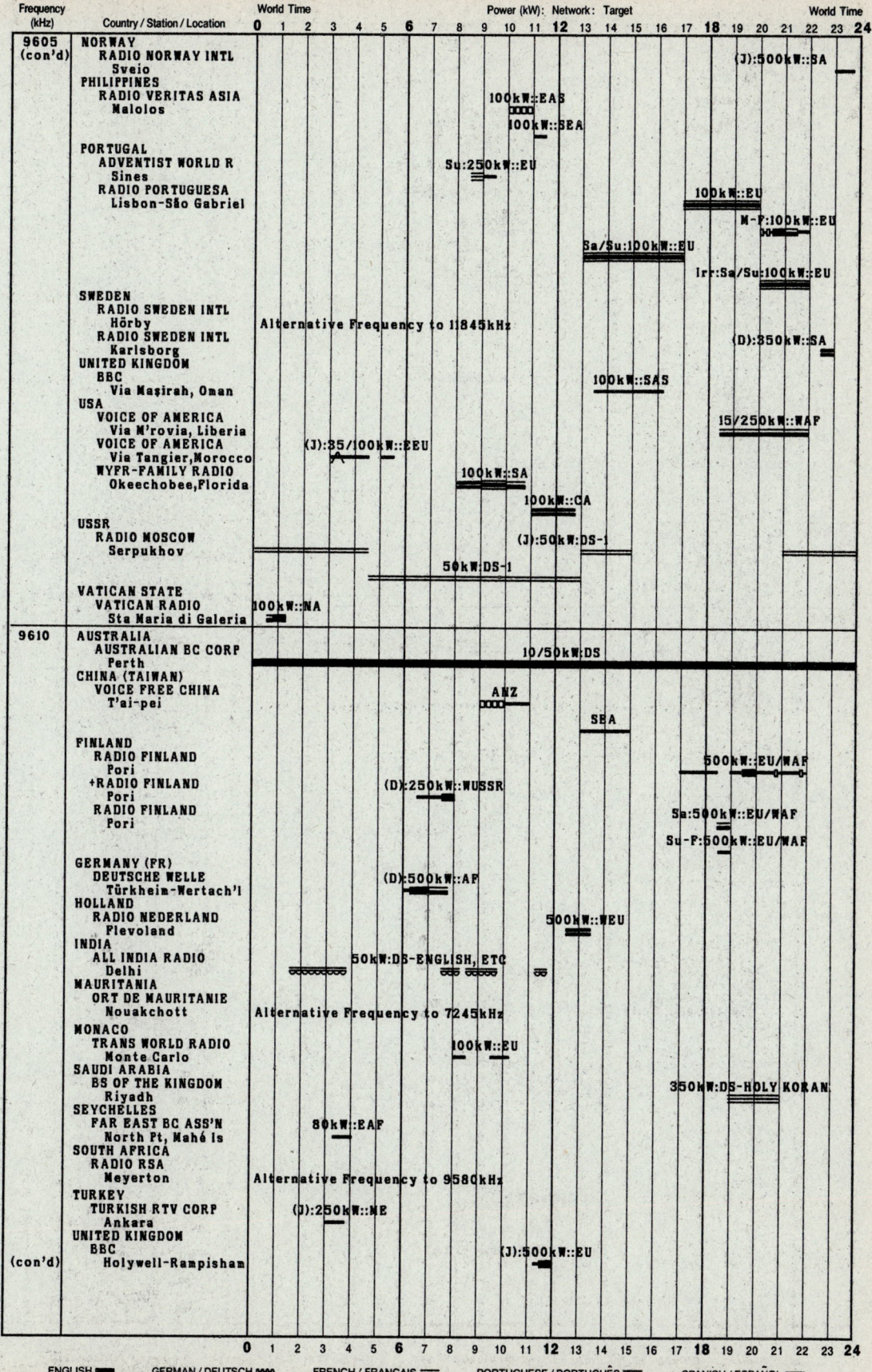

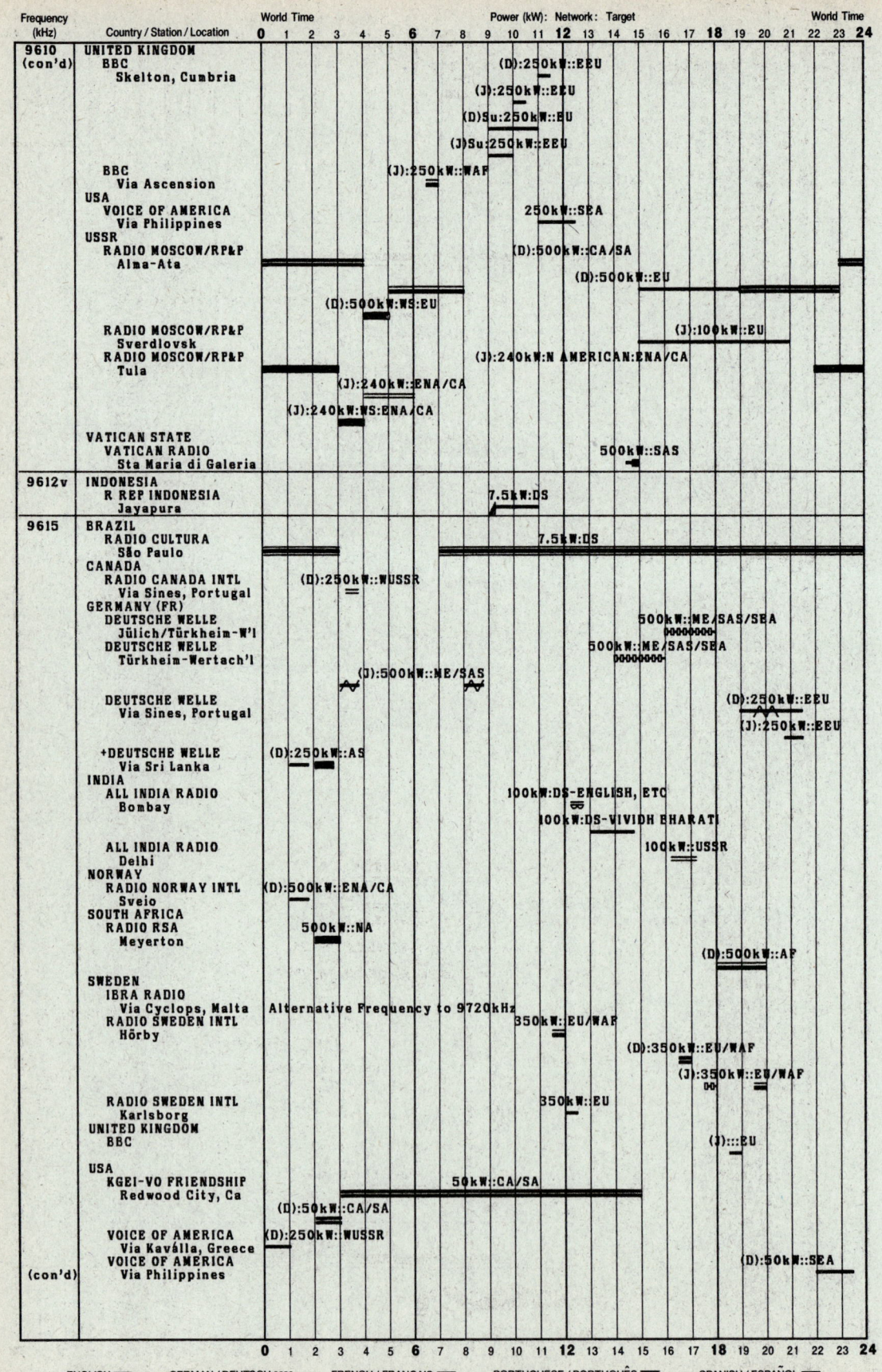

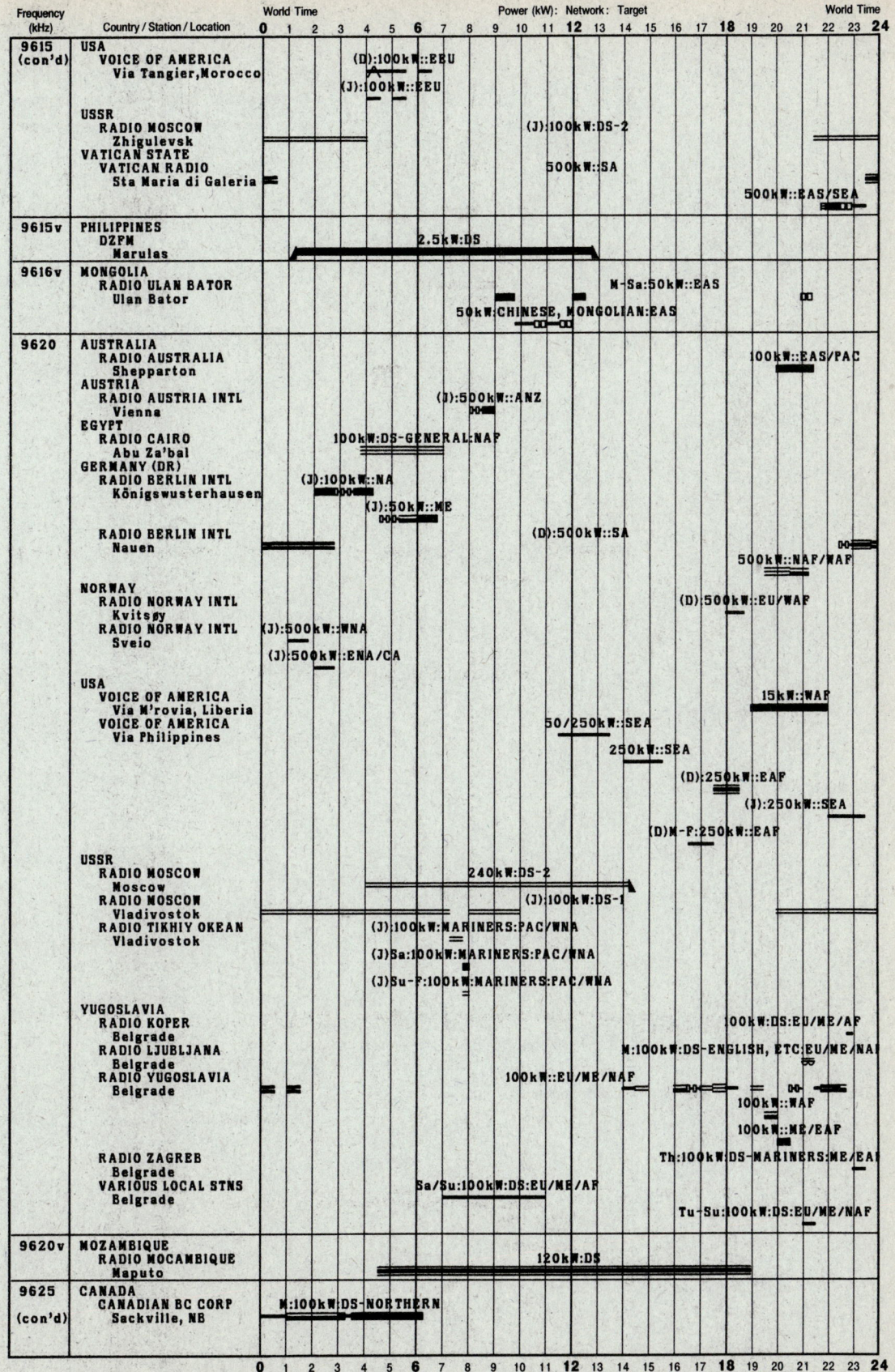

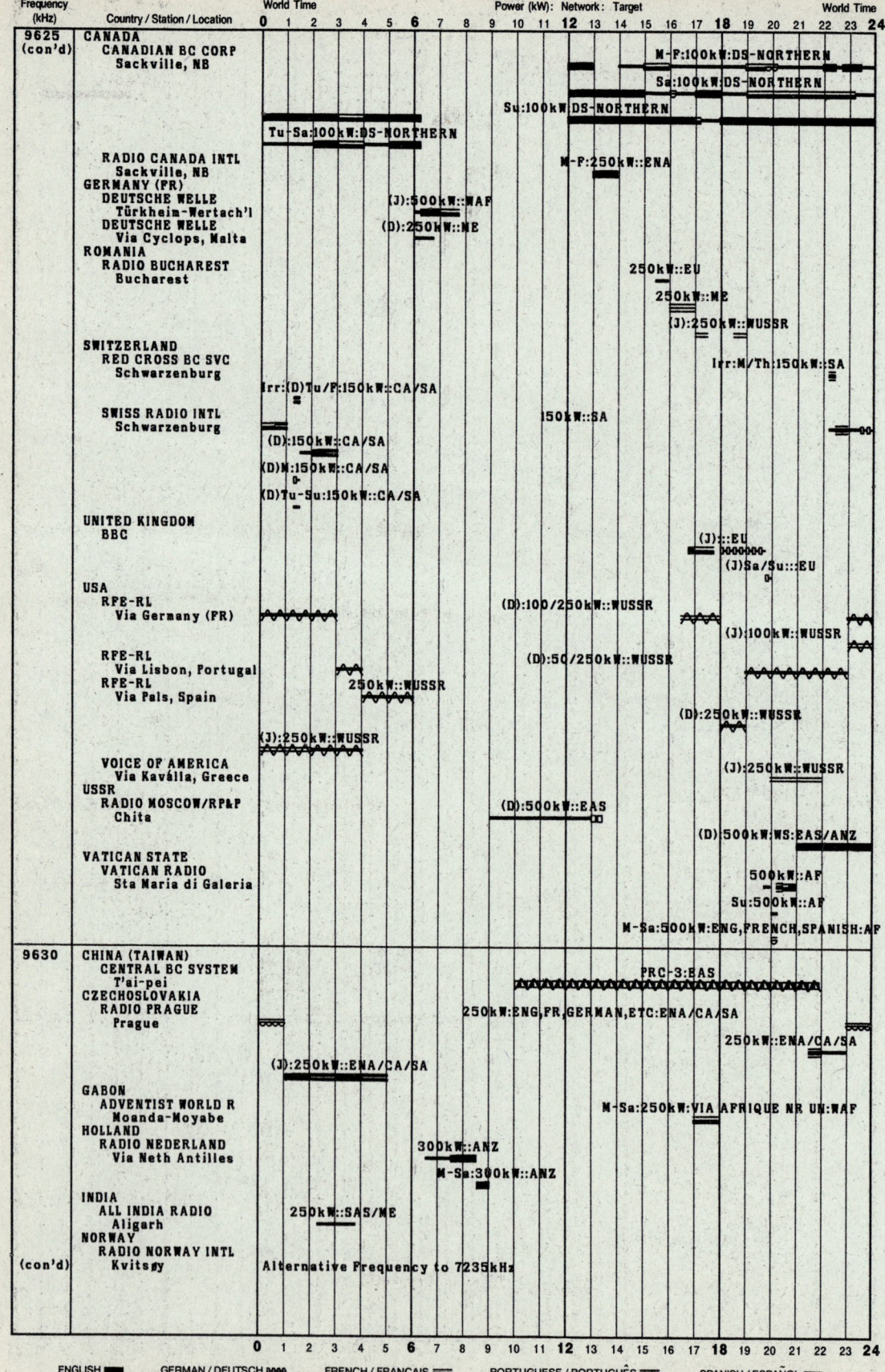

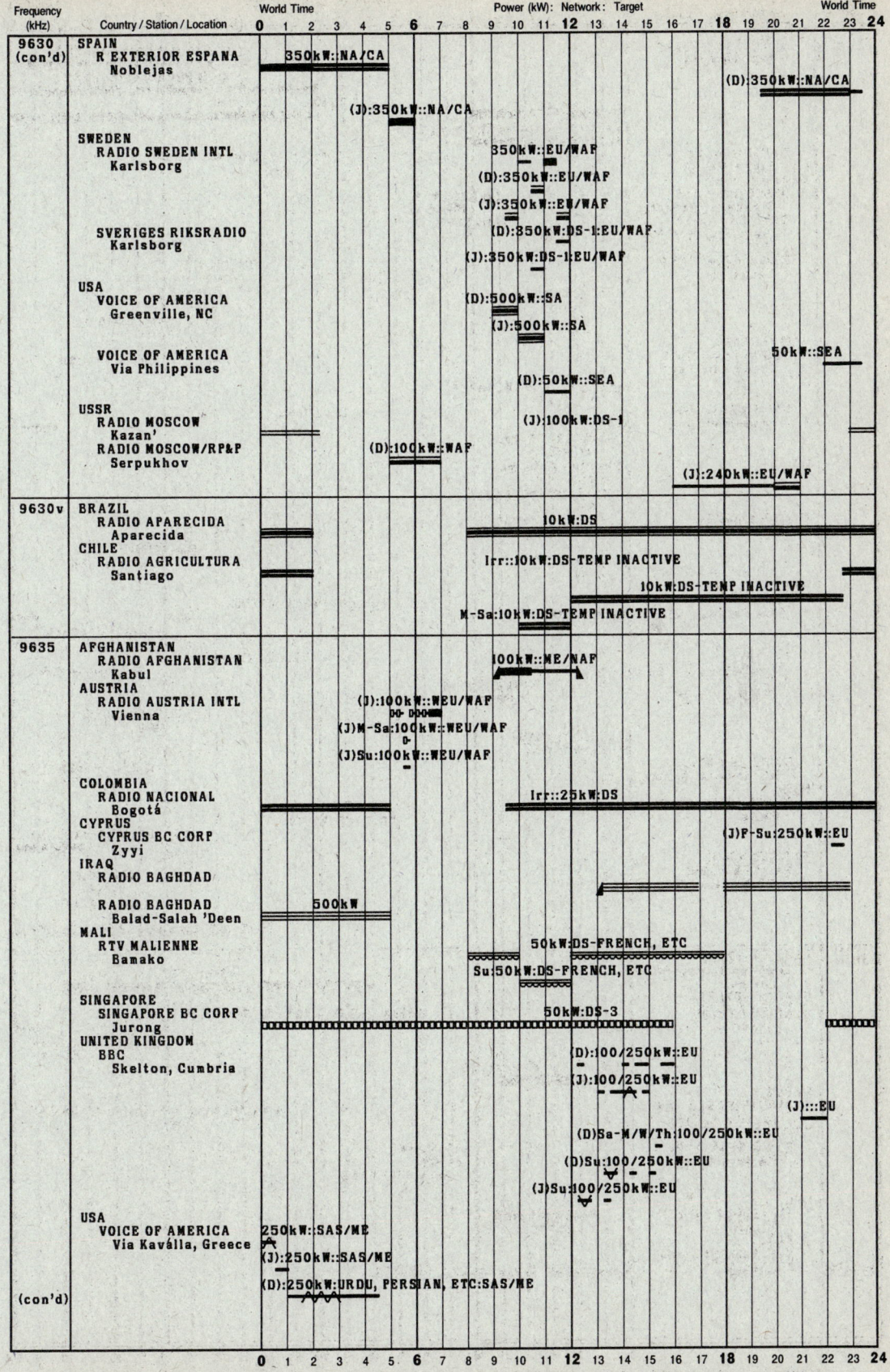

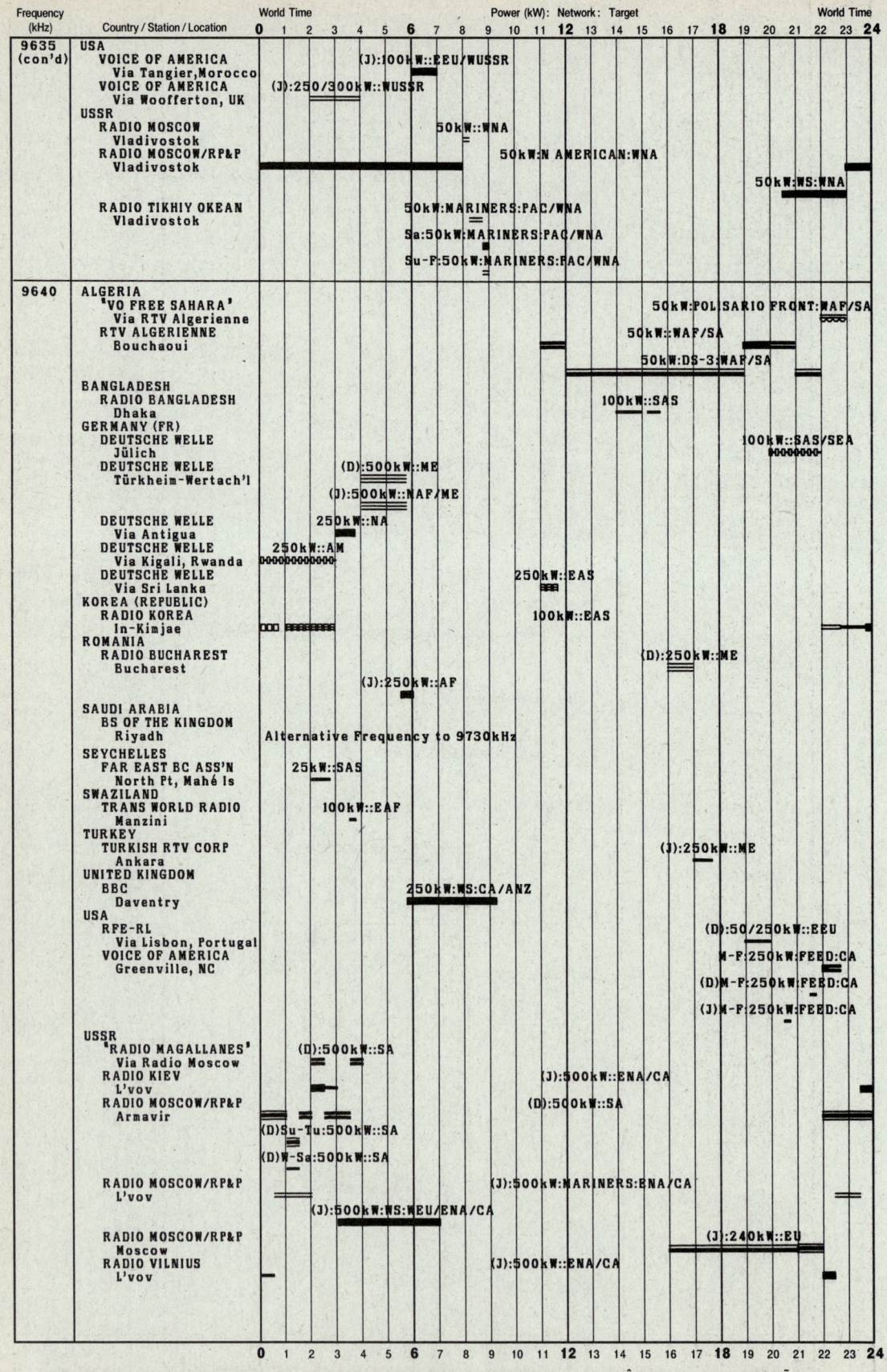

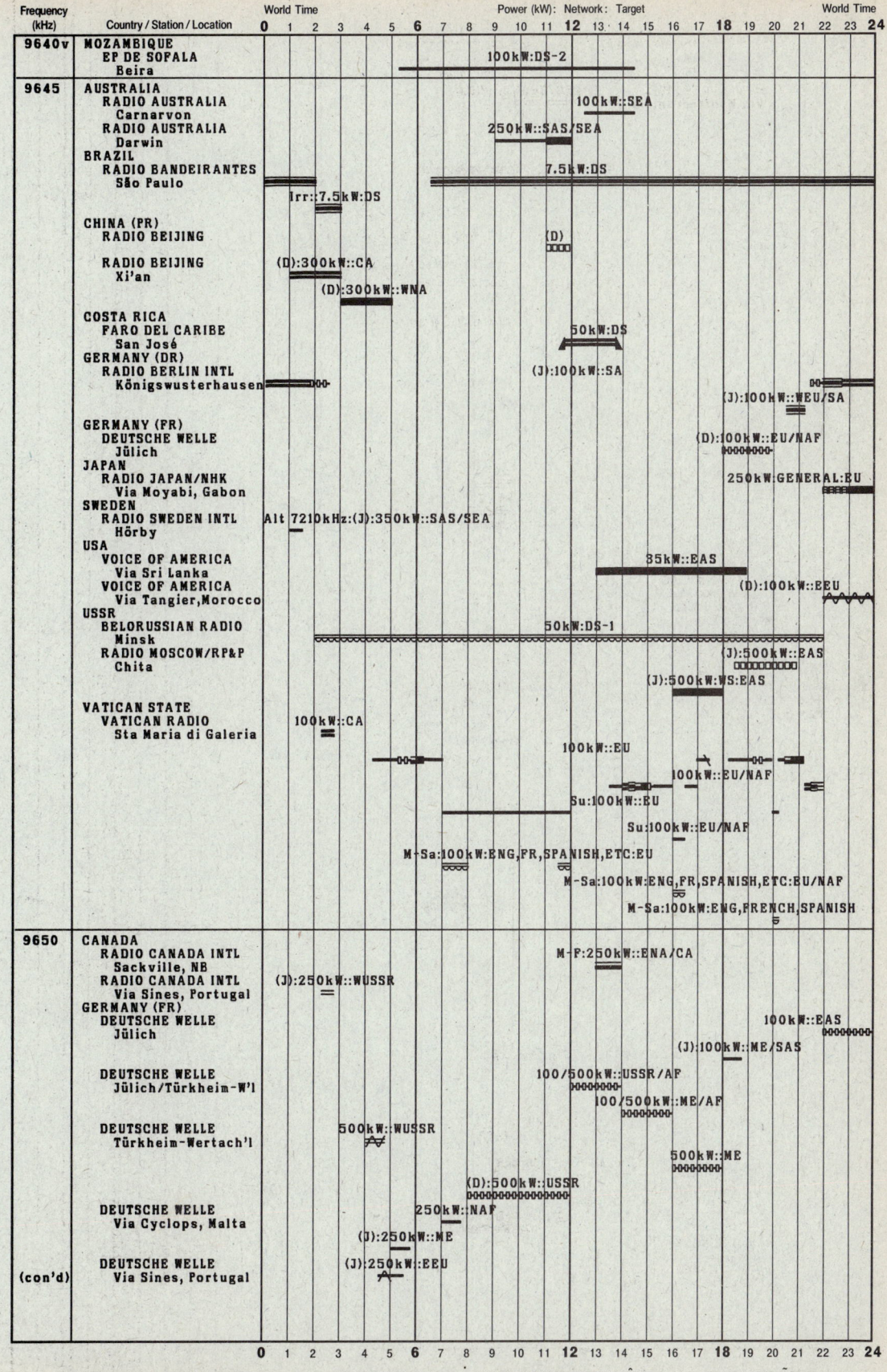

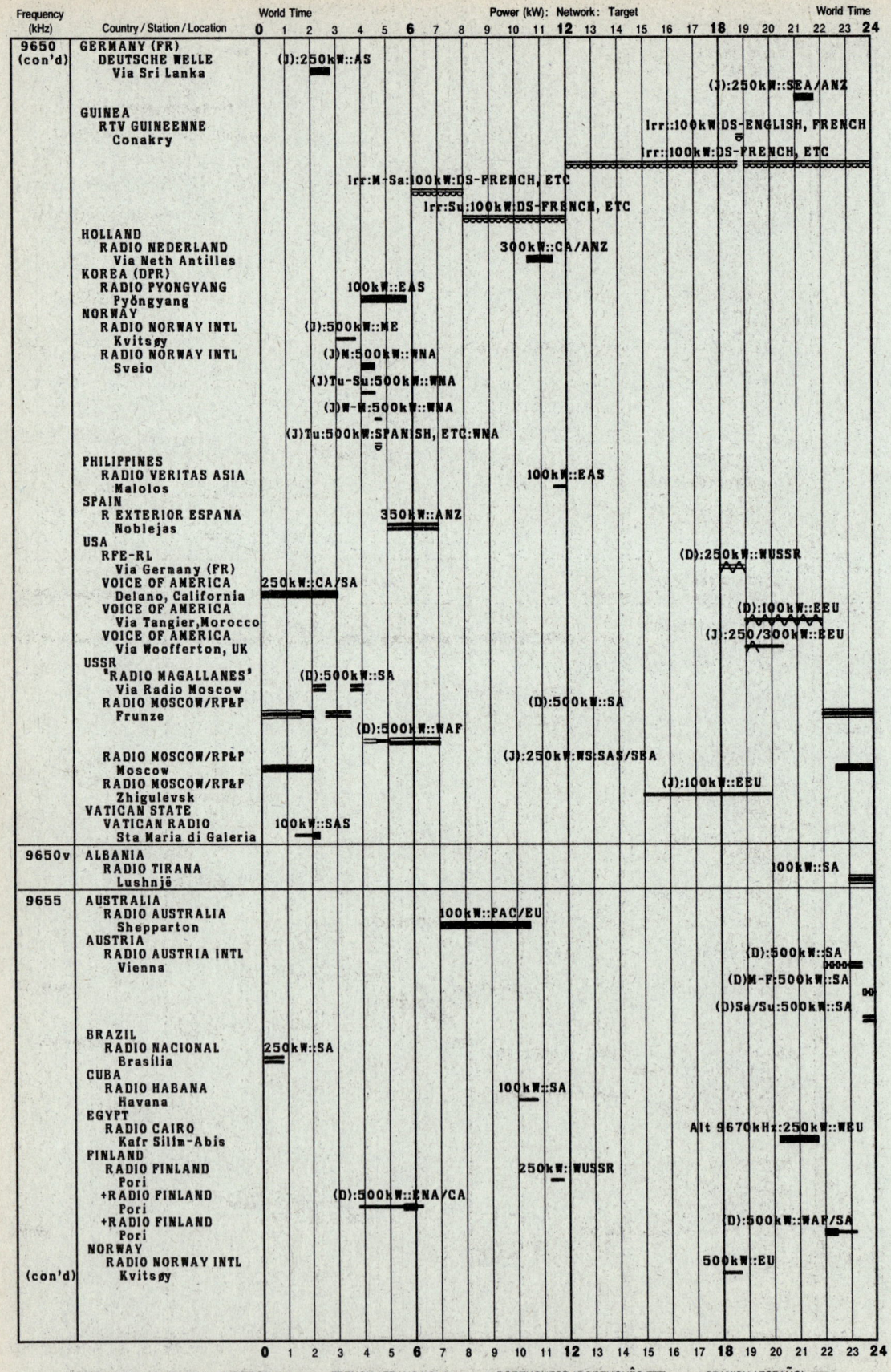

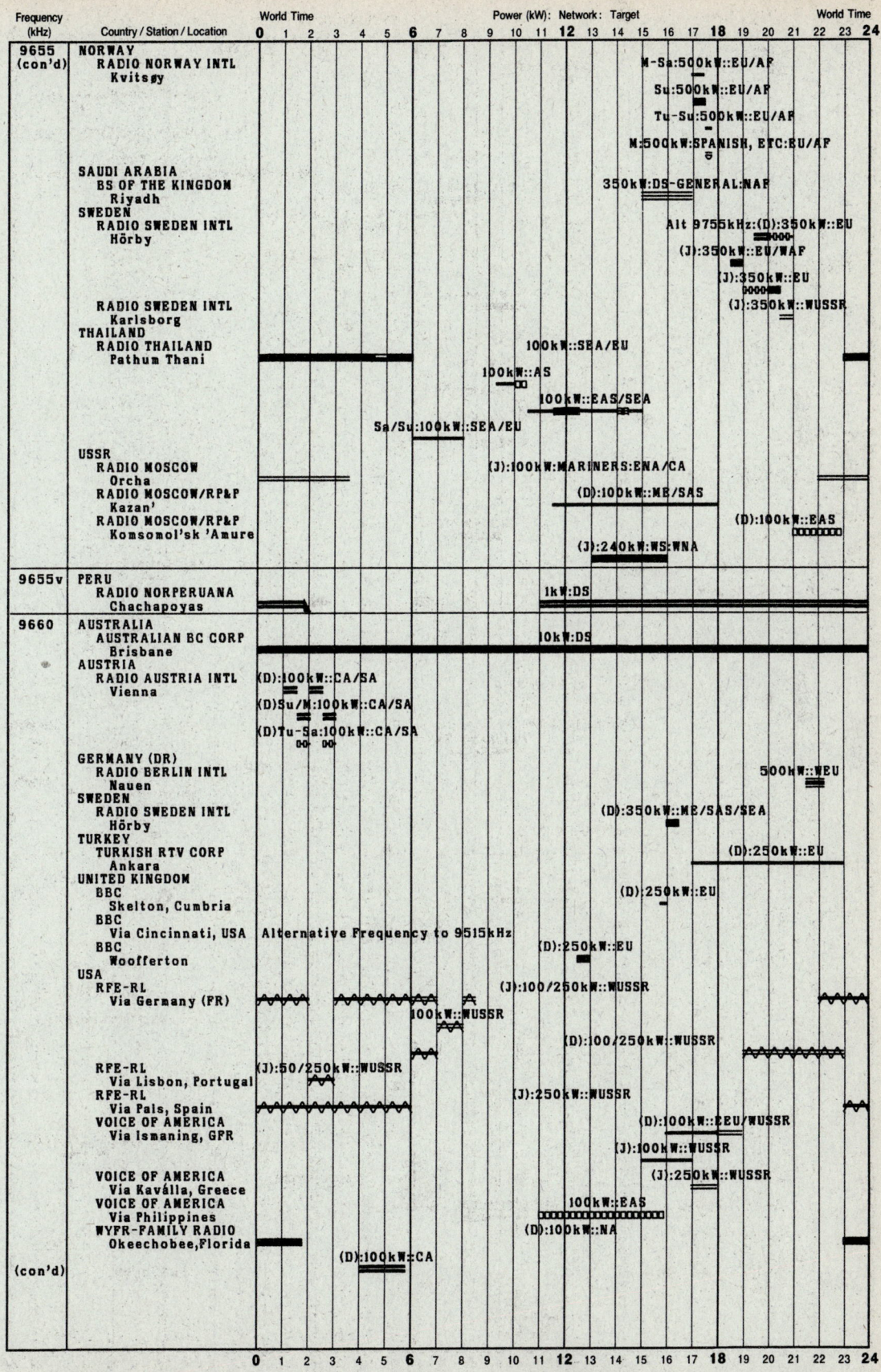

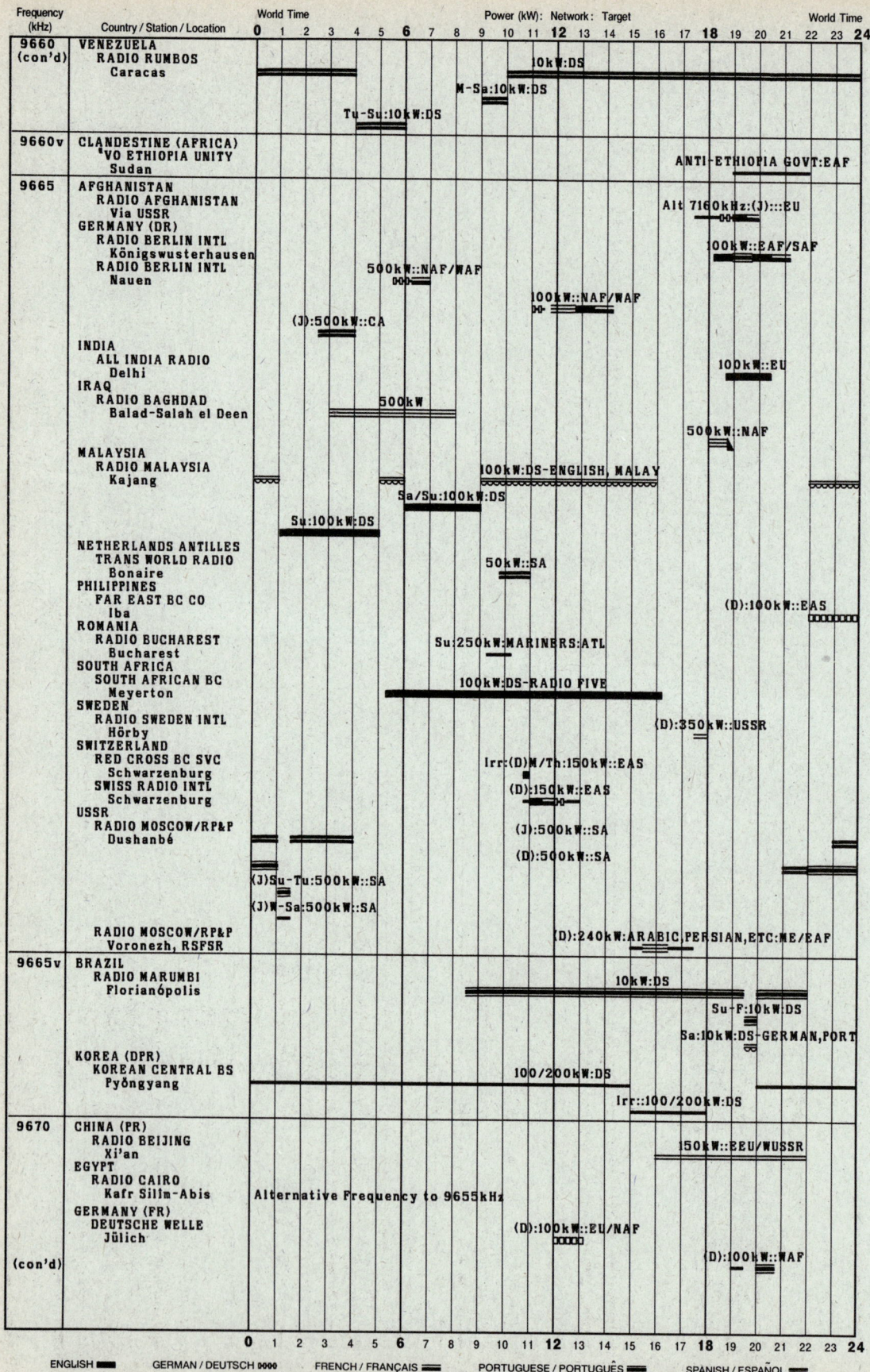

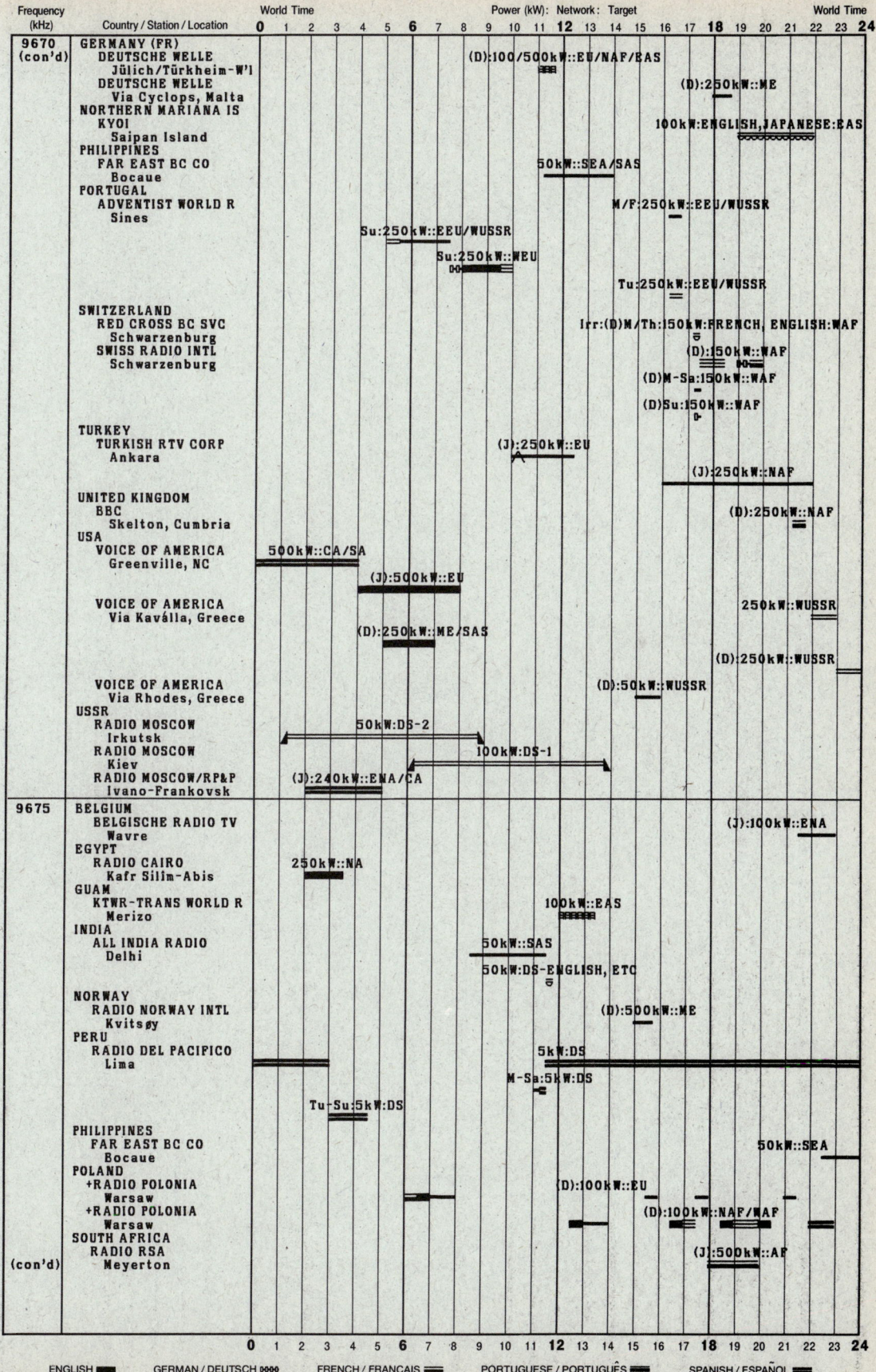

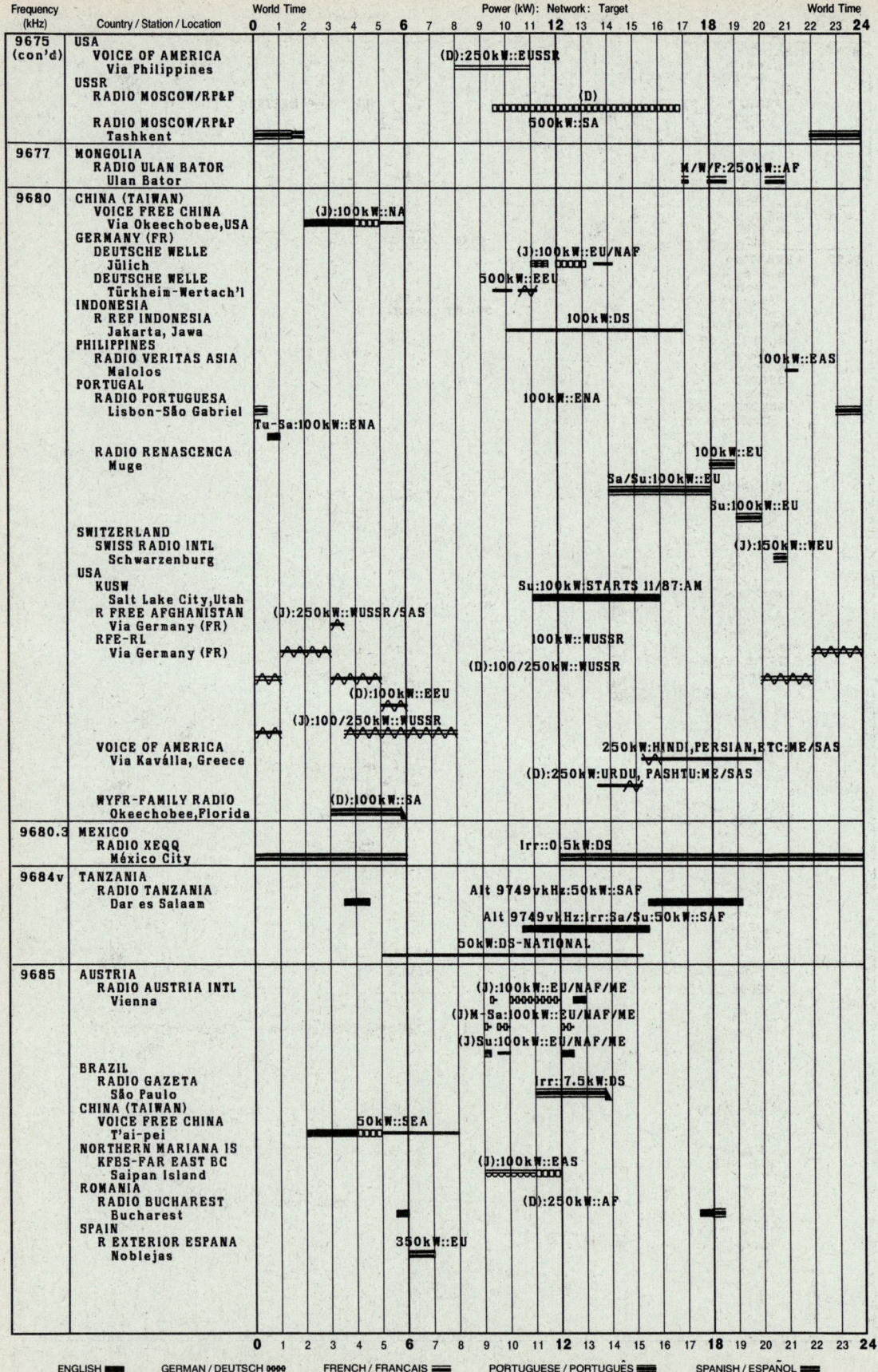

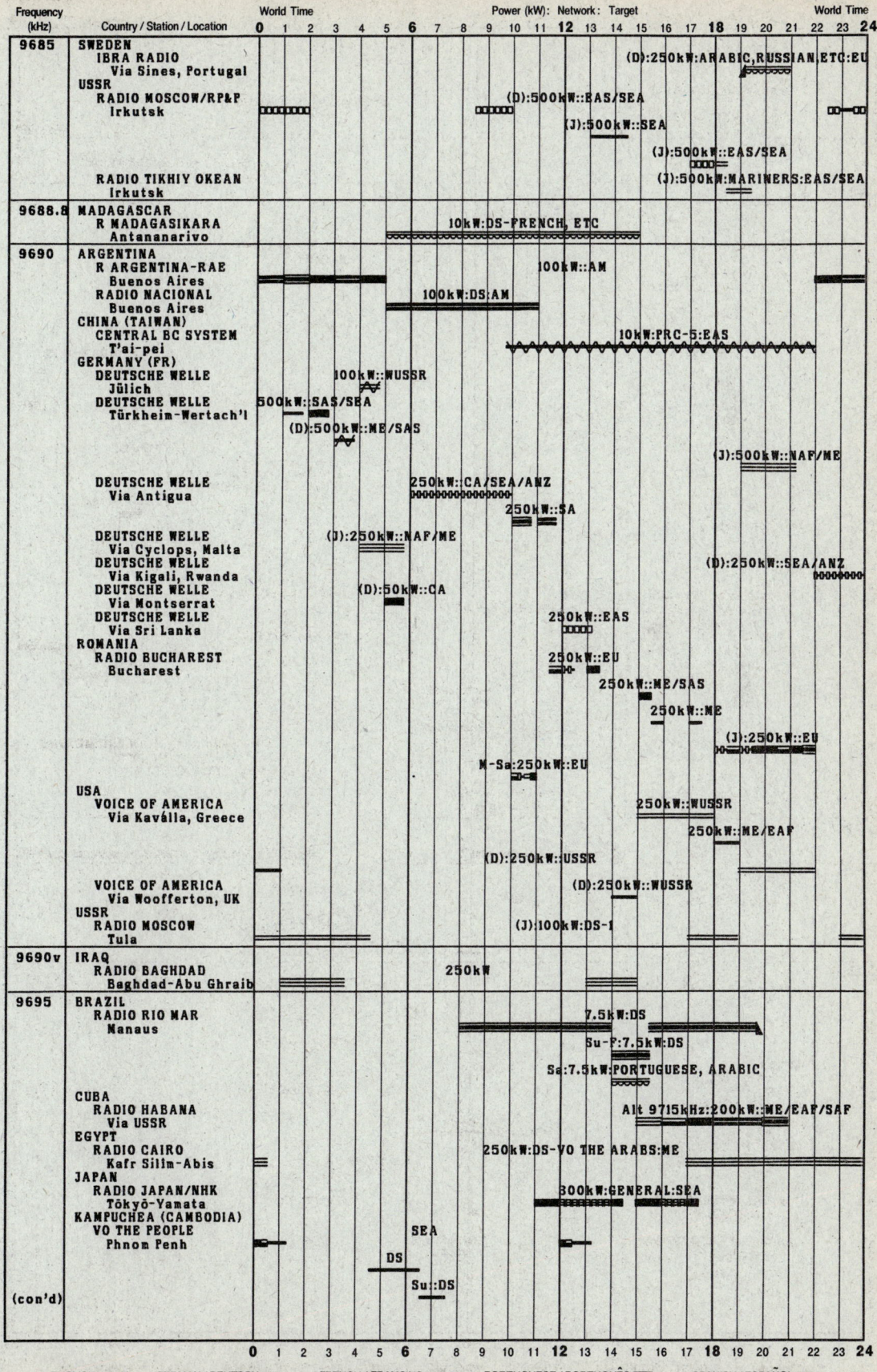

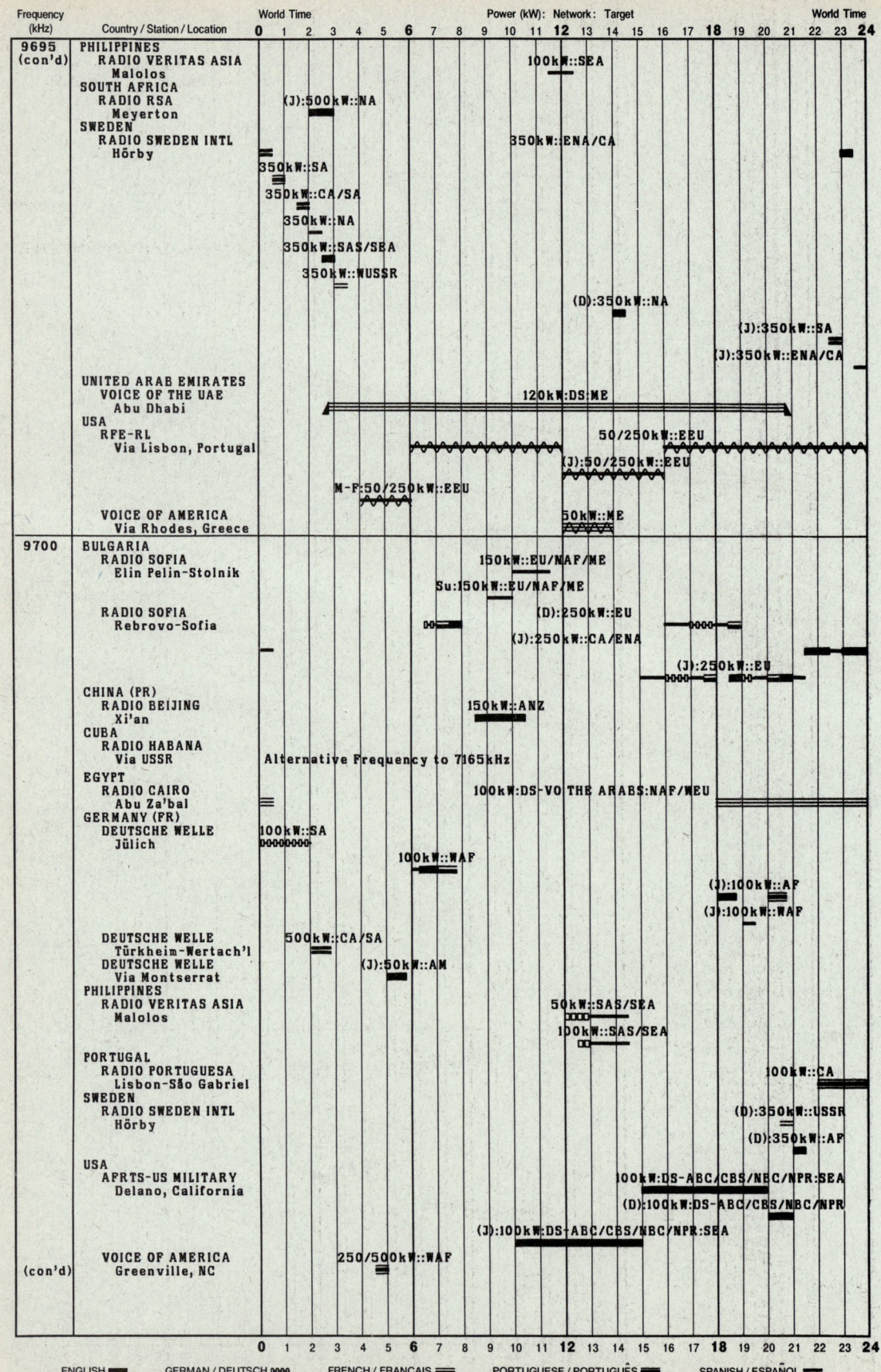

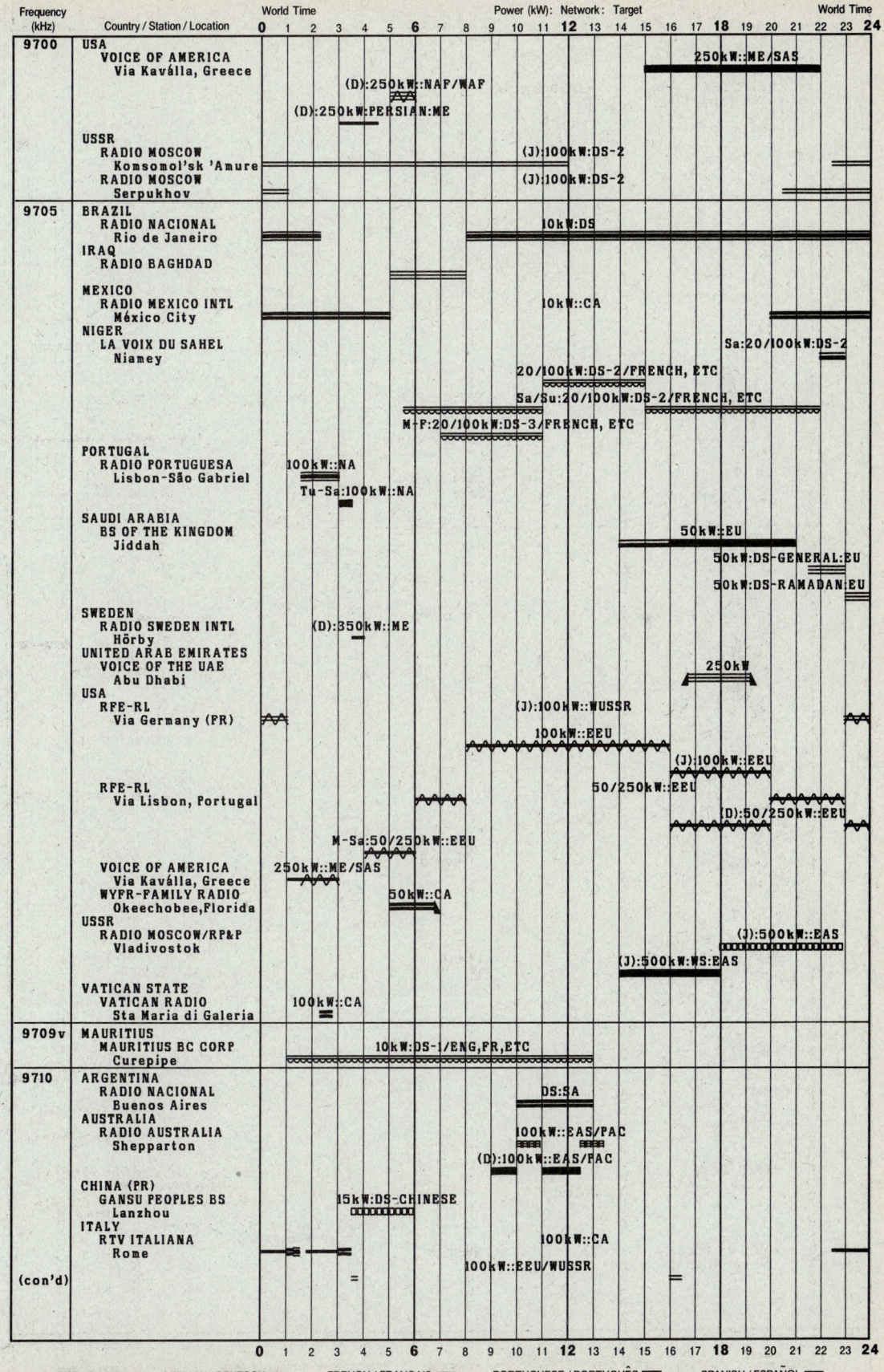

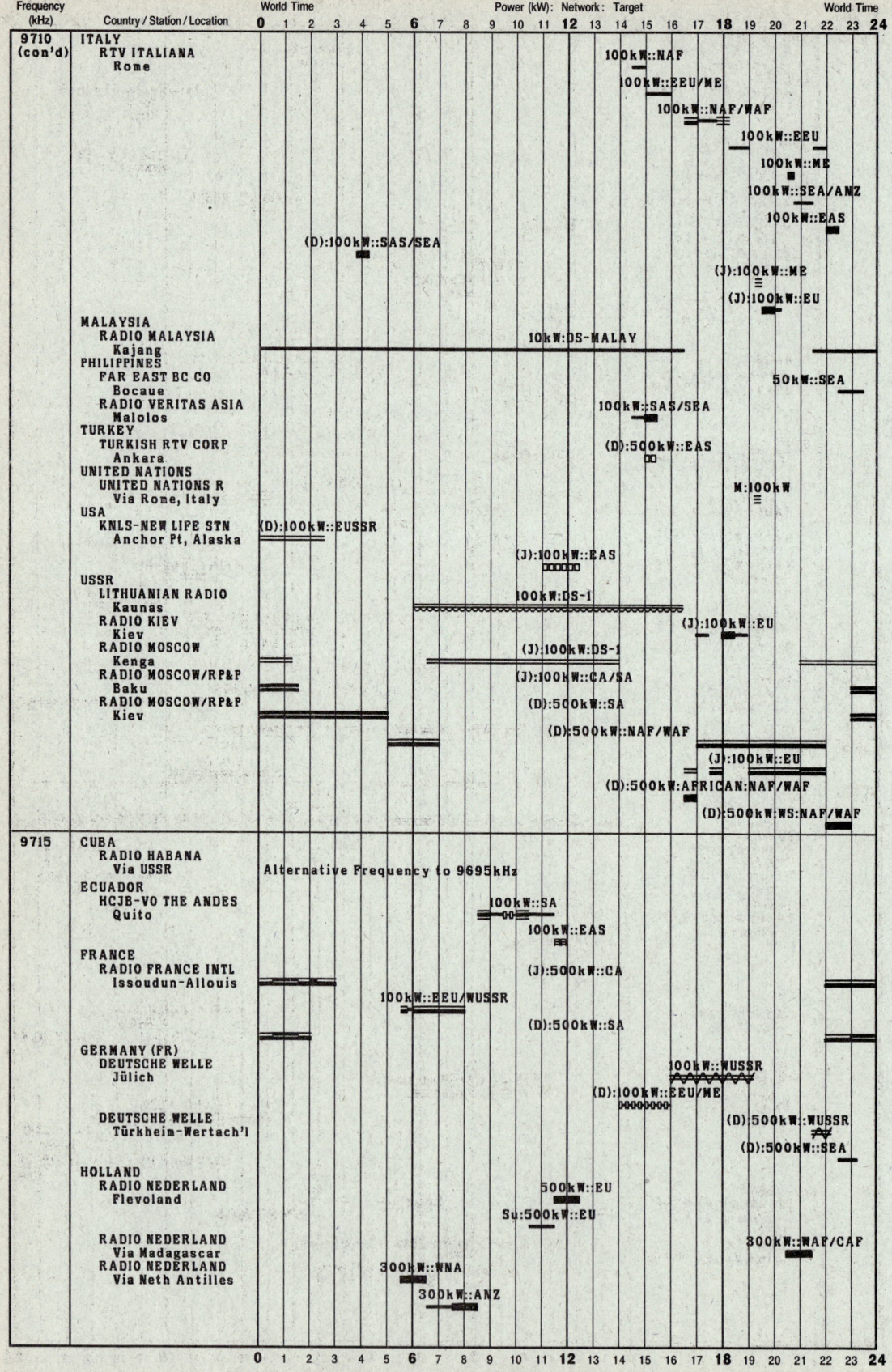

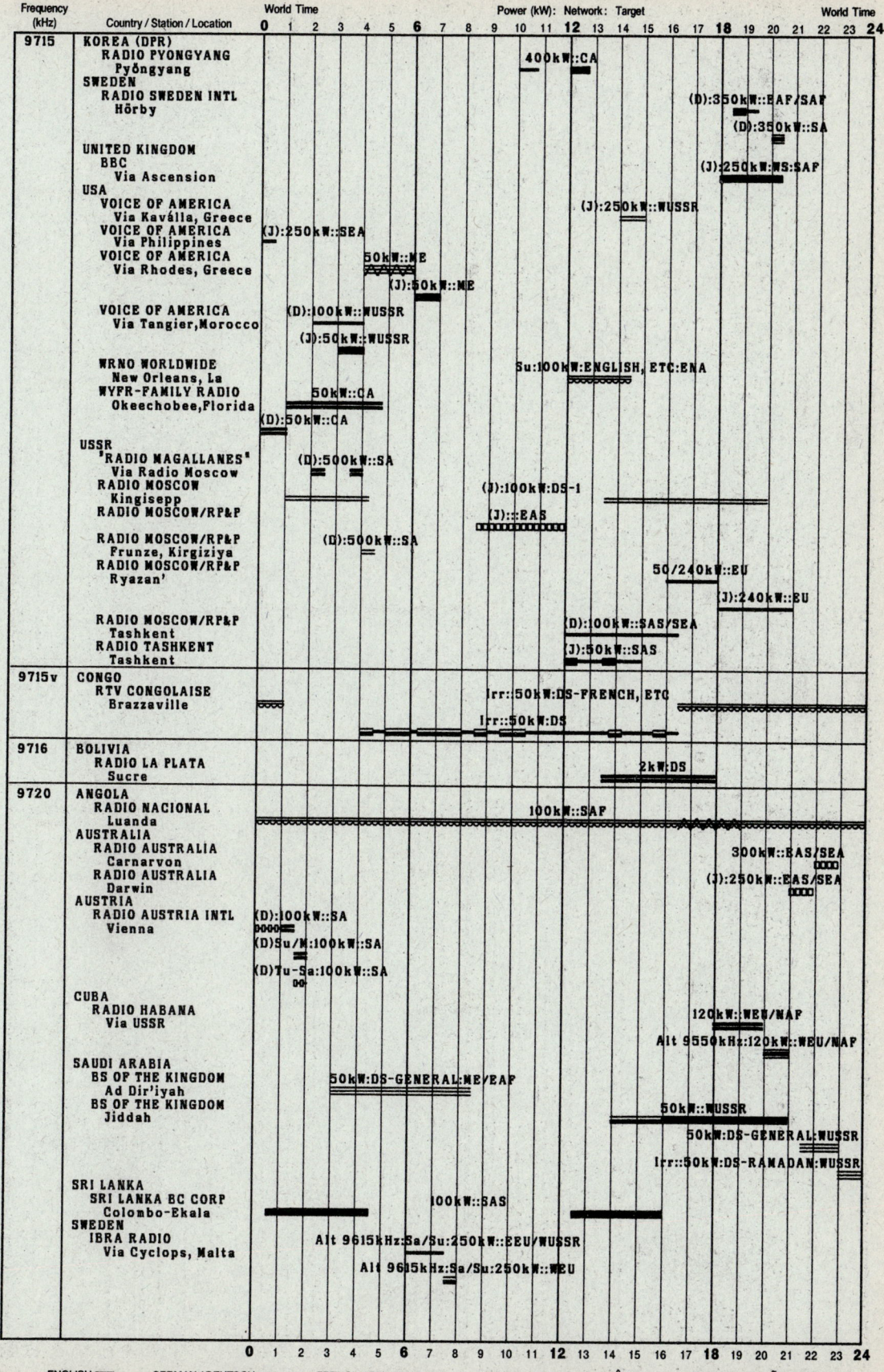

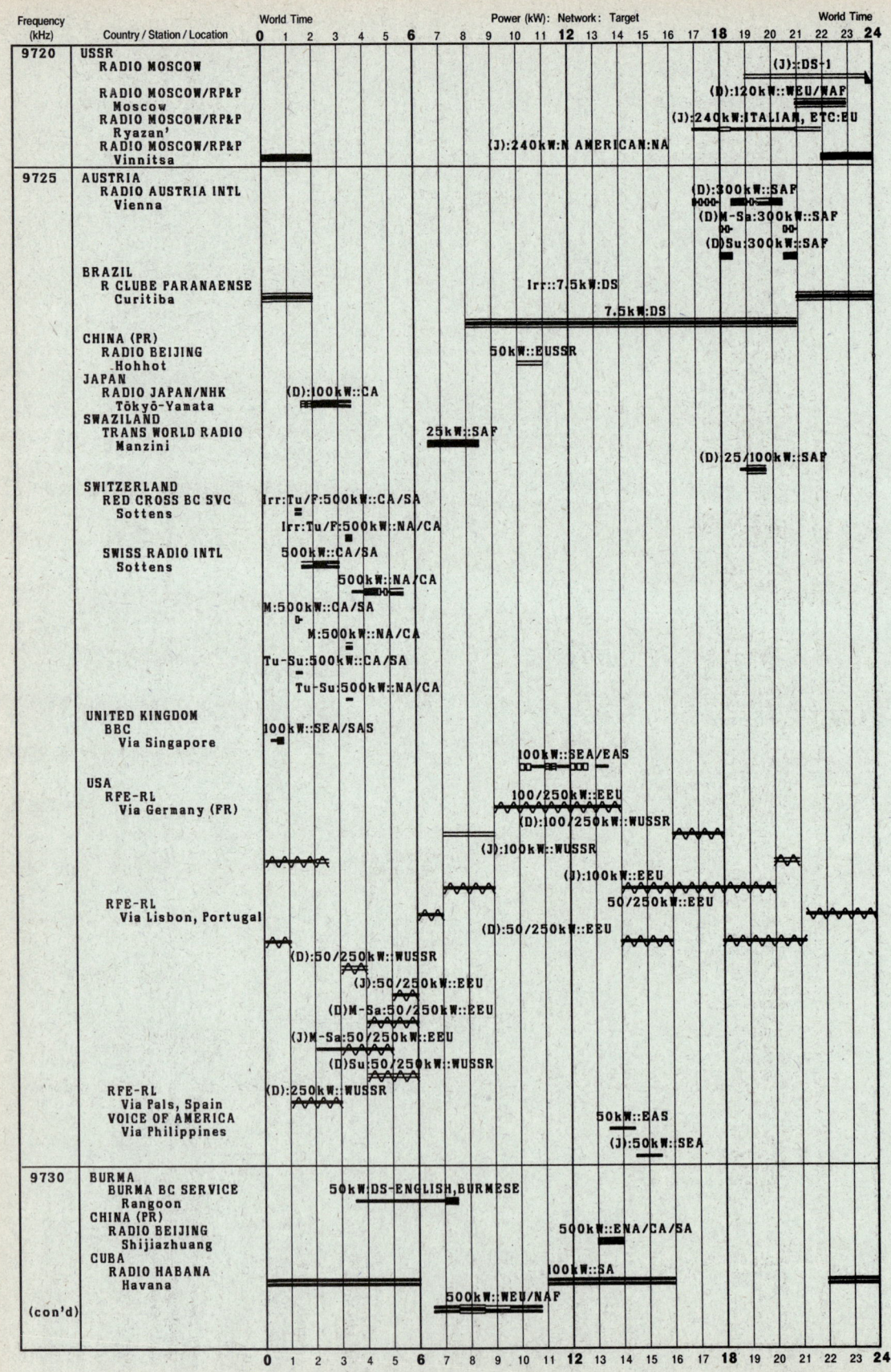

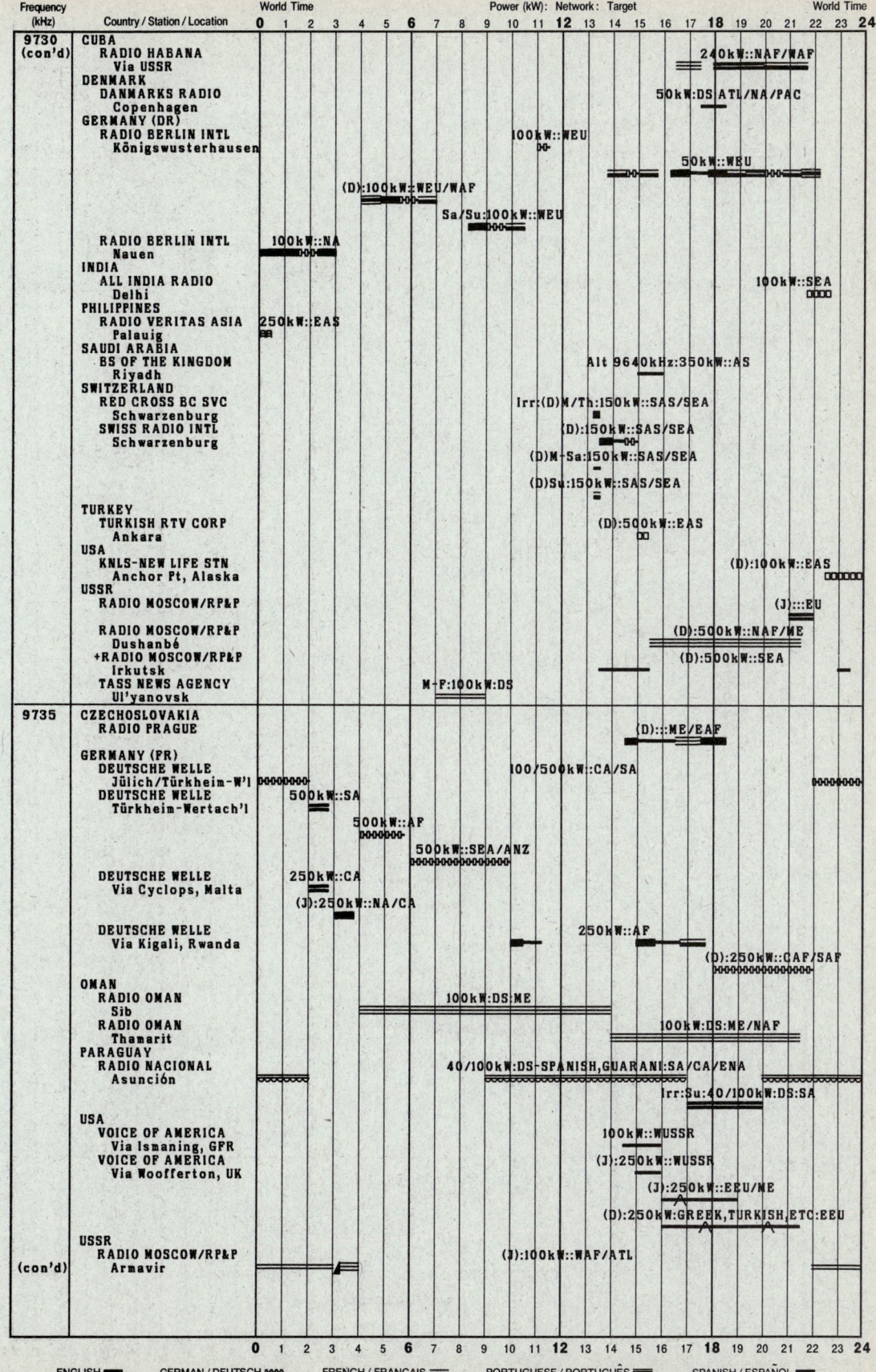

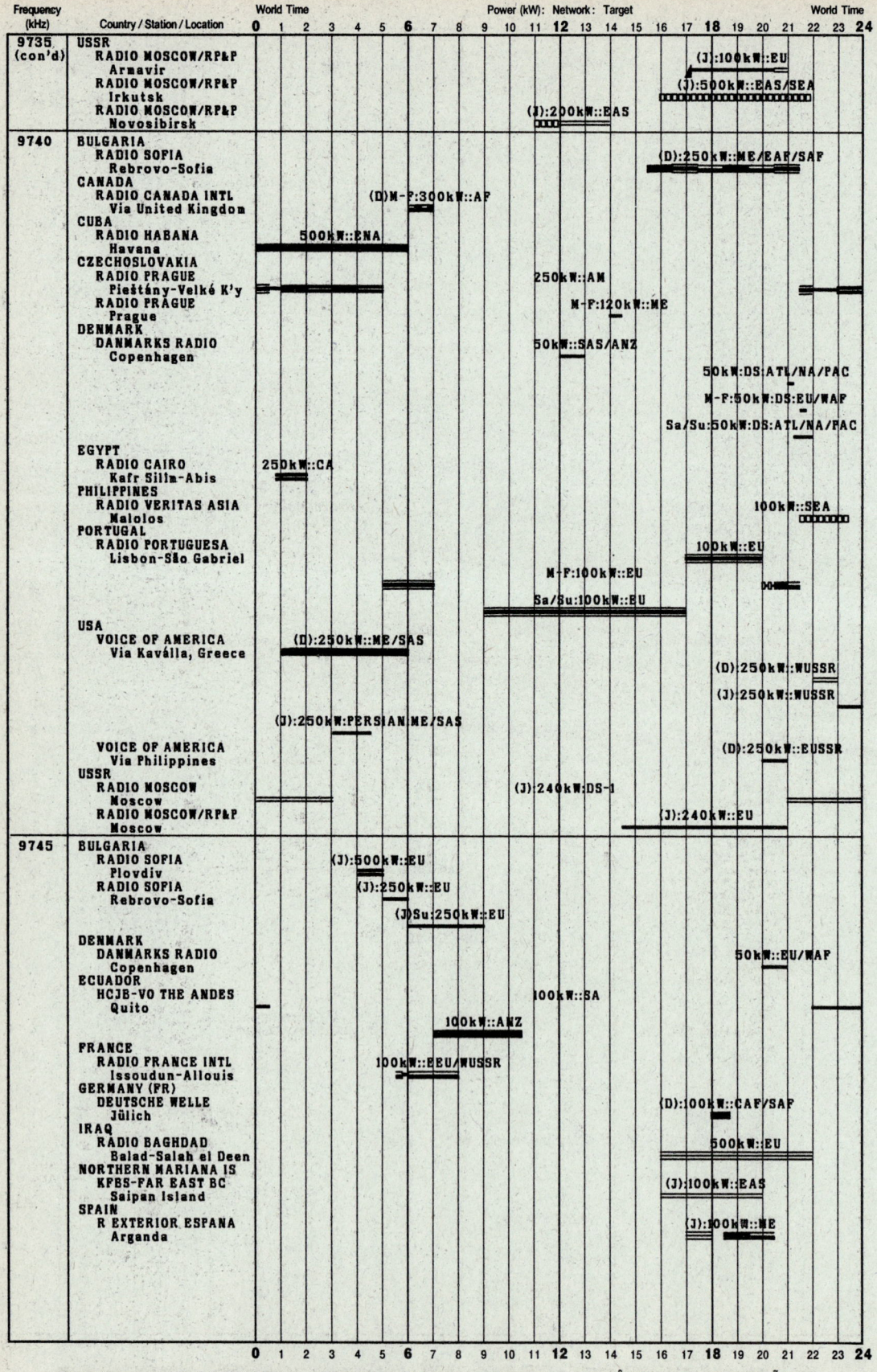

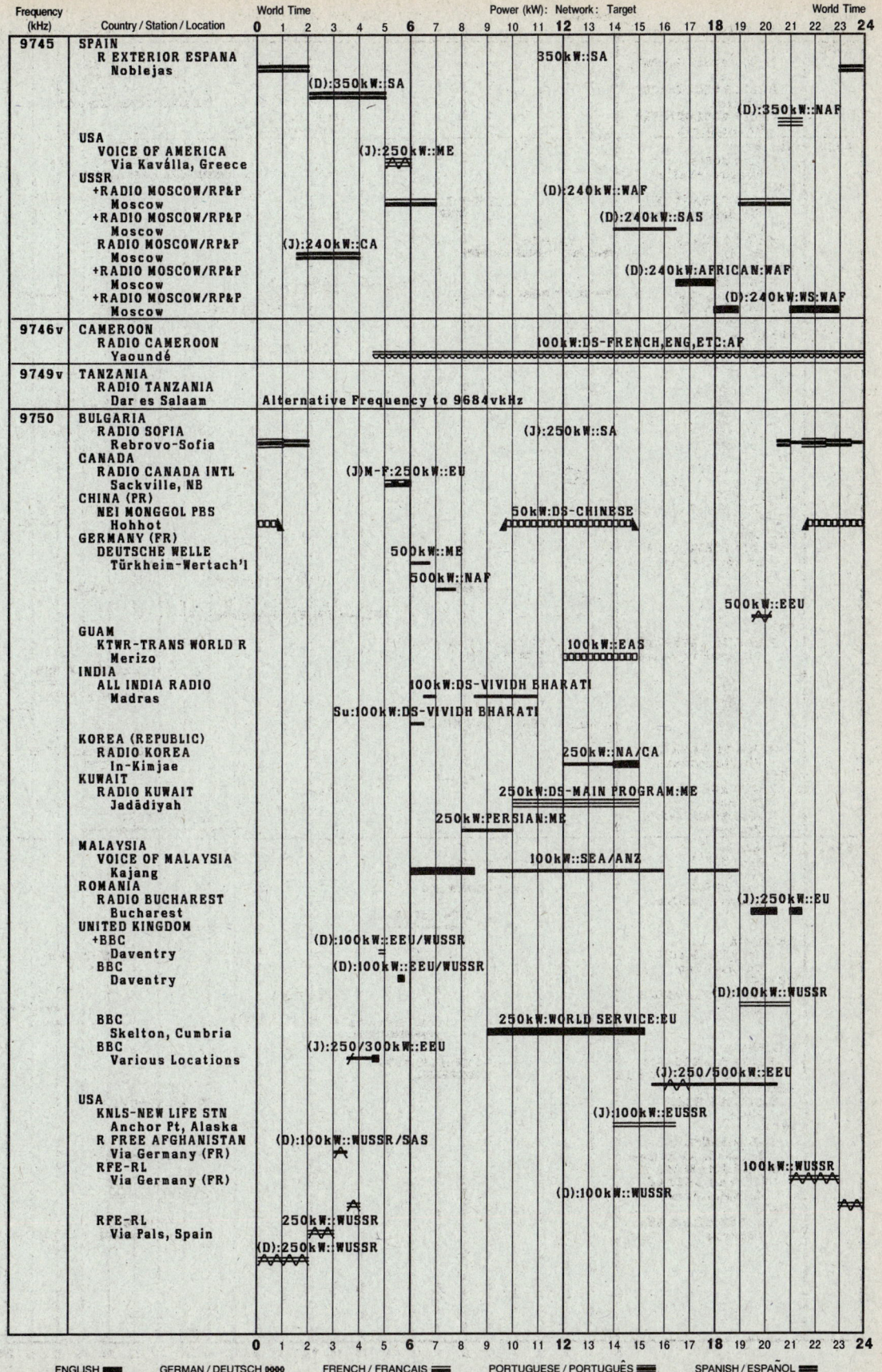

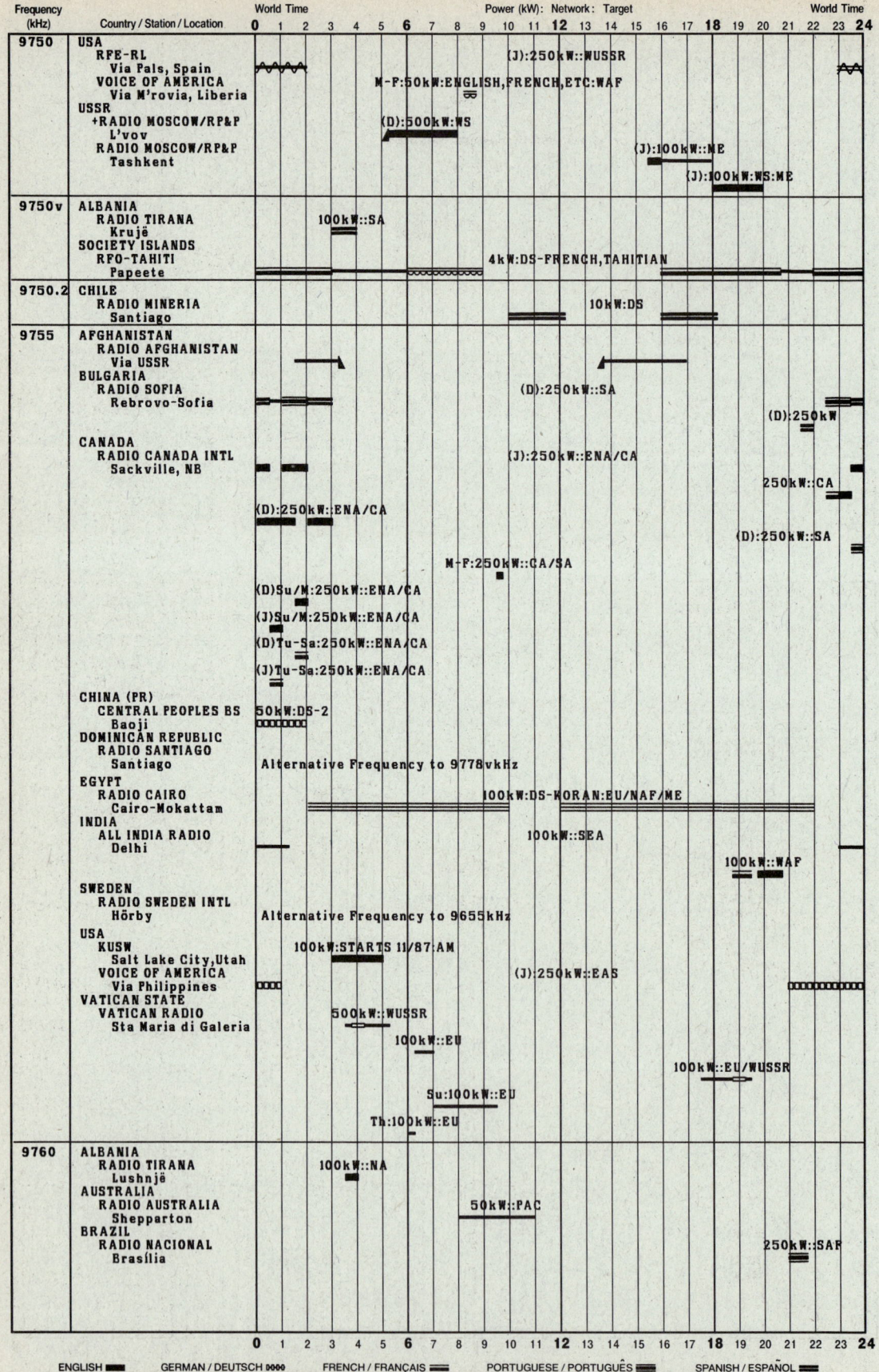

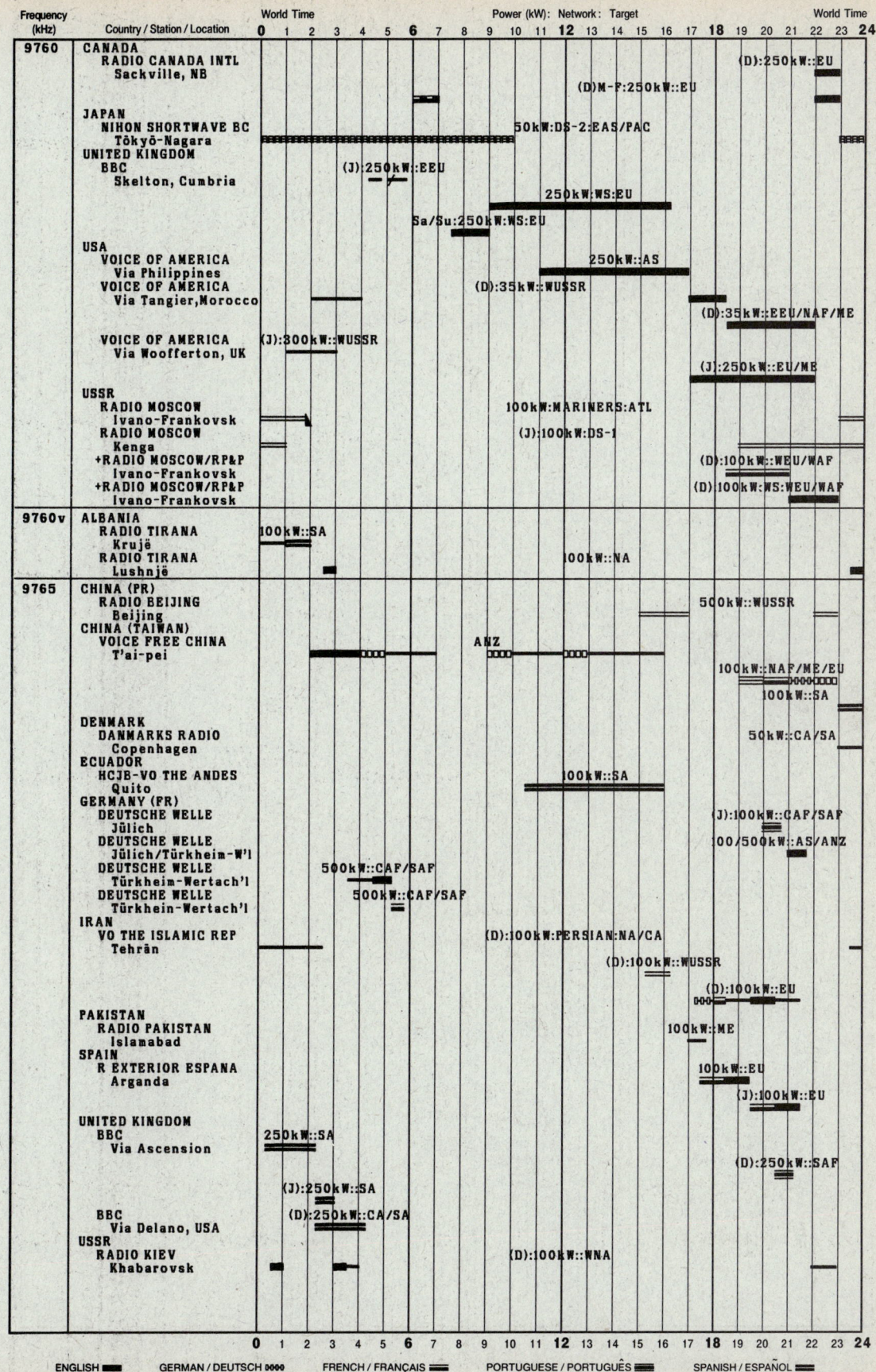

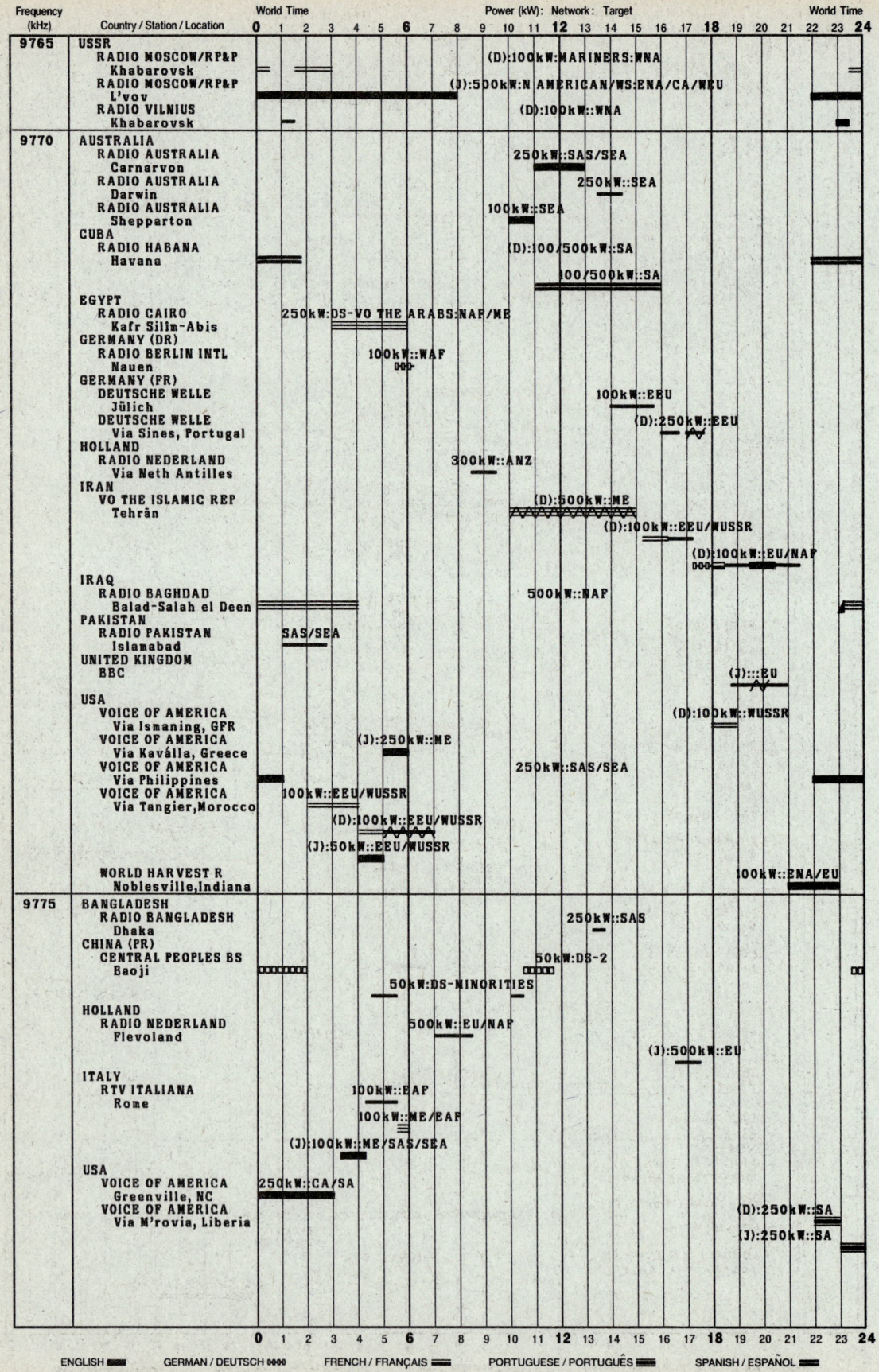

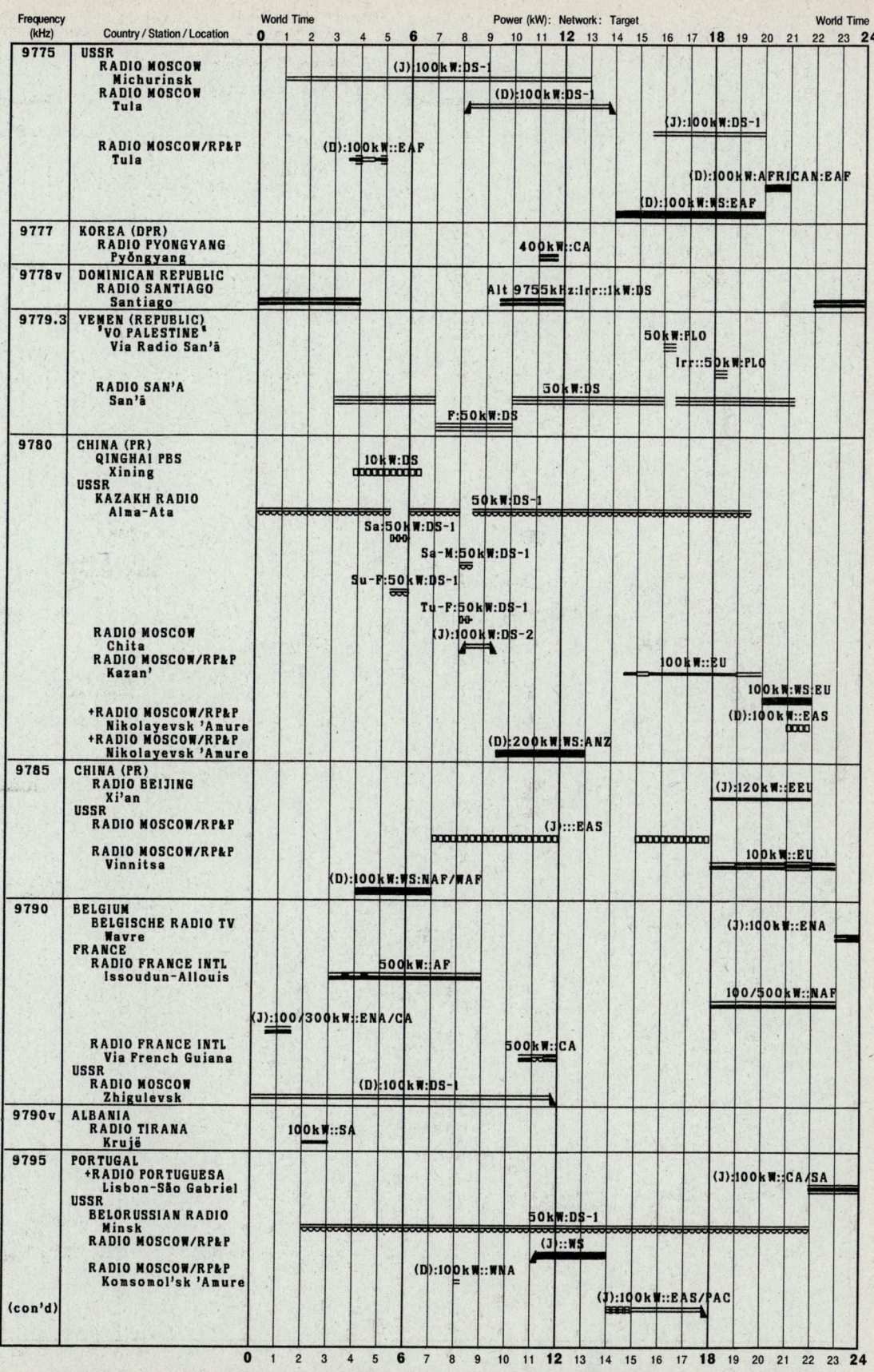

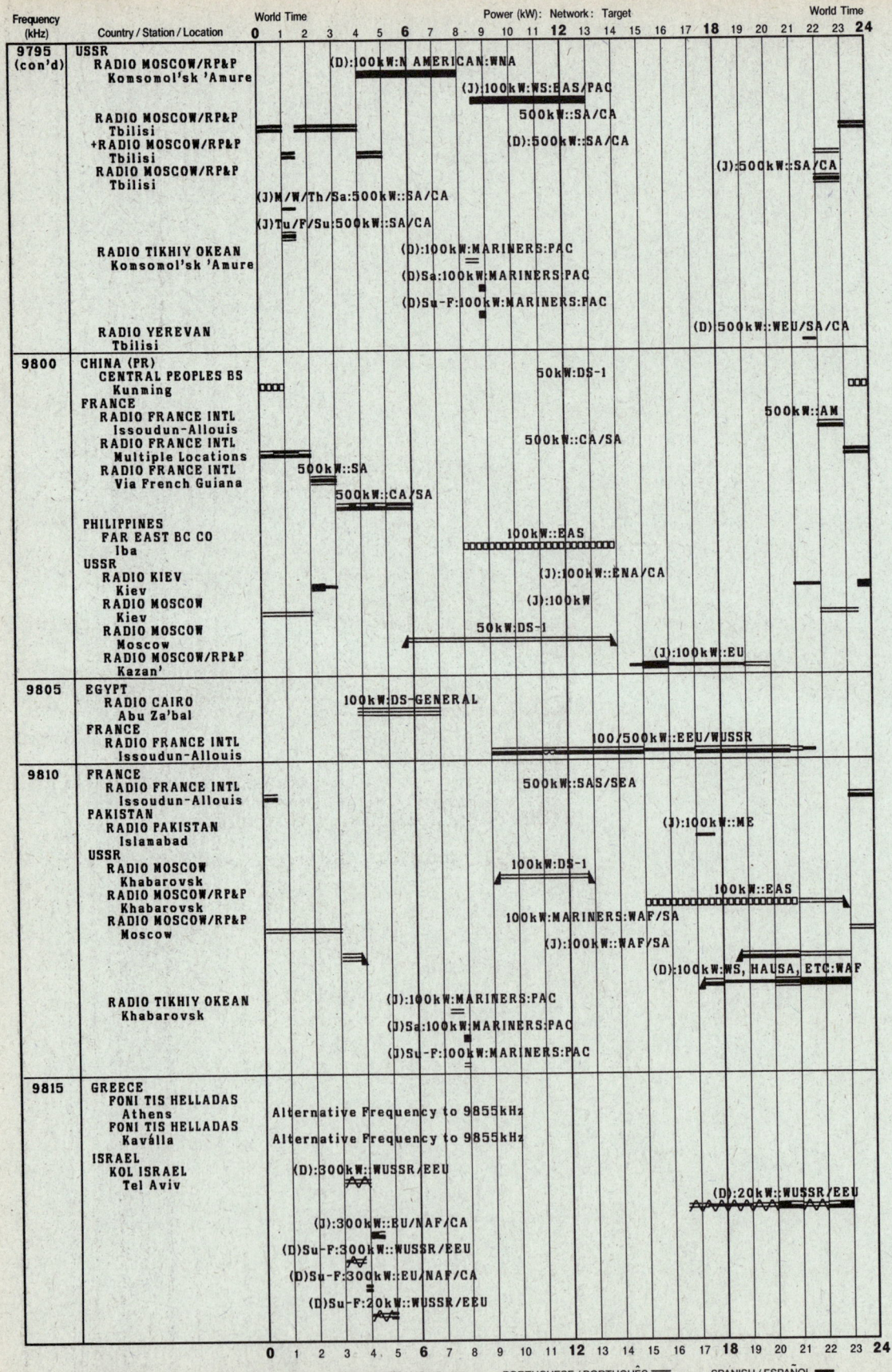

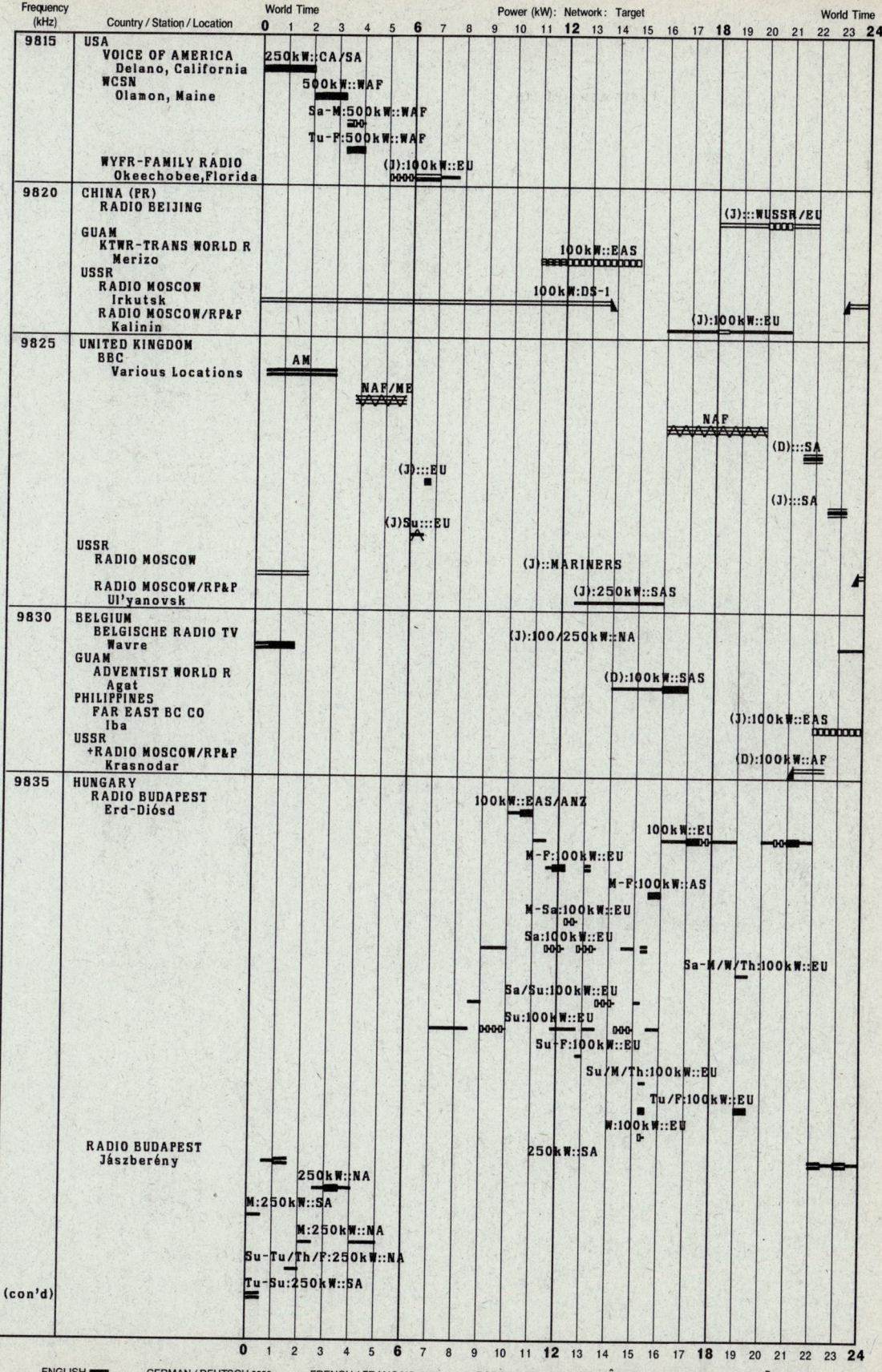

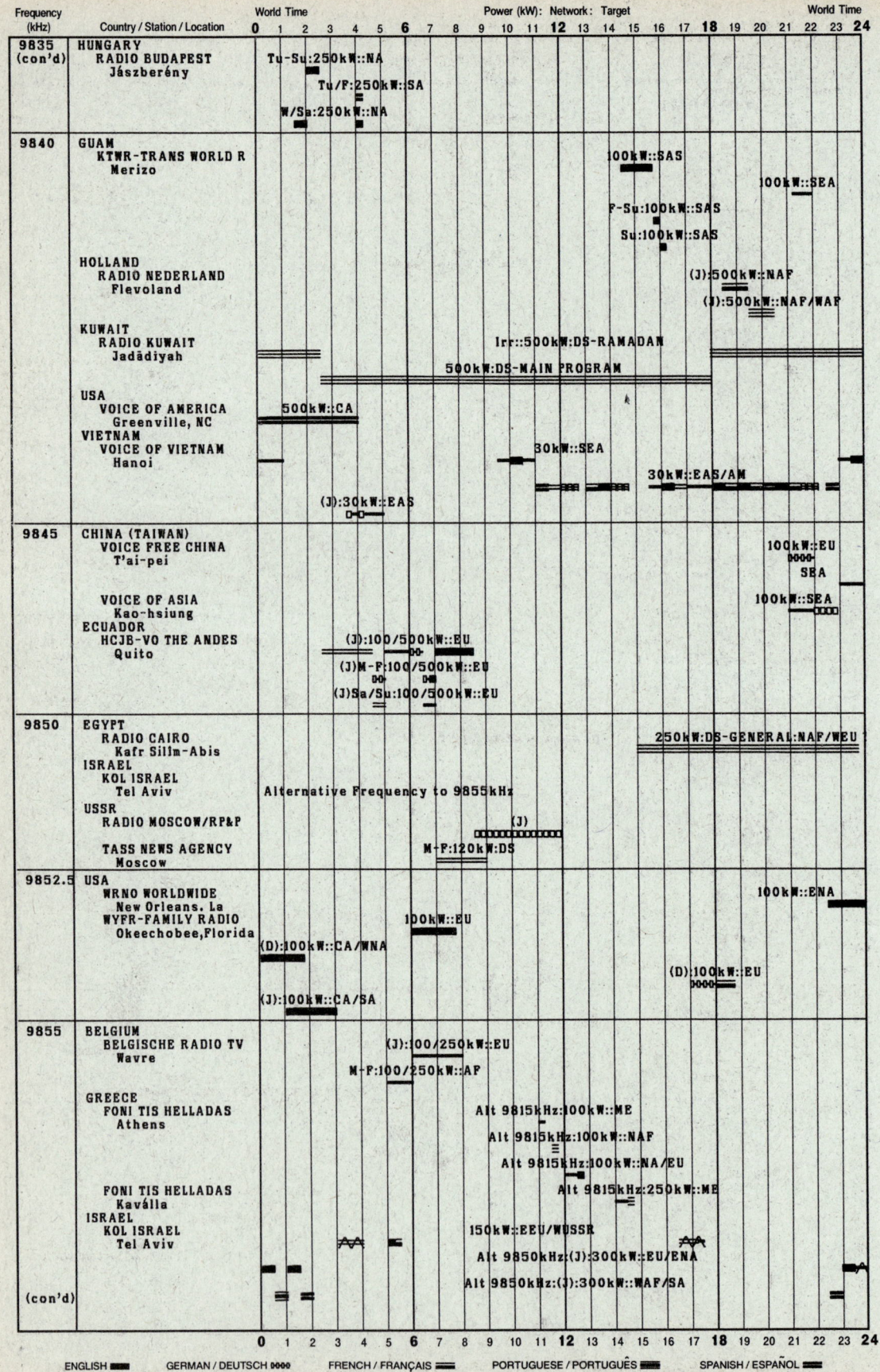

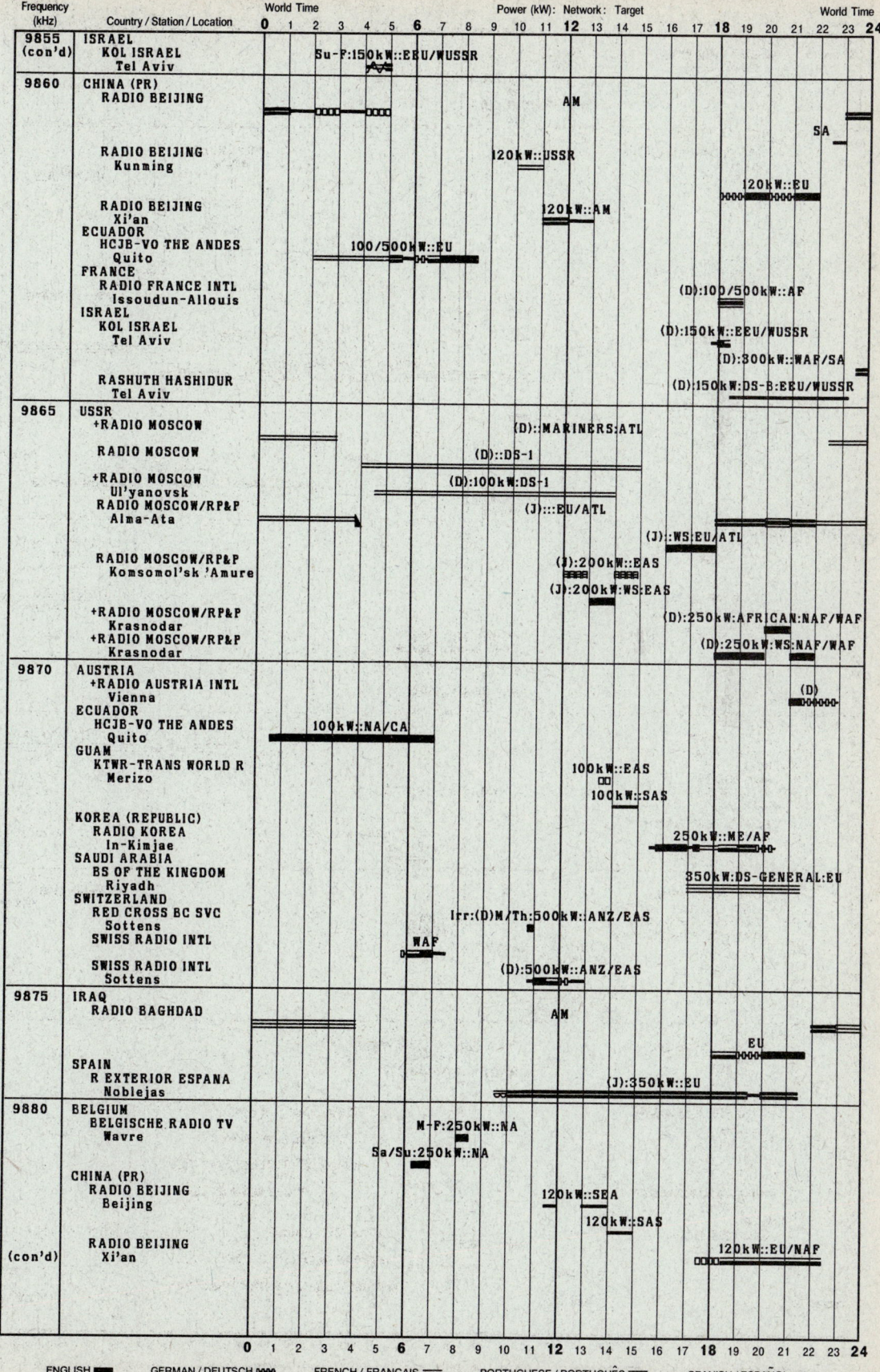

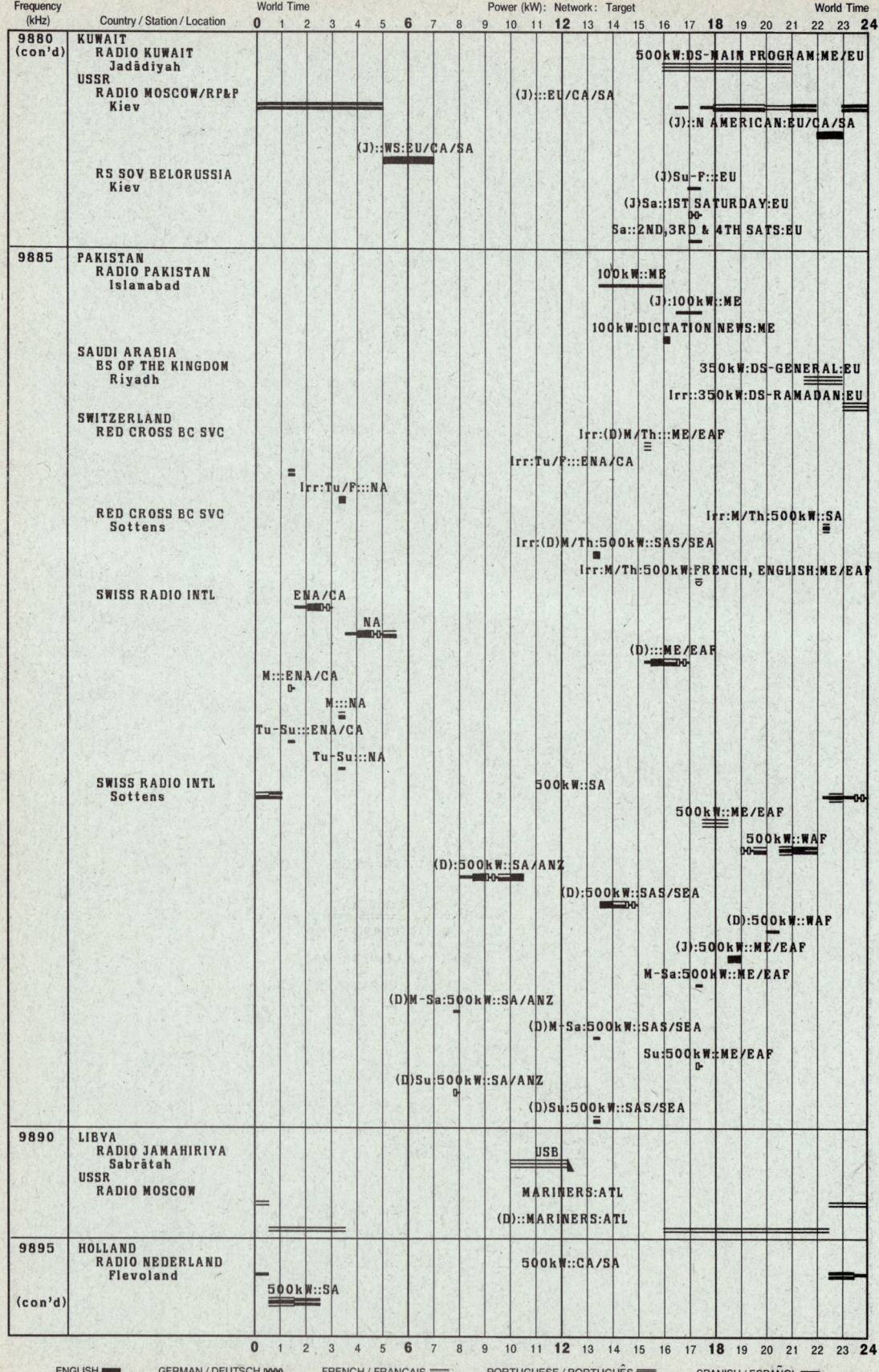

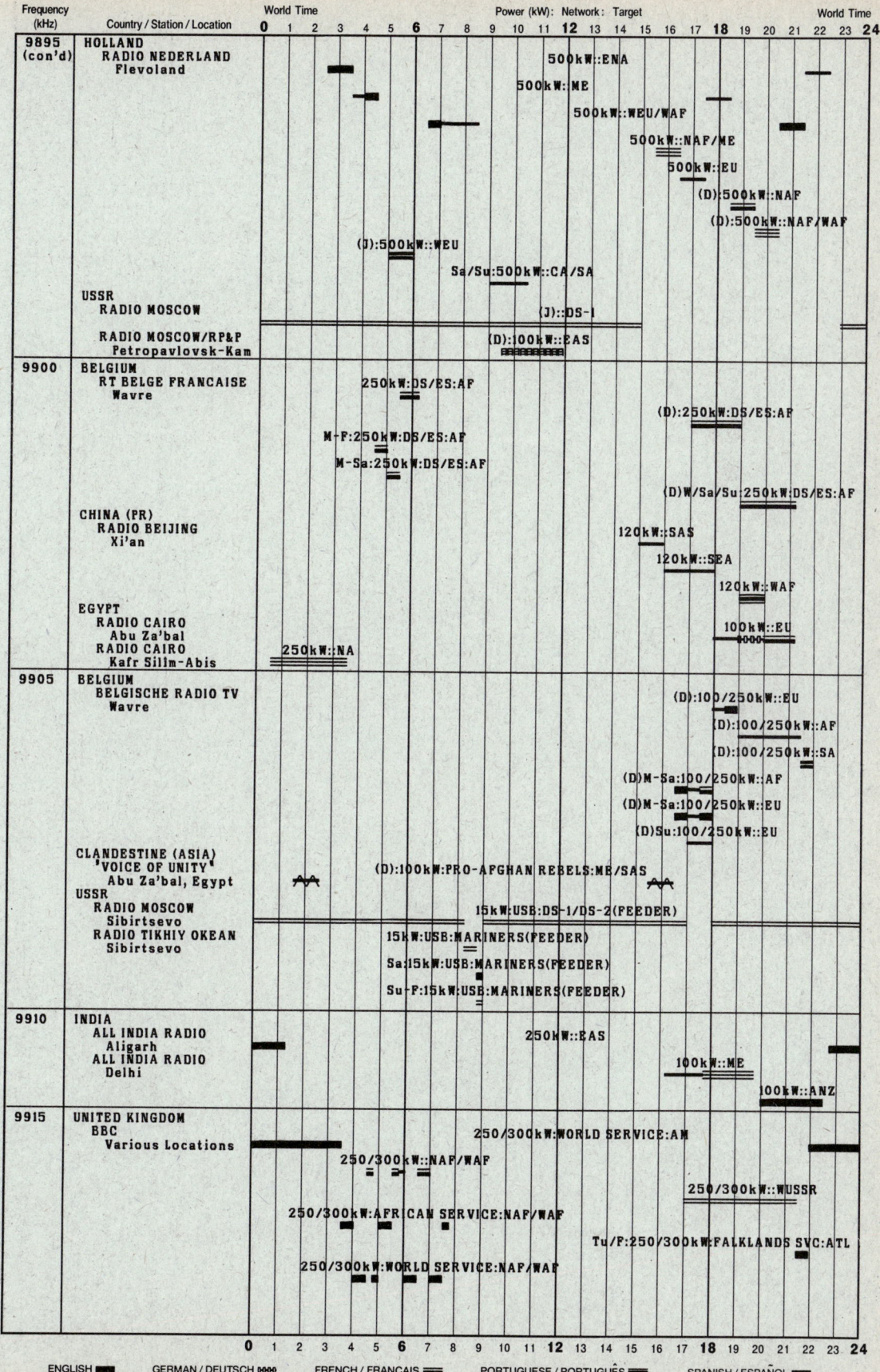

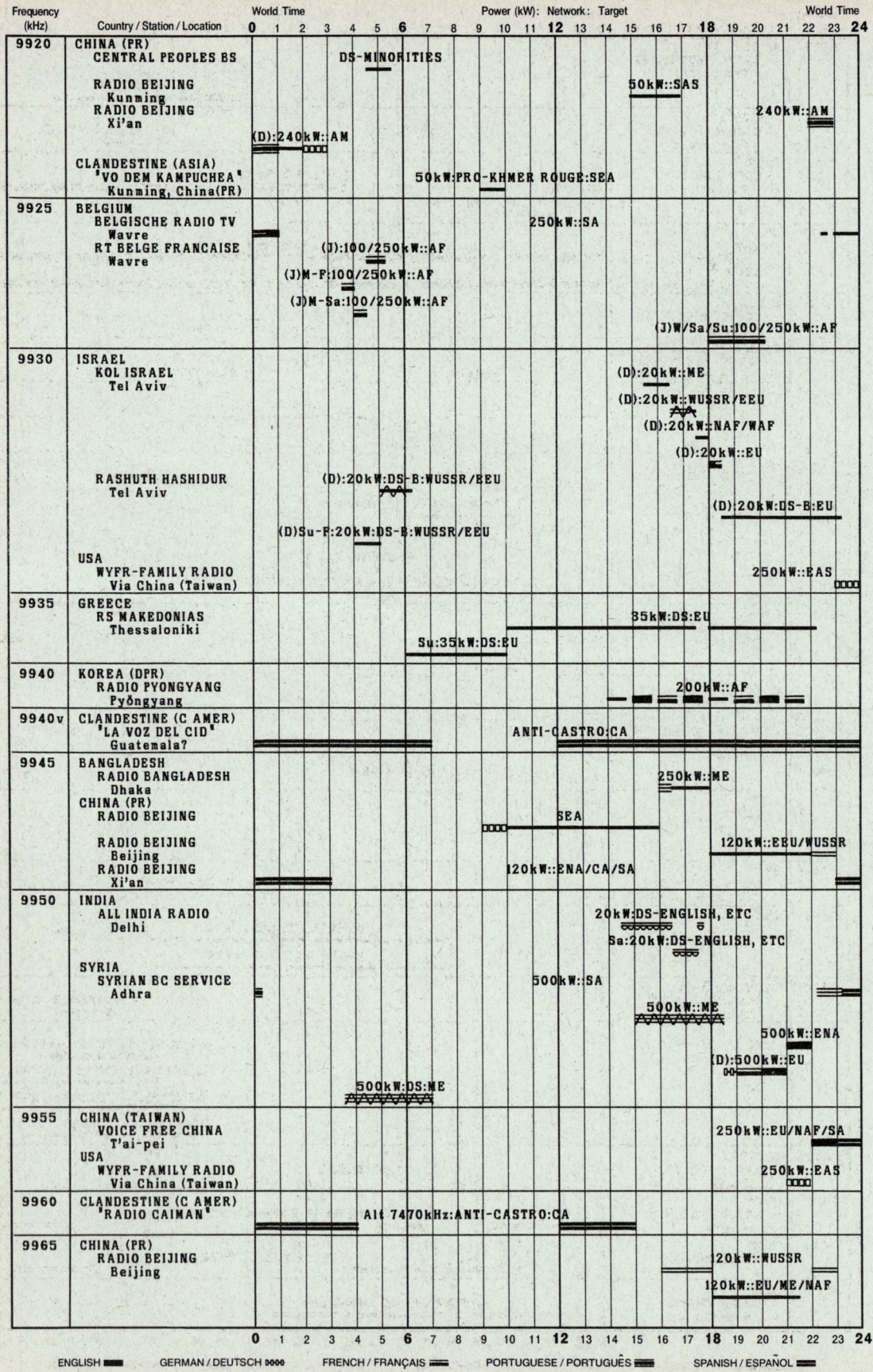

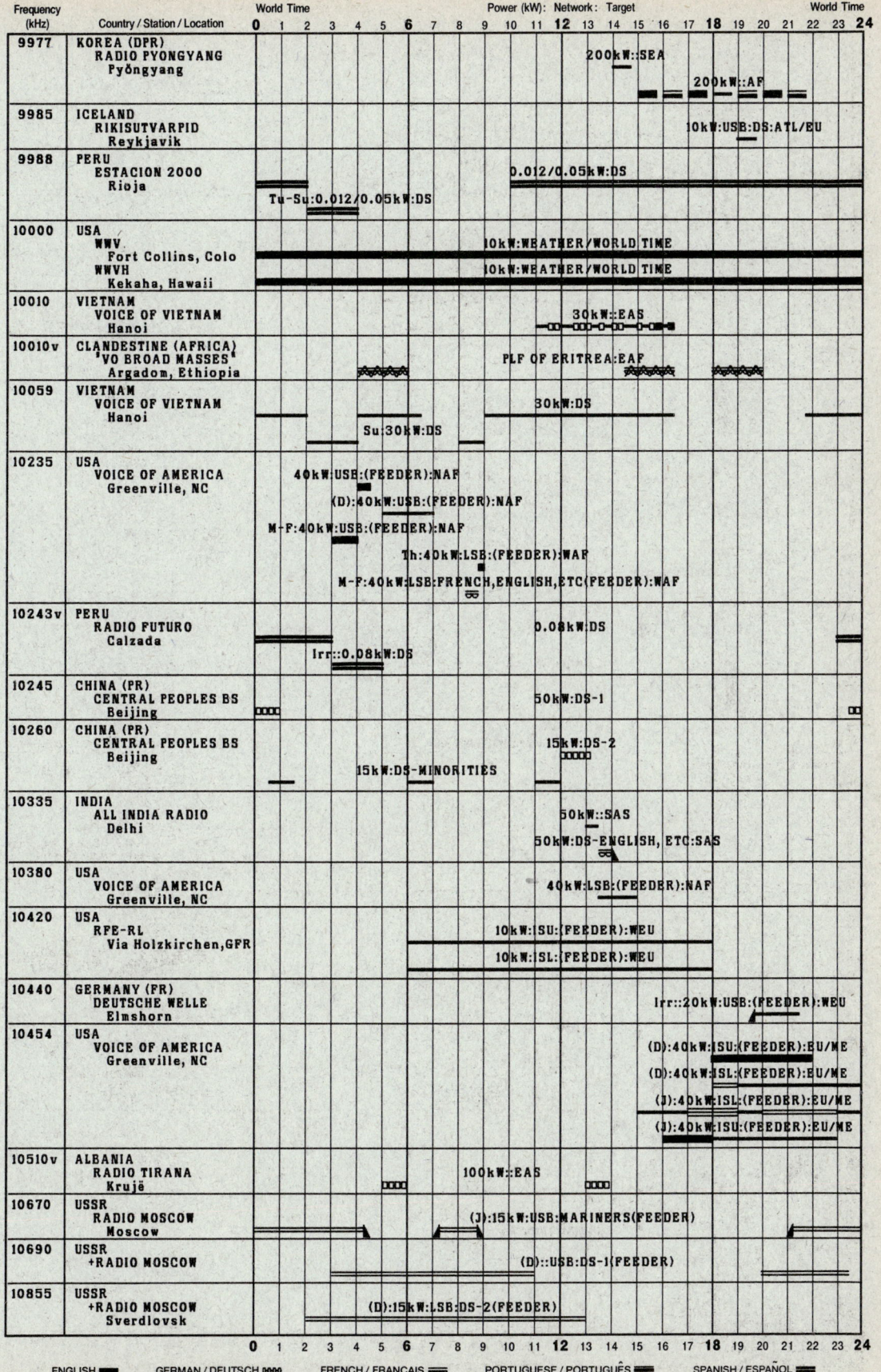

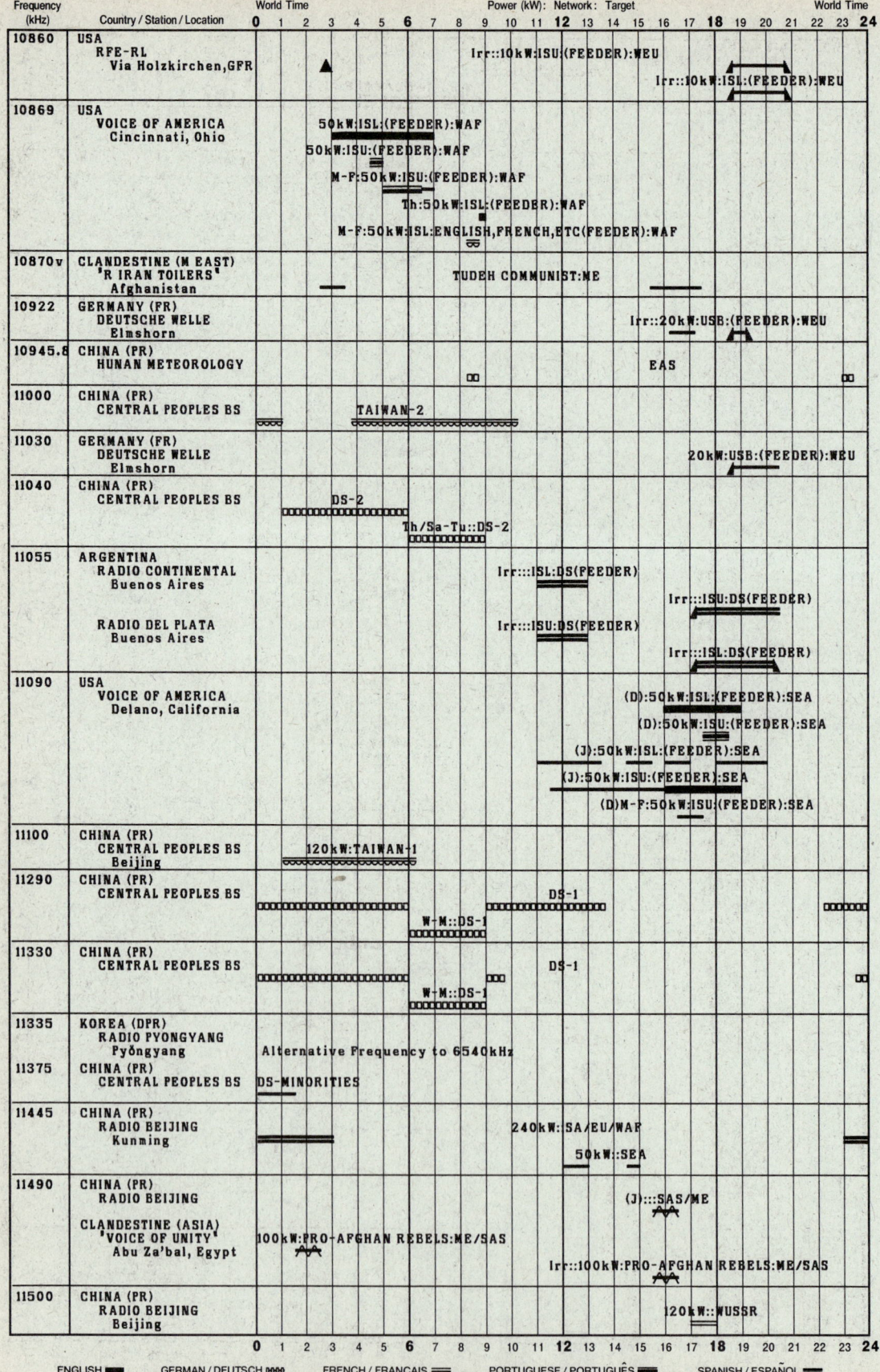

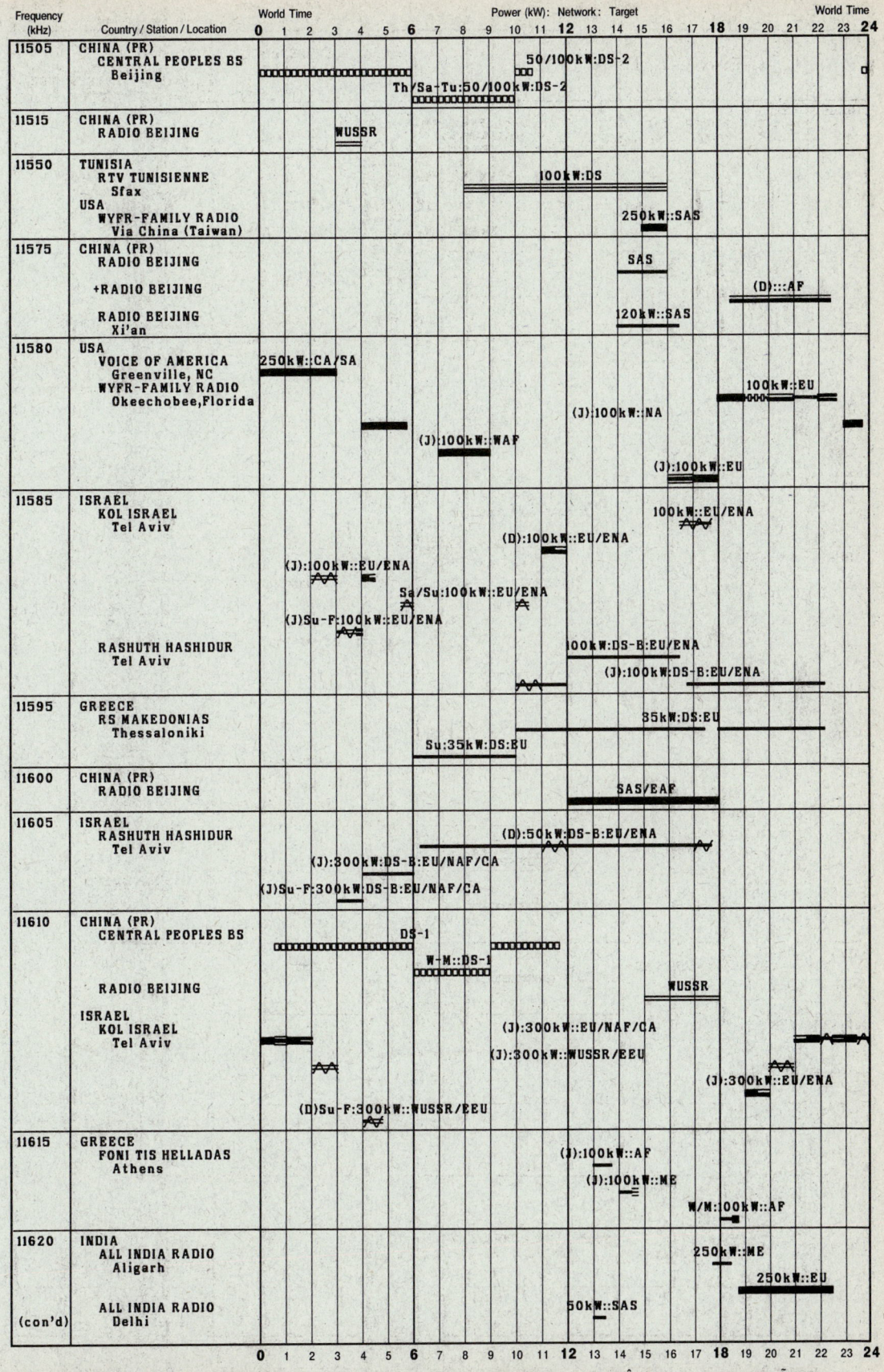

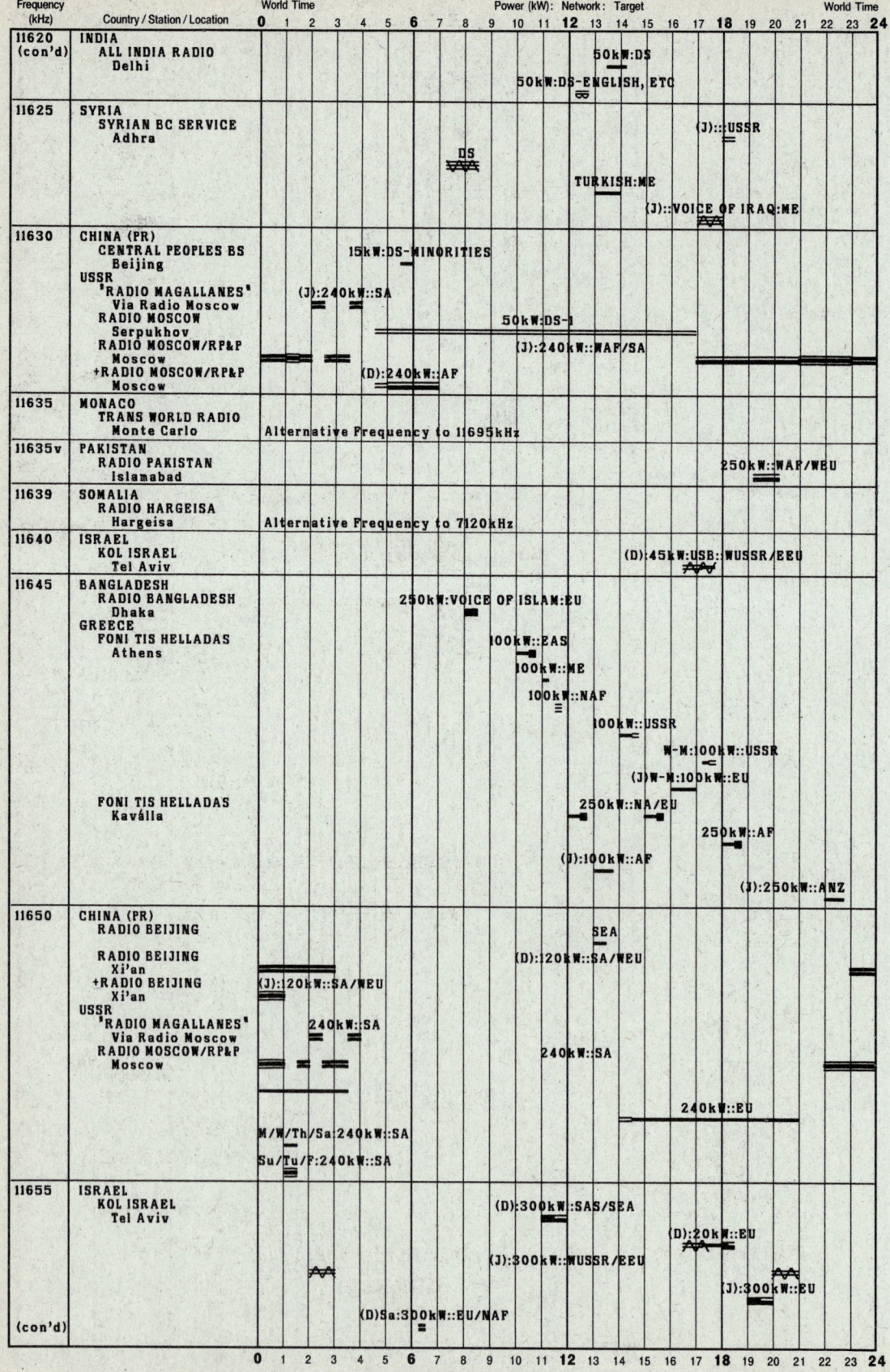

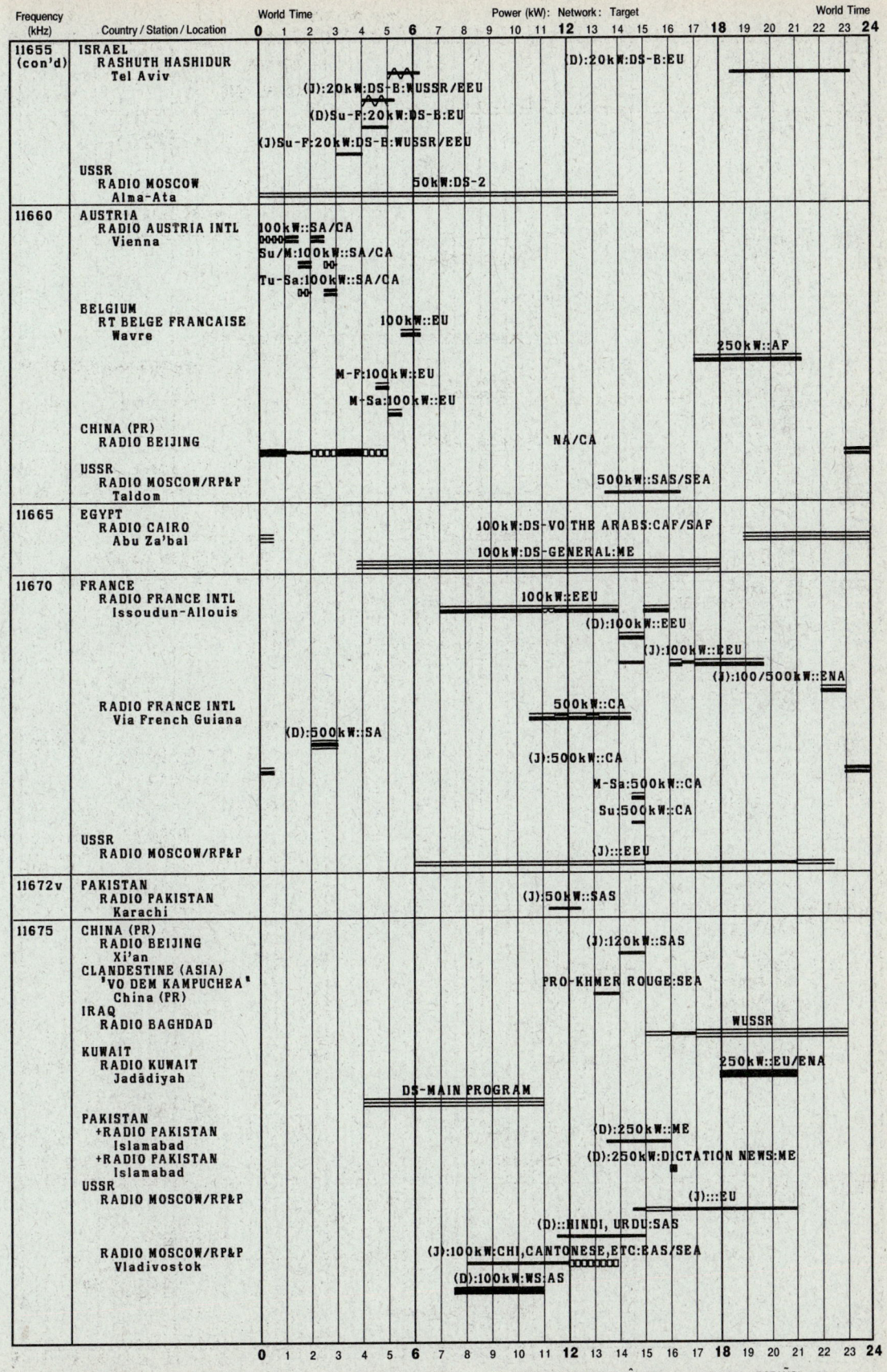

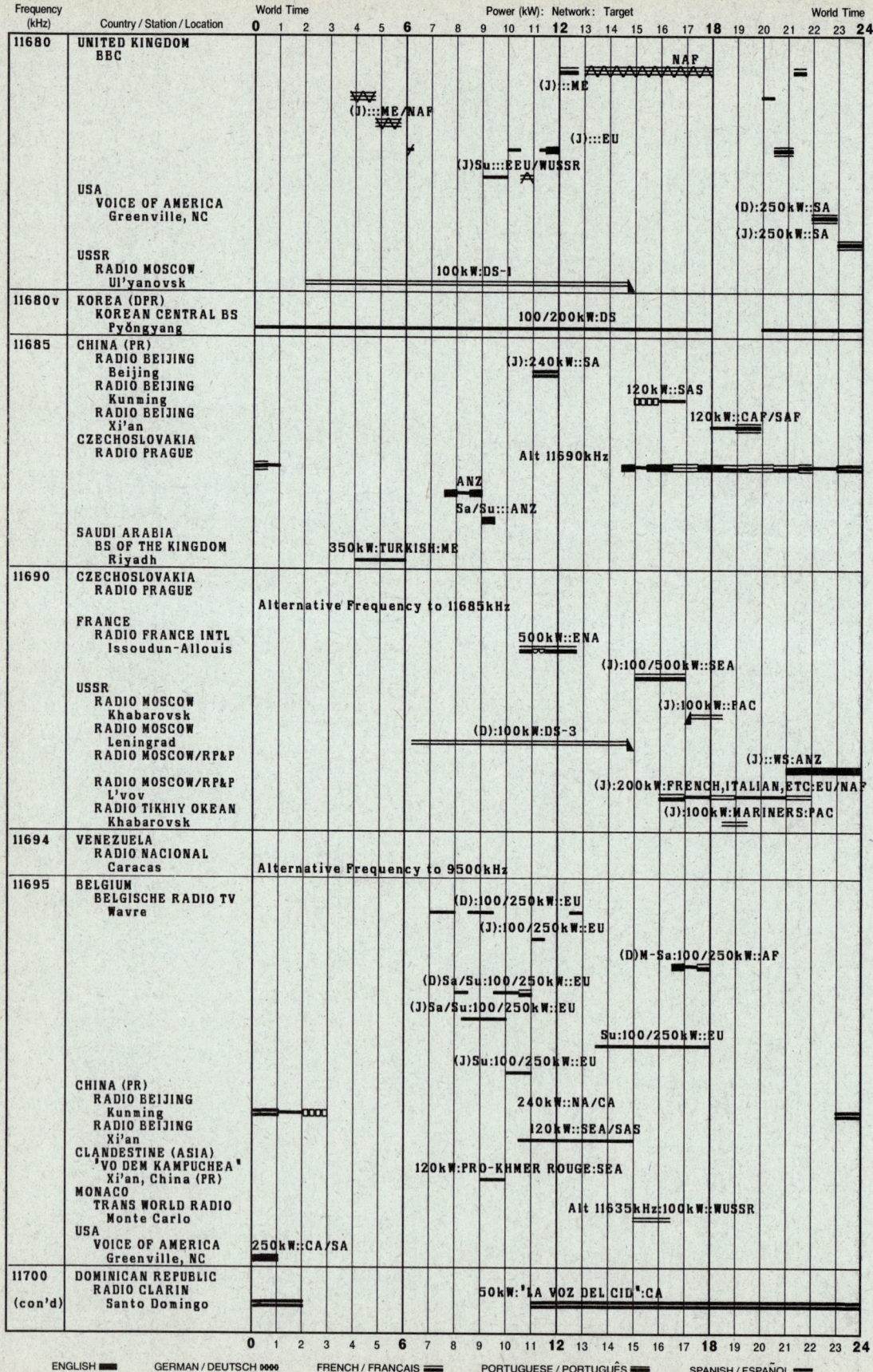

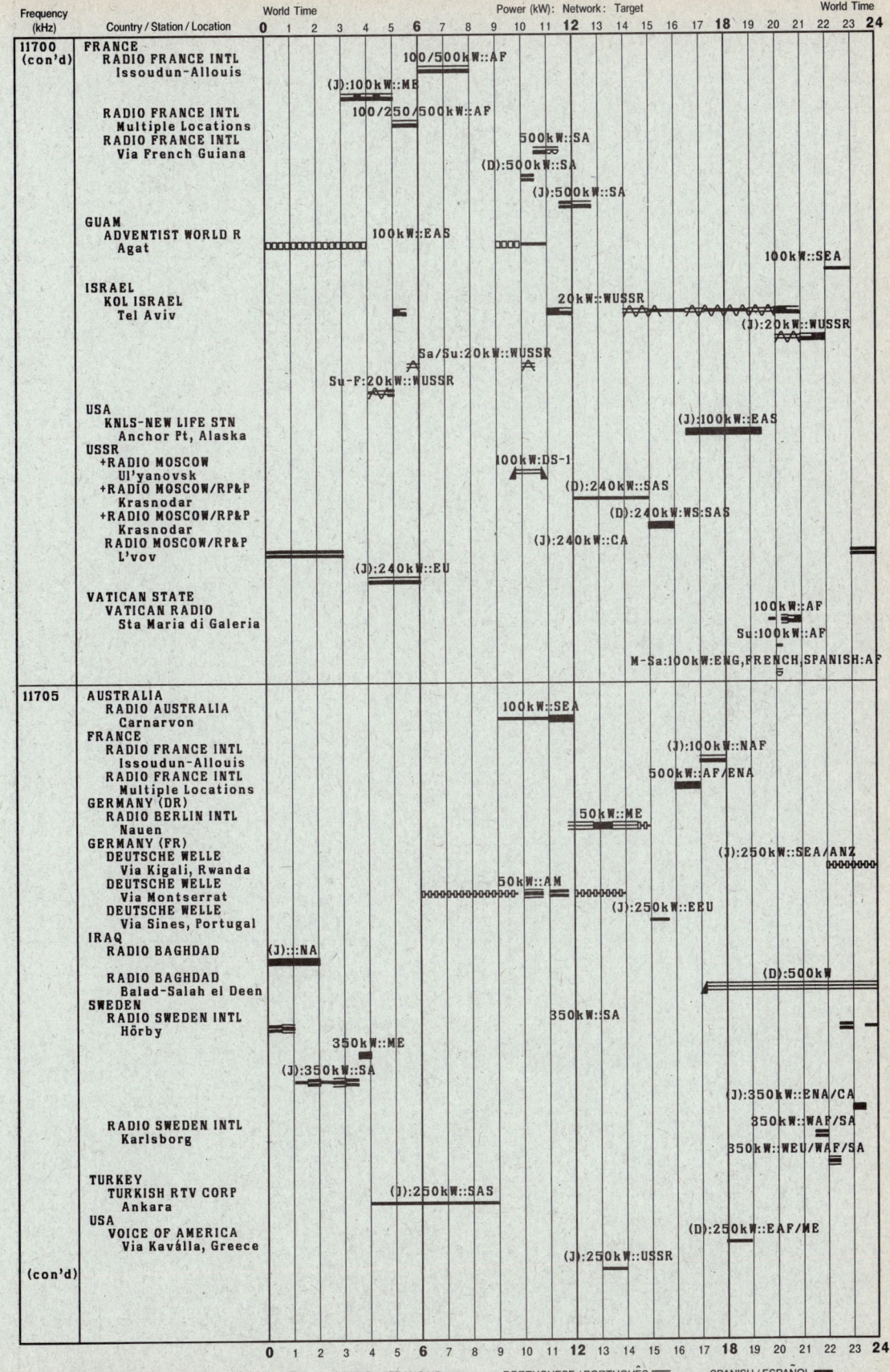

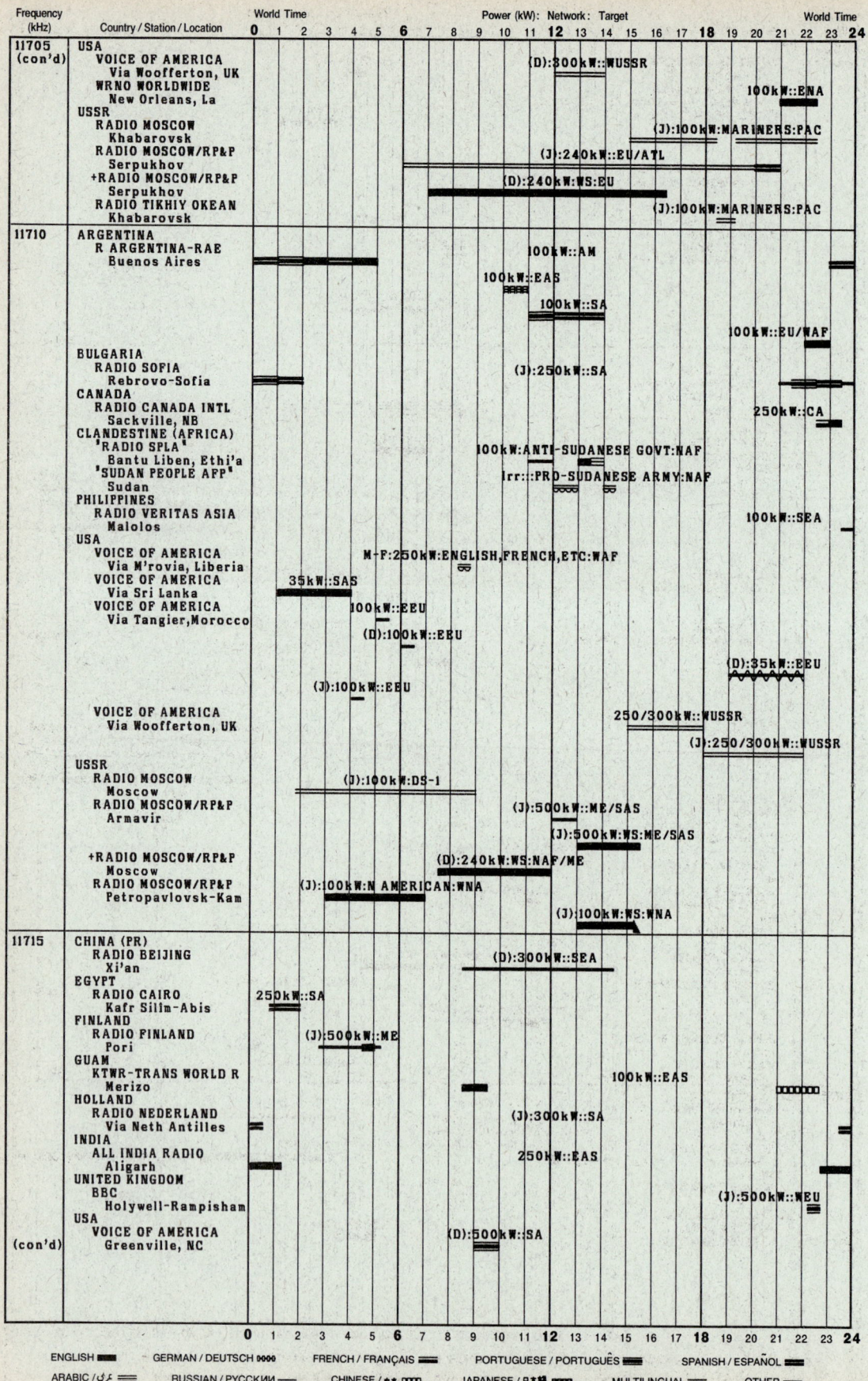

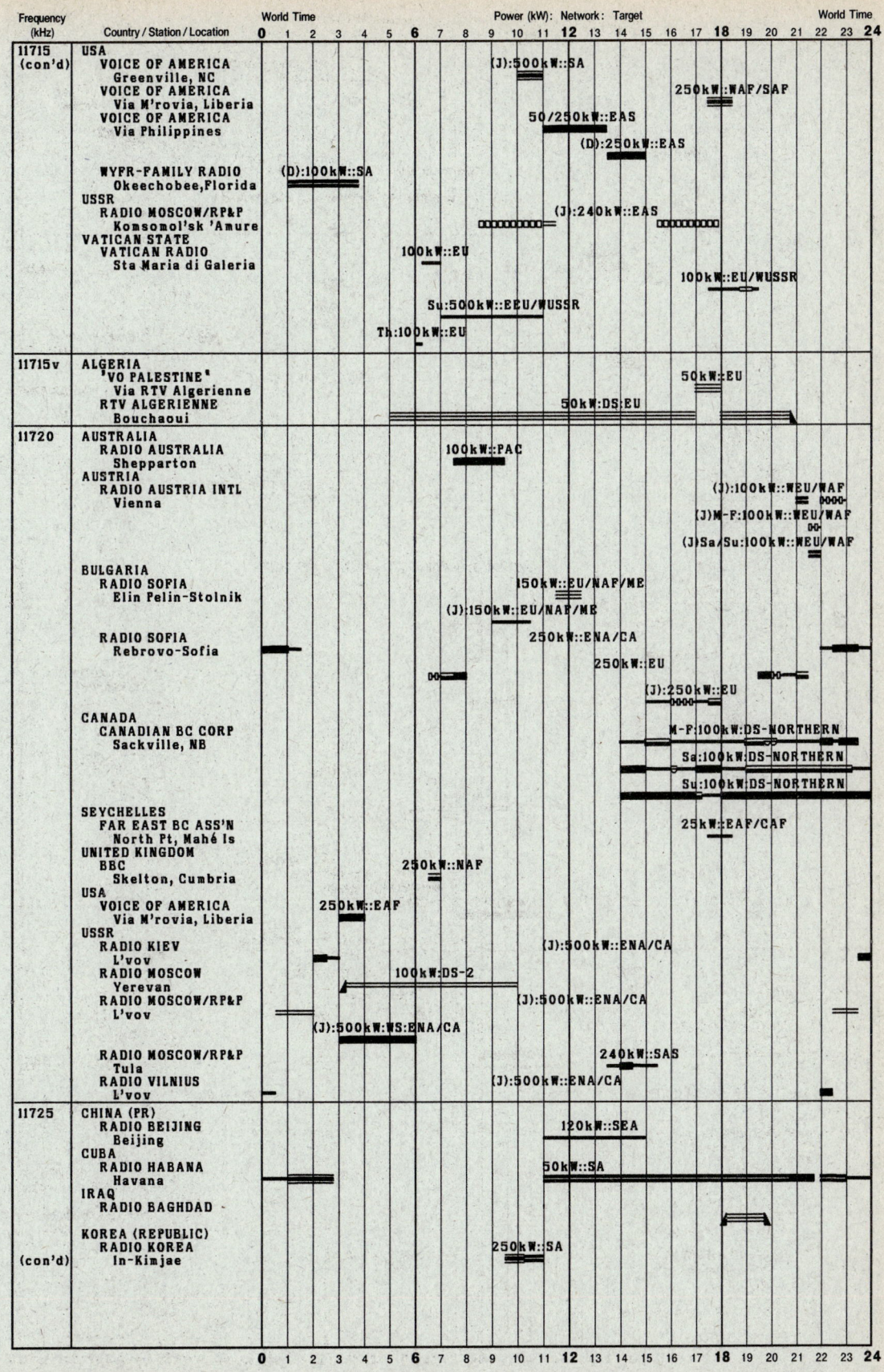

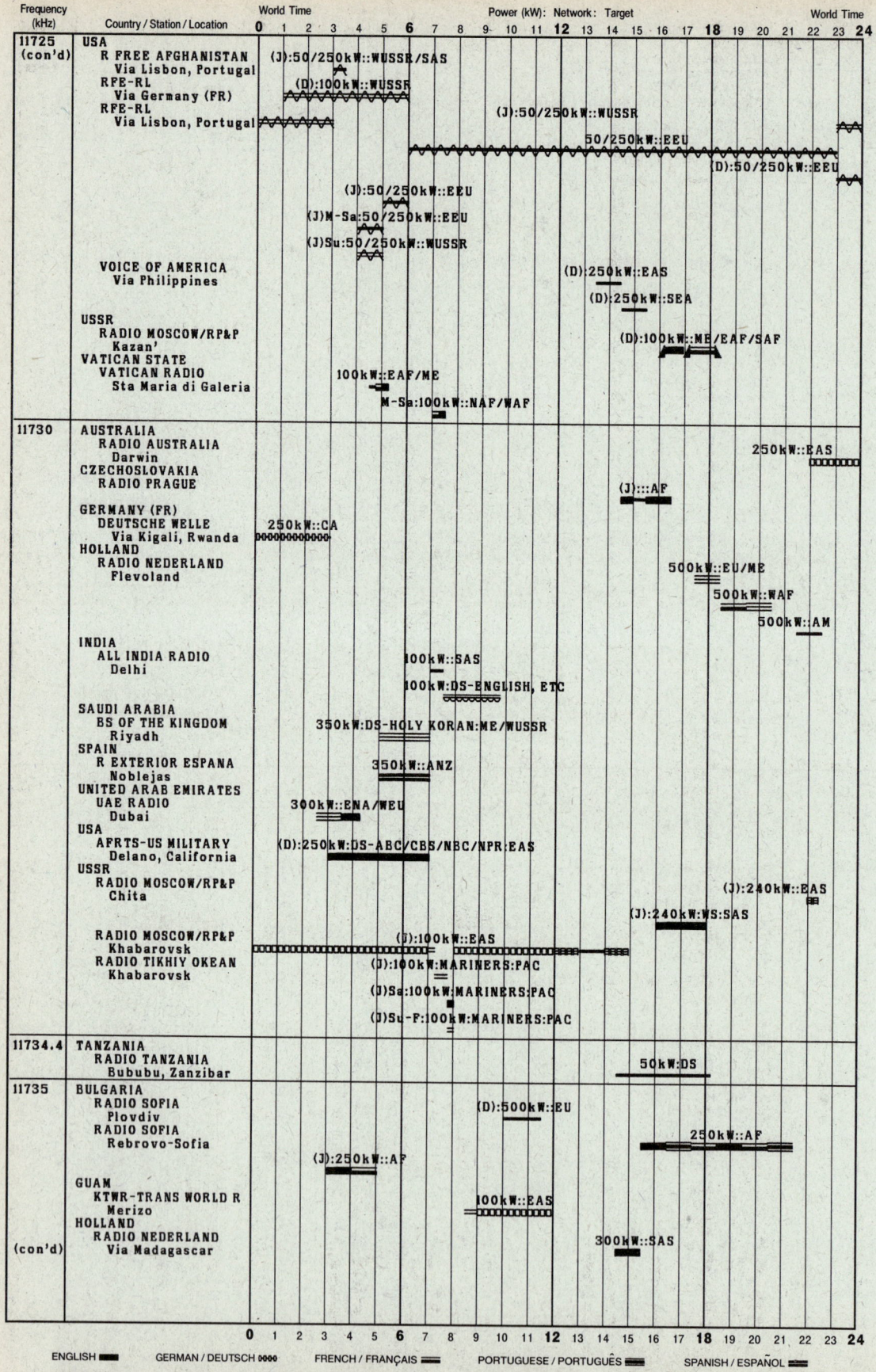

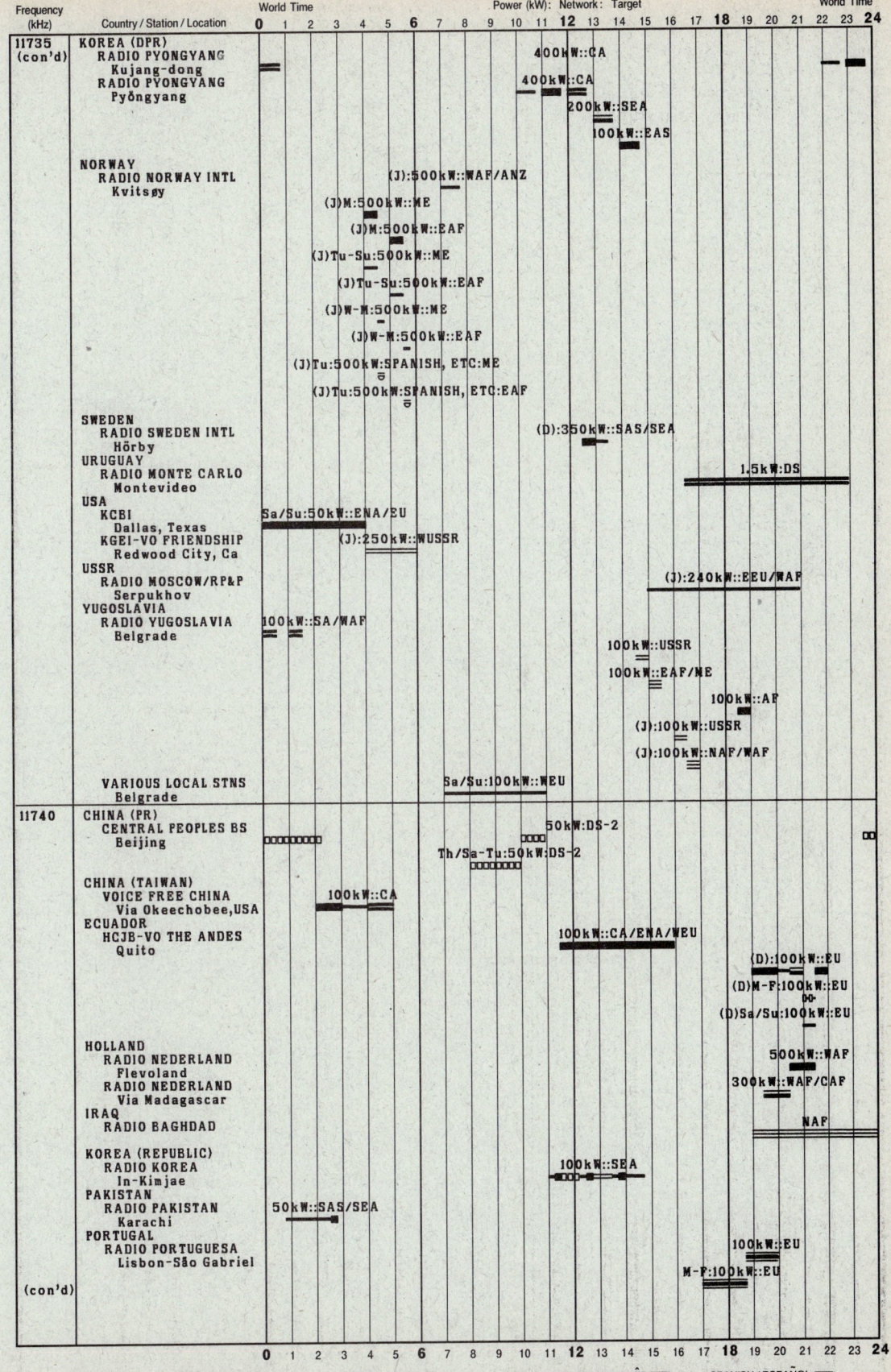

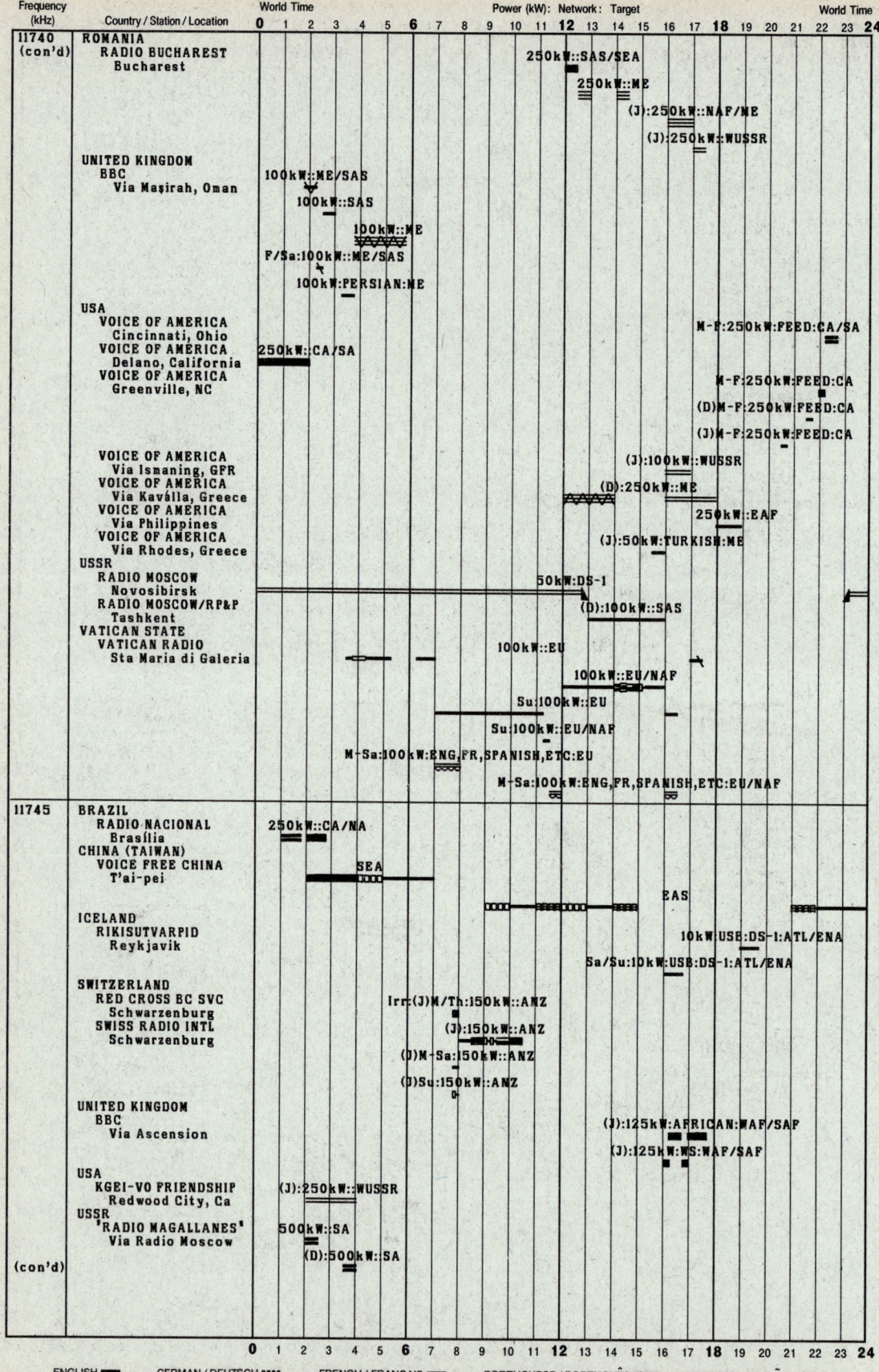

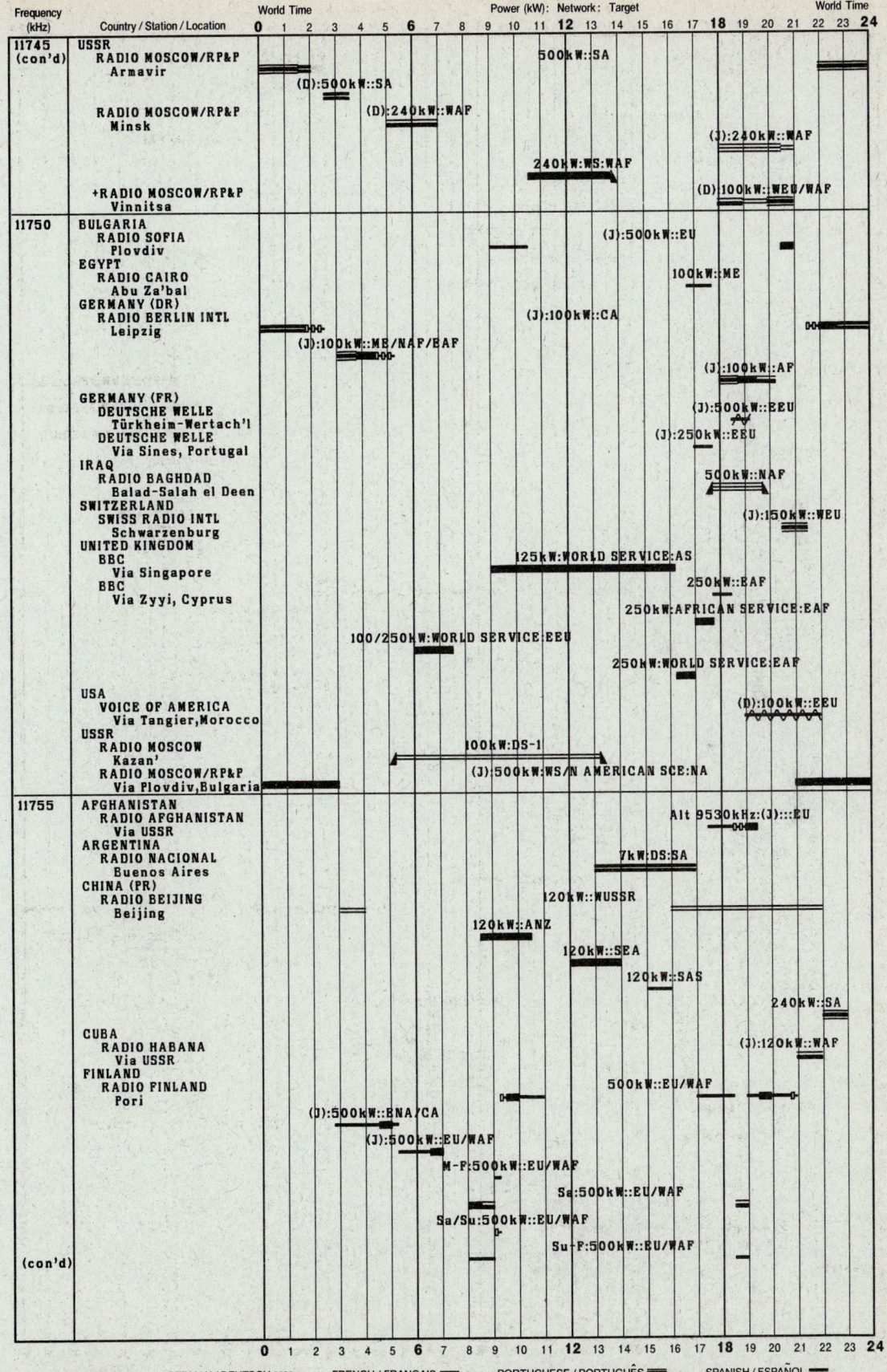

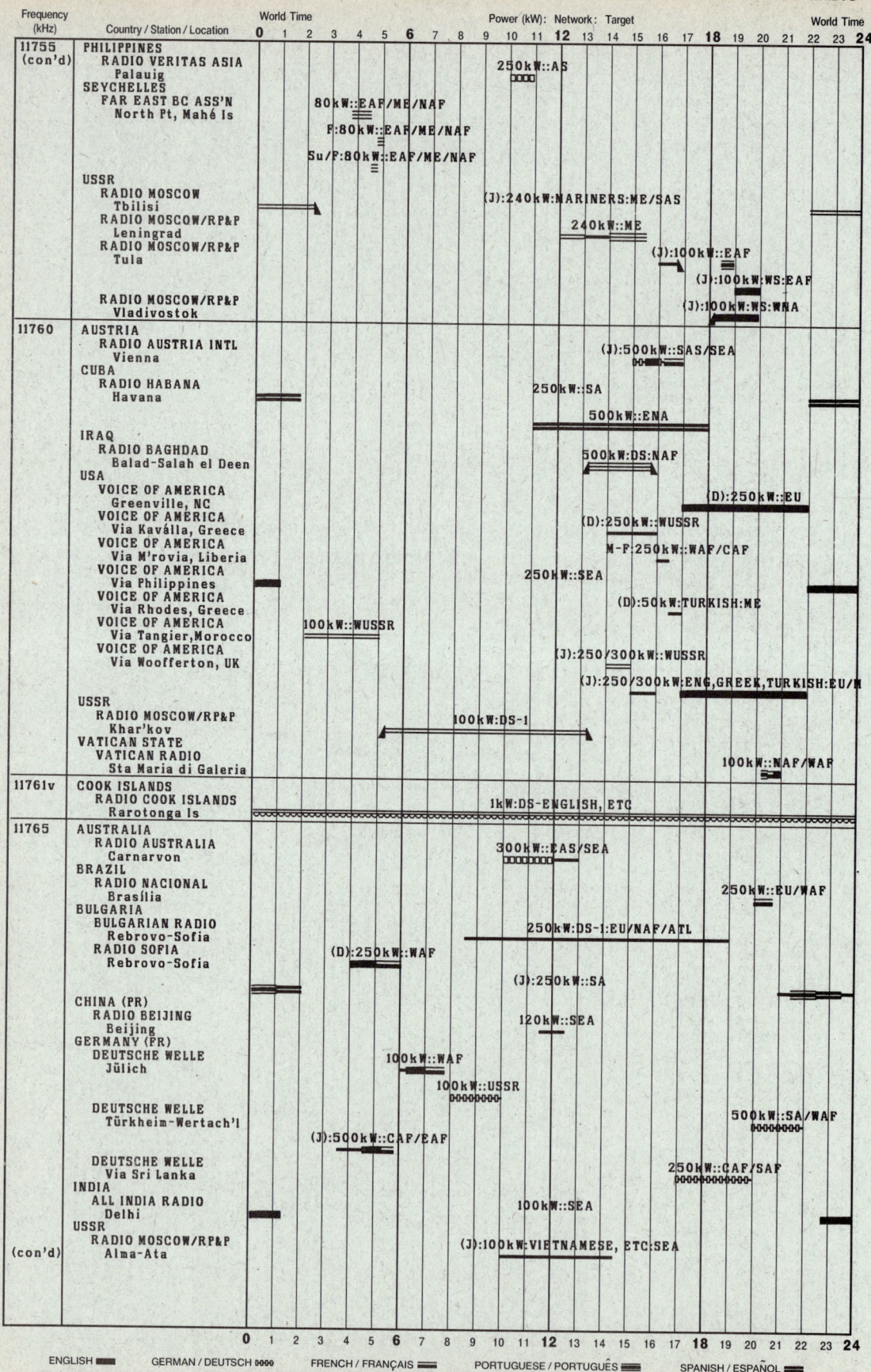

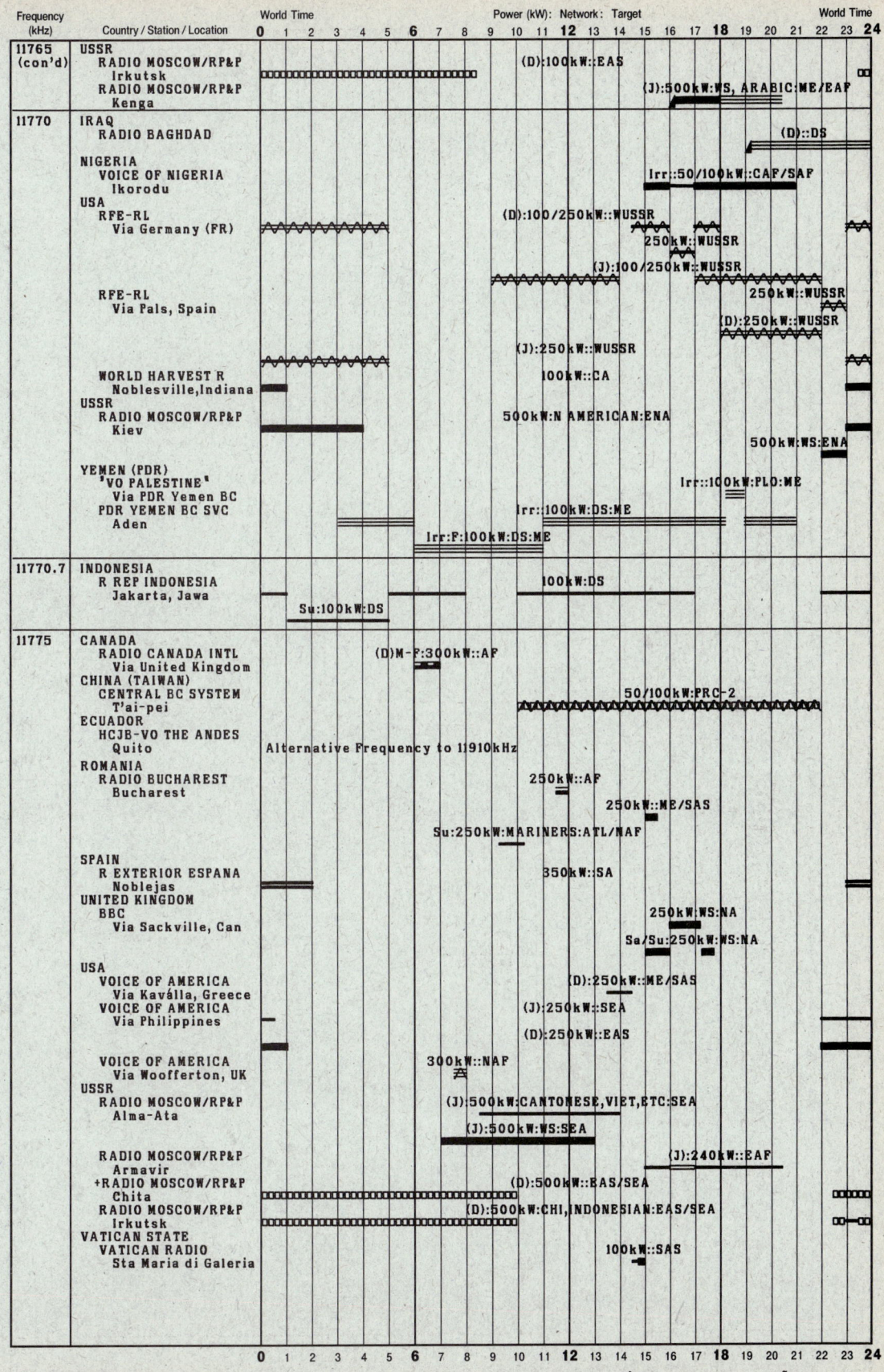

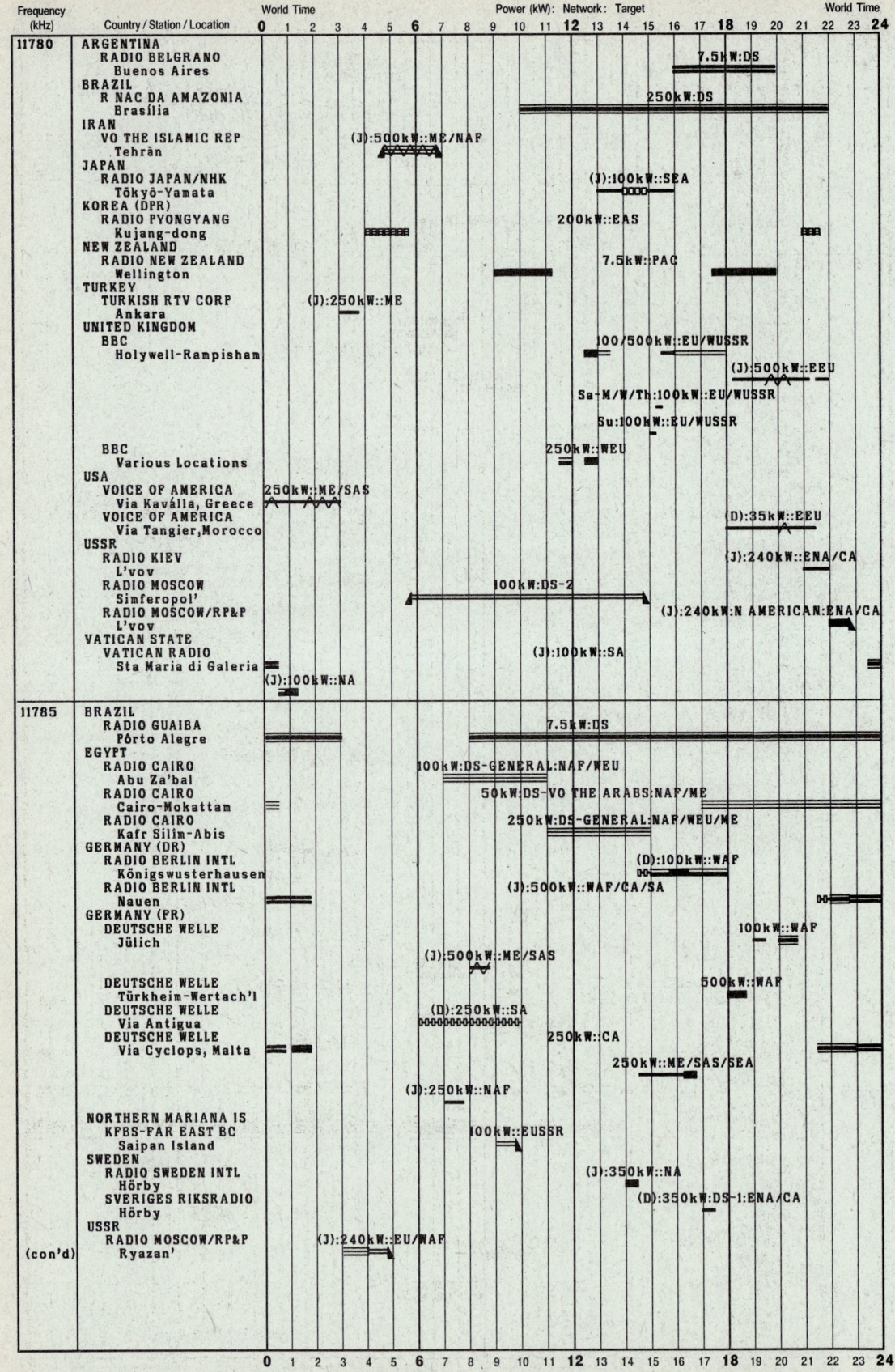

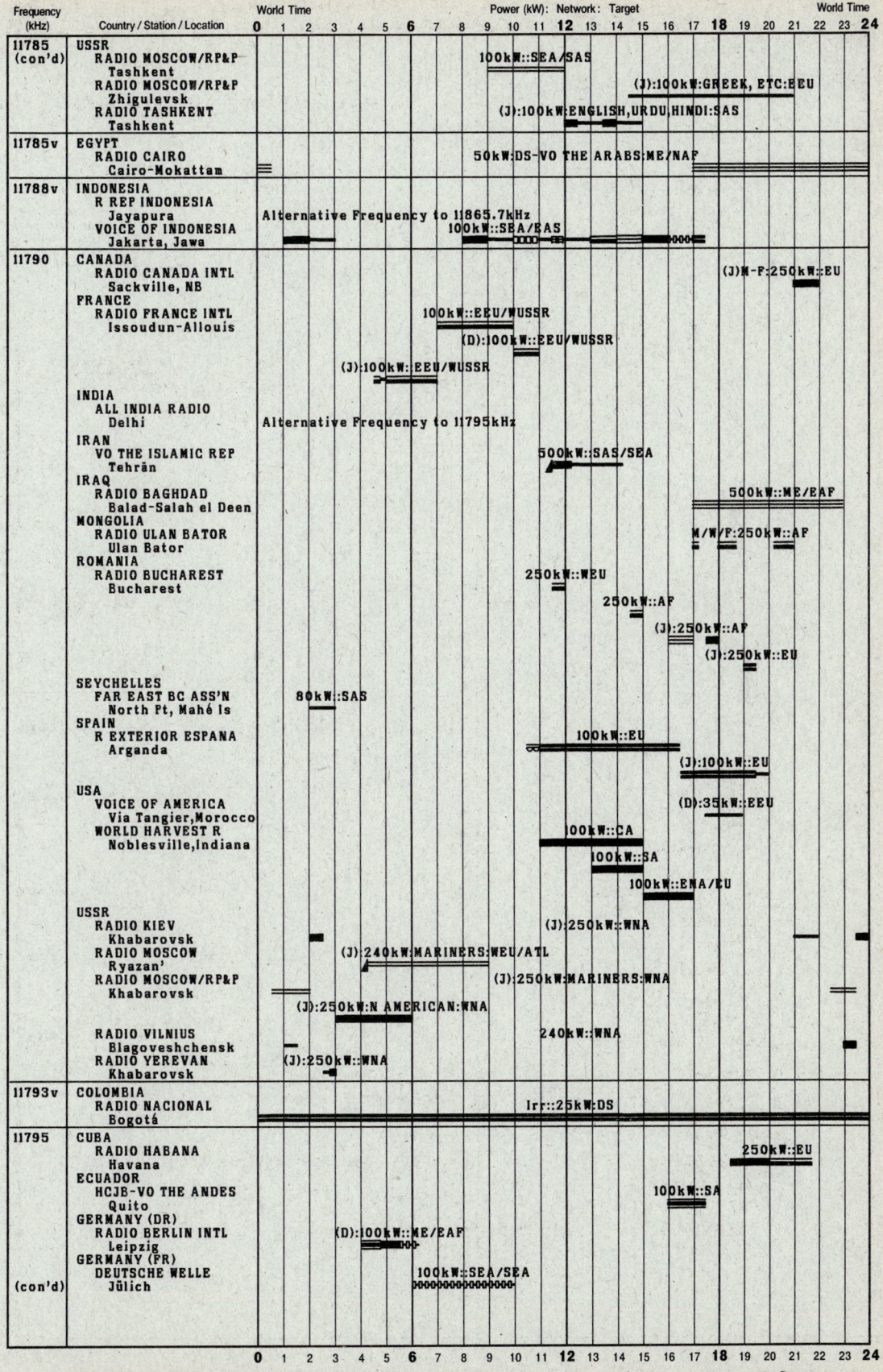

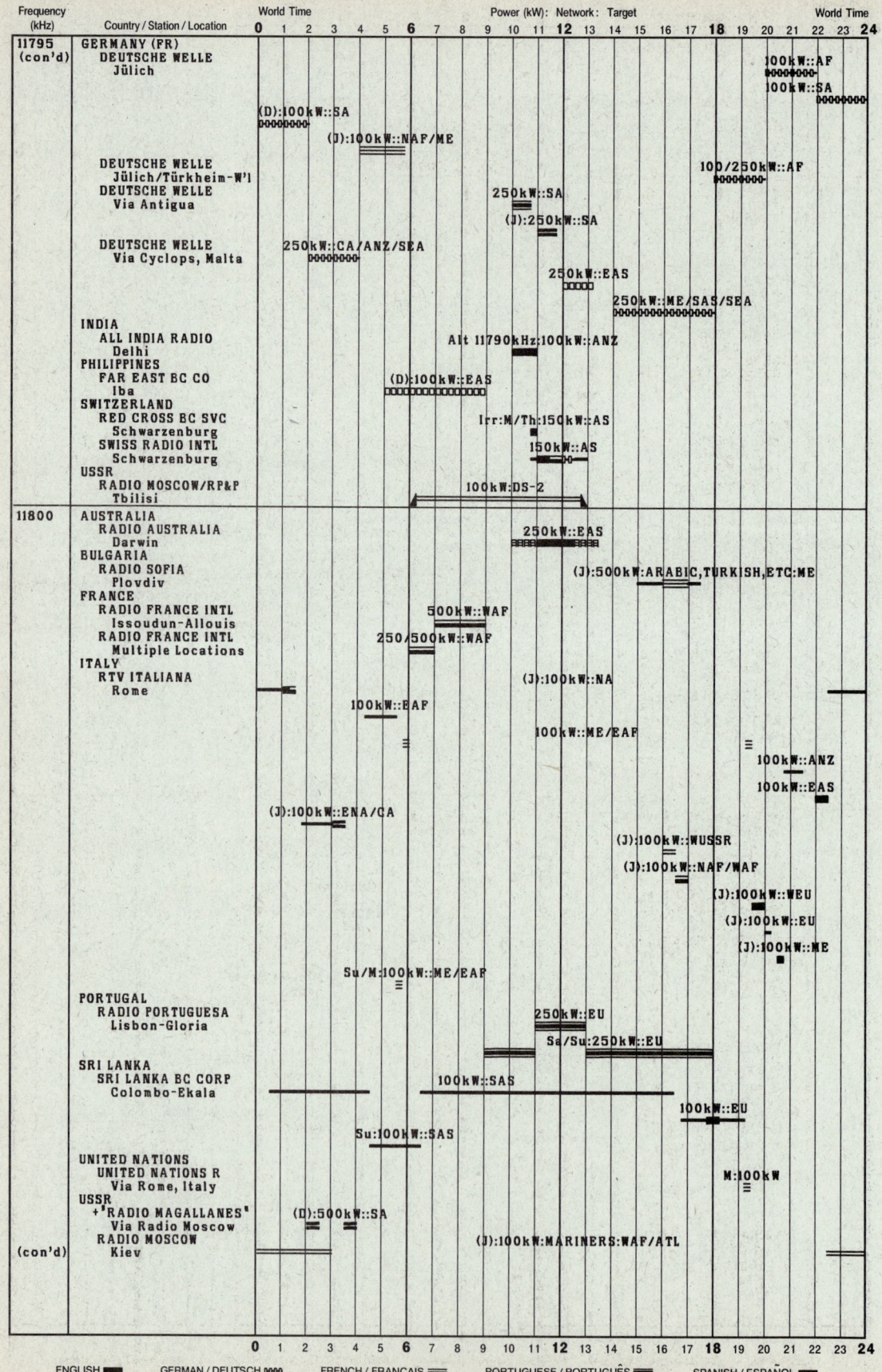

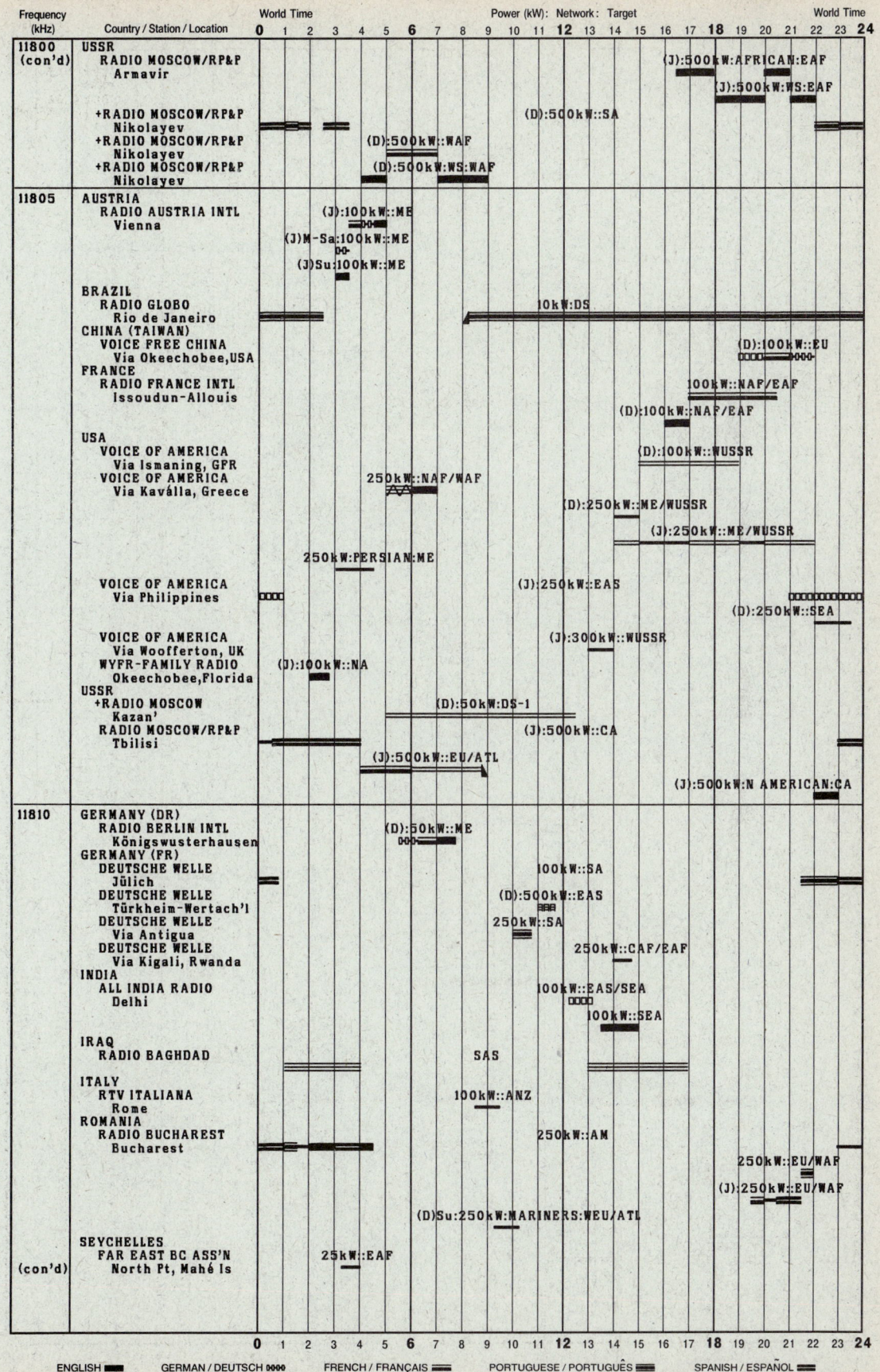

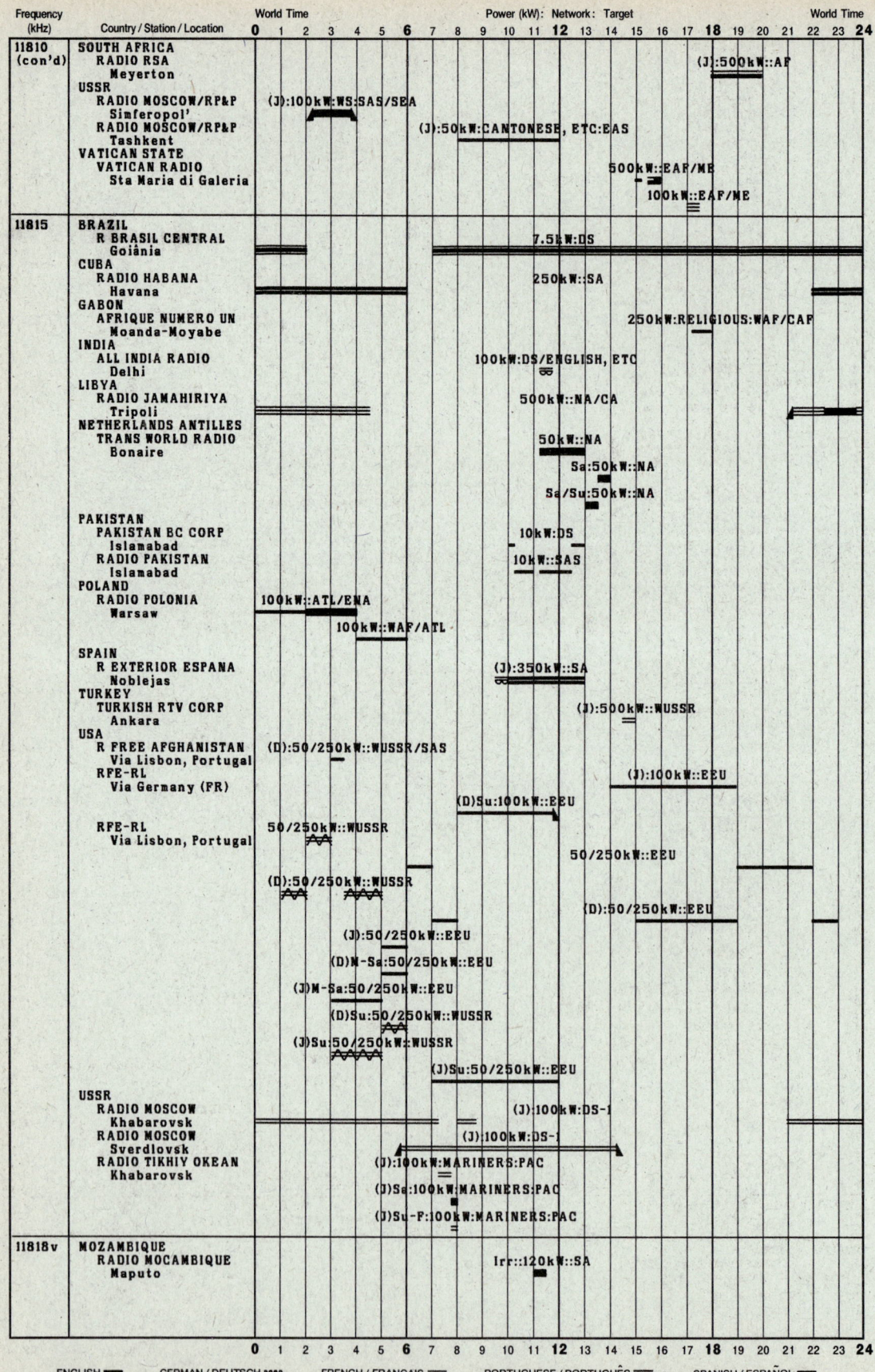

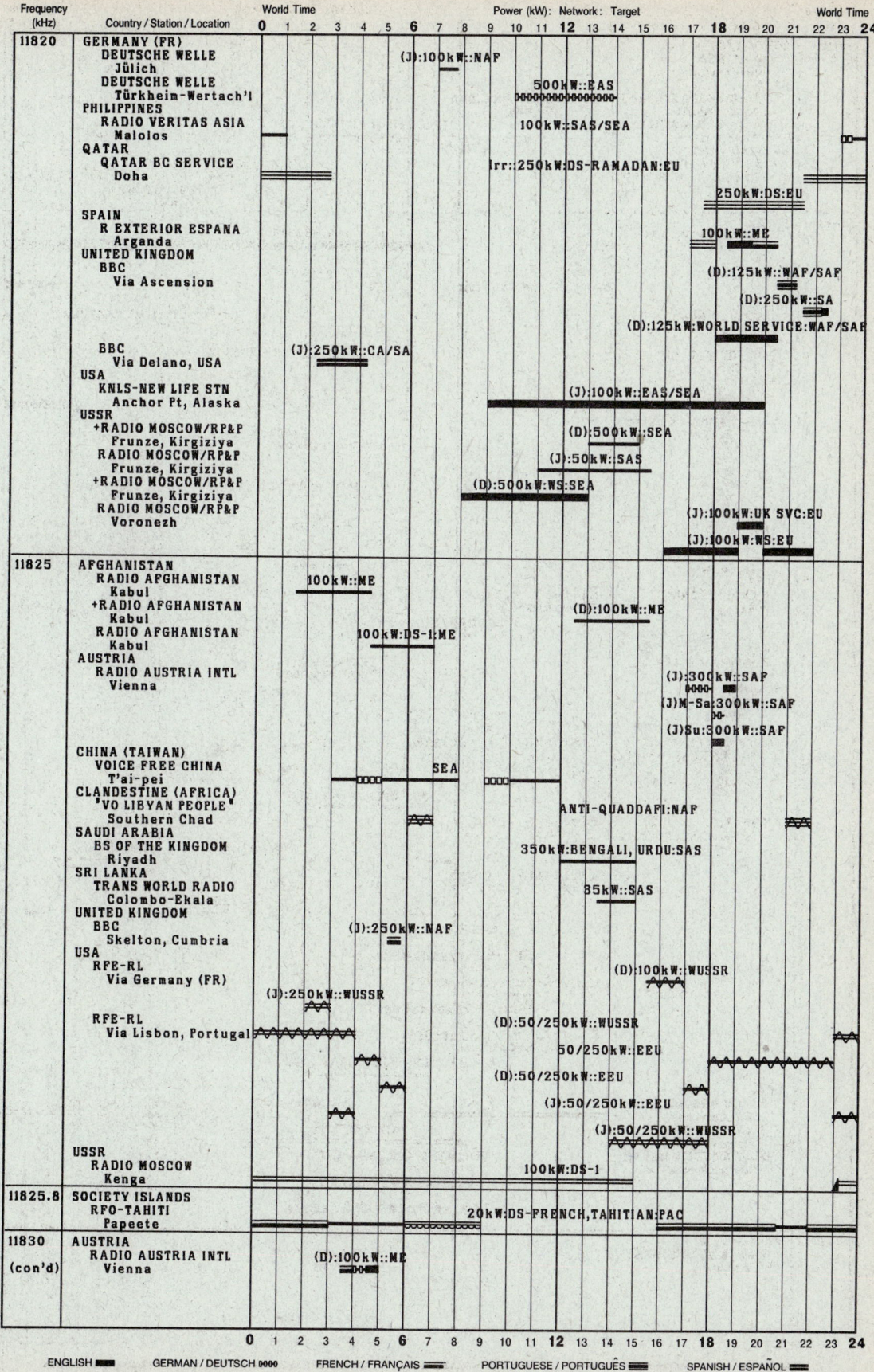

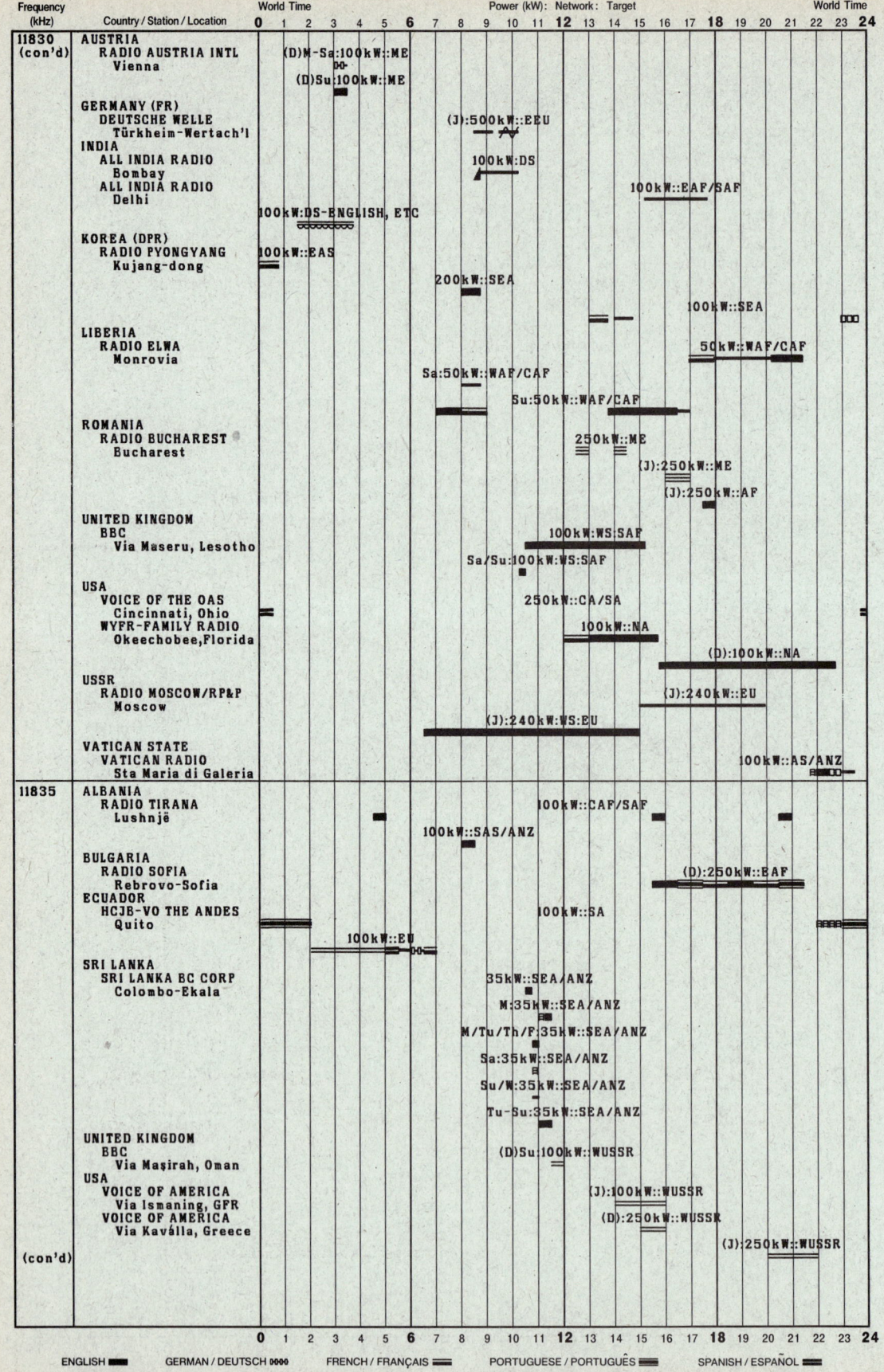

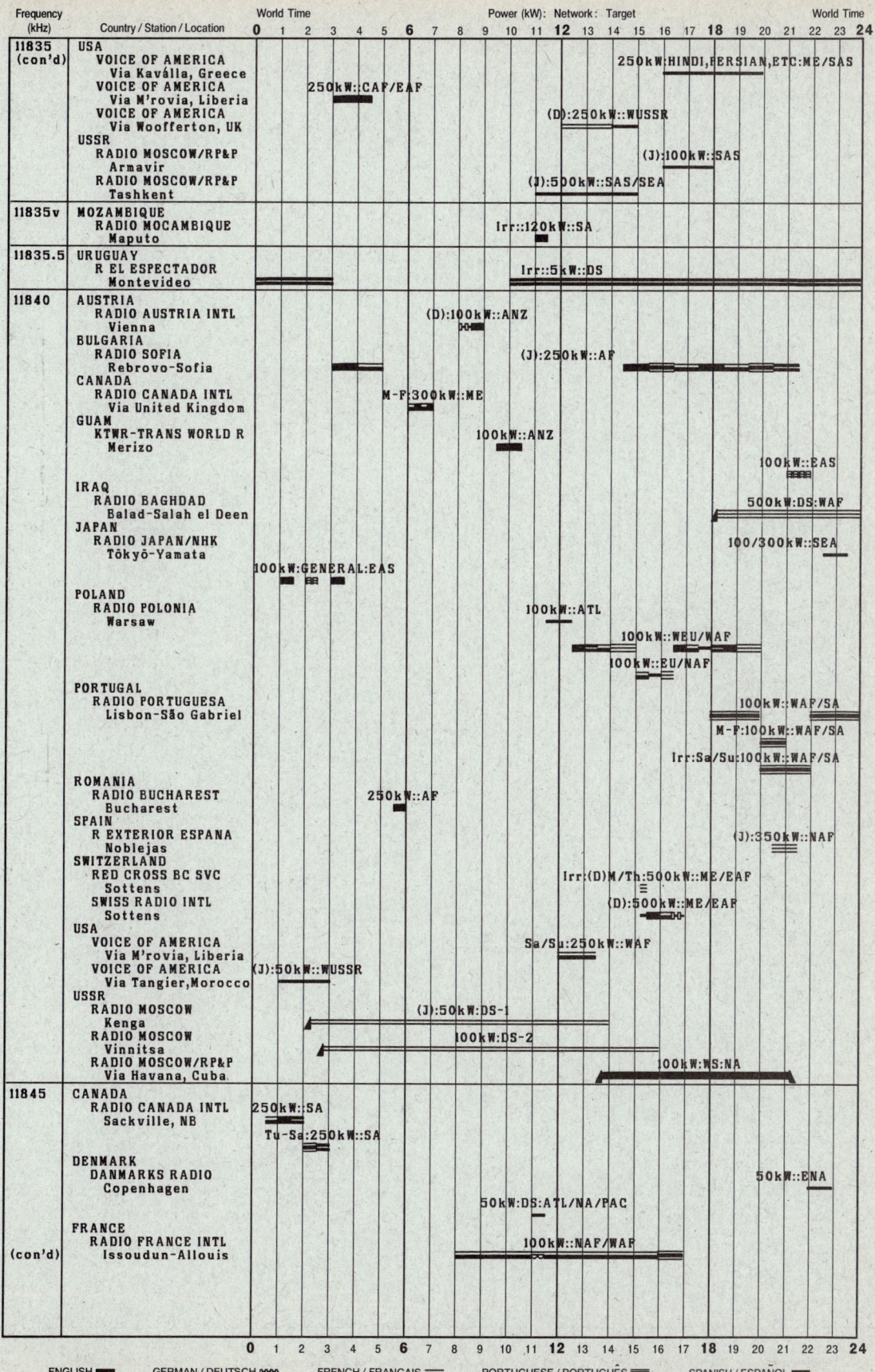

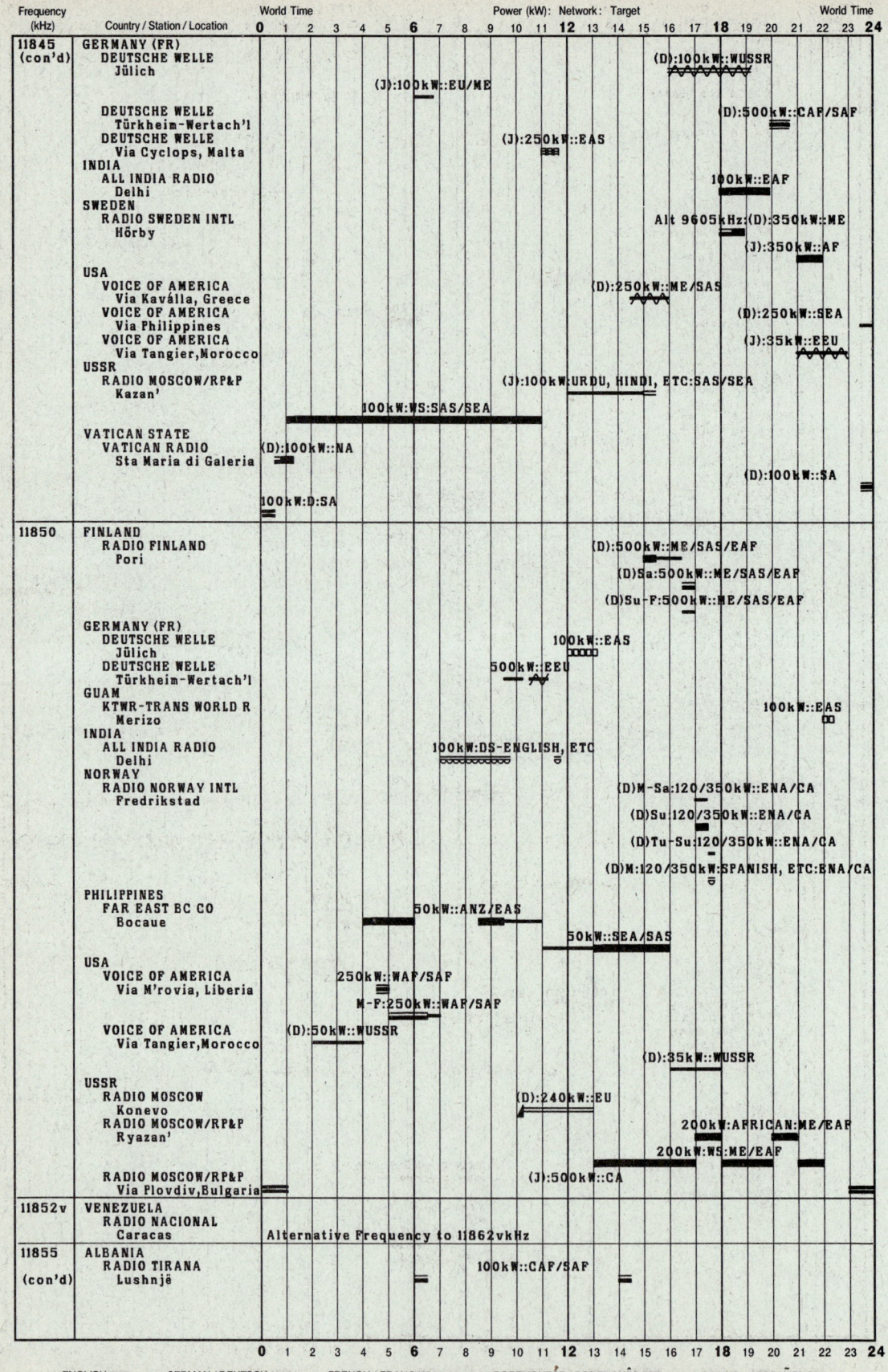

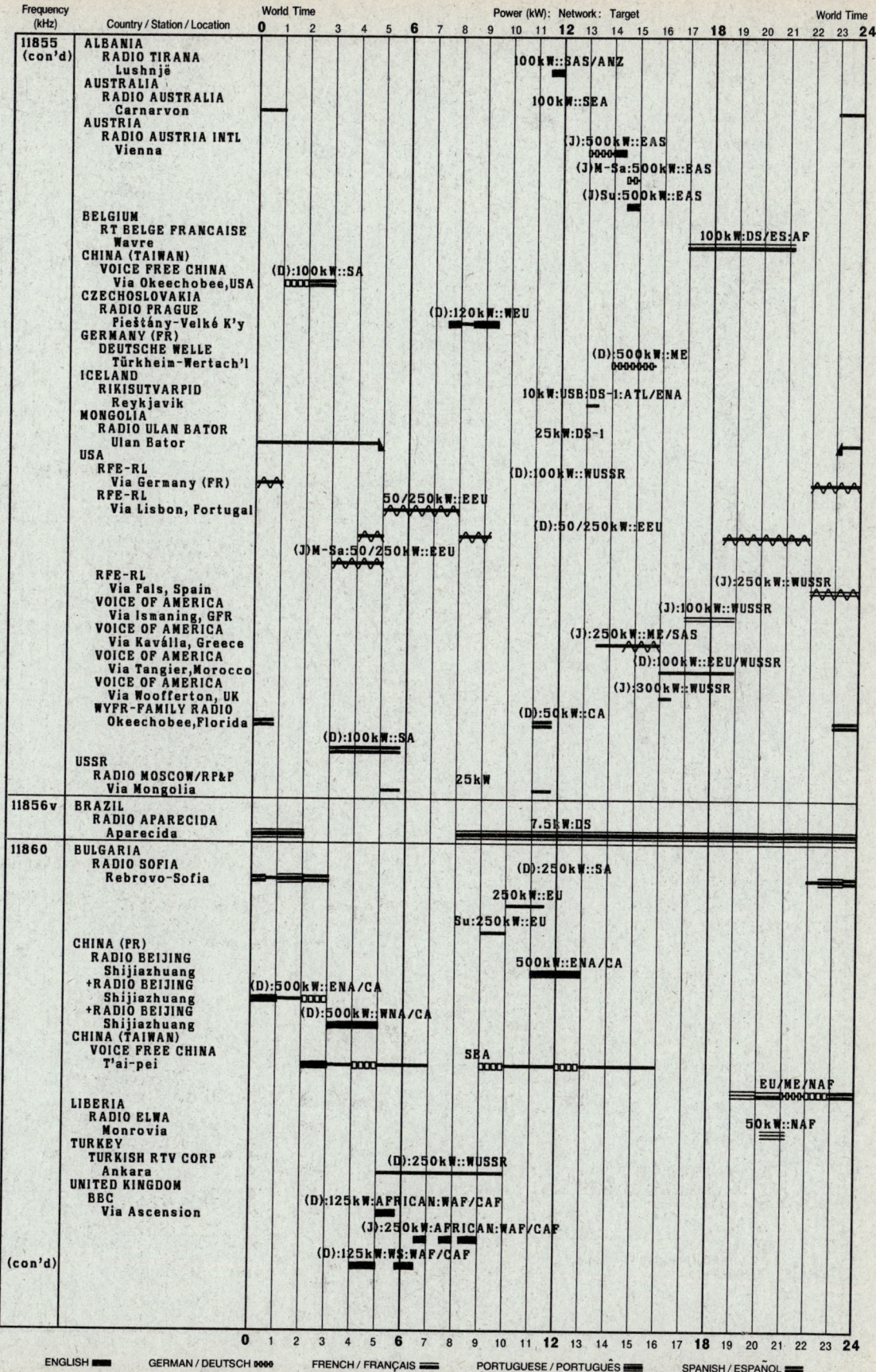

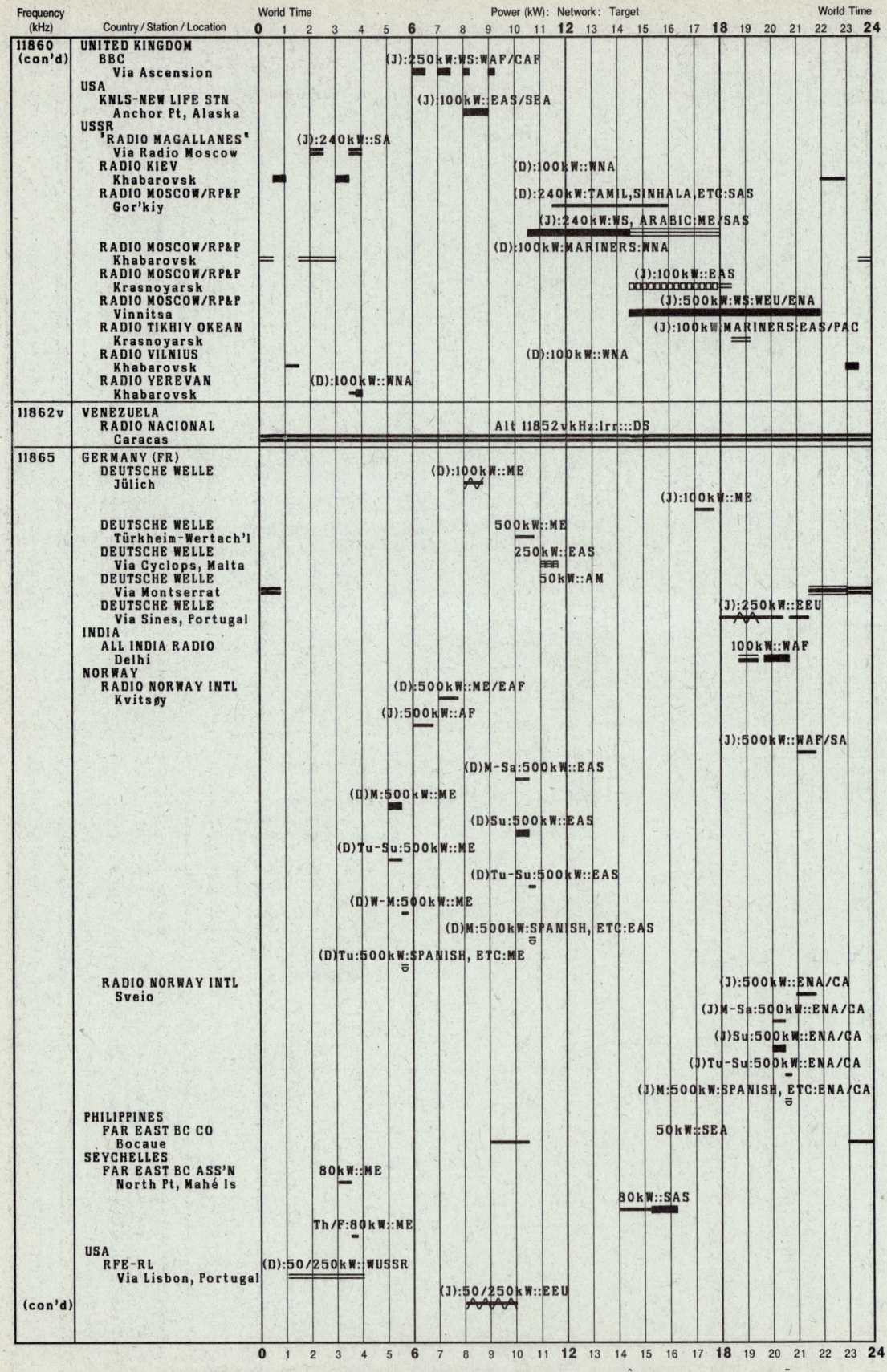

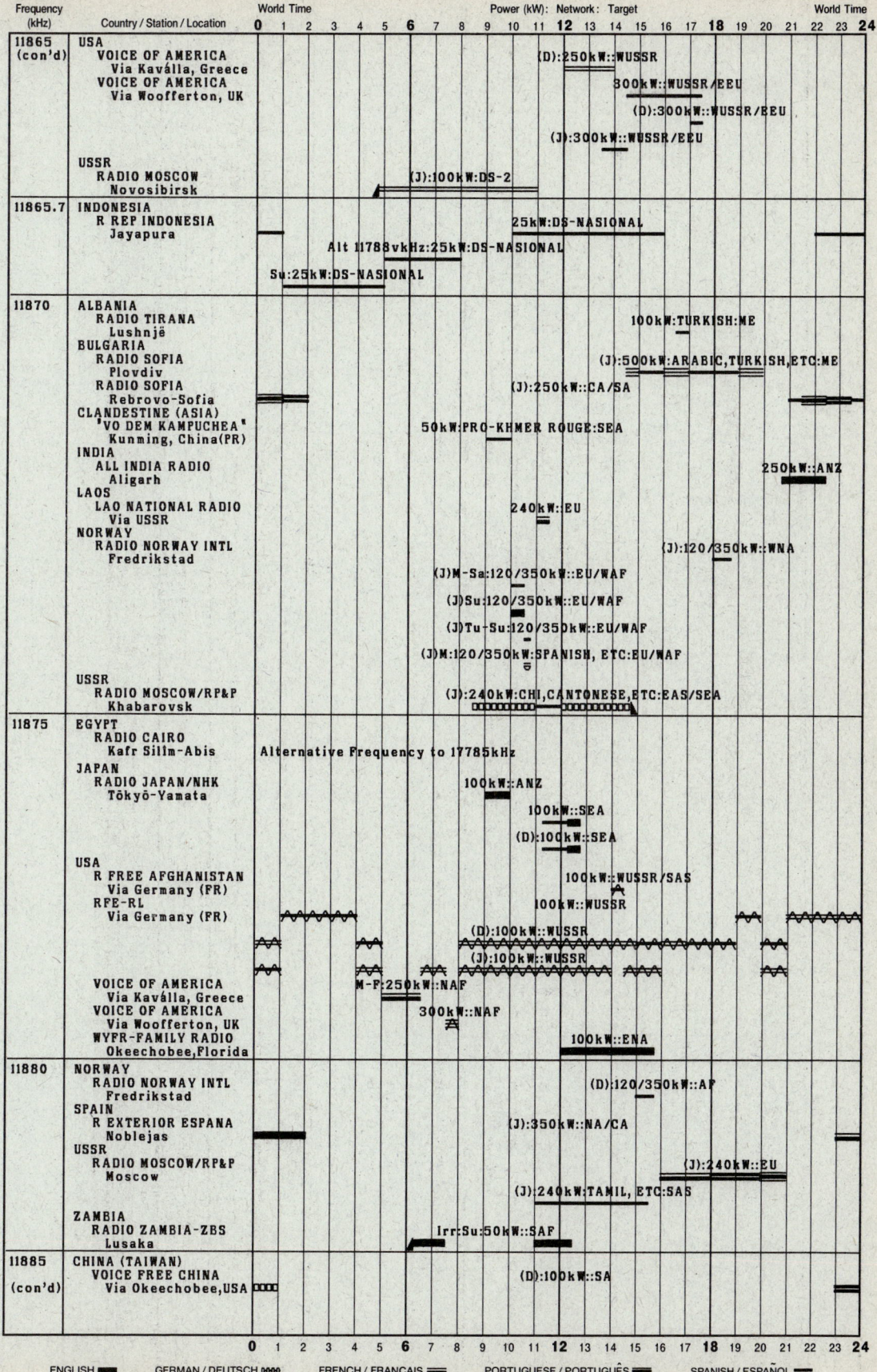

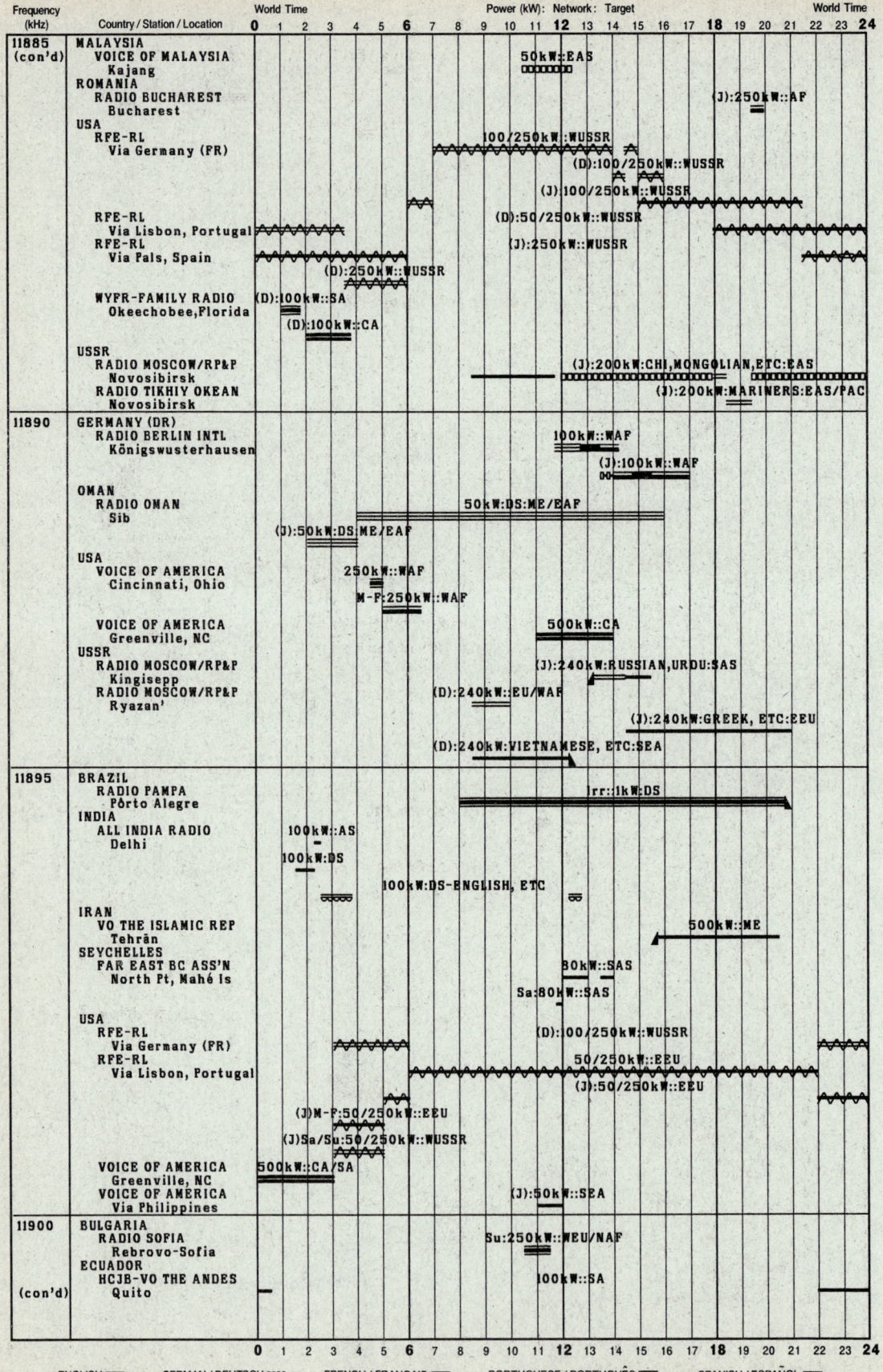

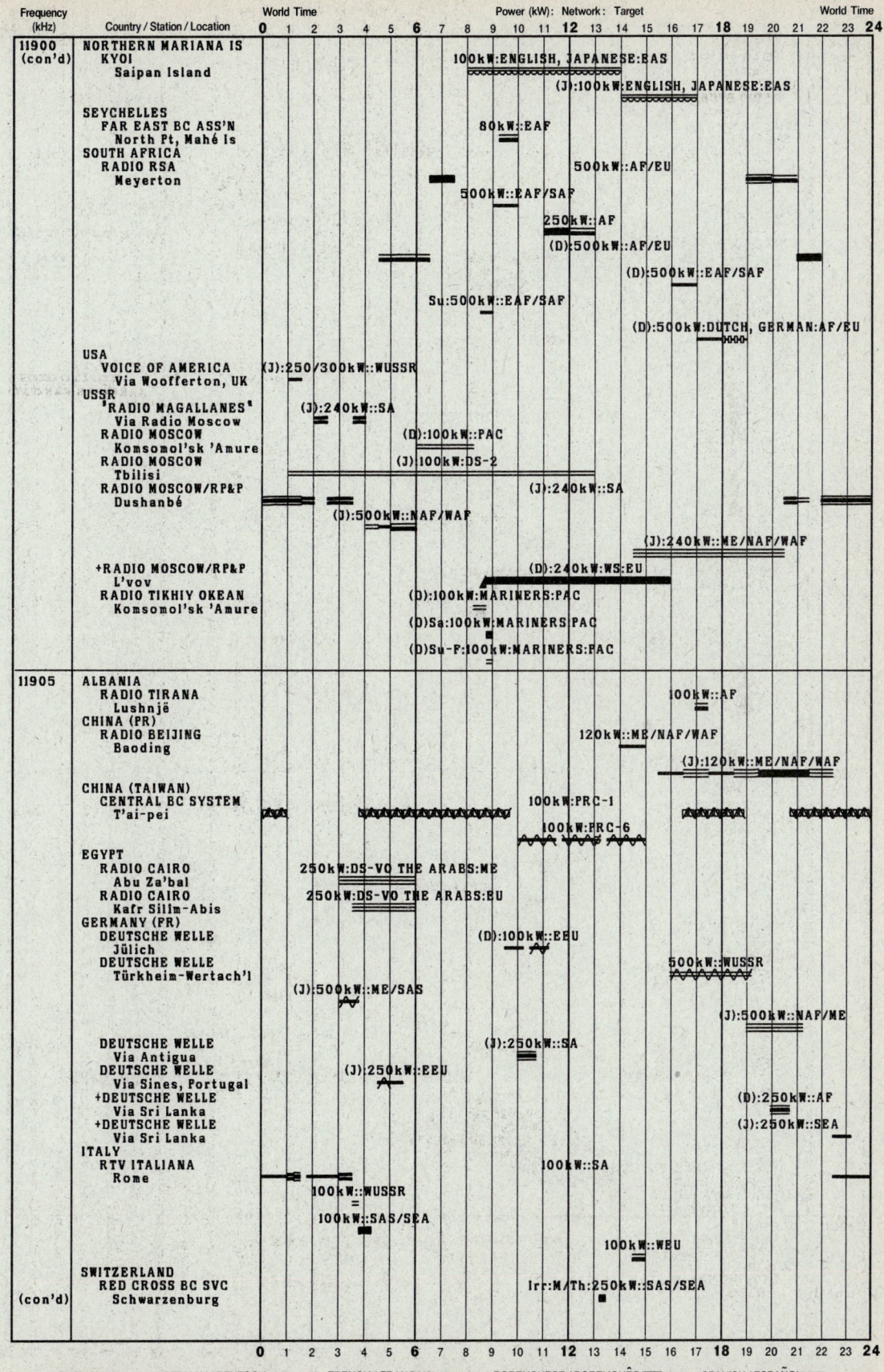

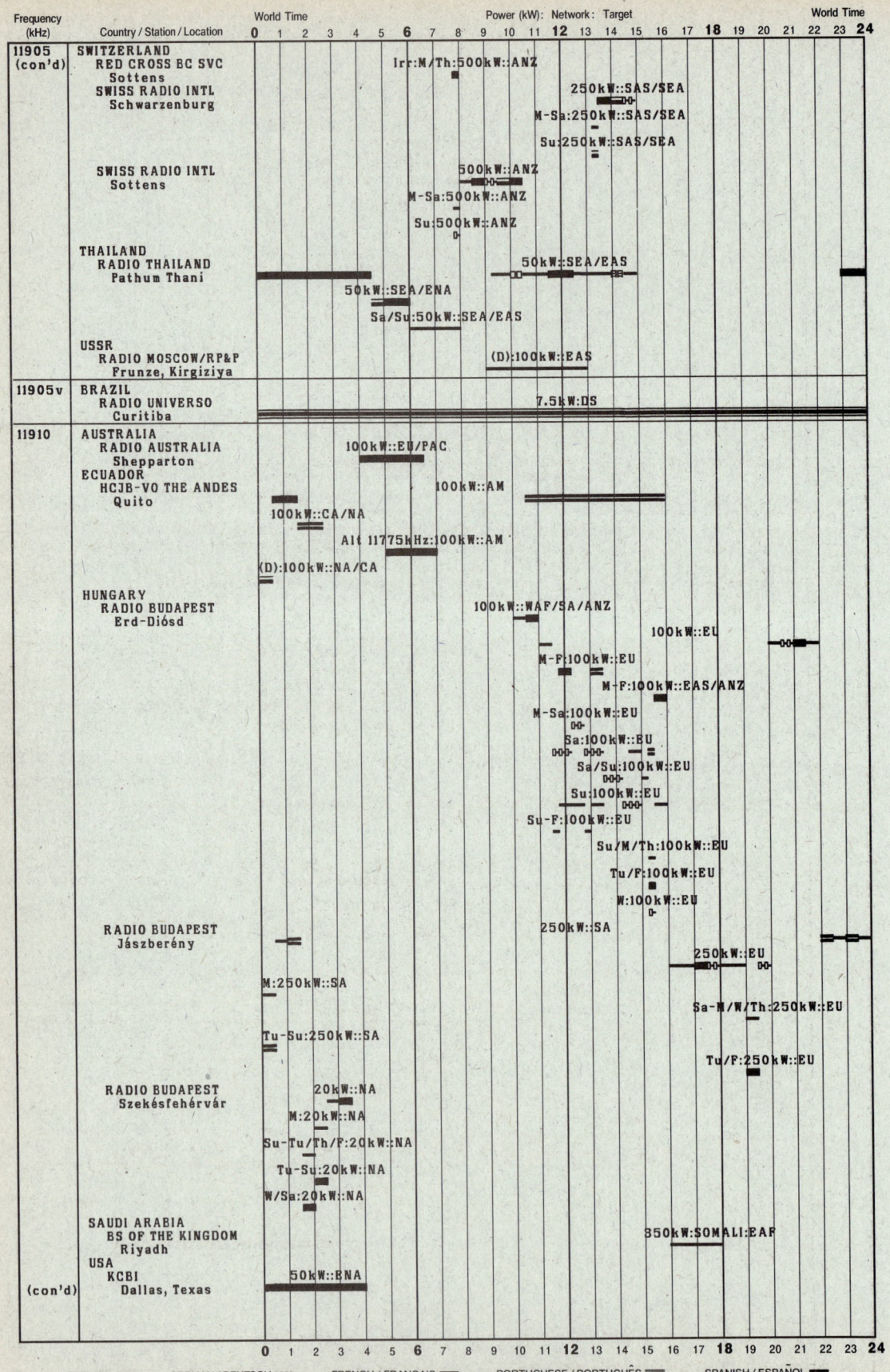

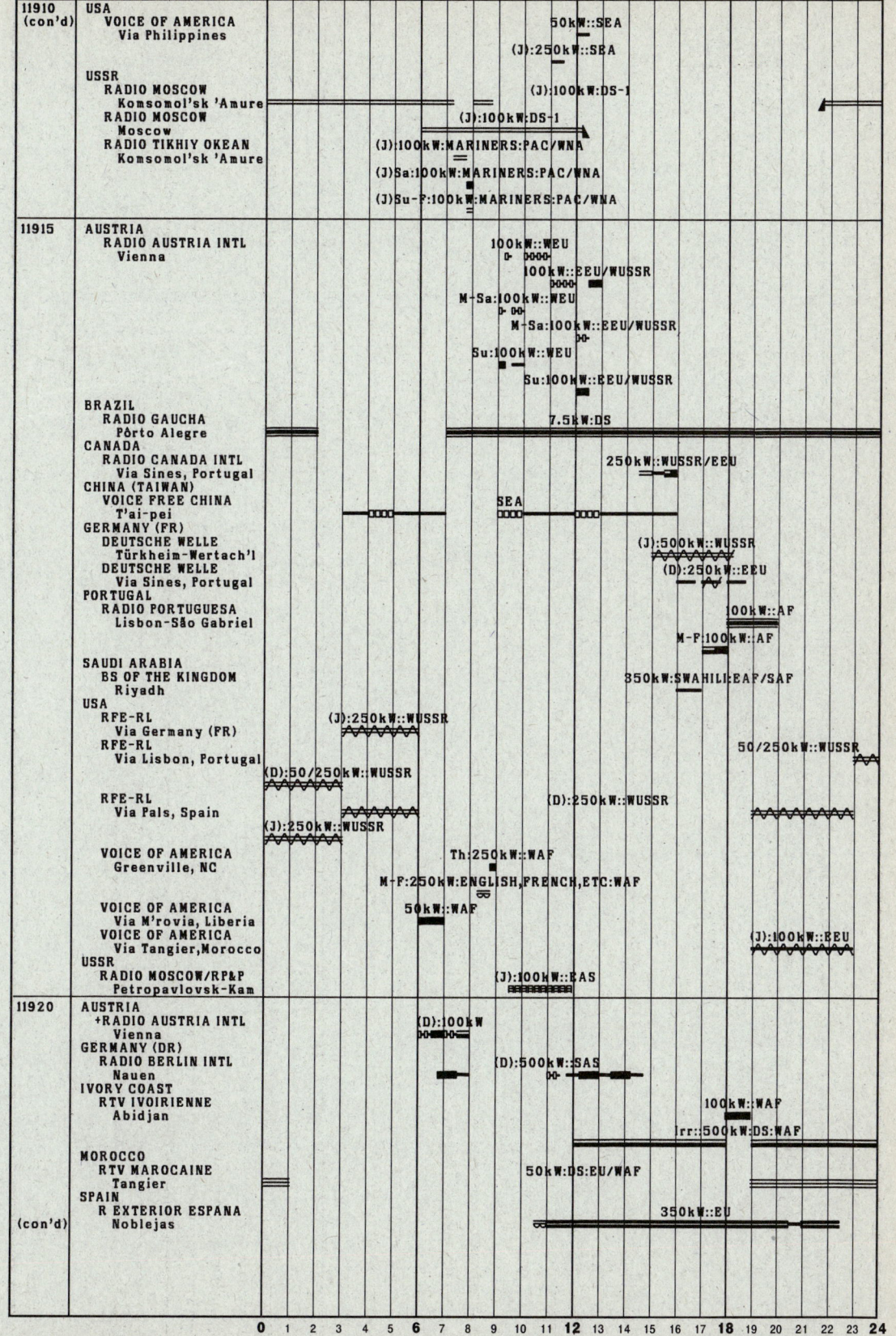

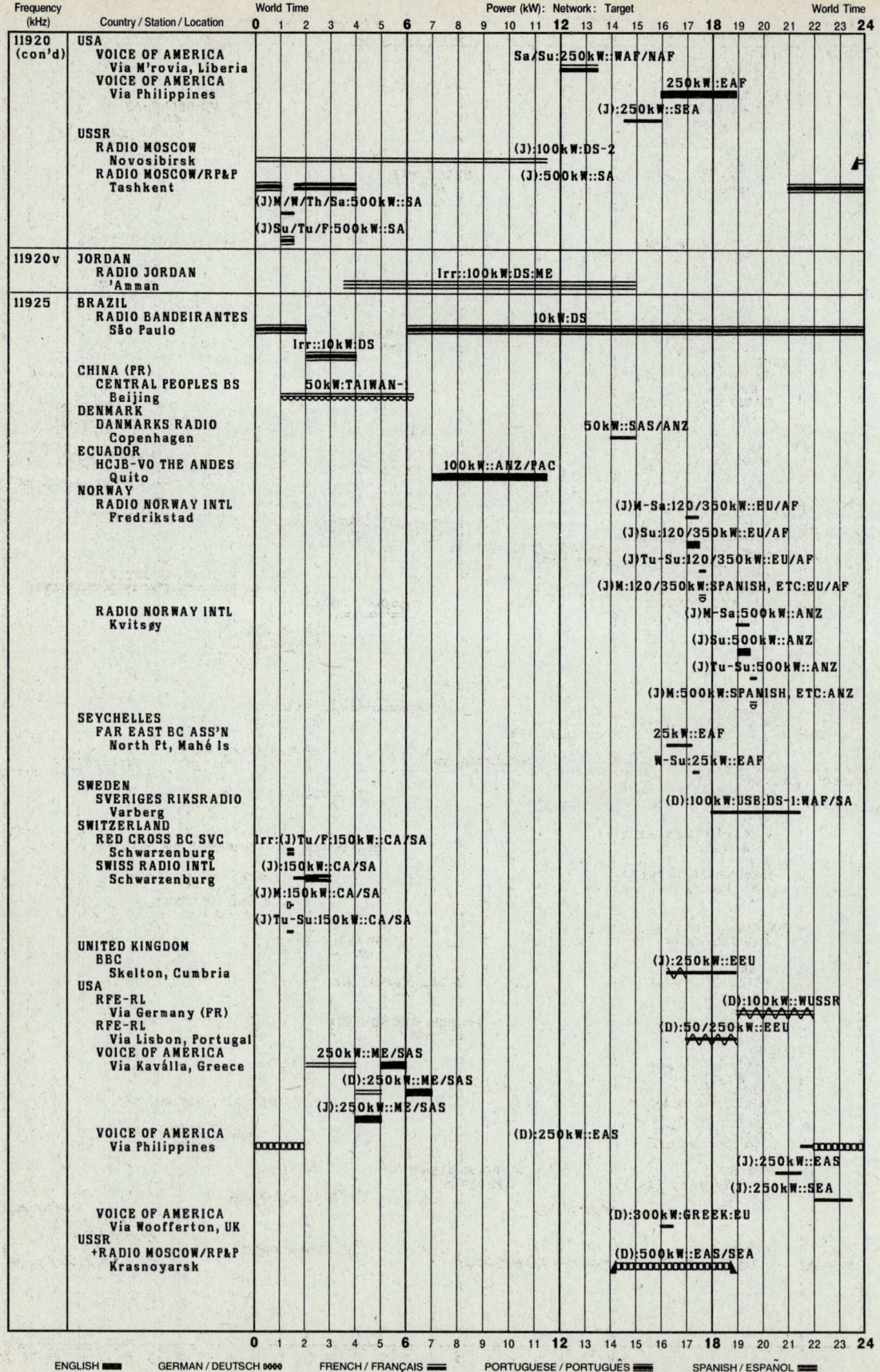

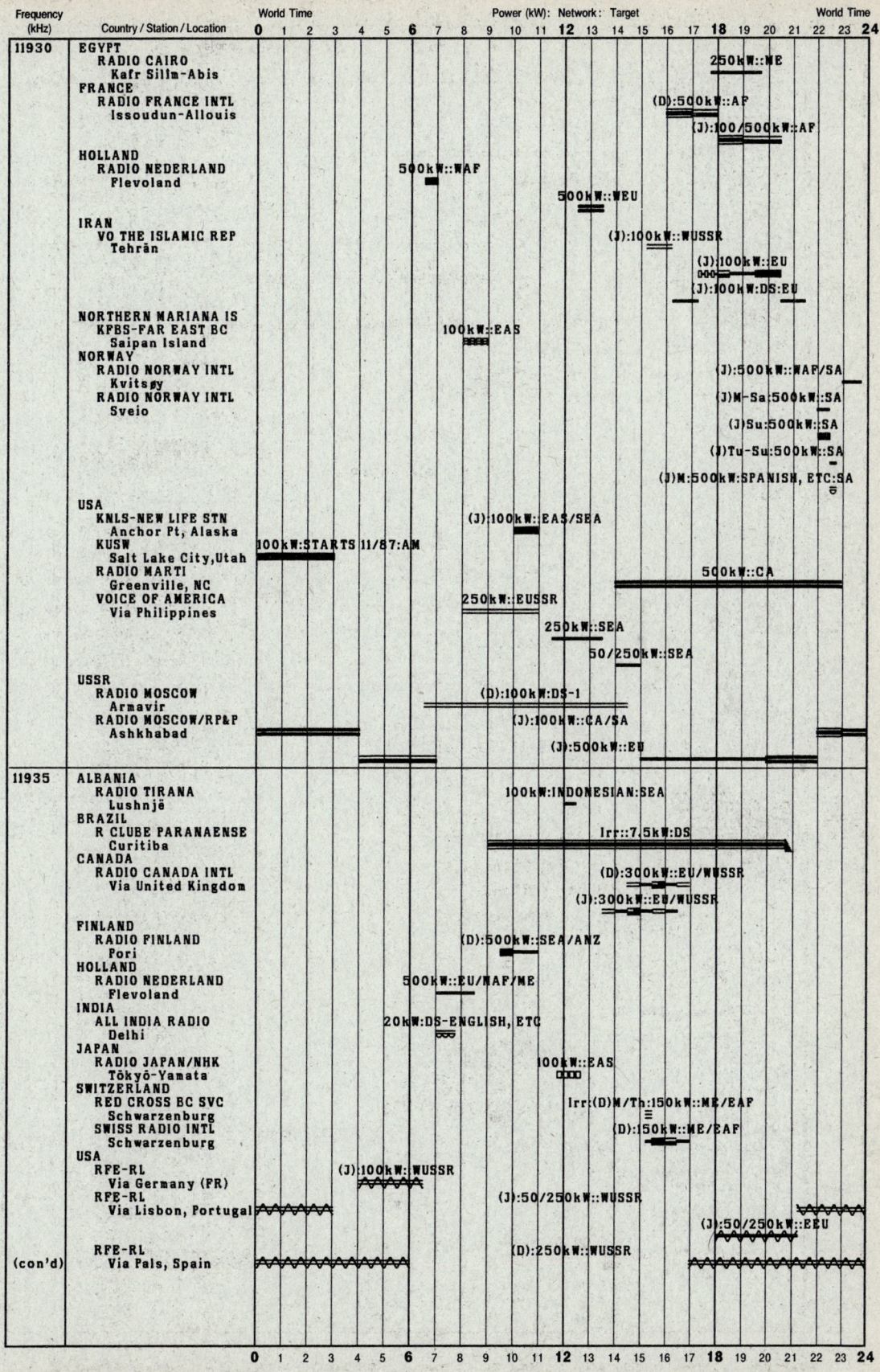

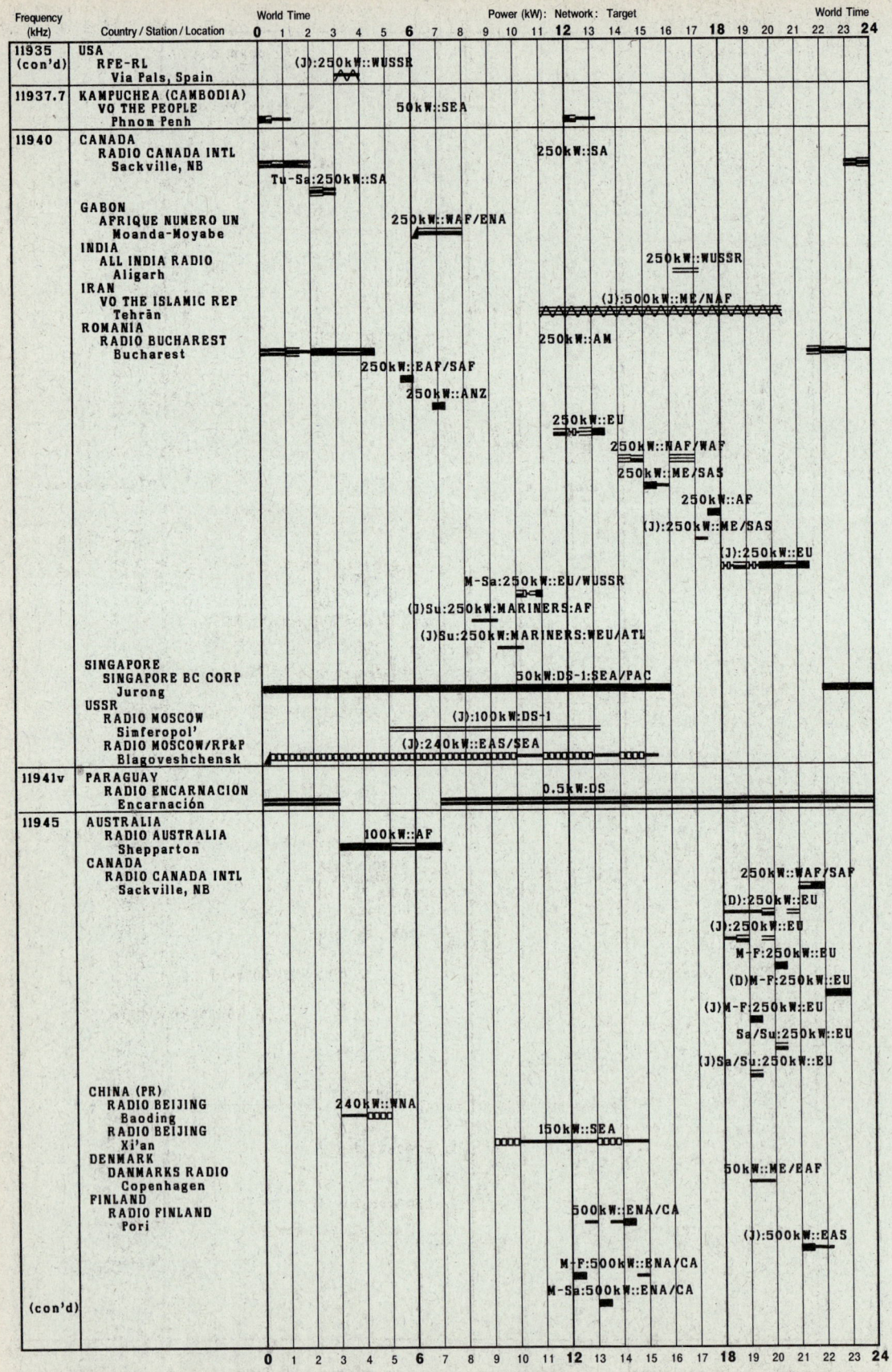

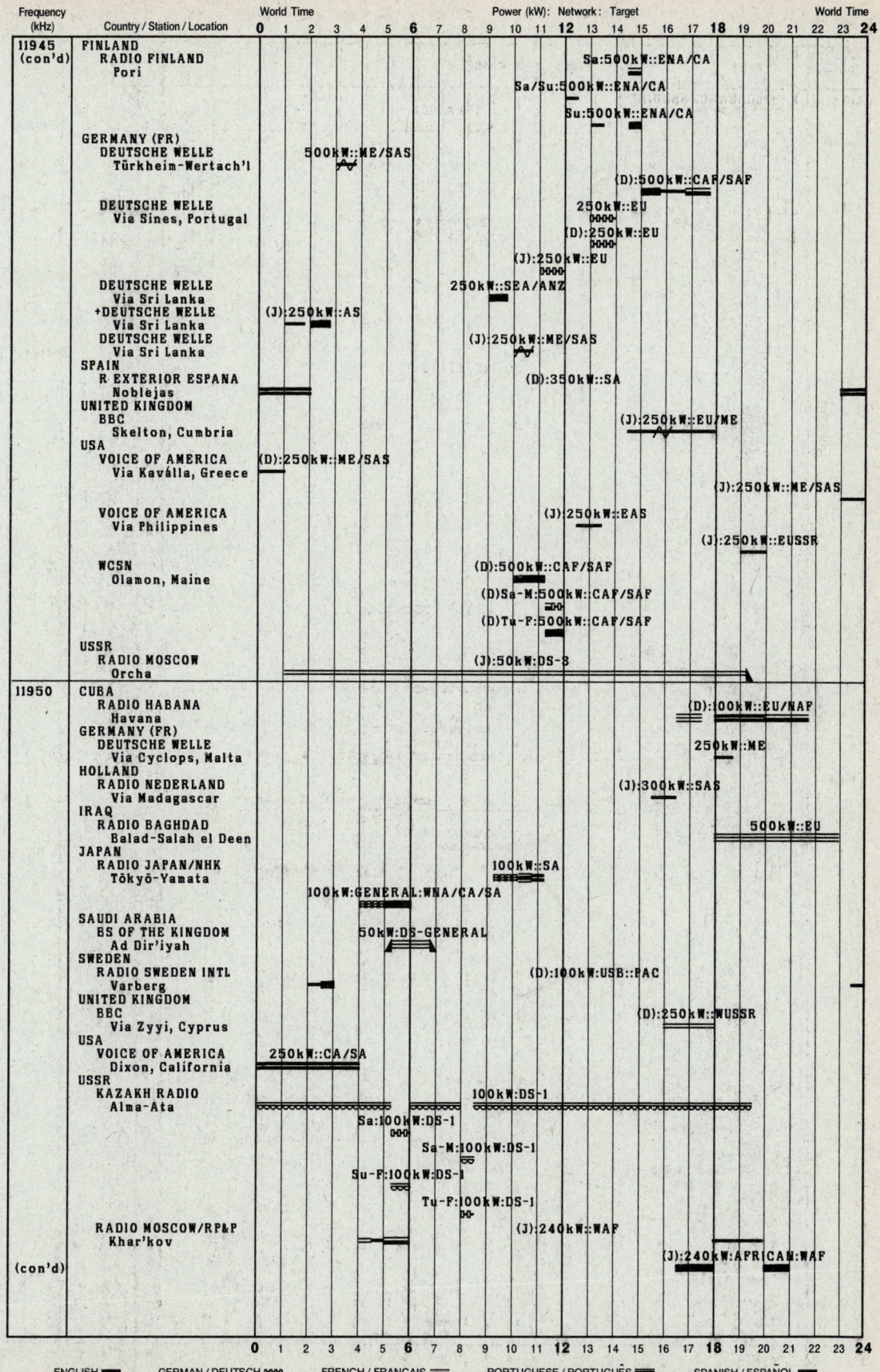

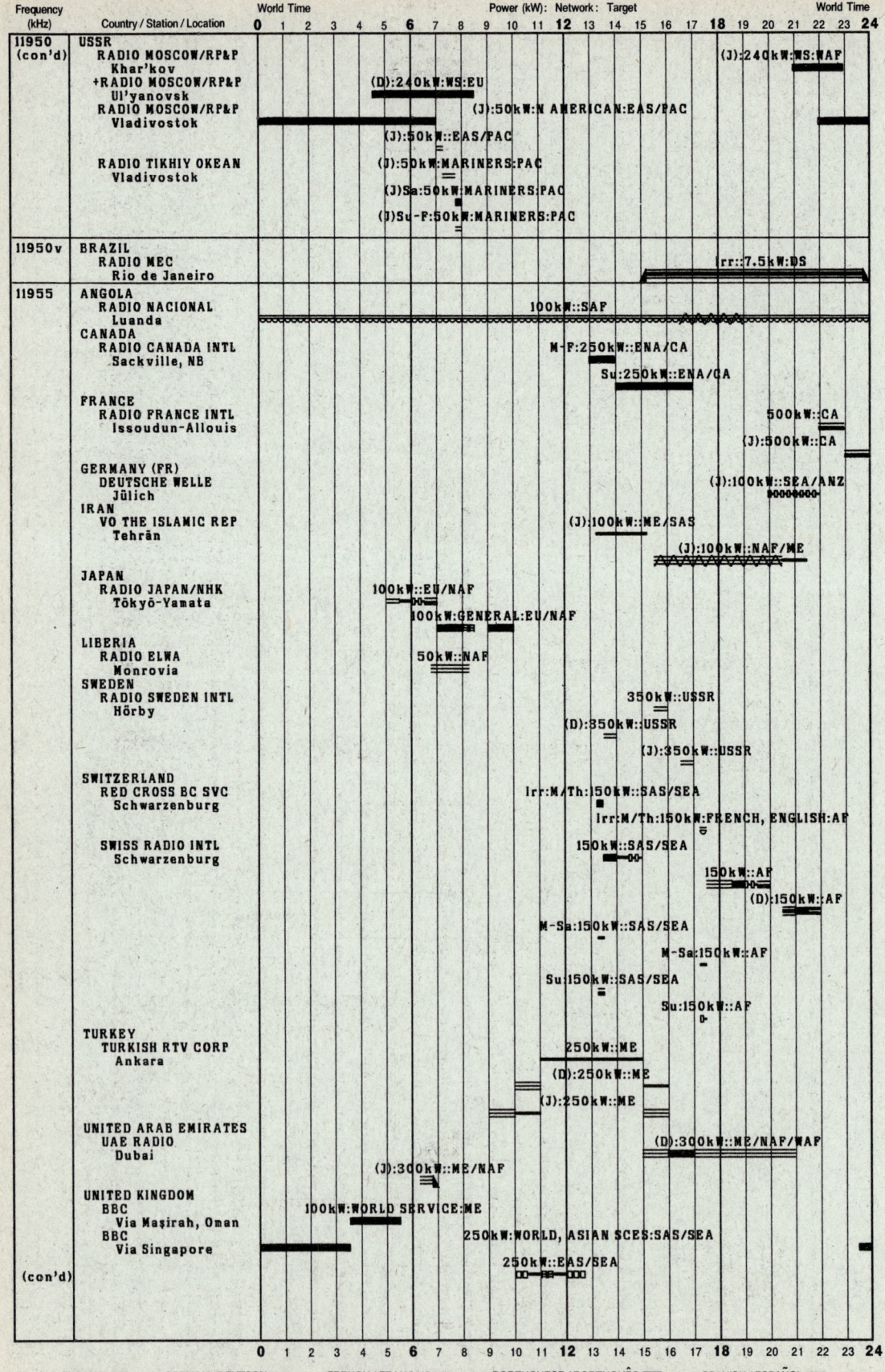

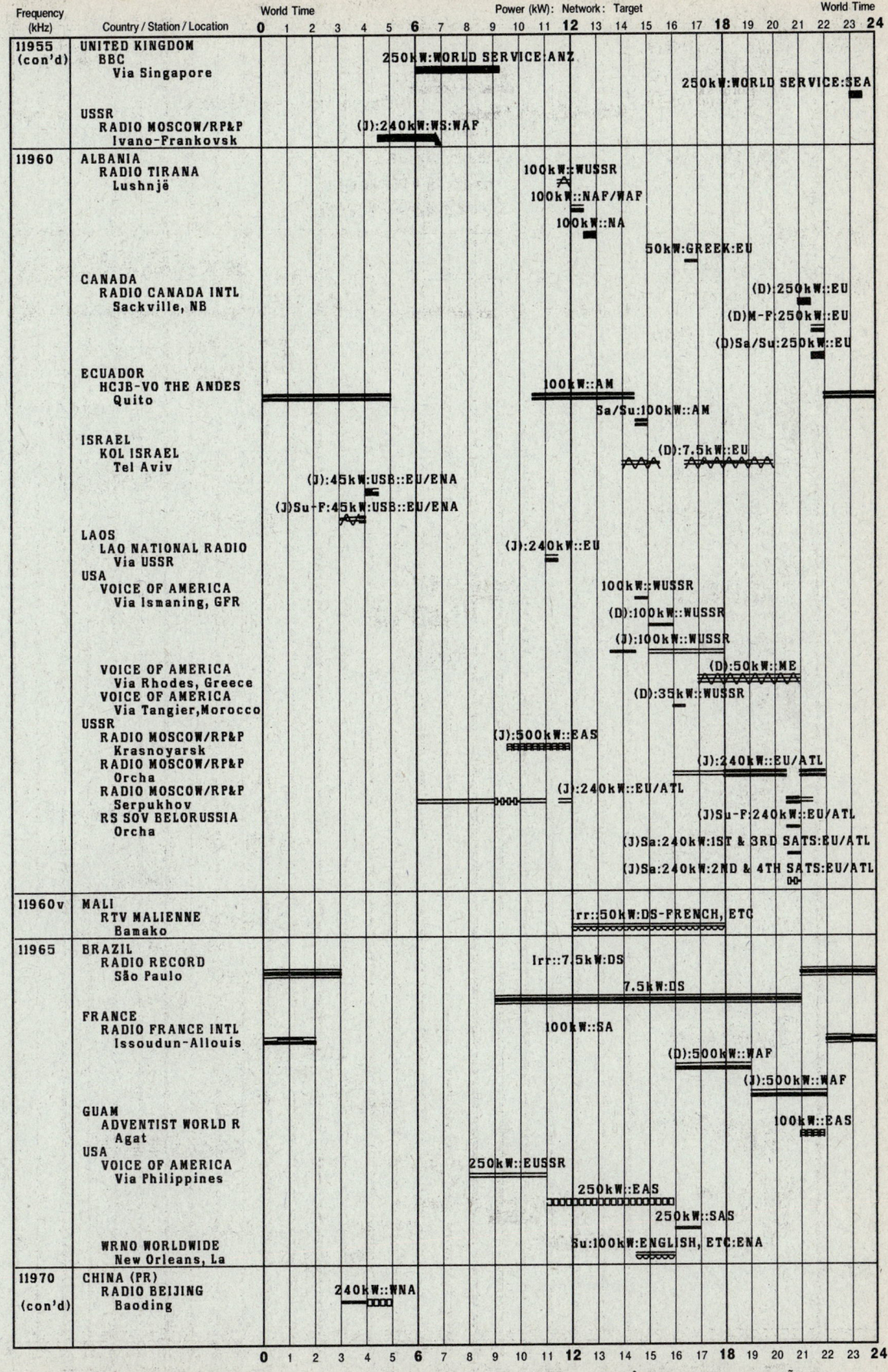

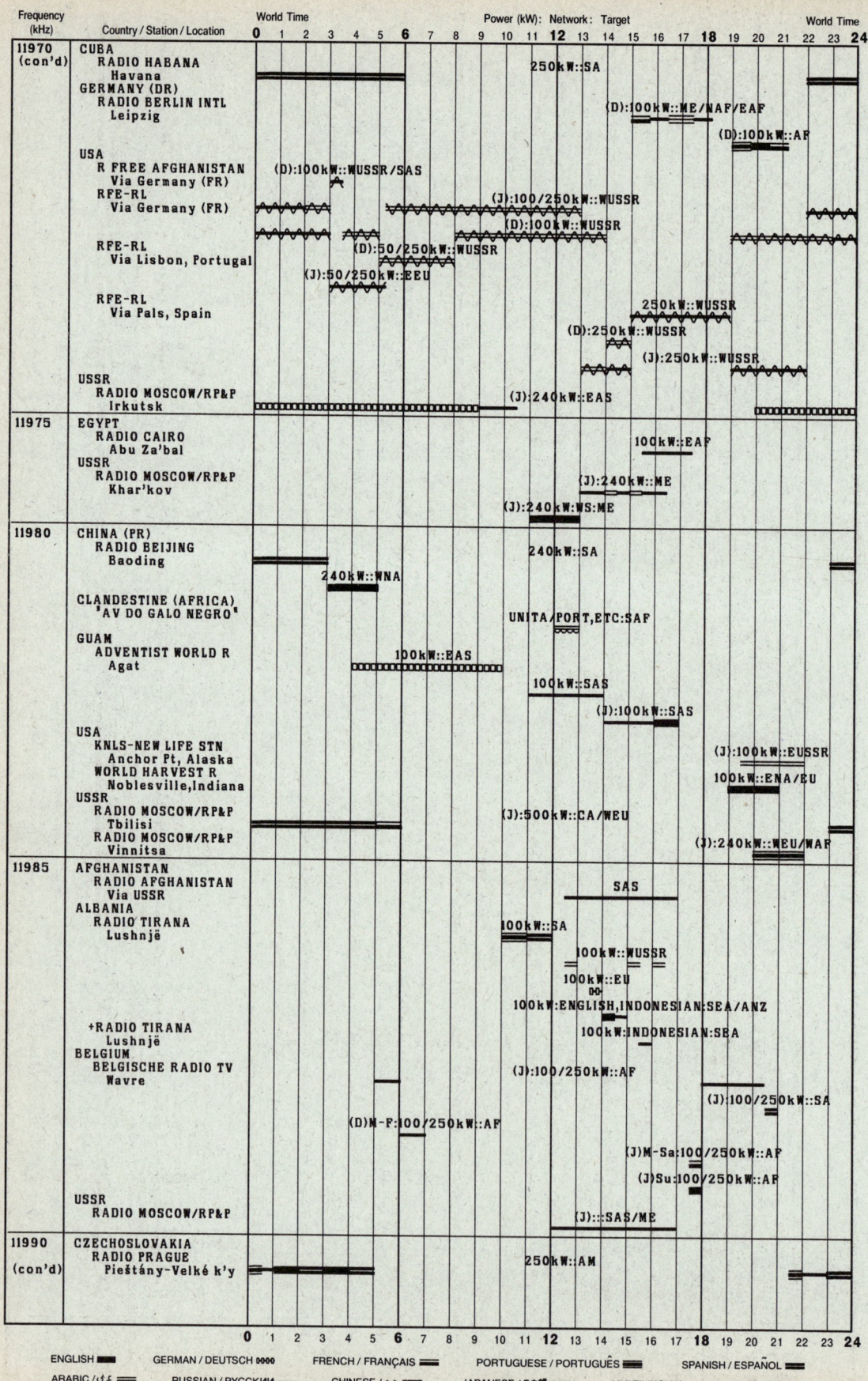

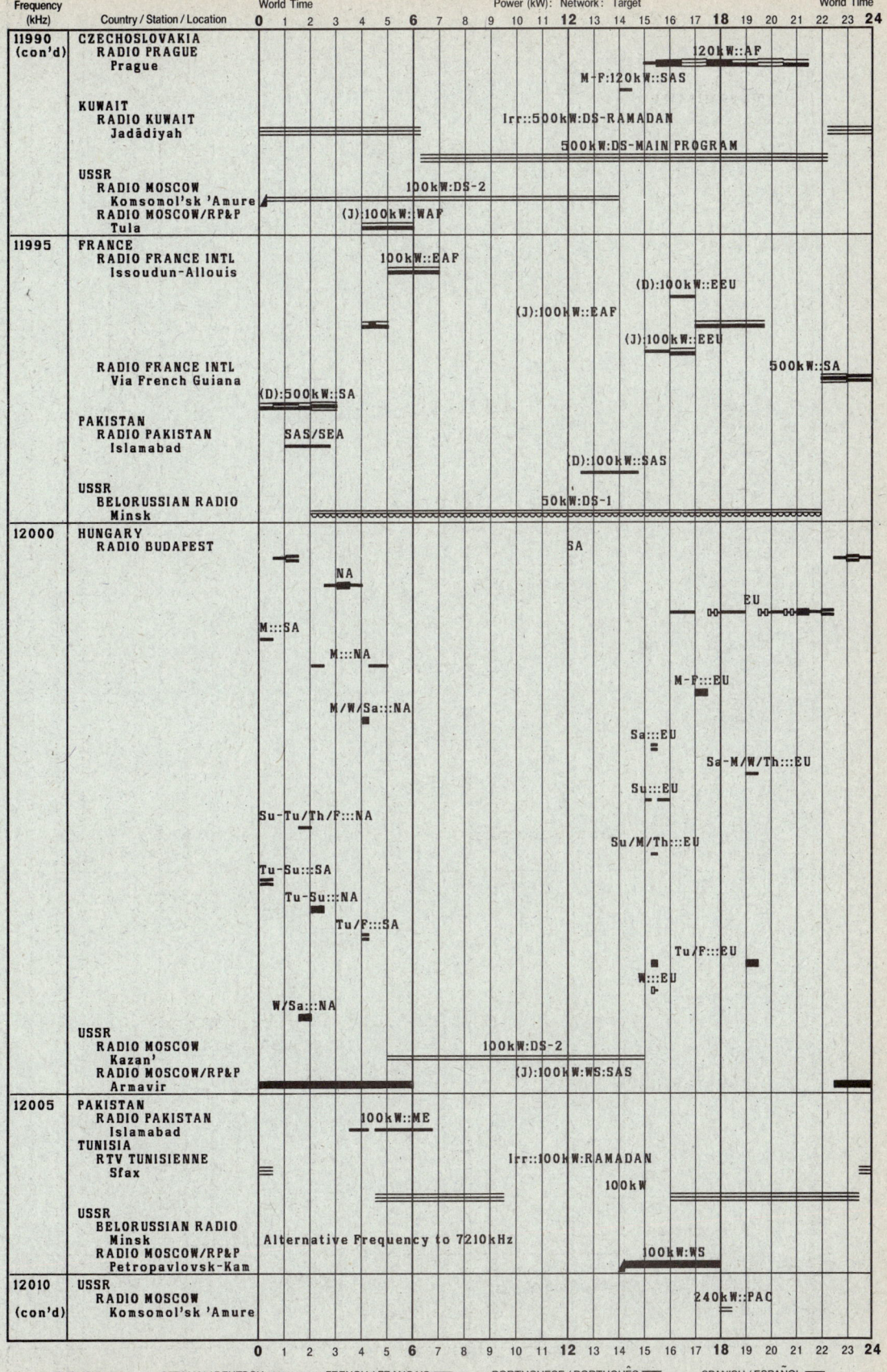

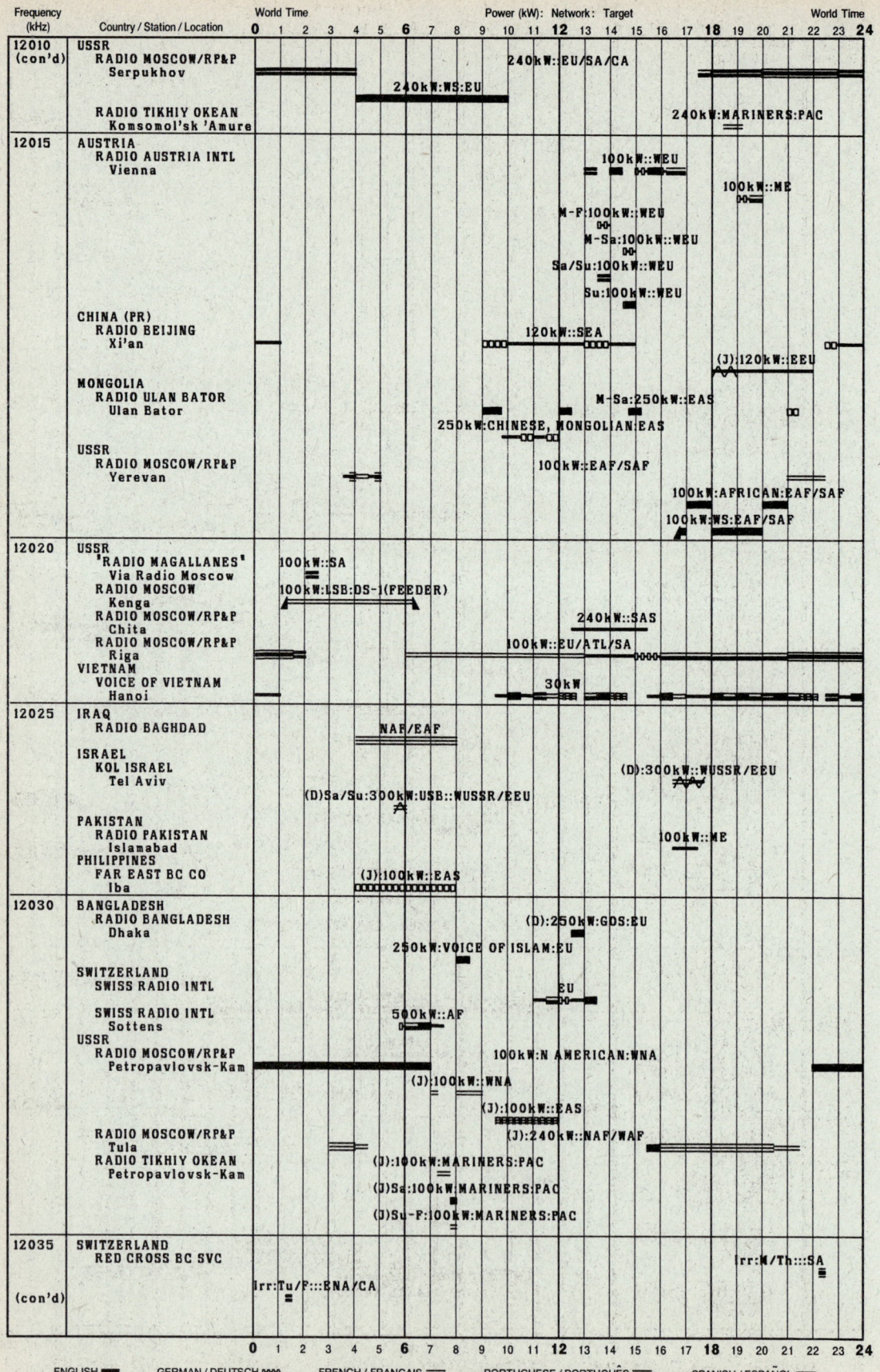

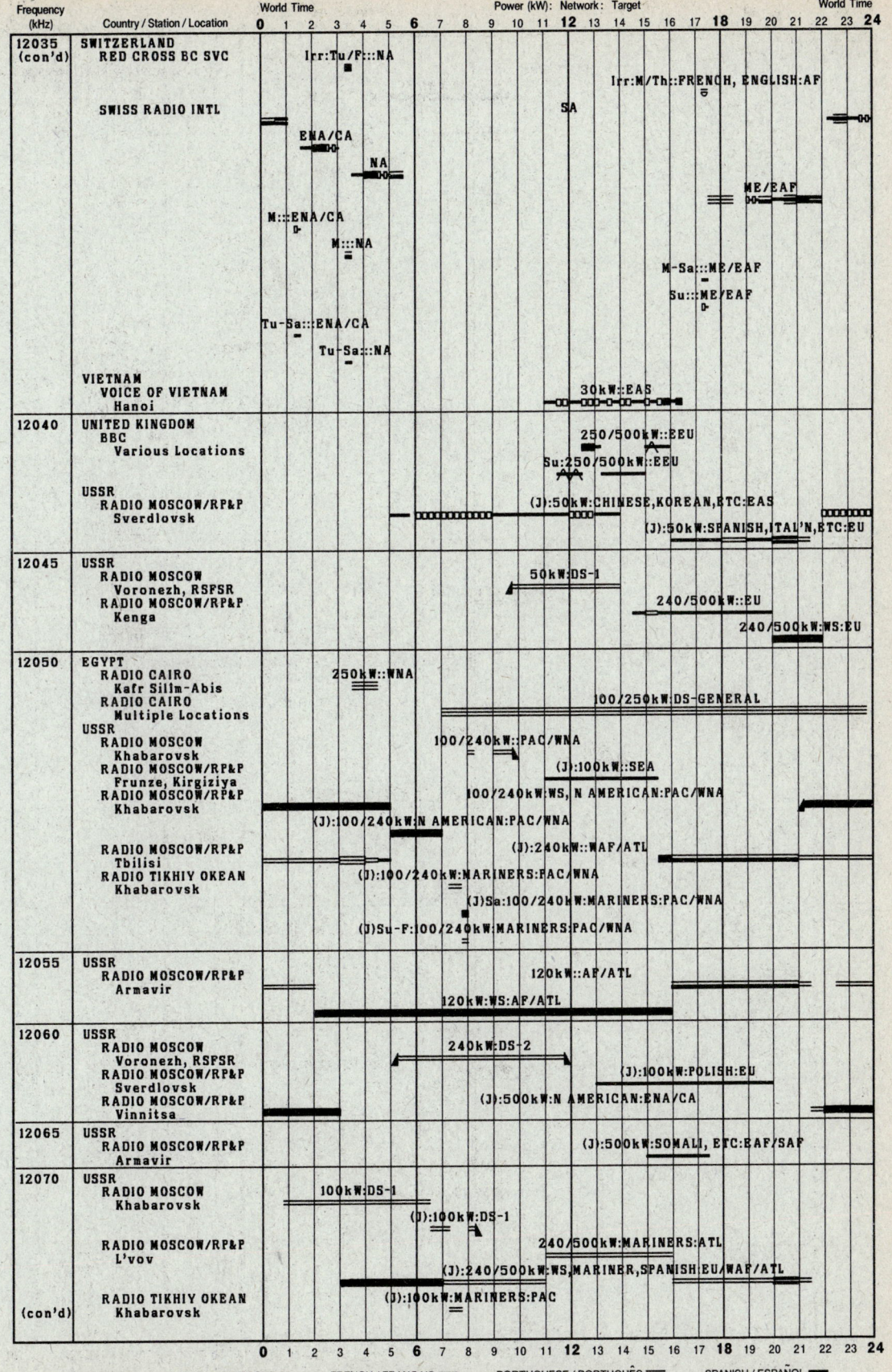

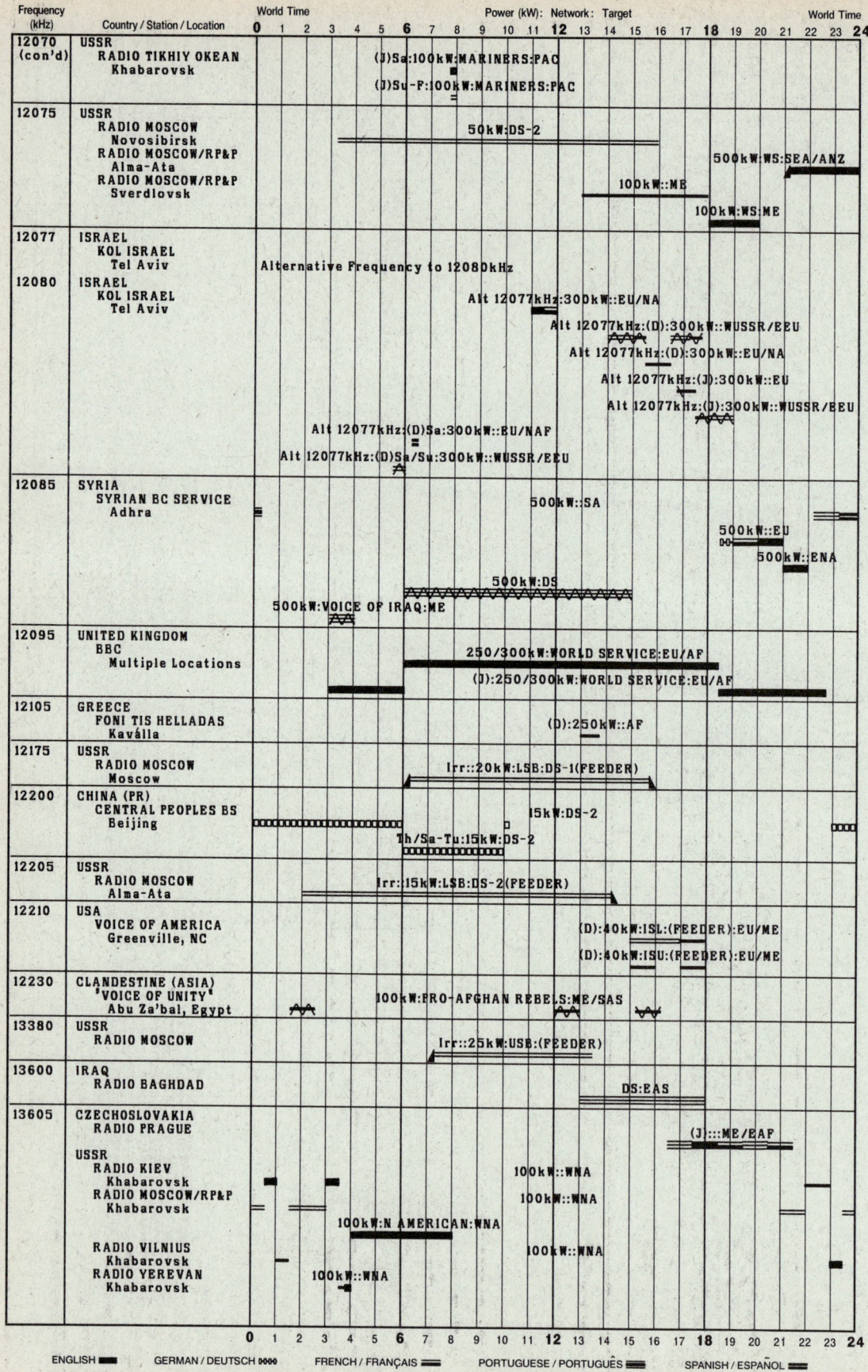

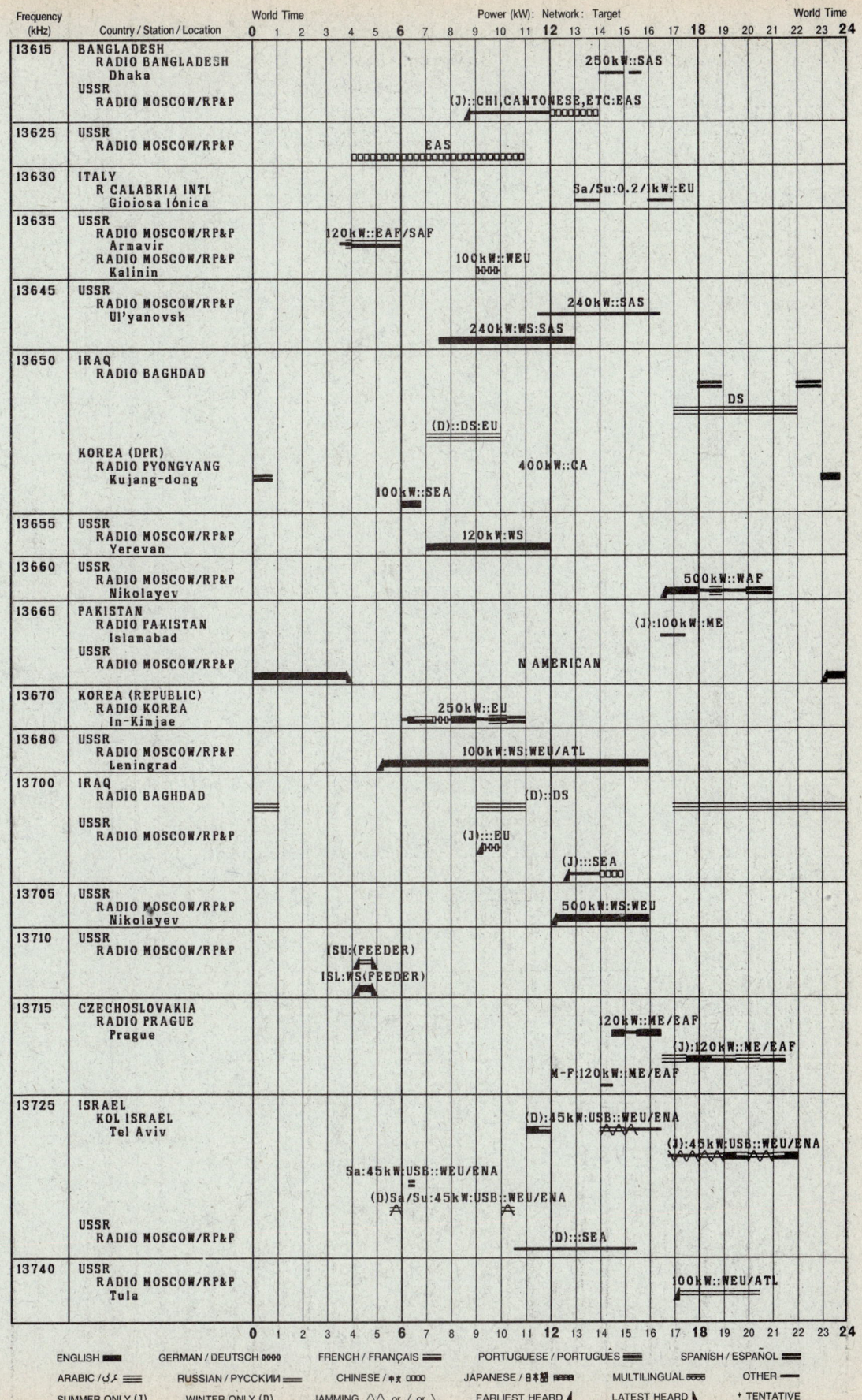

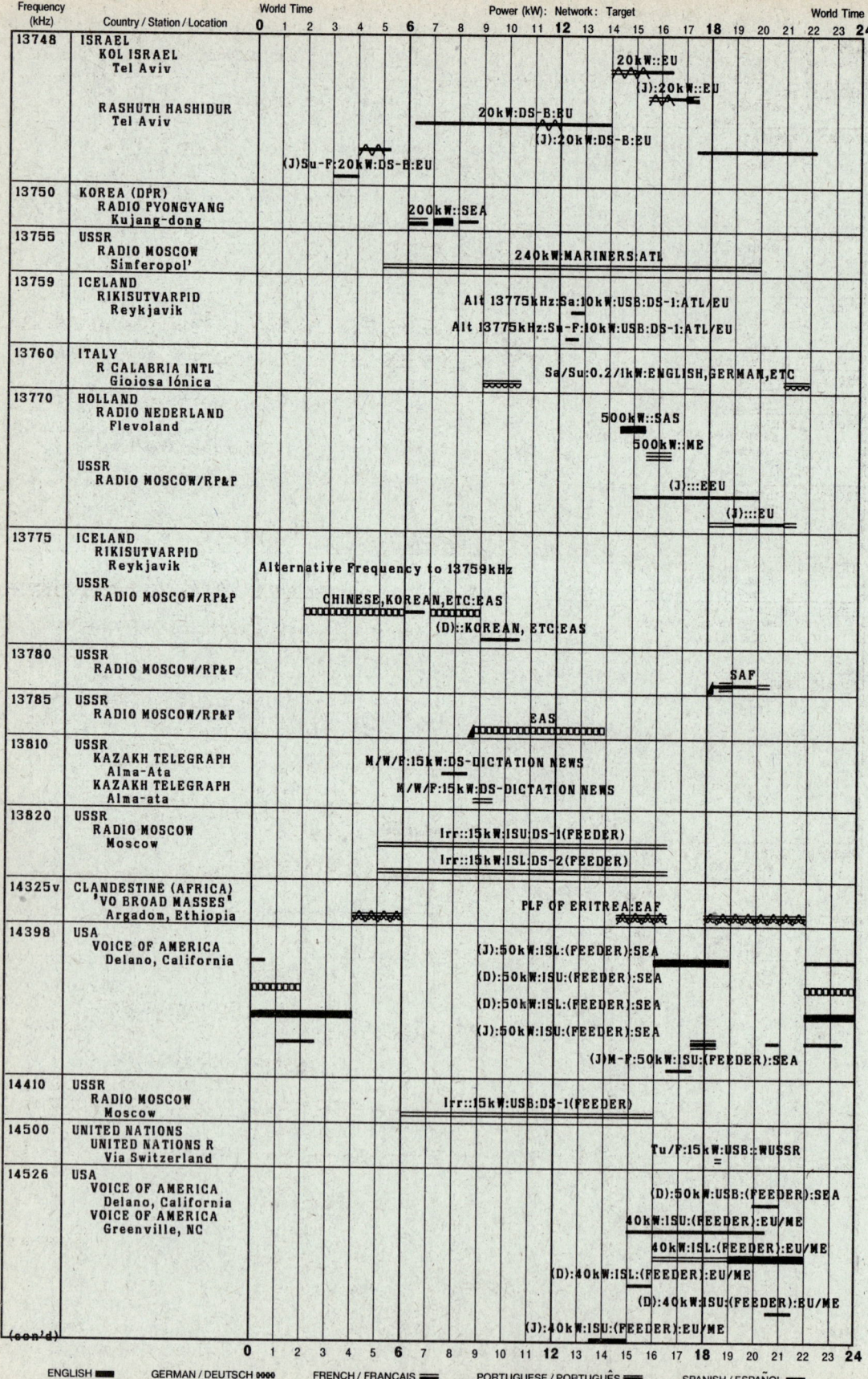

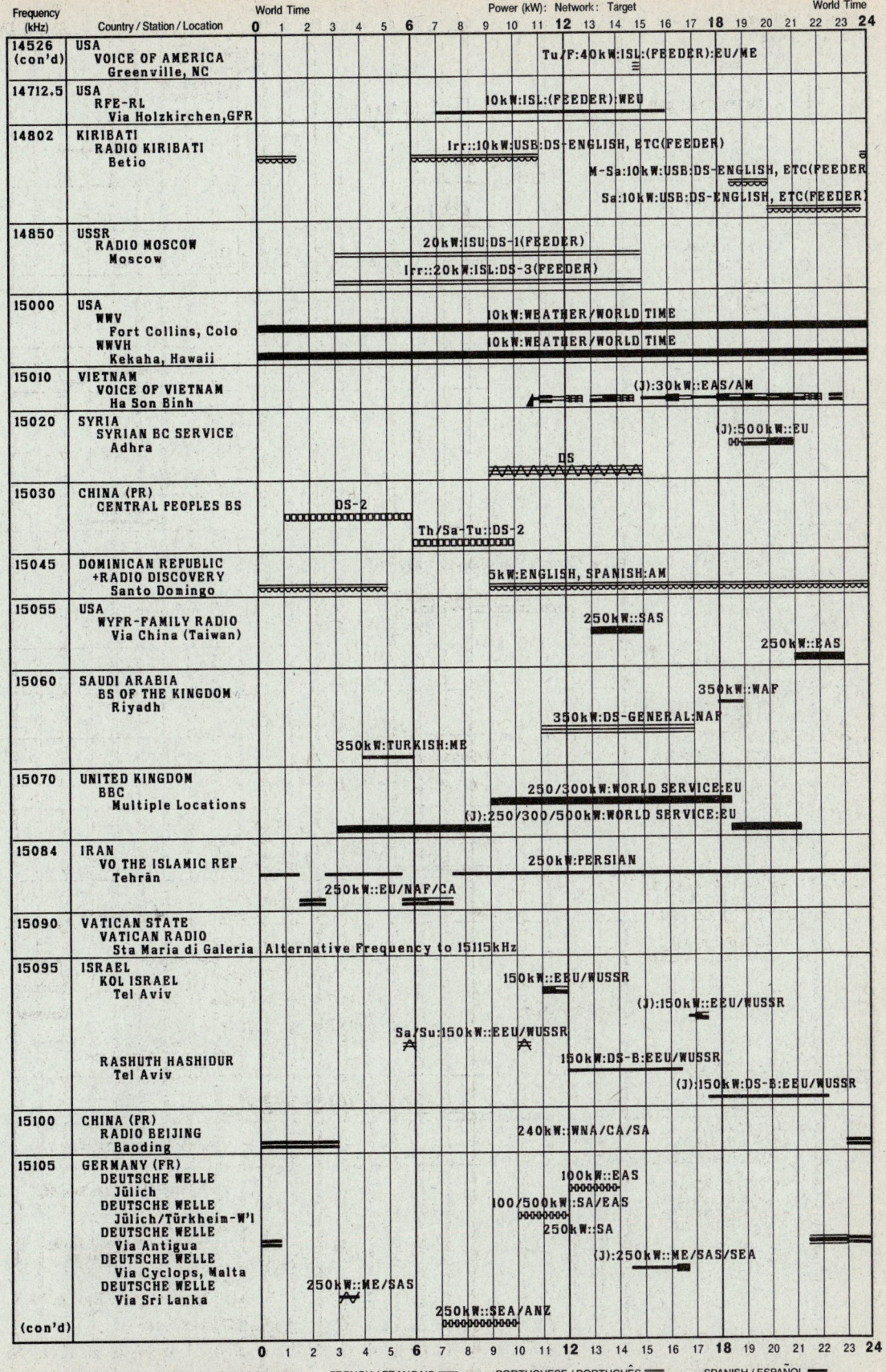

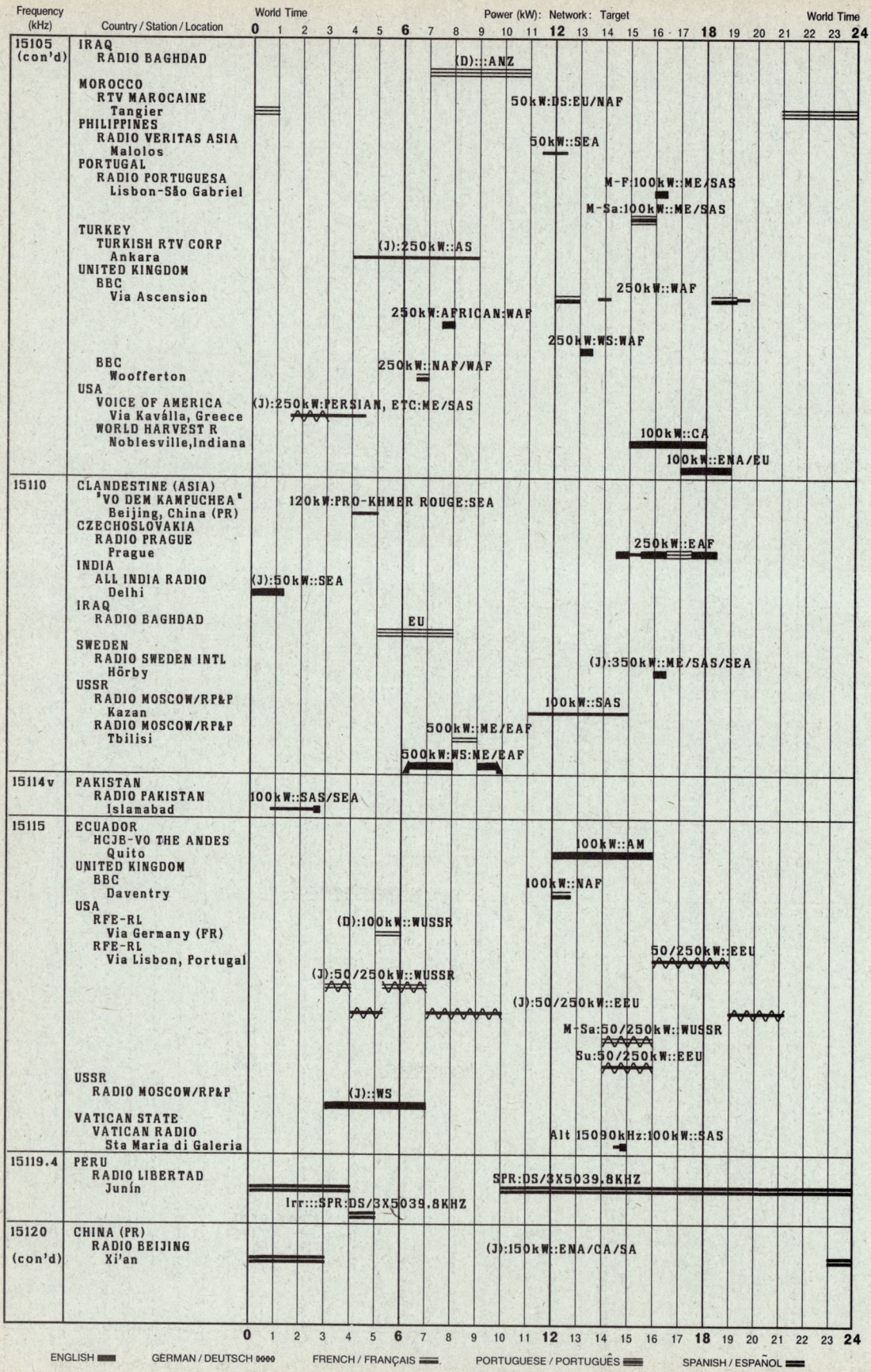

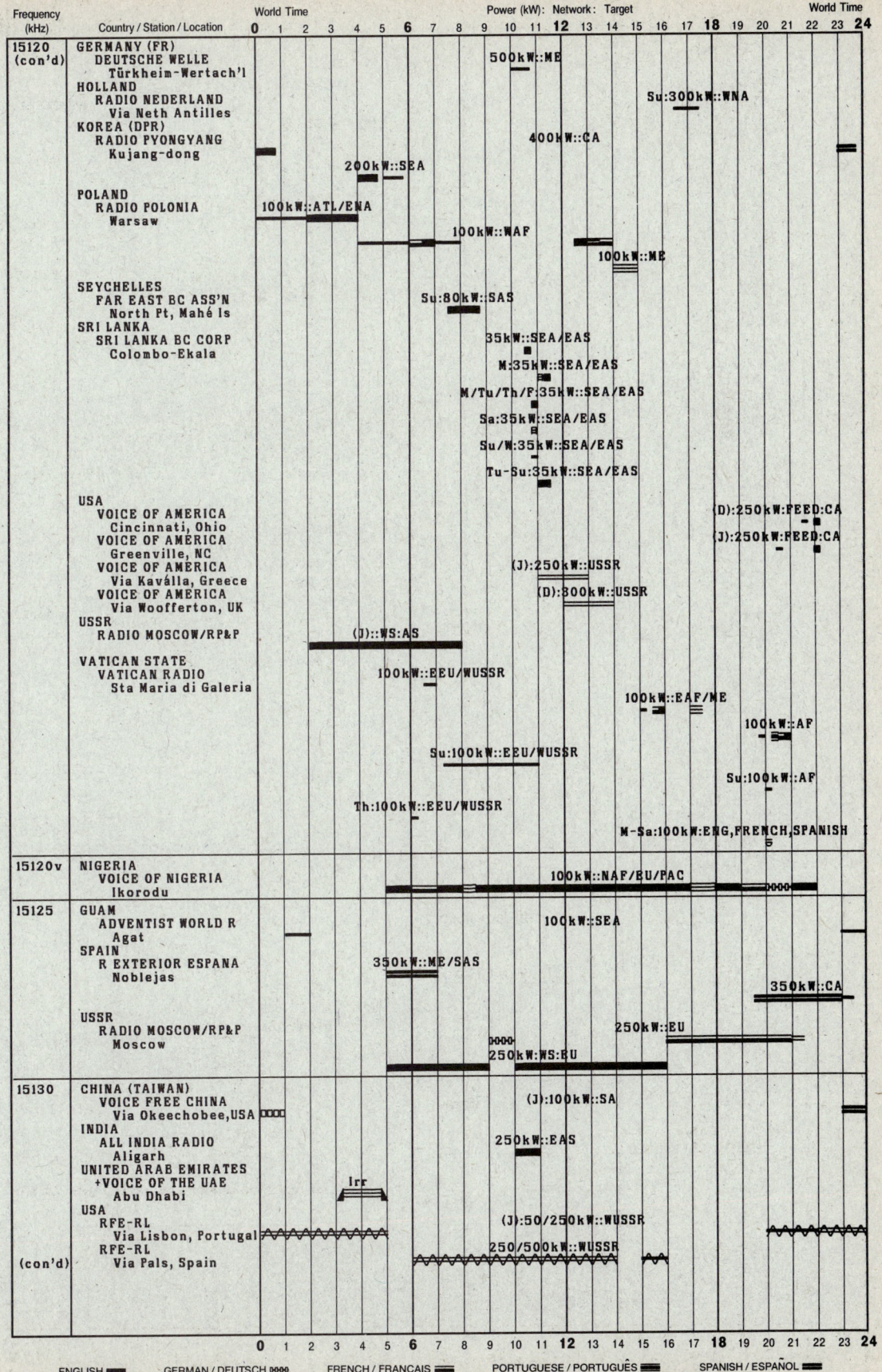

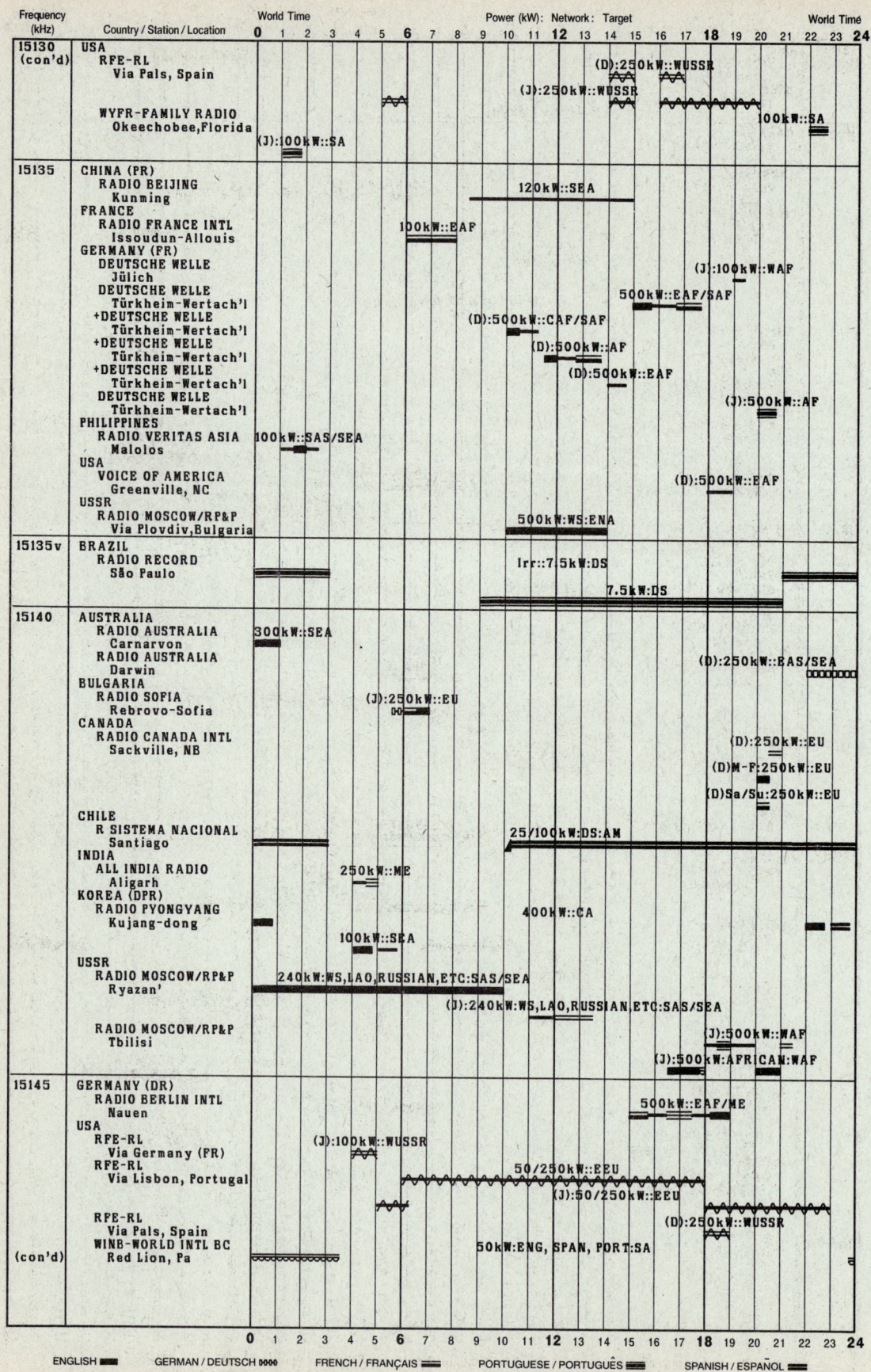

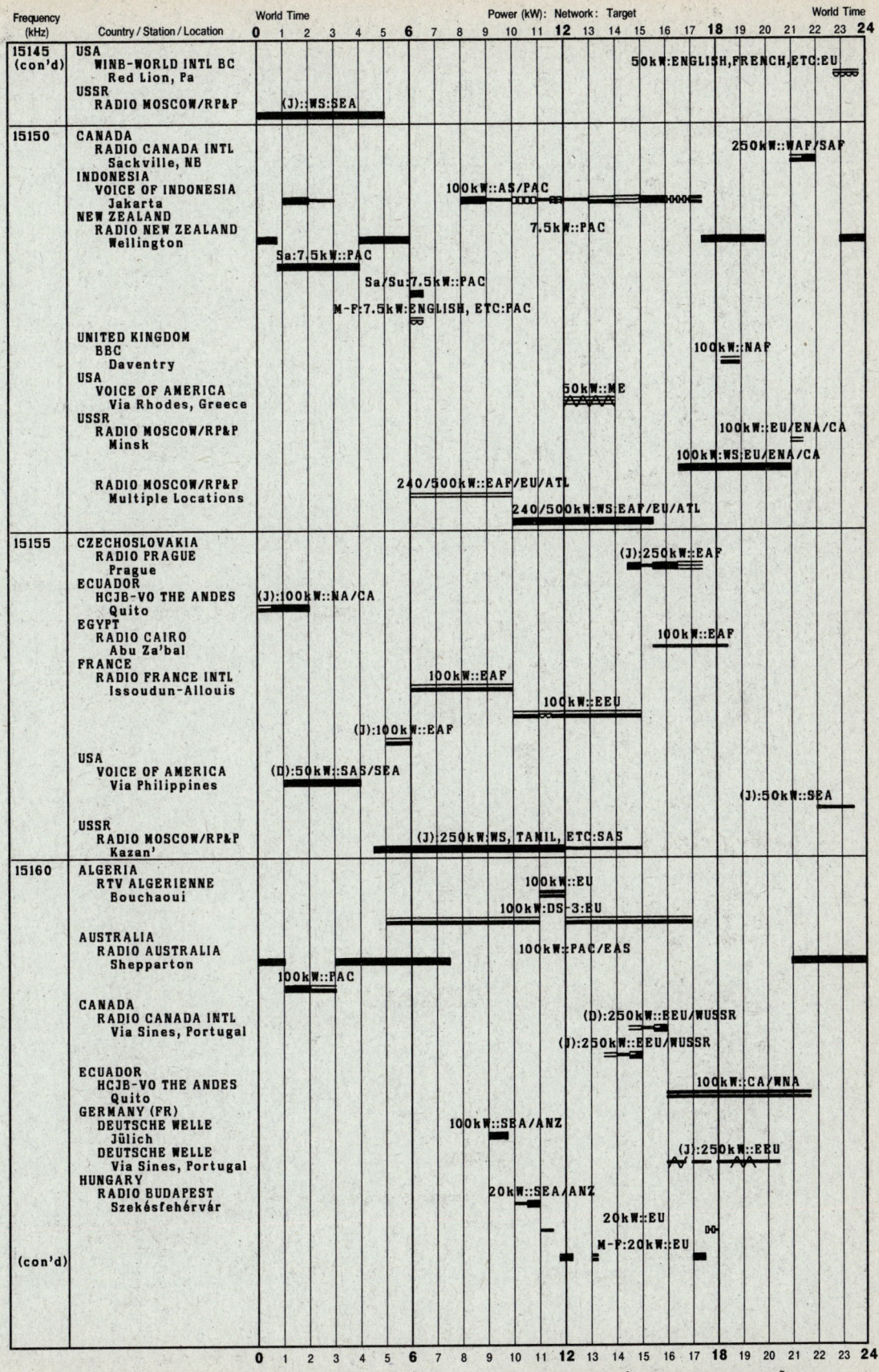

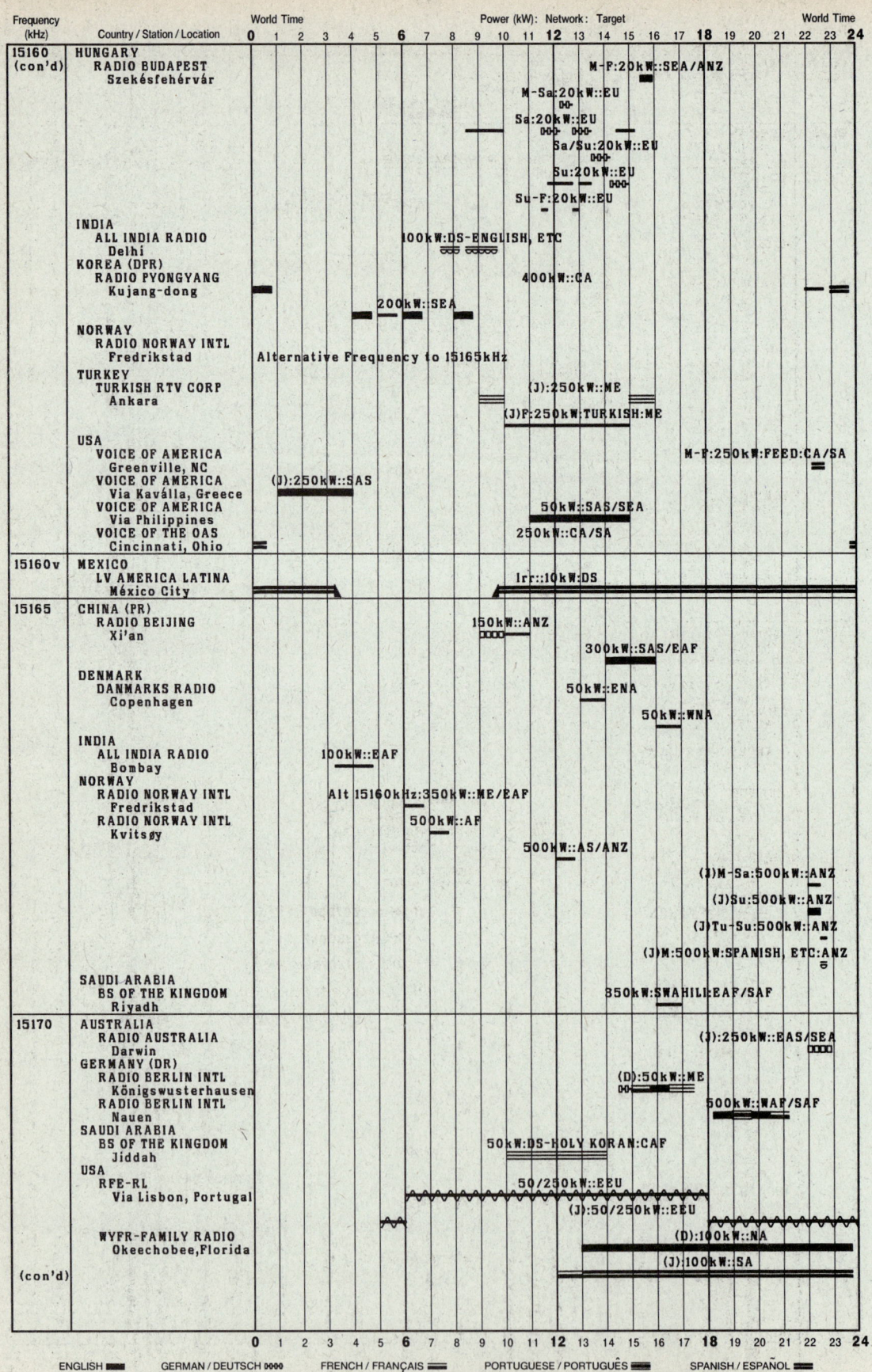

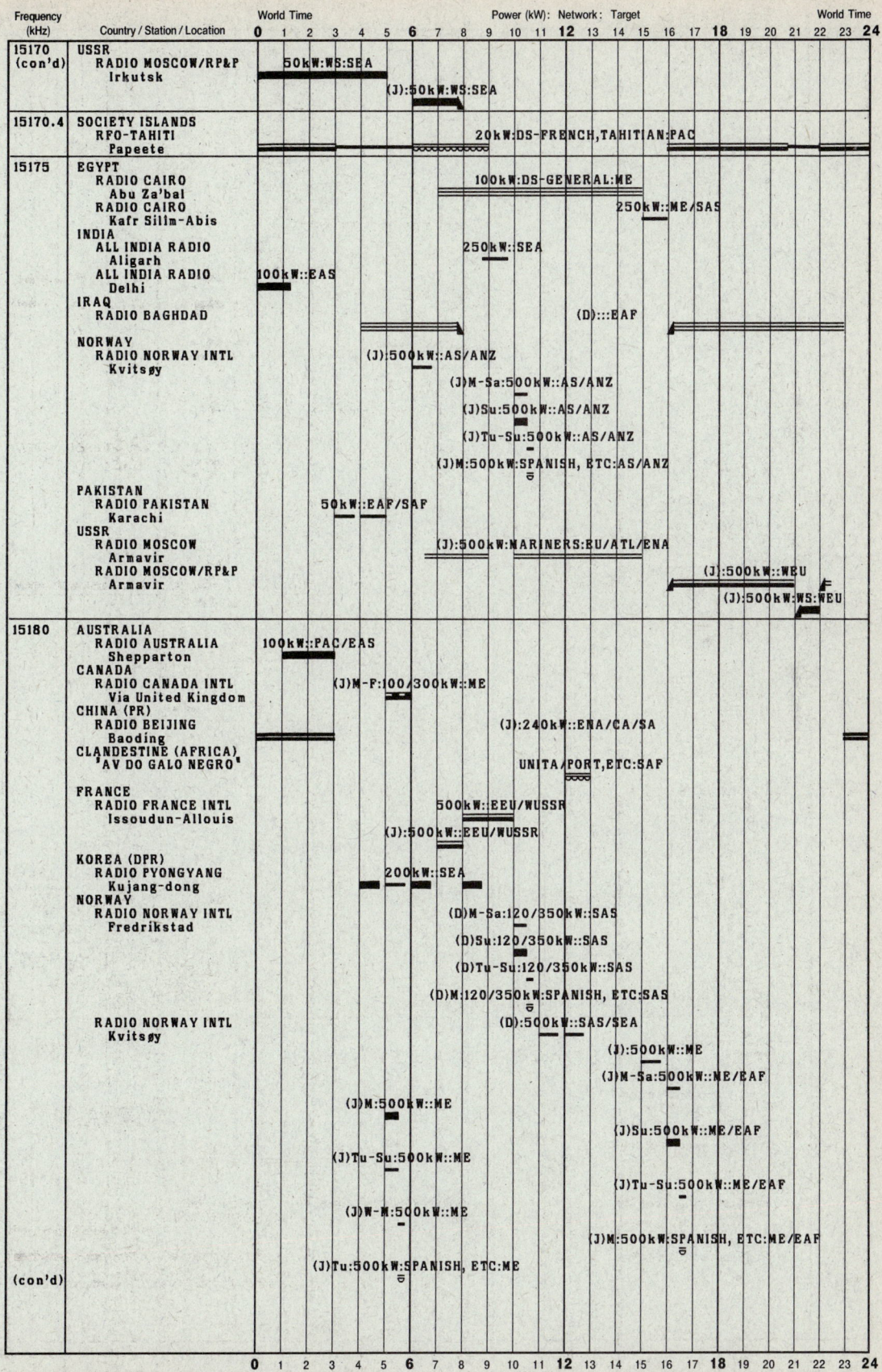

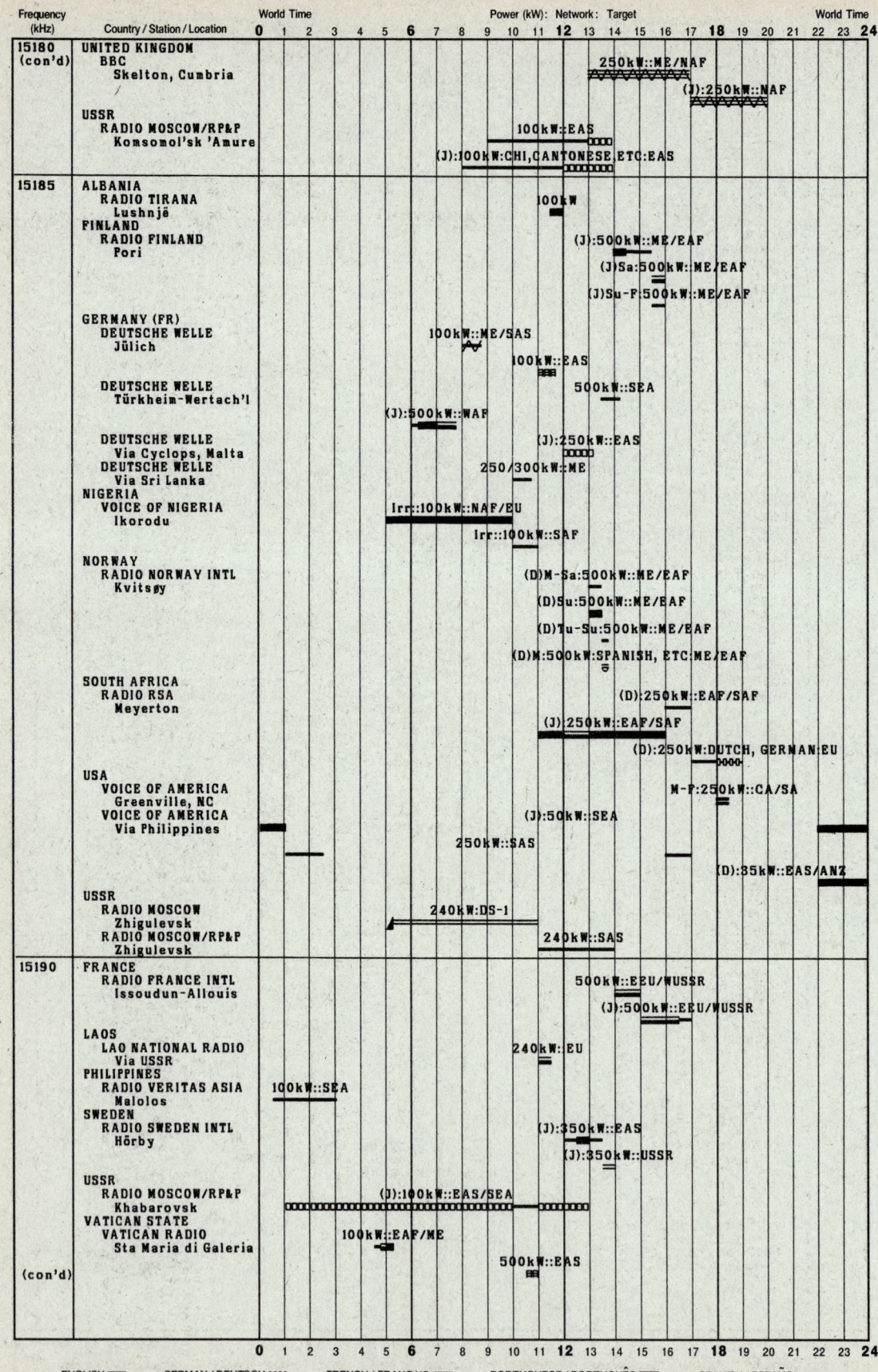

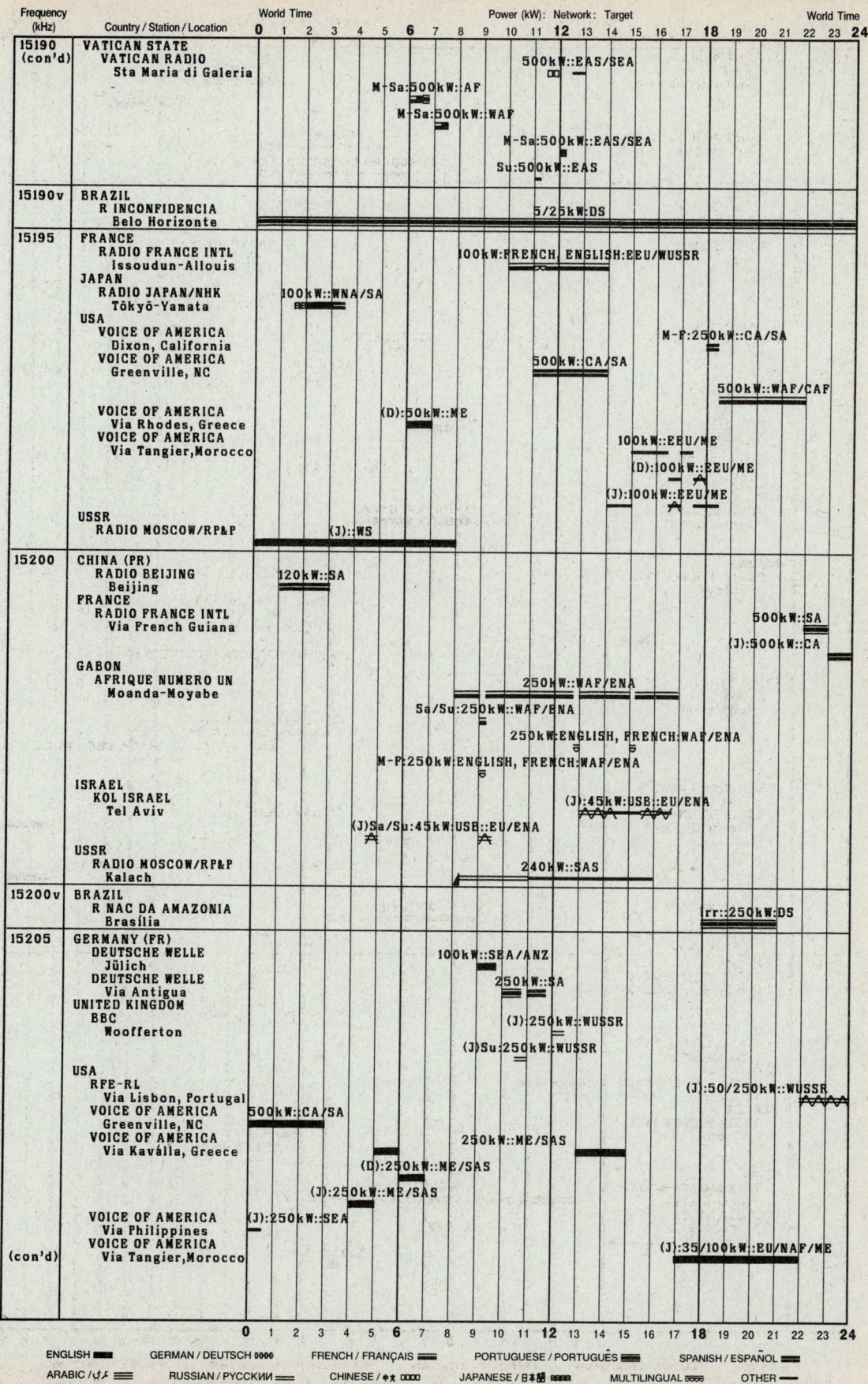

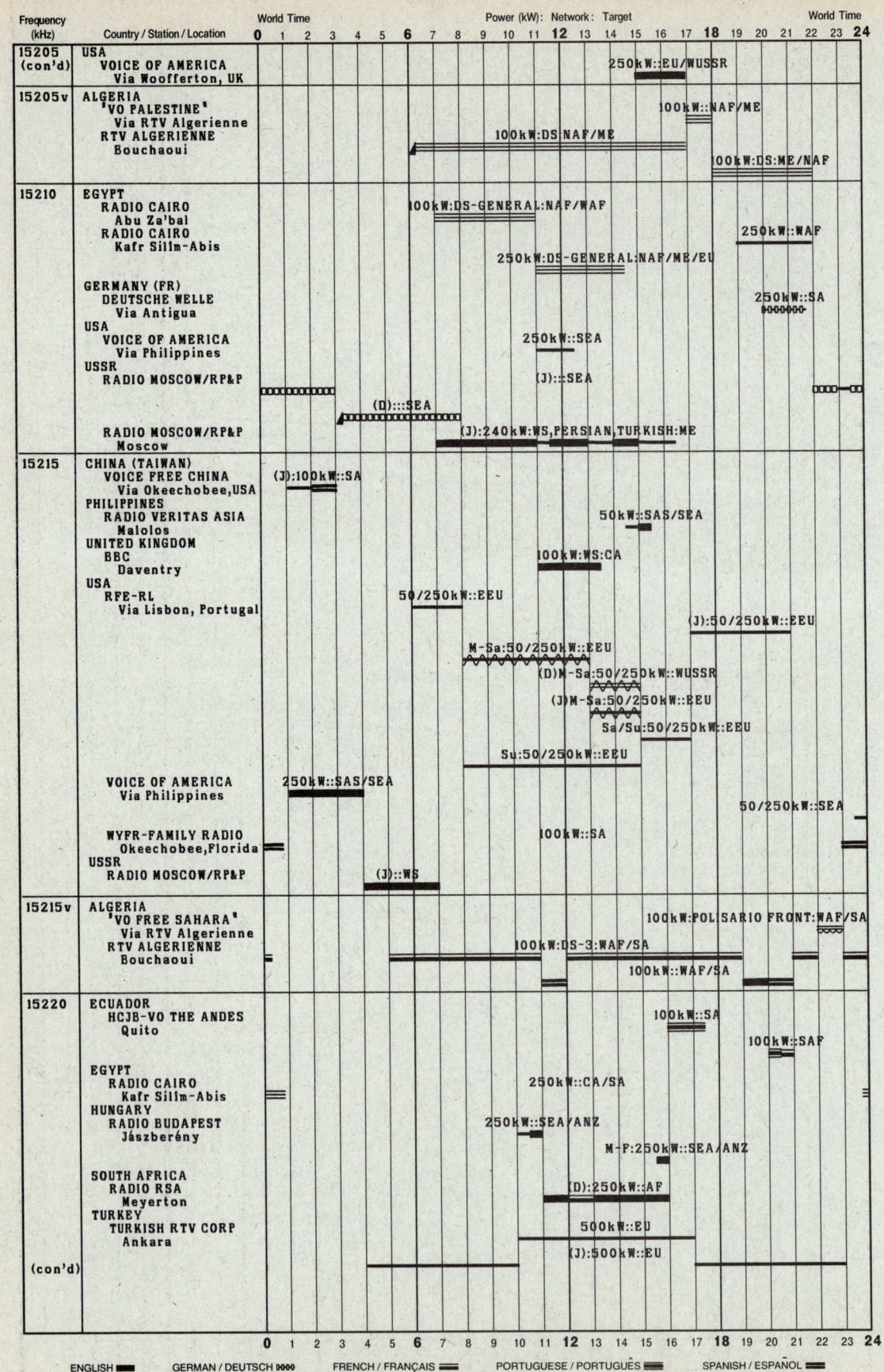

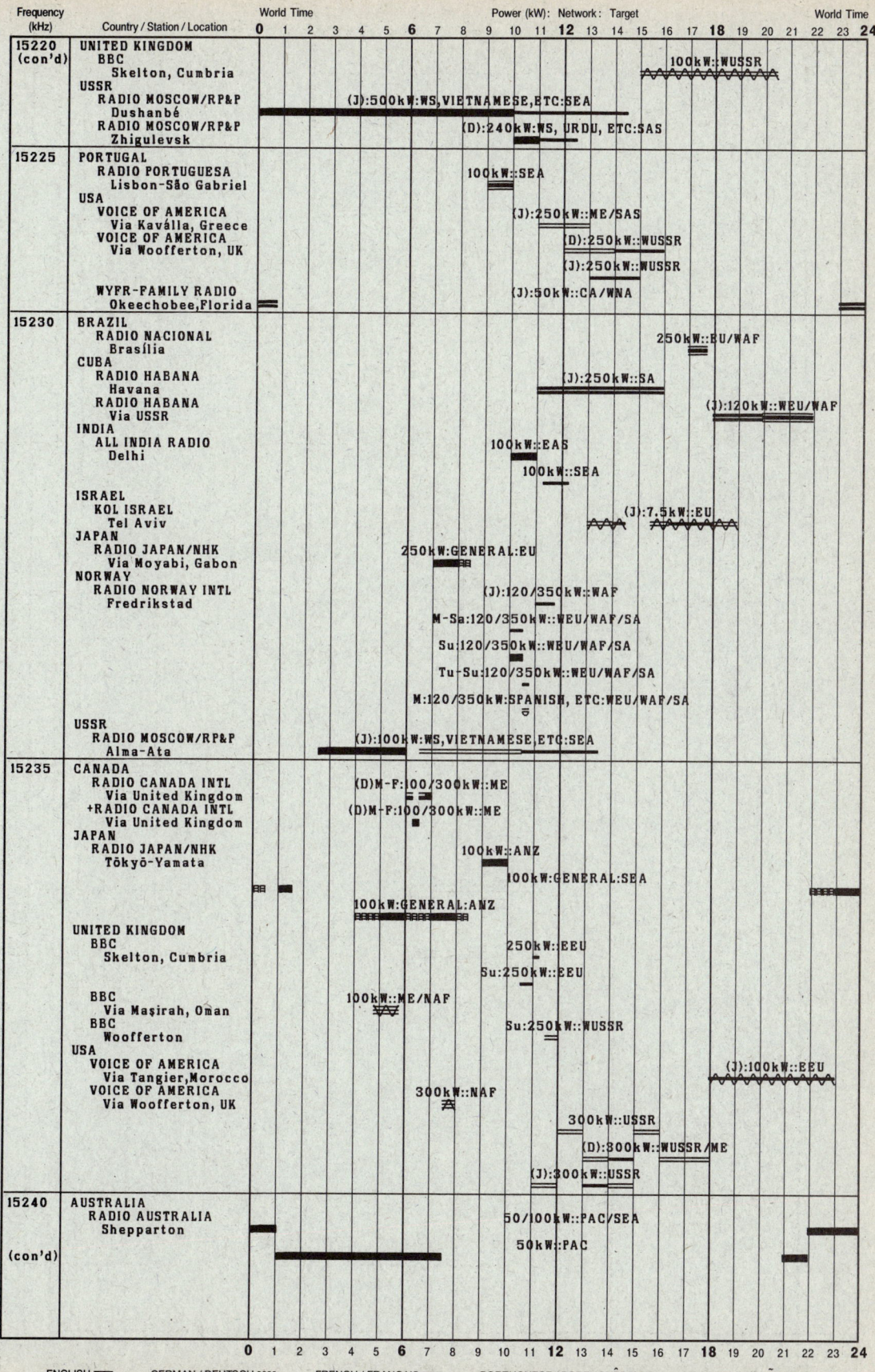

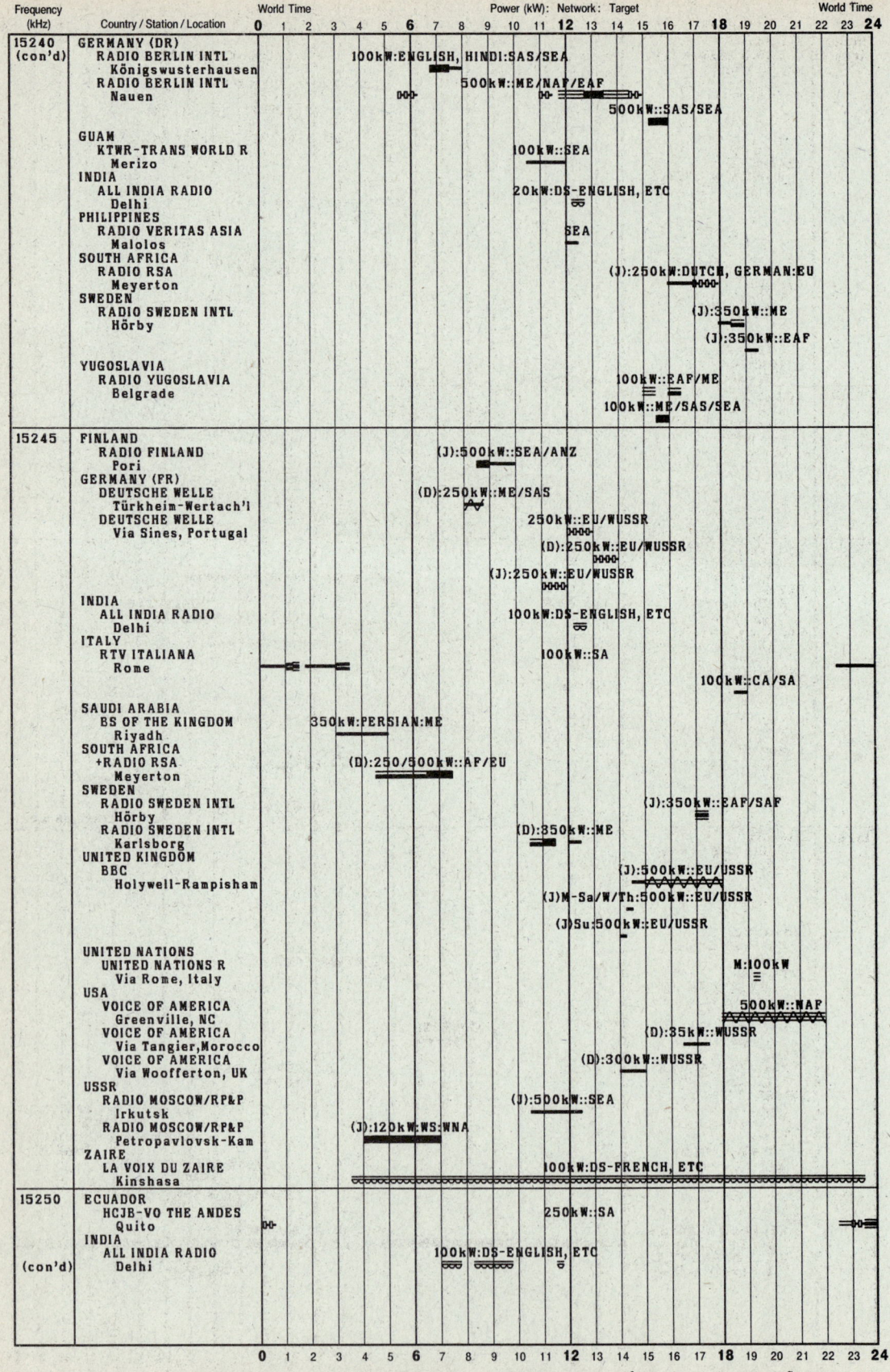

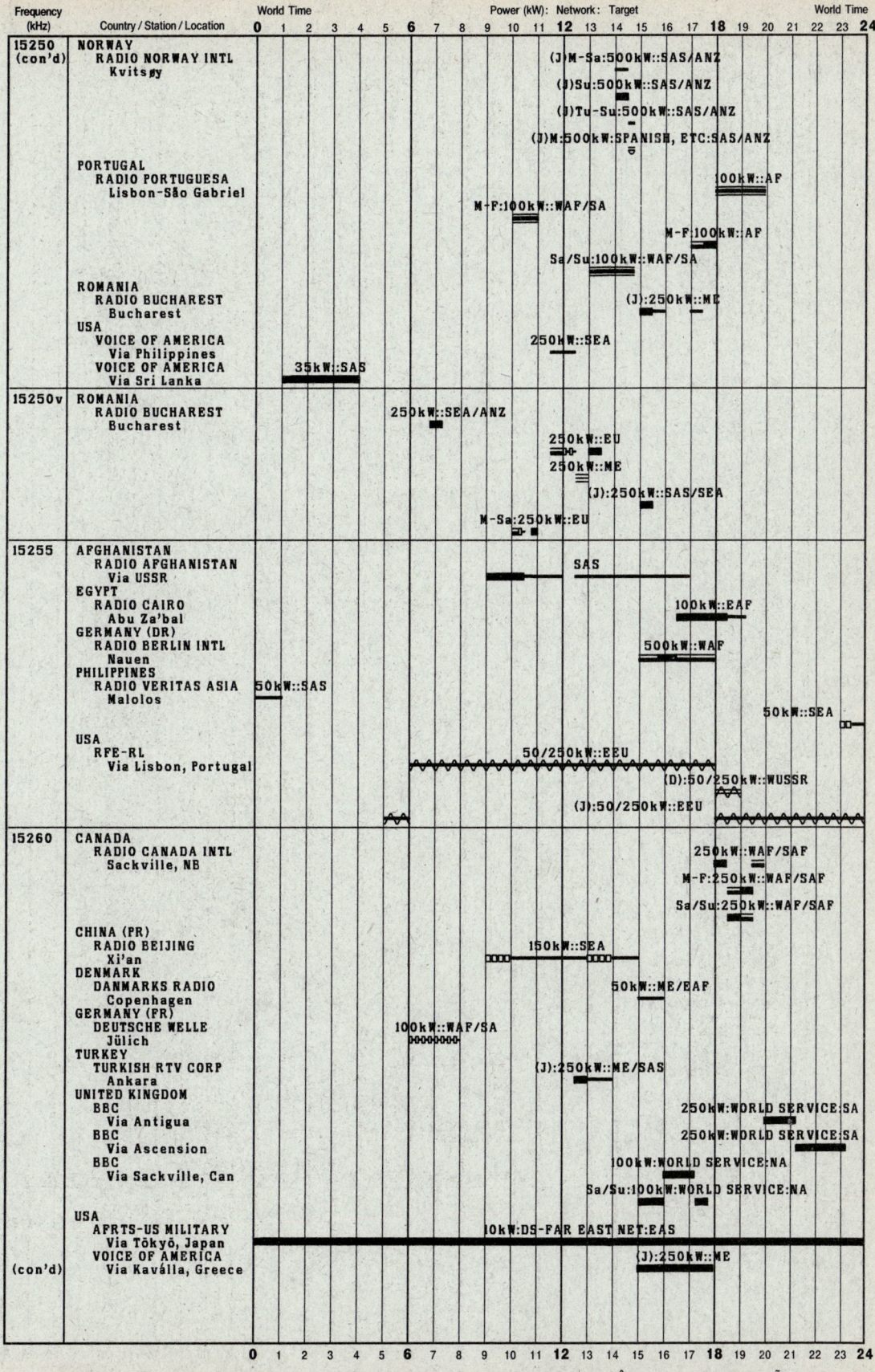

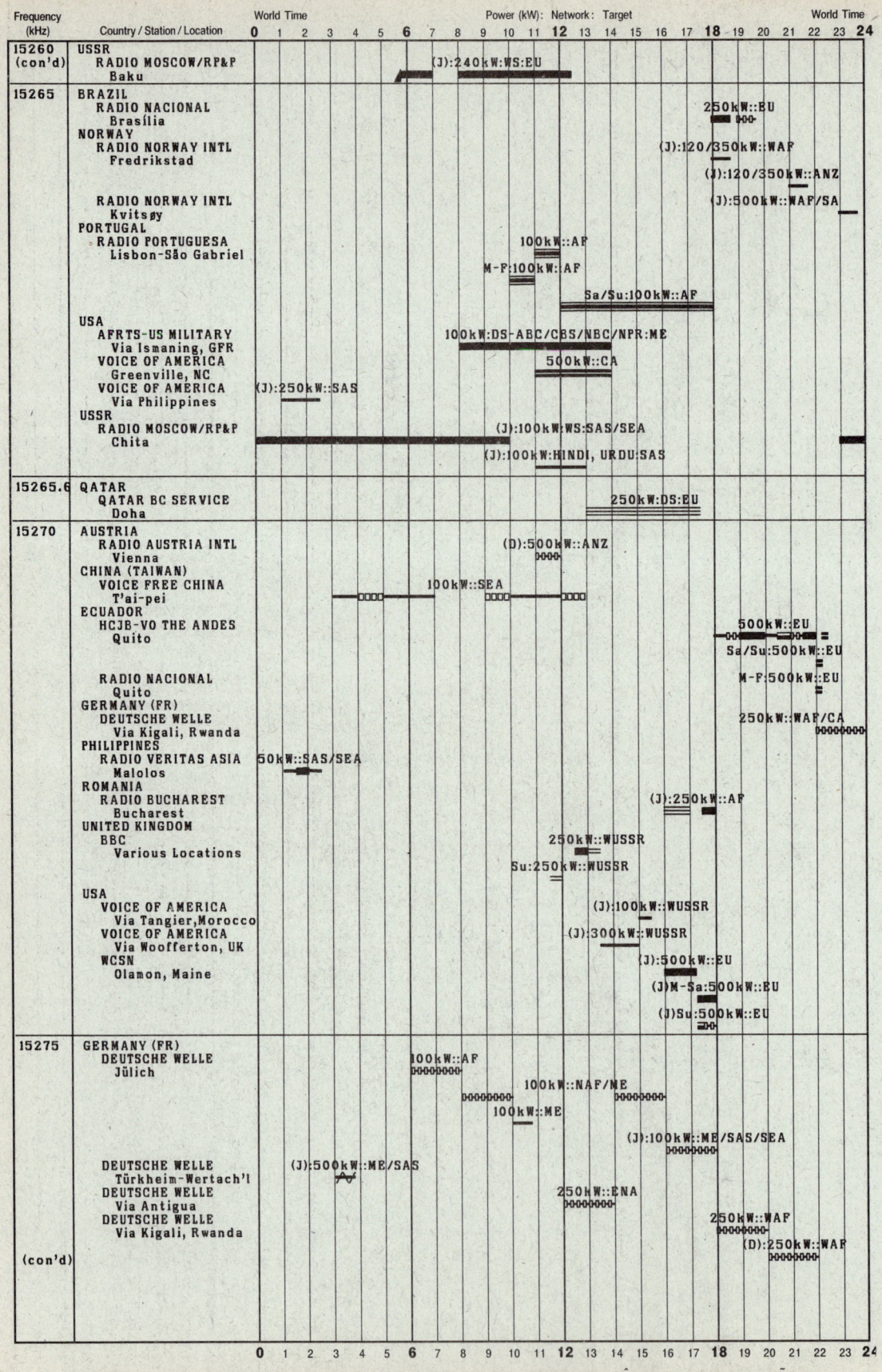

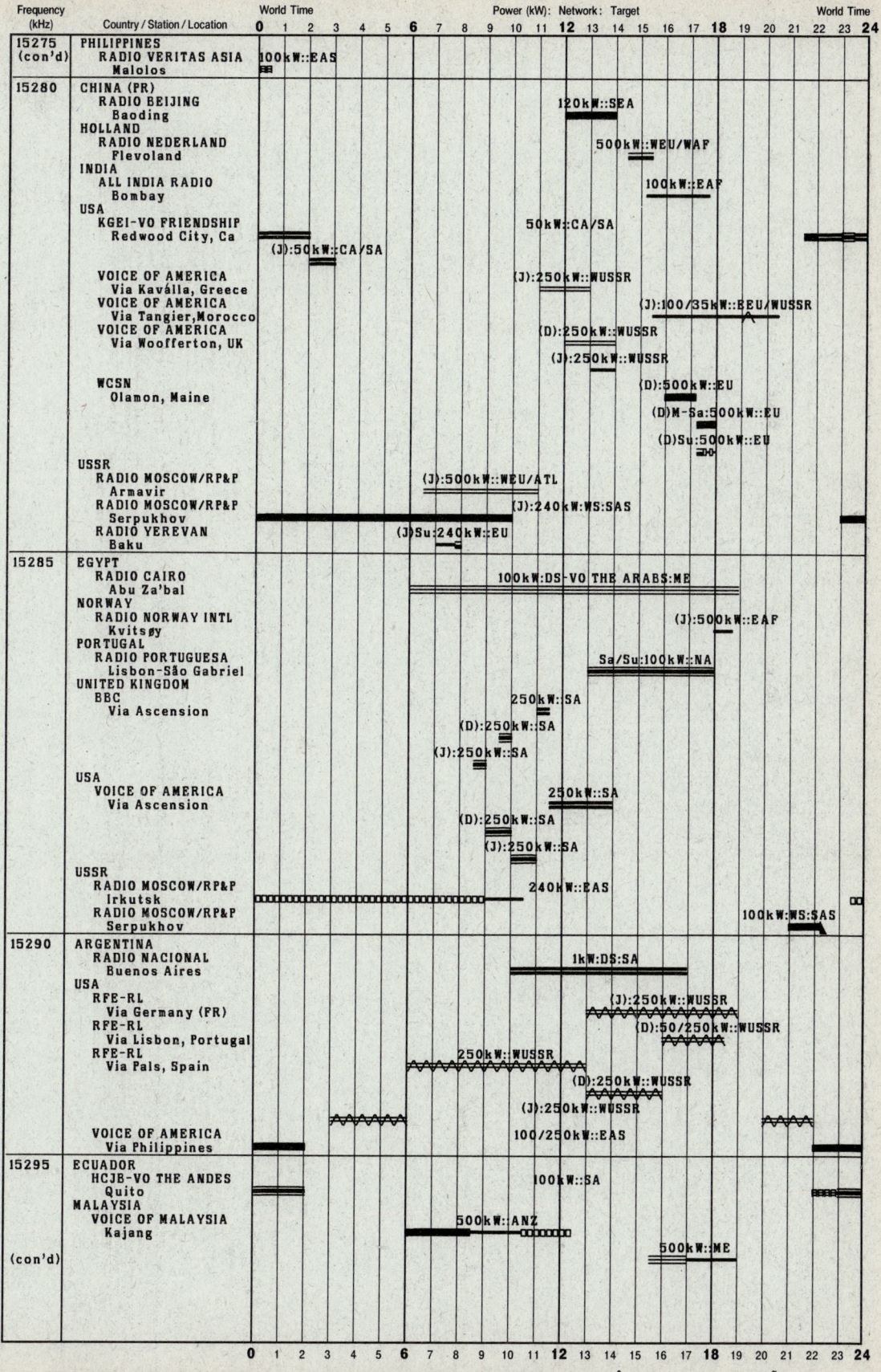

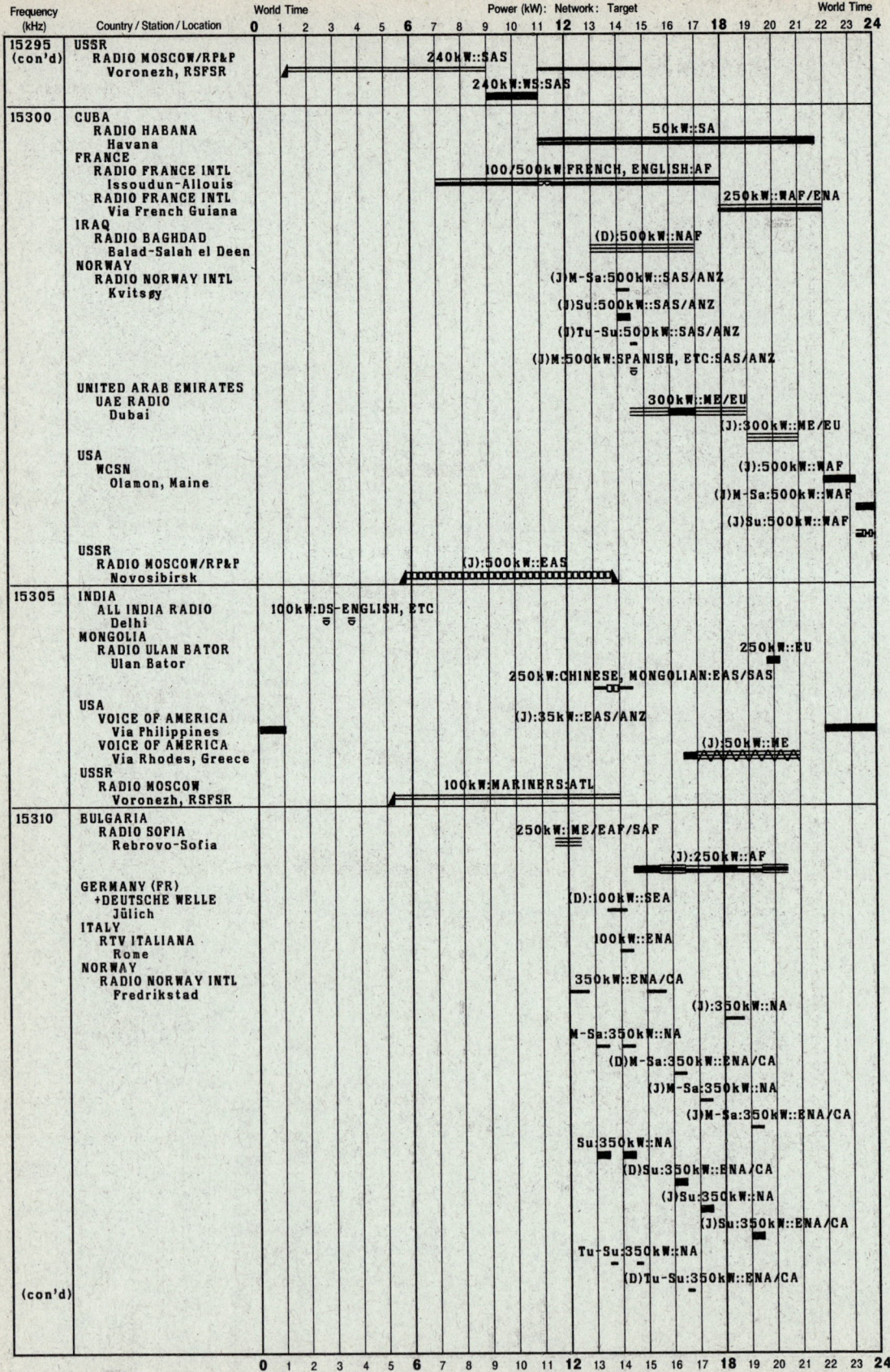

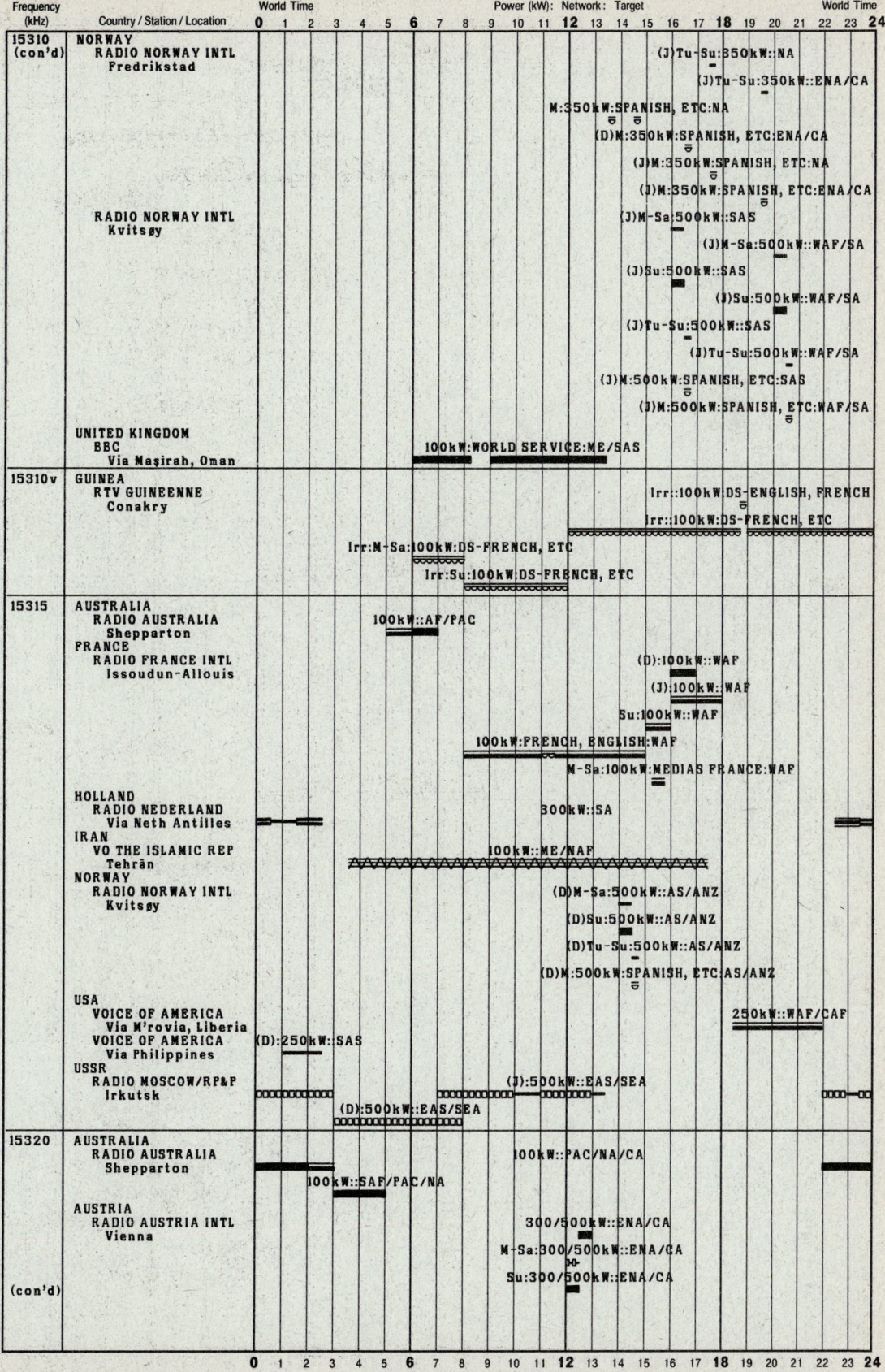

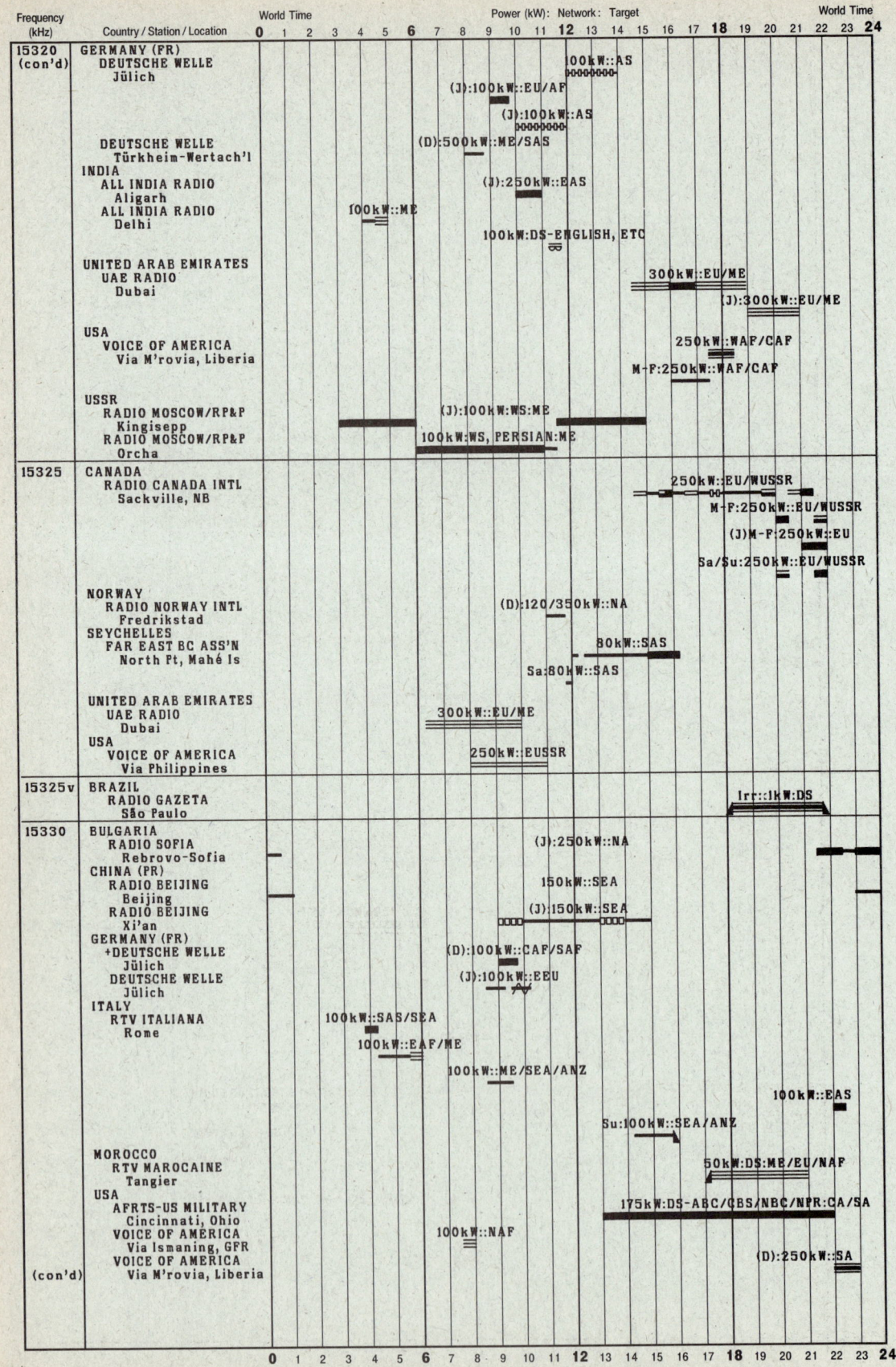

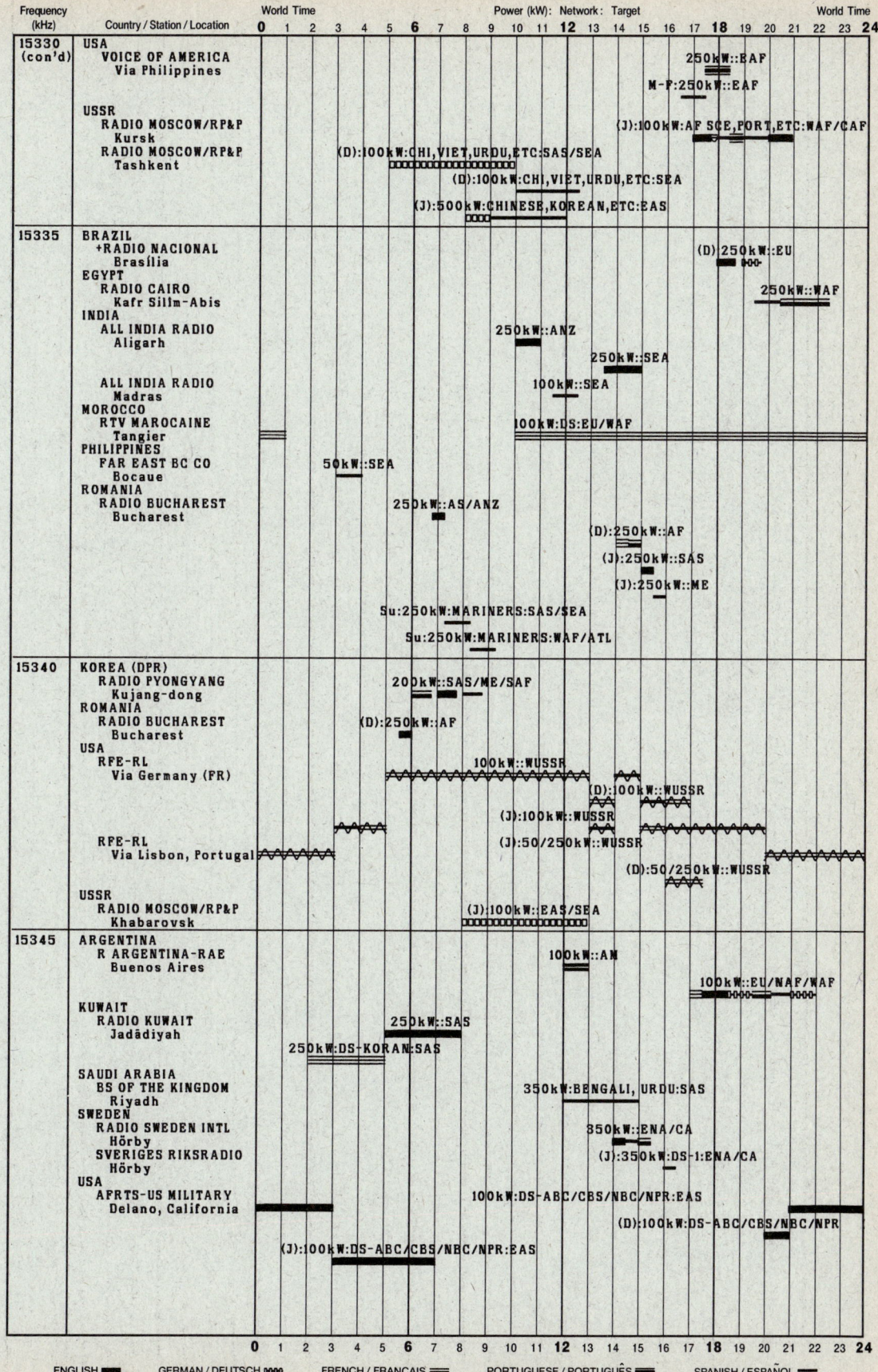

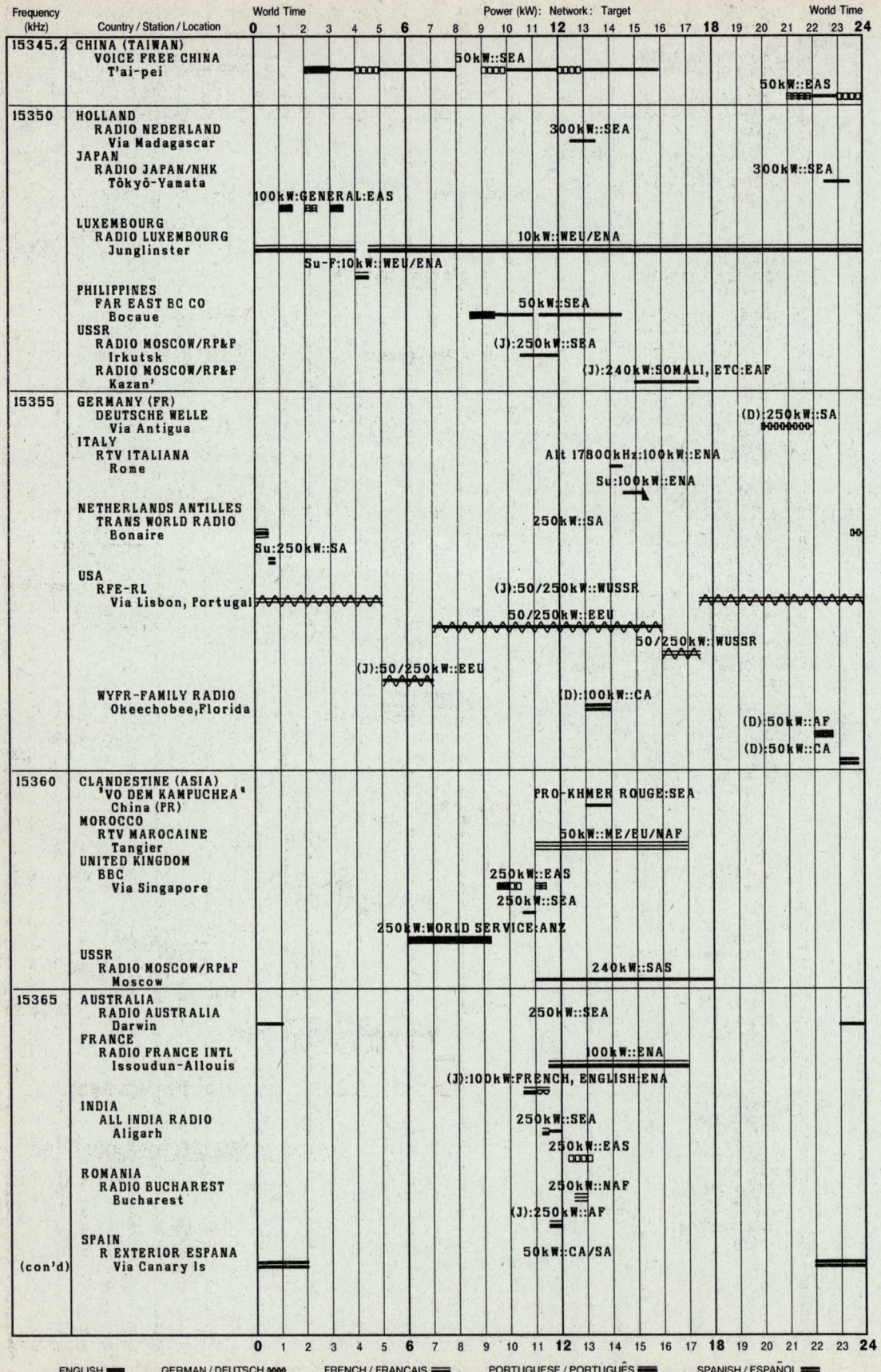

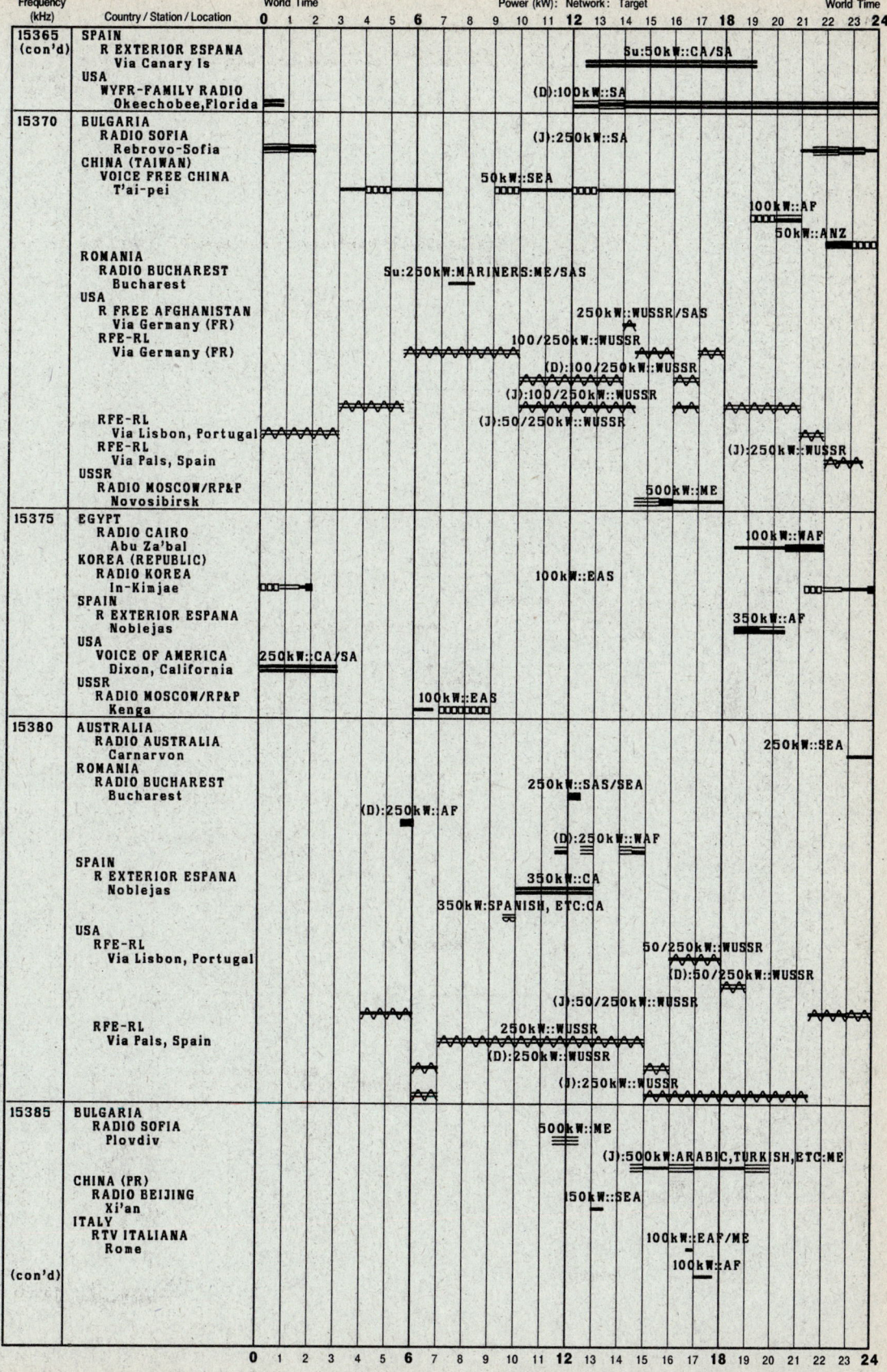

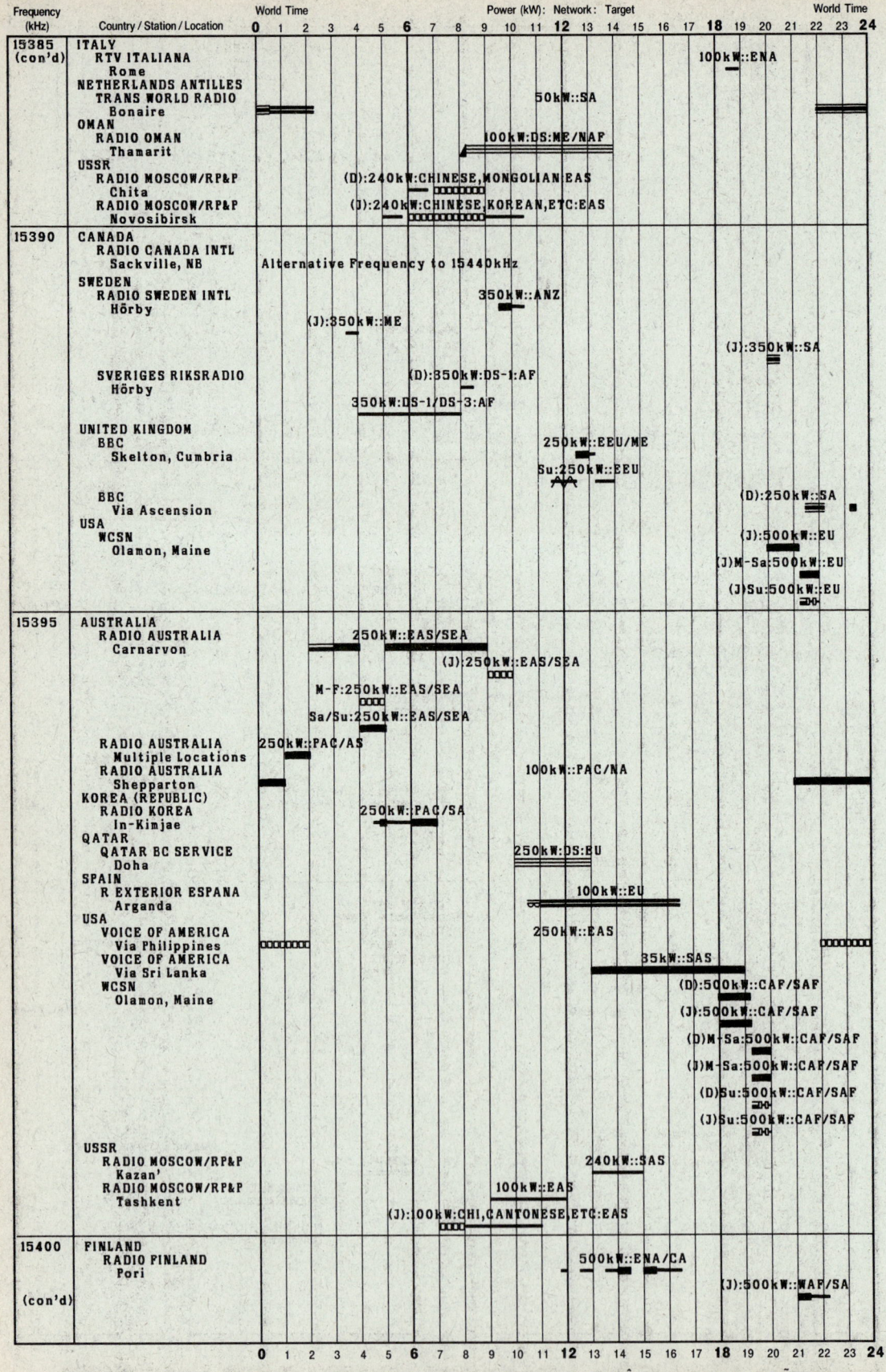

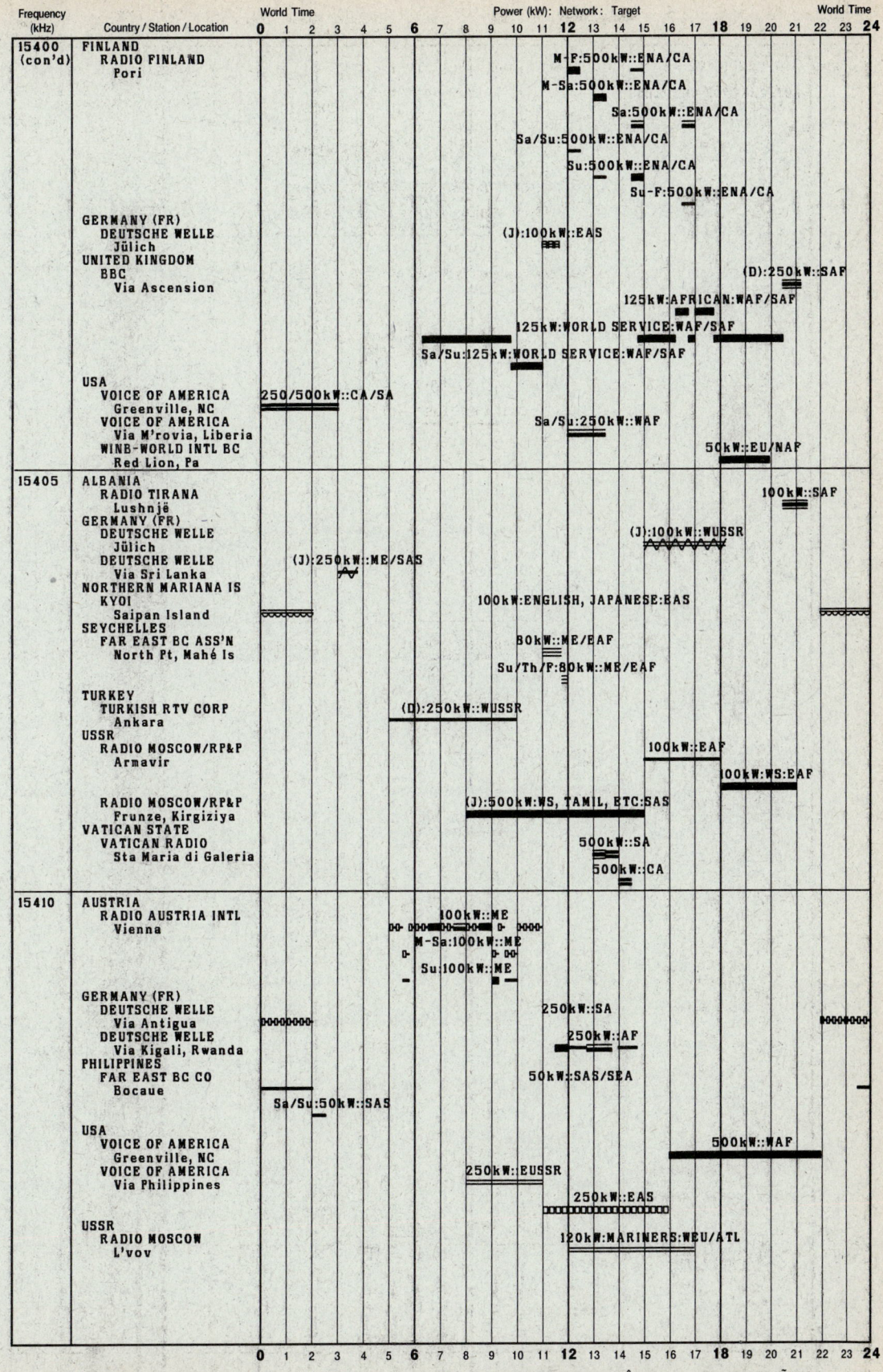

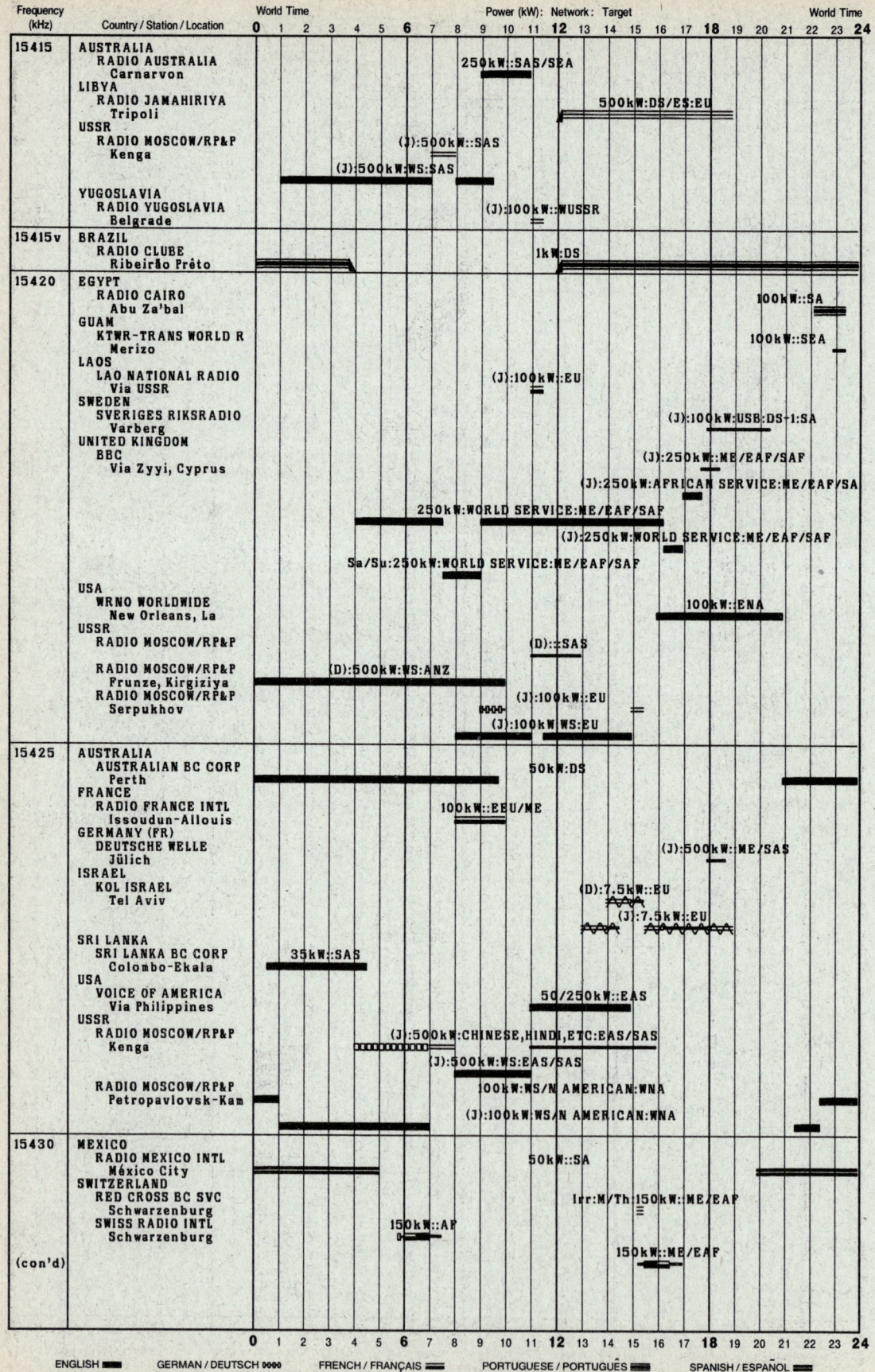

WORLDSCAN

15430–15445 kHz

Frequency (kHz)	Country / Station / Location	Broadcast Schedule
15430 (con'd)	USA — AFRTS-US MILITARY, Cincinnati, Ohio	175kW:DS-ABC/CBS/NBC/NPR:ATL/WEU/AF
	VOICE OF AMERICA, Via Philippines	250kW::EUSSR
15435	AFGHANISTAN — RADIO AFGHANISTAN, Via USSR	ME/SAS
	CHINA (PR) — RADIO BEIJING, Beijing	240kW::SA; 240kW::WUSSR
	FRANCE — RADIO FRANCE INTL, Via French Guiana	500kW::SA; 500kW::CA/SA; 500kW::CA; M-Sa:500kW::CA; Su:500kW::CA
	GERMANY (FR) — DEUTSCHE WELLE, Via Sri Lanka	(J):250kW::ME/SAS
	SAUDI ARABIA — BS OF THE KINGDOM, Jiddah	50kW:DS-GENERAL:EU
	SWEDEN — SVERIGES RIKSRADIO, Varberg	100kW:USB:DS-1:AF
	TANZANIA — RADIO TANZANIA, Dar es Salaam	Irr:::AF
	USA — R FREE AFGHANISTAN, Via Biblis, GFR	(J)Tu/Th/Sa:100kW::WUSSR/SAS
	VOICE OF AMERICA, Via Kaválla, Greece	250kW::SAS/ME; (J):250kW::SAS/ME
	USSR — RADIO MOSCOW/RP&P, Frunze, Kirgiziya	500kW::SEA/EAS; (J):240kW::SEA/EAS
15440	CANADA — RADIO CANADA INTL, Sackville, NB	M-F:250kW::CA/SA; Alt 15390kHz:Su:250kW::CA/SA
	CHINA (PR) — RADIO BEIJING, Kunming	120kW::ANZ
	CHINA (TAIWAN) — VOICE FREE CHINA, Via Okeechobee, USA	(J):100kW::EU
	GERMANY (DR) — RADIO BERLIN INTL, Königswusterhausen	(J):100kW:ENGLISH, HINDI:SAS/SEA
	GERMANY (FR) — DEUTSCHE WELLE, Via Sines, Portugal	(J):250kW::EEU
	IRAQ — RADIO BAGHDAD, Balad-Salah el Deen	500kW::NAF
	UNITED KINGDOM — BBC, Via Zyyi, Cyprus	250kW::EAF; 250kW:AFRICAN SERVICE:EAF
	USA — WYFR-FAMILY RADIO, Okeechobee, Florida	100kW::EU; (J):100kW::SA
	USSR — RADIO MOSCOW/RP&P, Ryazan'	120/500kW::ME/EAF; 120/500kW::AFRICAN:ME/EAF; 120/500kW::WS:ME/EAF
15445 (con'd)	CHINA (PR) — RADIO BEIJING, Jinhua	500kW::WNA; 500kW::ENA/CA

Legend: ENGLISH, GERMAN / DEUTSCH, FRENCH / FRANÇAIS, PORTUGUESE / PORTUGUÊS, SPANISH / ESPAÑOL, ARABIC / ﻋربي, RUSSIAN / РУССКИЙ, CHINESE / 中文, JAPANESE / 日本語, MULTILINGUAL, OTHER

SUMMER ONLY (J) · WINTER ONLY (D) · JAMMING ∧∧ or / or \ · EARLIEST HEARD ◢ · LATEST HEARD ◣ · † TENTATIVE

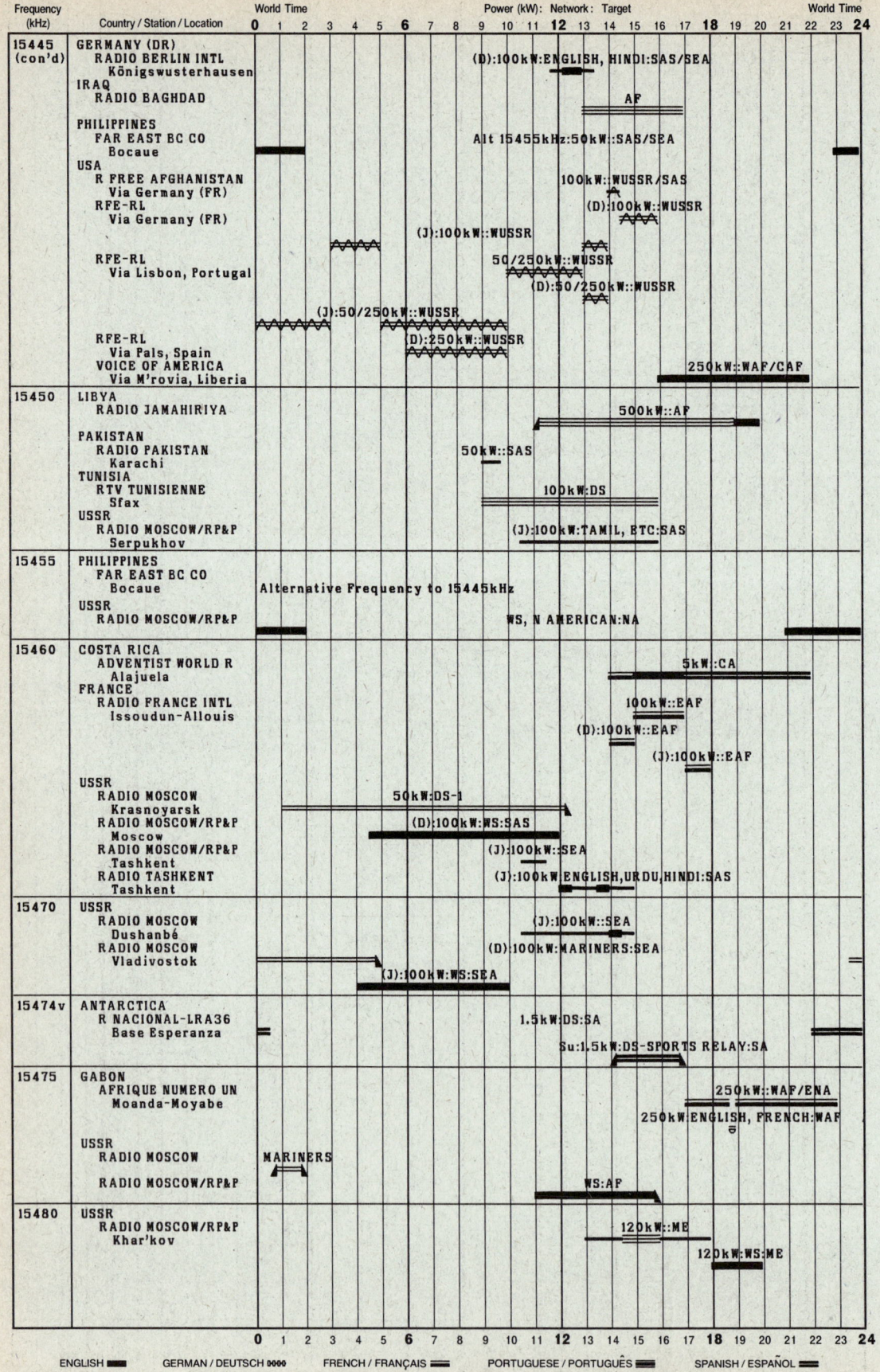

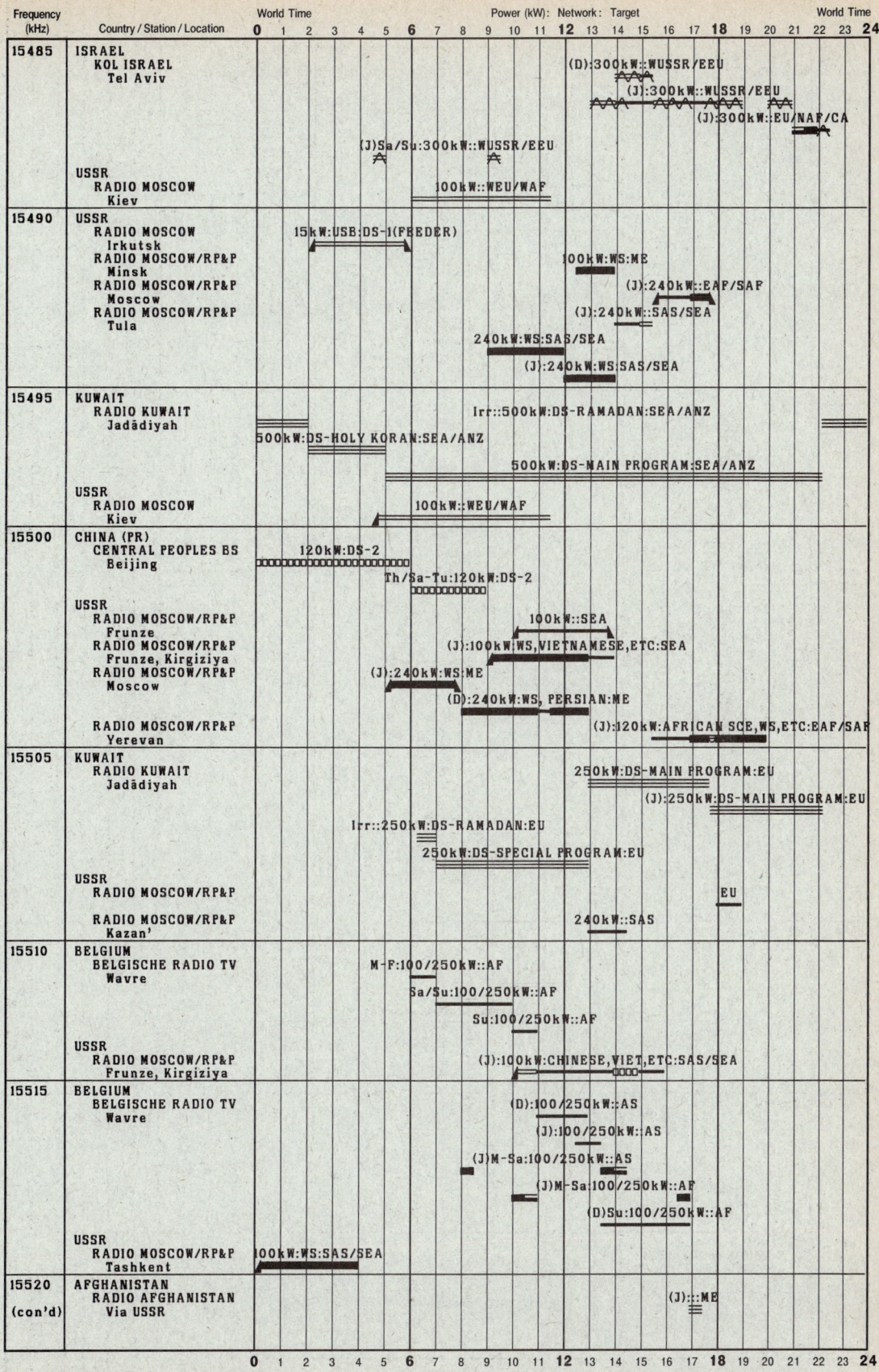

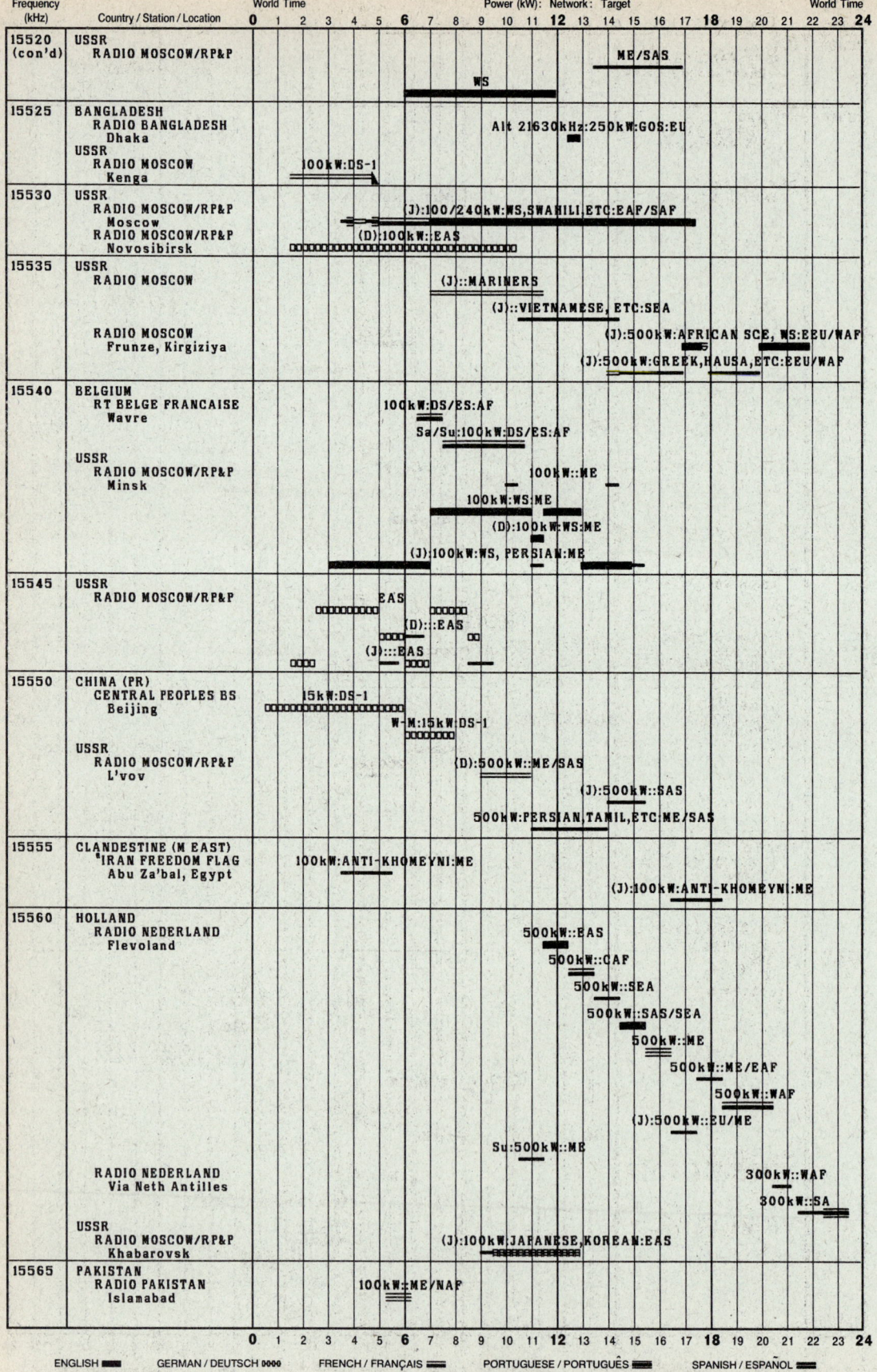

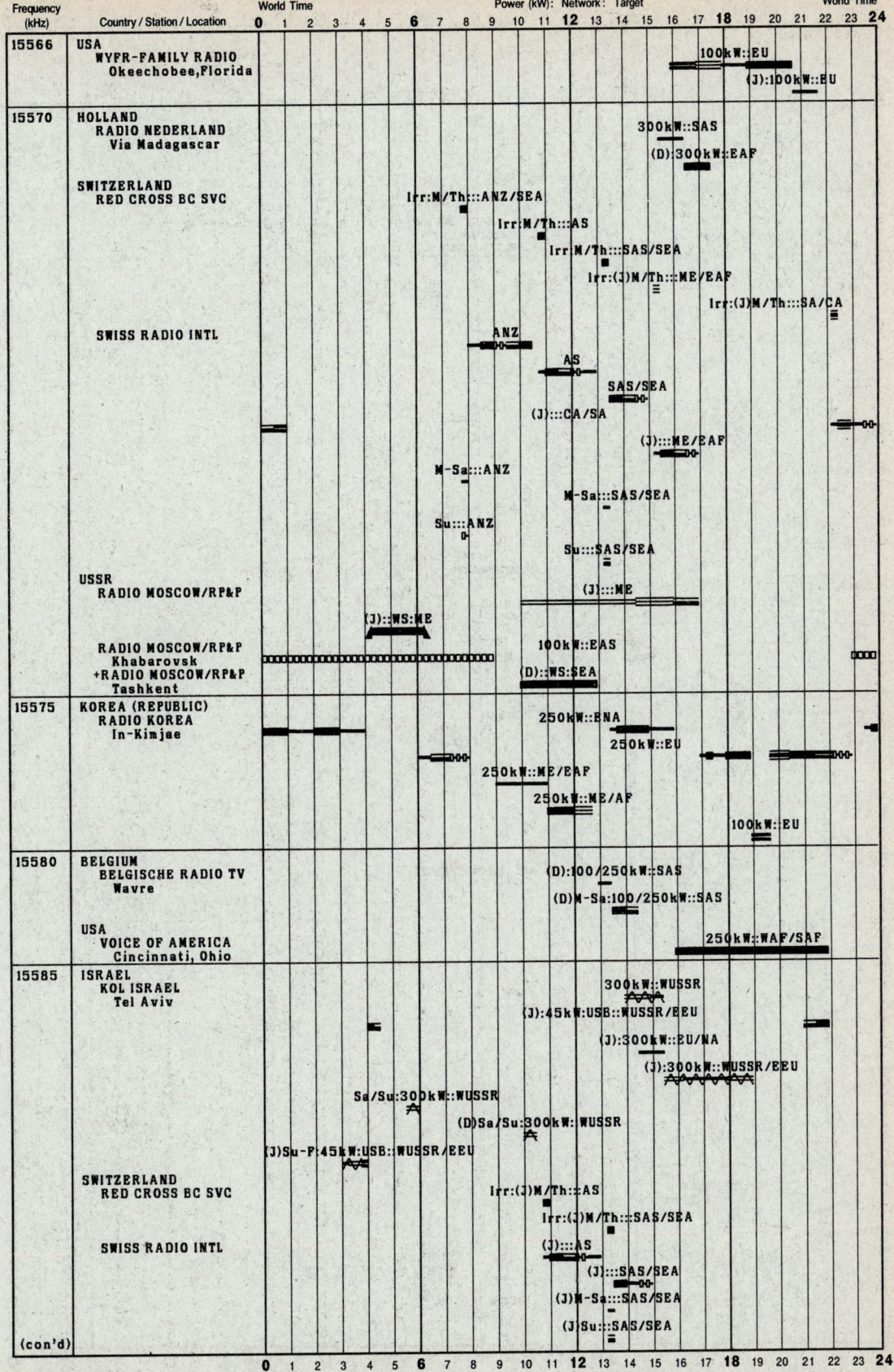

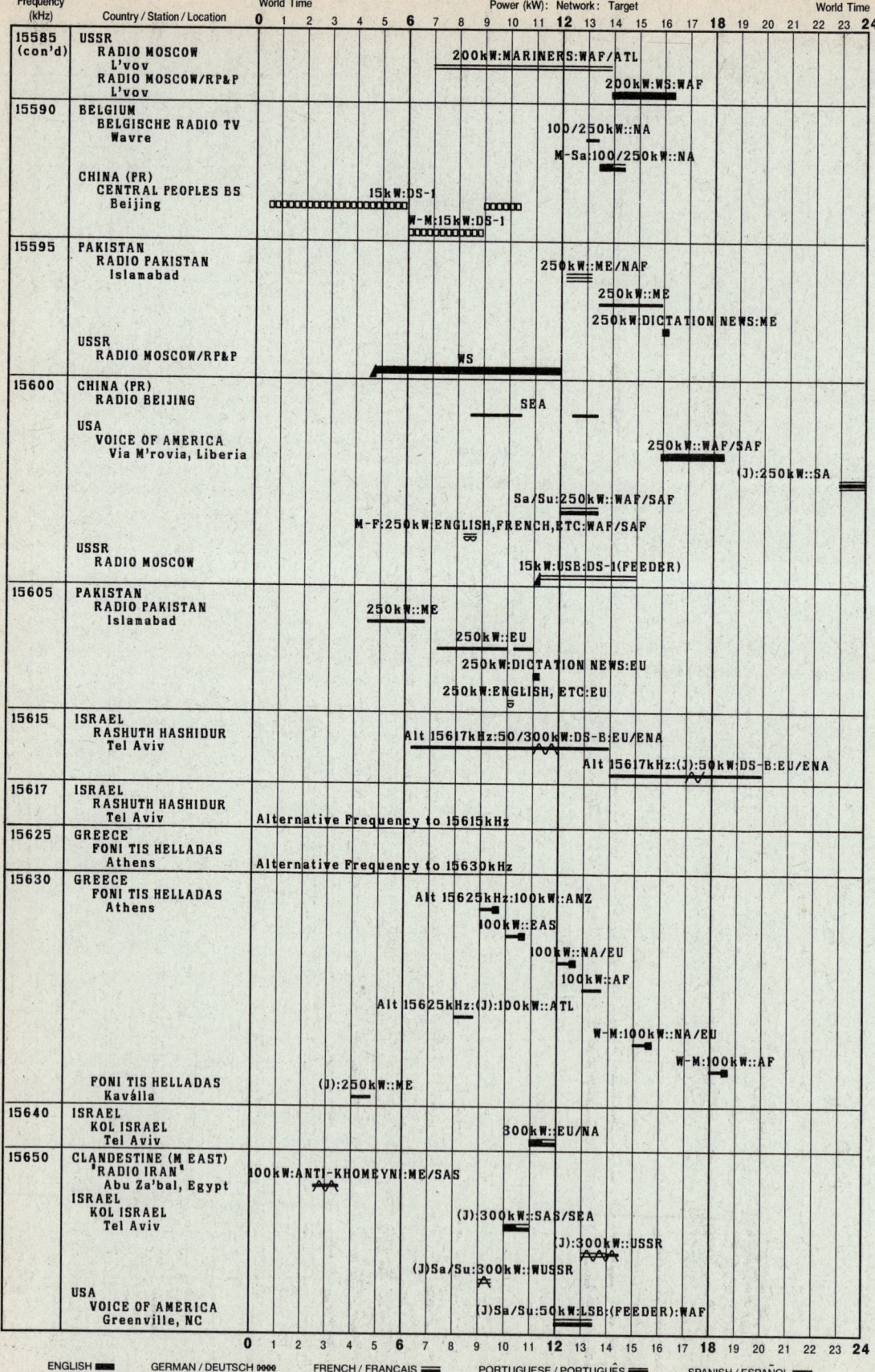

WORLDSCAN

369   15670–17605 kHz

Frequency (kHz)	Country / Station / Location	Power (kW) : Network : Target (World Time 0–24)
15670	CHINA (PR) — CENTRAL PEOPLES BS — Kunming	50kW:DS-MINORITIES
15685	CLANDESTINE (ASIA) — "VOICE OF UNITY" — Abu Za'bal, Egypt	100kW:PRO-AFGHAN REBELS:ME/SAS
15710	CHINA (PR) — CENTRAL PEOPLES BS — Beijing	10kW:TAIWAN-1
15880	CHINA (PR) — CENTRAL PEOPLES BS — Beijing	15kW:TAIWAN-2
16065	USA — RFE-RL — Via Holzkirchen, GFR	10kW:ISL:(FEEDER):WEU 10kW:ISU:(FEEDER):WEU
16230	ALBANIA — RADIO TIRANA — Lushnjë	50kW::EAS
16330	USSR — RADIO MOSCOW — Moscow	Irr::20kW:SU:DS-2(FEEDER) Irr::20kW:ISL:DS-3(FEEDER)
17387	INDIA — ALL INDIA RADIO — Delhi	100kW::EAS
17490	CLANDESTINE (ASIA) — "VOICE OF UNITY" — Abu Za'bal, Egypt	(J):100kW:PRO-AFGHAN REBELS:ME/SAS
17533	CHINA (PR) — RADIO BEIJING	SA
17555	ISRAEL — KOL ISRAEL — Tel Aviv	20kW::EEU/WUSSR (J):20kW::ME (J):20kW::EEU/WUSSR
	RASHUTH HASHIDUR — Tel Aviv	20kW:DS-B:EEU/WUSSR
17565	GREECE — FONI TIS HELLADAS — Athens	100kW::ANZ (D):100kW::AF W-M:100kW::NA/EU
17570	SWITZERLAND — RED CROSS BC SVC SWISS RADIO INTL	Irr:(J)M/Th:::ME/EAF (J):::ME/EAF
17575	HOLLAND — RADIO NEDERLAND — Via Madagascar	300kW::SEA (D):300kW::SEA (D):300kW::SAS
17580	BELGIUM — RT BELGE FRANCAISE — Wavre	250kW:DS/ES:AF Sa/Su:250kW:DS/ES:AF
	USSR — RADIO MOSCOW — Moscow	15kW:LSB:DS-2(FEEDER)
17595	BELGIUM — BELGISCHE RADIO TV — Wavre	Alt 17600kHz:100/250kW::AF (J):100/250kW::AF Alt 17600kHz:(D)M-F:100/250kW::EAS (J)M-Sa:100/250kW::AF (J)Su:100/250kW::AF
	MOROCCO — RTV MAROCAINE — Tangier	50kW:DS
17600	BELGIUM — BELGISCHE RADIO TV — Wavre	Alternative Frequency to 17595kHz
17605 (con'd)	CHINA (PR) — CENTRAL PEOPLES BS — Beijing	15kW:DS-1

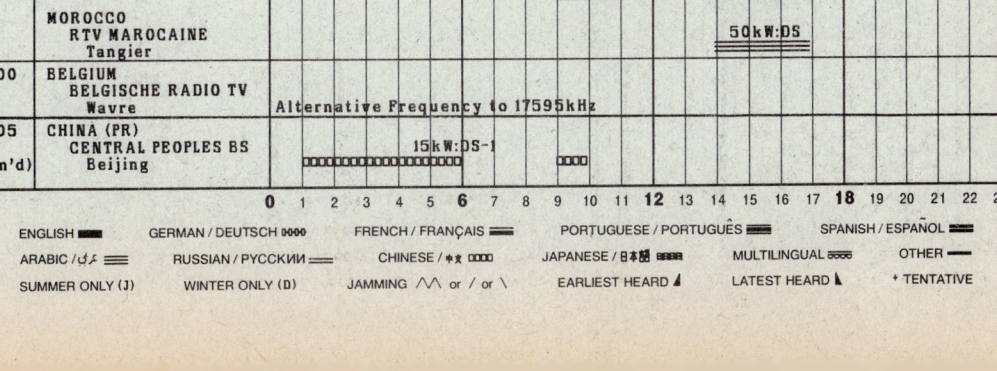

ENGLISH ▬▬   GERMAN / DEUTSCH ∞∞   FRENCH / FRANÇAIS ═══   PORTUGUESE / PORTUGUÊS ▬▬   SPANISH / ESPAÑOL ▬▬
ARABIC / العربية ═══   RUSSIAN / РУССКИЙ ═══   CHINESE / 中文 ∞∞   JAPANESE / 日本語 ▬▬   MULTILINGUAL ∞∞   OTHER ▬▬
SUMMER ONLY (J)   WINTER ONLY (D)   JAMMING ∧∧ or / or \   EARLIEST HEARD ◢   LATEST HEARD ◣   + TENTATIVE

17605–17660v kHz

Frequency (kHz)	Country / Station / Location	Schedule
17605 (con'd)	**CHINA (PR)** CENTRAL PEOPLES BS, Beijing	M-W:15kW:DS-1 (06–09)
	HOLLAND RADIO NEDERLAND, Flevoland	500kW::ME (09–11); 500kW::AF (10–12); (J):500kW::SEA (11–13); Su:500kW::ME (09–11); (D):300kW::SA (10–12)
	RADIO NEDERLAND Via Neth Antilles	(00–01); 300kW::NAF (16–19); 300kW::WAF (18–21); 300kW::SA (19–22); Su:500kW::SA (15–17); Su:300kW::NA (16–18); (23–24)
17610	**TUNISIA** RTV TUNISIENNE, Sfax	100kW:DS (08–13)
17620	**FRANCE** RADIO FRANCE INTL, Issoudun-Allouis	(J):100/500kW::AF (15–18); Su:100/500kW::AF (08–11); 100/500kW:FRENCH, ENGLISH:AF (08–13); M-Sa:100/500kW:MEDIAS FRANCE:AF (07–13)
	ISRAEL KOL ISRAEL, Tel Aviv	(J):300kW::SEA/ANZ (04–07)
	RASHUTH HASHIDUR, Tel Aviv	(J):300kW:DS-E:EU/NA (05–08)
17630	**IRAQ** RADIO BAGHDAD	(06–15)
	ISRAEL KOL ISRAEL, Tel Aviv	45kW:USB::WUSSR/EEU (11–14); (J):45kW:USB::WUSSR/EEU (14–19); (D):300kW::SEA/ANZ (05–08); Sa/Su:45kW:USB::EEU/WUSSR (09–12); (D)Sa/Su:300kW::USSR (06–08); (J)Sa/Su:40kW:USB::WUSSR/EEU (05–07)
17635	**CHINA (PR)** CENTRAL PEOPLES BS, Kunming	50kW:DS-MINORITIES (06–10)
17640	**PAKISTAN** RADIO PAKISTAN, Karachi	50kW::SEA (08–09)
	USA VOICE OF AMERICA, Greenville, NC	500kW::WAF/SAF (17–21)
	WCSN, Olamon, Maine	(J):500kW::CAF/SAF (09–13); (J)Sa-M:500kW::CAF/SAF (10–12); (J)Tu-F:500kW::CAF/SAF (11–13)
	WYFR-FAMILY RADIO, Okeechobee, Florida	(D):100kW::EU (16–18); (J):100kW::EU (17–19)
17650	**CHINA (PR)** RADIO BEIJING, Kunming	50kW::SA (02–05)
17653v	**BANGLADESH** RADIO BANGLADESH, Dhaka	(J):250kW:GOS:EU (12–14)
17655	**AFGHANISTAN** RADIO AFGHANISTAN, Via USSR	(J):::ME/SAS (07–11)
17660v	**PAKISTAN** RADIO PAKISTAN, Islamabad	250kW::ME (09–11); 250kW::EU (10–13); 100kW:DICTATION NEWS:SEA (03–07)
(con'd)		

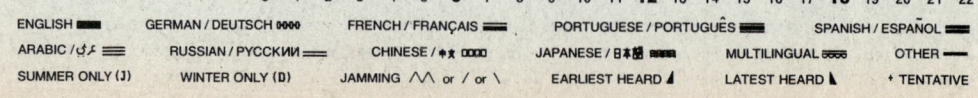

ENGLISH ▬ GERMAN / DEUTSCH ◊◊◊◊ FRENCH / FRANÇAIS ≡ PORTUGUESE / PORTUGUÊS ≡ SPANISH / ESPAÑOL ≡
ARABIC / ﻋﺮﺑﻲ ≡ RUSSIAN / РУССКИЙ ≡ CHINESE / 中文 ◊◊◊◊ JAPANESE / 日本語 ▬▬ MULTILINGUAL ◊◊◊◊ OTHER ▬
SUMMER ONLY (J) WINTER ONLY (D) JAMMING ∧∧ or / or \ EARLIEST HEARD ▲ LATEST HEARD ▼ + TENTATIVE

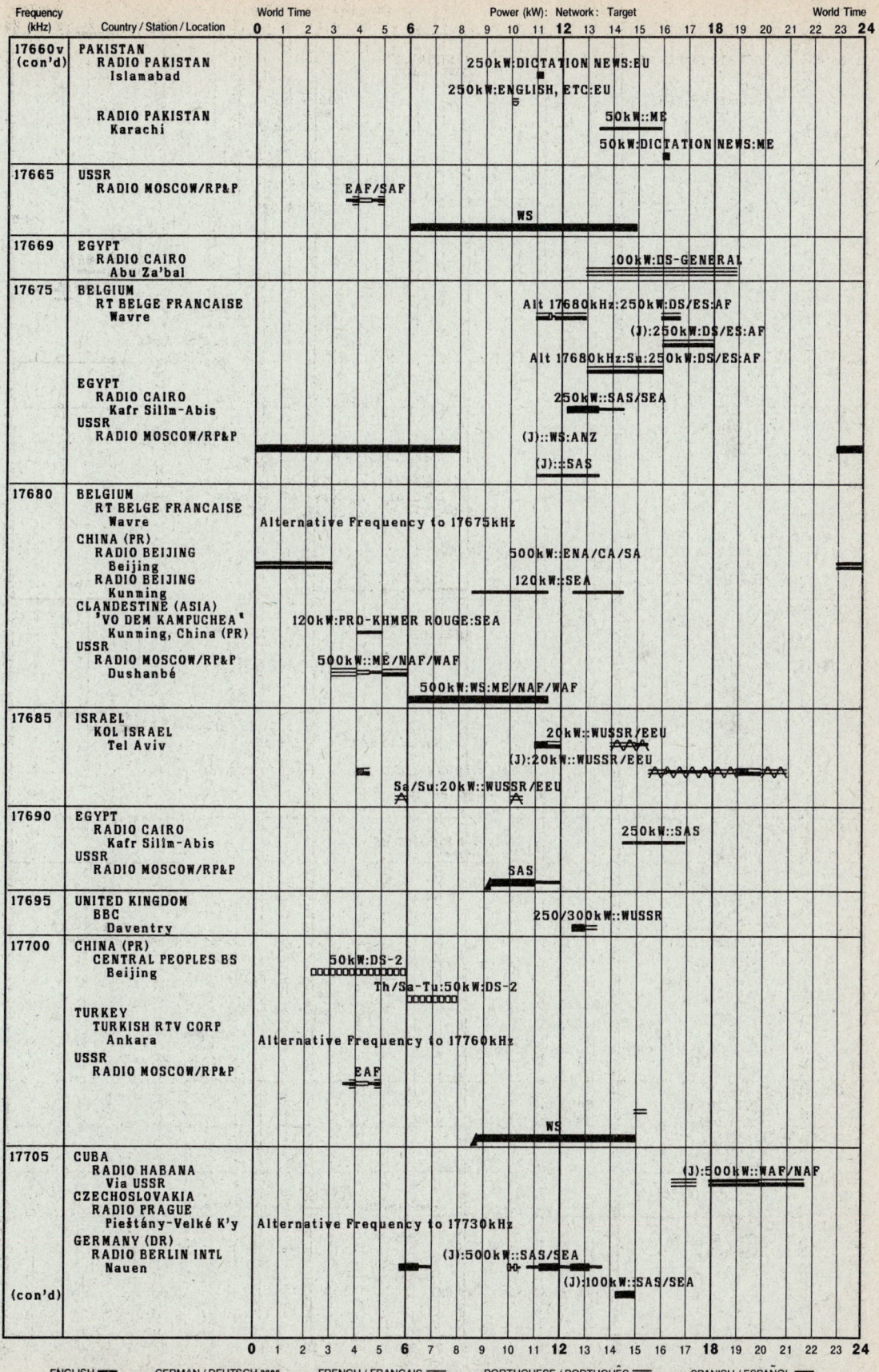

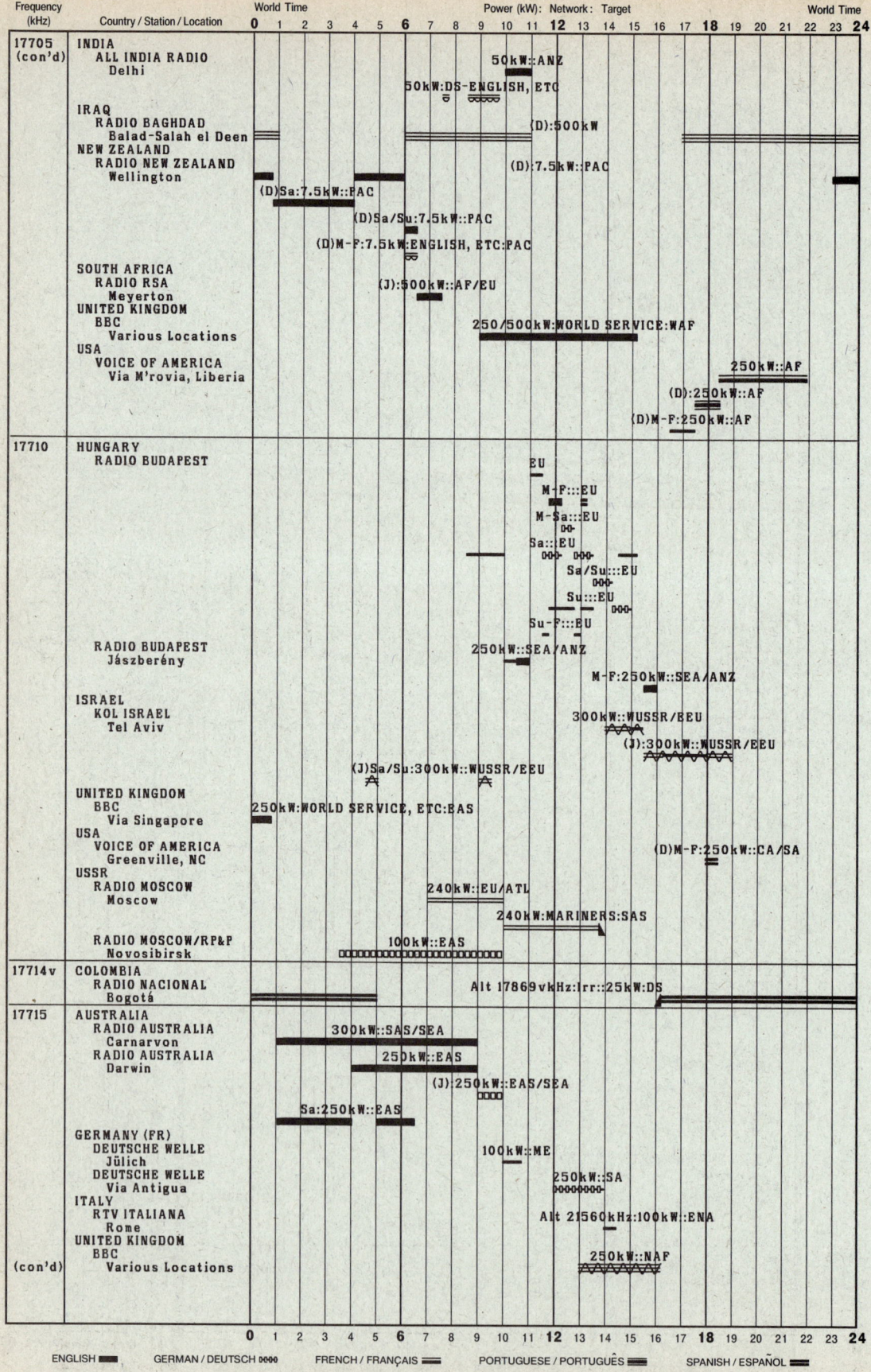

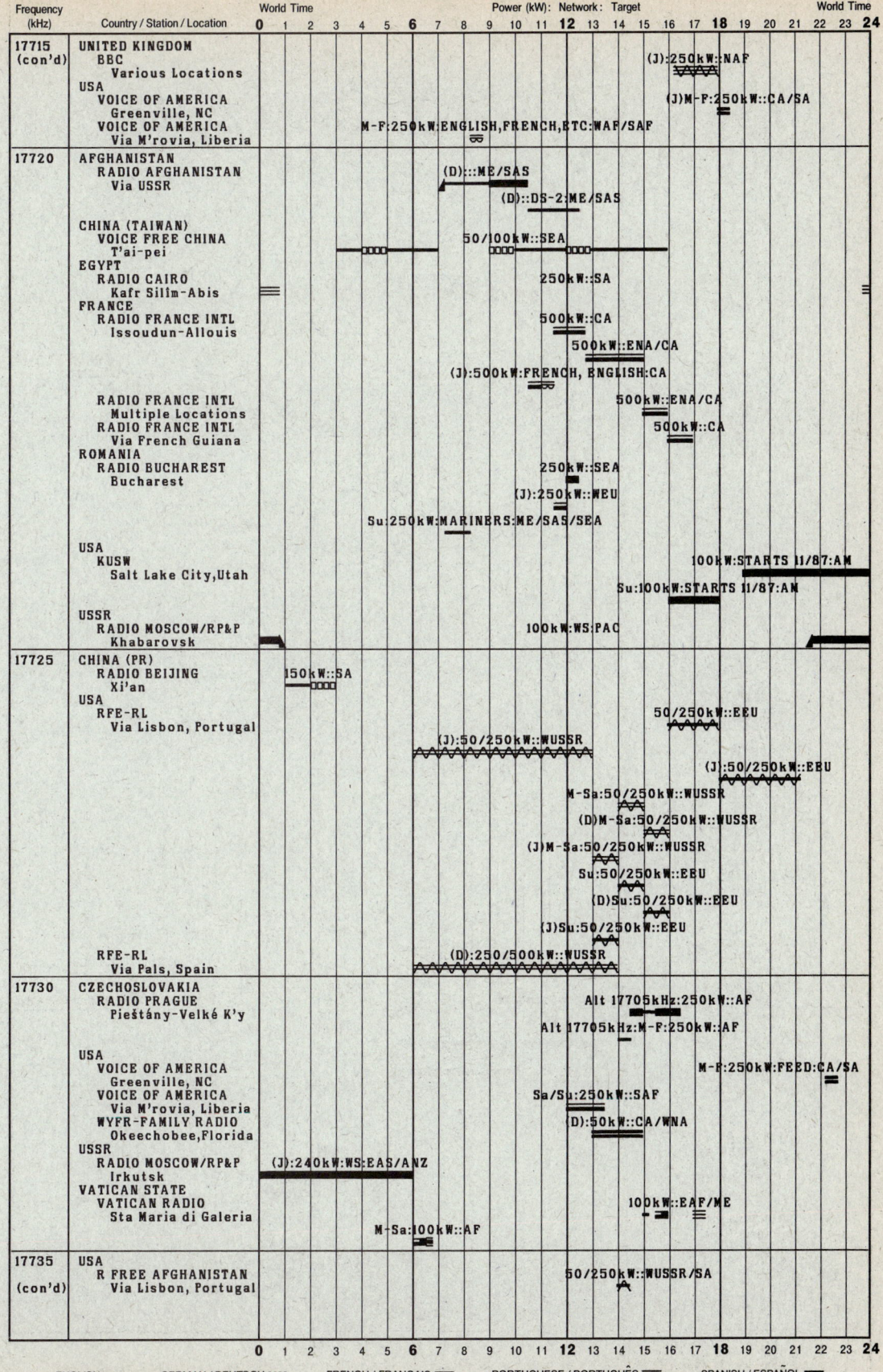

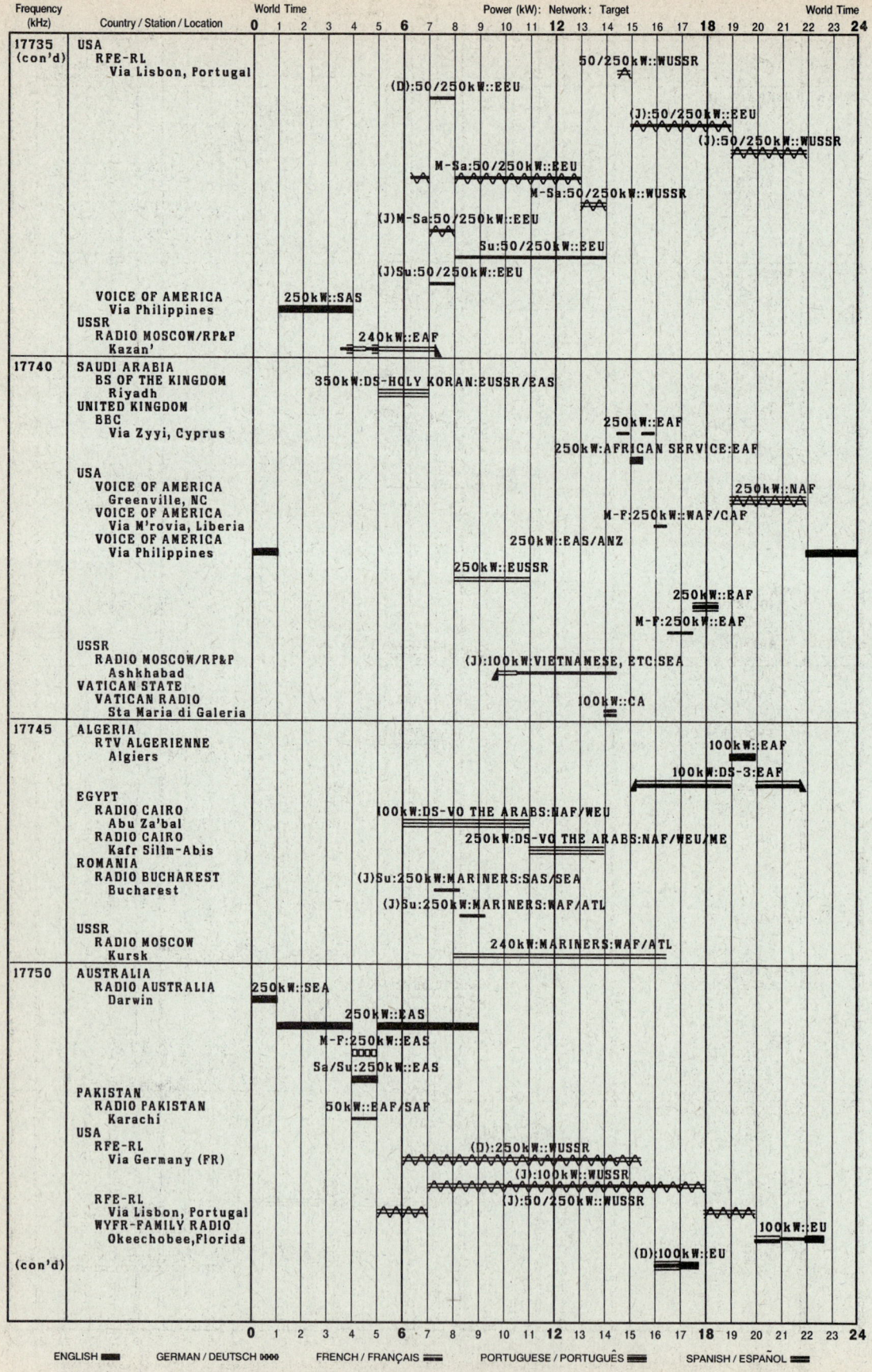

WORLDSCAN — 17750–17775 kHz

Frequency (kHz)	Country / Station / Location	Schedule (World Time 0–24) — Power (kW): Network: Target
17750 (con'd)	**USA** — WYFR-FAMILY RADIO, Okeechobee, Florida	(J):100kW::EU (approx 20–23)
17755	**GERMANY (DR)** — RADIO BERLIN INTL, Königswusterhausen	50kW::ME/NAF (≈12–14); (J):50kW::ME/NAF (≈13–14)
	JAPAN — RADIO JAPAN/NHK, Tōkyō-Yamata	300kW:GENERAL:WNA/CA (≈19–21)
	SURINAME — R SURINAME INTL, Via Brasília, Brazil	Alt 17758.2kHz:M-F:250kW:EU ; Alt 17758.2kHz:M-F:250kW:ENGLISH, ETC
17758.2	**SURINAME** — R SURINAME INTL, Via Brasília, Brazil	Alternative Frequency to 17755 kHz
17760	**NORWAY** — RADIO NORWAY INTL, Kvitsøy	Alternative Frequency to 17770 kHz
	TURKEY — TURKISH RTV CORP, Ankara	Alt 17700kHz:(J):500kW::NA
	USA — RFE-RL, Via Germany (FR)	100kW::WUSSR (≈5–12); (D):100kW::WUSSR; (J):100kW::WUSSR
17765	**GERMANY (FR)** — DEUTSCHE WELLE, Jülich	100kW::CAF/EAF ; 100kW::AF ; 100kW::EAF/SAF
	USA — VOICE OF AMERICA, Dixon, California	M-F:250kW::CA/SA
	USSR — RADIO MOSCOW/RP&P, Tula	500kW::SEA
17770	**EGYPT** — RADIO CAIRO, Kafr Sillm-Abis	Alternative Frequency to 21465 kHz
	IRAQ — RADIO BAGHDAD, Balad-Salah el Deen	500kW::AF ; (D):500kW::AF
	NORWAY — RADIO NORWAY INTL, Kvitsøy	(J):500kW::WAF/SA ; Alt 17760kHz:M-Sa:500kW::ME/EAF ; Alt 17760kHz:Su:500kW::ME/EAF ; Alt 17760kHz:Tu-Su:500kW::ME/EAF ; Alt 17760kHz:M:500kW:SPANISH, ETC:ME/EAF
	SPAIN — R EXTERIOR ESPANA, Noblejas	350kW::ME
	SWEDEN — SVERIGES RIKSRADIO, Varberg	100kW:USB:DS-1:SEA/ANZ ; Sa/Su:100kW:USB:DS-1:ME/SAS
	USA — RFE-RL, Via Lisbon, Portugal ; RFE-RL, Via Pals, Spain	(D):50/250kW::WUSSR ; (J):250kW::WUSSR
17775	**NORTHERN MARIANA IS** — KYOI, Saipan Island	100kW:ENGLISH, JAPANESE:EAS ; (D):100kW:ENGLISH, JAPANESE:EAS
	UNITED ARAB EMIRATES — UAE RADIO, Dubai	300kW::ME/NAF/EU ; 300kW::EU
(con'd)	**USA** — KVOH-VOICE OF HOPE, Rancho Simi, Ca	50kW::CA

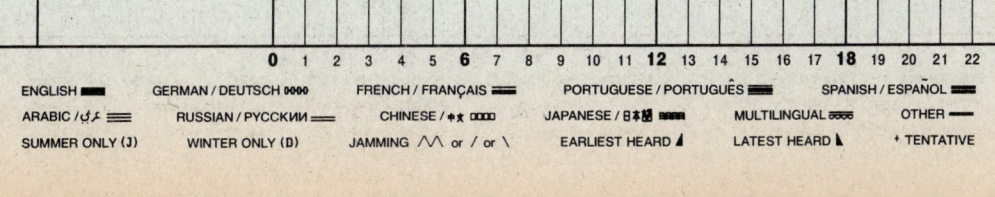

ENGLISH ▬ GERMAN / DEUTSCH ◊◊◊◊ FRENCH / FRANÇAIS ≡ PORTUGUESE / PORTUGUÊS ≣ SPANISH / ESPAÑOL ≋
ARABIC / عربي ≡ RUSSIAN / РУССКИЙ ═ CHINESE / 中文 □□□□ JAPANESE / 日本語 ▦▦▦ MULTILINGUAL ⚌⚌⚌ OTHER ─
SUMMER ONLY (J) WINTER ONLY (D) JAMMING ∧∧∧ or / or \ EARLIEST HEARD ◀ LATEST HEARD ▶ ✦ TENTATIVE

Frequency (kHz)	Country / Station / Location	Transmission details
17775 (con'd)	**USA** VOICE OF AMERICA, Greenville, NC	M-F:250kW:FEED:CA/SA (22–23)
	USSR RADIO MOSCOW, Khabarovsk	(J):240kW:DS-1:PAC/ANZ (6–9)
	RADIO MOSCOW/RP&P, Frunze, Kirgiziya	(J):100kW:WS, HINDI, ETC:SAS/SEA (9–15); (J):500kW:WS, PORT, SWAHILI:EAF (4–9)
	RADIO TIKHIY OKEAN, Khabarovsk	(J):240kW:MARINERS:PAC/ANZ (5–7); (J)Sa:240kW:MARINERS:PAC/ANZ; (J)Su-F:240kW:MARINERS:PAC/ANZ
17780	**GERMANY (FR)** DEUTSCHE WELLE, Jülich	(J):100kW::ME/SAS (6–8)
	DEUTSCHE WELLE, Türkheim-Wertach'l	500kW::SEA/ANZ (7–8)
	INDIA ALL INDIA RADIO, Aligarh	250kW::SEA (9–10)
	ITALY RTV ITALIANA, Rome	100kW::SEA/ANZ (6–7); 100kW::EAF (16–17); 100kW::CA (17–18)
	NORTHERN MARIANA IS KYOI, Saipan Island	(J):100kW:ENGLISH, JAPANESE:EAS (4–7)
	SOUTH AFRICA RADIO RSA, Meyerton	(D):500kW::AF/EU (6–8); (J):500kW::AF/EU (11–15); Su:500kW::AF/EU (7–9)
	USA VOICE OF AMERICA, Via Tangier, Morocco	(D):35kW::WUSSR (13–15)
	VOICE OF AMERICA, Via Woofferton, UK	(J):300kW::WUSSR (13–15)
17785	**EGYPT** RADIO CAIRO, Kafr Silim-Abis	Alt 11875kHz:250kW::SAF (15–17)
	FRANCE RADIO FRANCE INTL, Issoudun-Allouis	M-Sa:500kW:MEDIAS FRANCE:AF (7–10)
	INDIA ALL INDIA RADIO, Aligarh	250kW::ME (4–6)
	JAPAN RADIO JAPAN/NHK, Via Moyabi, Gabon	250kW:GENERAL:EU (13–15)
	SWEDEN RADIO SWEDEN INTL, Hörby	(J):350kW::SAS/SEA (12–13)
	USA VOICE OF AMERICA, Greenville, NC	250kW::WAF/SAF (16–22)
	VOICE OF AMERICA, Via Philippines	250kW::SAS (1–3)
	WYFR-FAMILY RADIO, Okeechobee, Florida	(D):100kW::CA (12–14)
	USSR RADIO MOSCOW/RP&P, Ivano-Frankovsk	240kW::SEA (8–10)
17790	**ECUADOR** HCJB-VO THE ANDES, Quito	500kW::EU (20–22); M-F:500kW::EU; Sa/Su:500kW::EU; M-F:500kW::EU
	RADIO NACIONAL, Quito	
	ROMANIA RADIO BUCHAREST, Bucharest	250kW::SEA/ANZ (5–6); Su:250kW::MARINERS:ME (6–7); Su:250kW::MARINERS:SAF (7–8)
(con'd)	**SOUTH AFRICA** RADIO RSA, Meyerton	(J):250kW:DUTCH, GERMAN:EU (15–17)

ENGLISH ▬ GERMAN / DEUTSCH ▭ FRENCH / FRANÇAIS ▬ PORTUGUESE / PORTUGUÊS ▬ SPANISH / ESPAÑOL ▬

ARABIC / ۞ ▬ RUSSIAN / РУССКИЙ ▬ CHINESE / 中文 ▭ JAPANESE / 日本語 ▬ MULTILINGUAL ▭ OTHER ▬

SUMMER ONLY (J)   WINTER ONLY (D)   JAMMING ∧∧ or / or \   EARLIEST HEARD ◢   LATEST HEARD ◣   + TENTATIVE

WORLDSCAN — 17790–17810 kHz

Frequency (kHz)	Country / Station / Location	Schedule (World Time 0–24) — Power (kW) :: Network : Target
17790 (con'd)	**UNITED KINGDOM** BBC, Holywell-Rampisham	Su:500kW::ME
	BBC, Via Ascension	250kW::WORLD SERVICE:SAF
17795	**AUSTRALIA** RADIO AUSTRALIA, Shepparton	(0–6, 22–24); 100kW::PAC/NA
	FRANCE RADIO FRANCE INTL, Issoudun-Allouis	100kW::EAF; Su:100kW::EAF; M-Sa:100kW::MEDIAS FRANCE:AF
	GERMANY (FR) DEUTSCHE WELLE, Türkheim-Wertach'l	(J):500kW::USSR
	ITALY RTV ITALIANA, Rome	Su:100kW::DS-1/DS-2:CA
	SEYCHELLES FAR EAST BC ASS'N, North Pt, Mahé Is	Su:80kW::SAS
	USSR RADIO MOSCOW/RP&P, Serpukhov	(J):100kW:VIETNAMESE, ETC:SEA
17800	**EGYPT** RADIO CAIRO, Kafr Silim-Abis	250kW::CAF/SAF
	FRANCE RADIO FRANCE INTL, Issoudun-Allouis	100kW::EAF; (D):100kW::EAF
	GERMANY (FR) DEUTSCHE WELLE, Via Kigali, Rwanda	250kW::SEA/ANZ; 250kW::WAF
	ITALY RTV ITALIANA, Rome	Alternative Frequency to 15355 kHz
	USA VOICE OF AMERICA, Cincinnati, Ohio	250kW::AF
17805	**CHINA (TAIWAN)** VOICE FREE CHINA, Via Okeechobee, USA	(J):100kW::SA
	INDIA ALL INDIA RADIO, Aligarh	250kW::EAF
	ROMANIA RADIO BUCHAREST, Bucharest	250kW::SEA/ANZ
	USA RFE-RL, Via Germany (FR)	(J):100kW::WUSSR
	RFE-RL, Via Lisbon, Portugal	50/250kW::EEU; (J):50/250kW::EEU; (J):50/250kW::WUSSR
	WYFR-FAMILY RADIO, Okeechobee, Florida	(J):100kW::SA
	USSR RADIO MOSCOW/RP&P, Tbilisi	500kW::SAS
17810	**GERMANY (FR)** DEUTSCHE WELLE, Via Antigua	125kW::AM
	DEUTSCHE WELLE, Via Cyclops, Malta	(J):250kW::ME/SAS
	JAPAN RADIO JAPAN/NHK, Tōkyō-Yamata	300kW:GENERAL:SEA/SA
	PAKISTAN RADIO PAKISTAN, Karachi	50kW::SAS
	UNITED KINGDOM BBC, Skelton, Cumbria	250kW::NAF/WAF
	BBC, Via Ascension	(D):250kW::SA; (J):250kW::SA
(con'd)	**USA** VOICE OF AMERICA, Cincinnati, Ohio	M-F:250kW::CA/SA

Legend: ENGLISH ▬ · GERMAN/DEUTSCH ◊◊◊◊ · FRENCH/FRANÇAIS ═ · PORTUGUESE/PORTUGUÊS ≡ · SPANISH/ESPAÑOL ≣ · ARABIC/عربي · RUSSIAN/РУССКИЙ · CHINESE/中文 · JAPANESE/日本語 · MULTILINGUAL · OTHER · SUMMER ONLY (J) · WINTER ONLY (D) · JAMMING ⋀⋀ or / or \ · EARLIEST HEARD ◢ · LATEST HEARD ◣ · † TENTATIVE

17810–17830 kHz

Frequency (kHz)	Country / Station / Location	Broadcast Schedule
17810 (con'd)	**USA** — VOICE OF AMERICA, Via Ascension	(D):250kW::SA (0900–1100)
	VOICE OF AMERICA, Via Ascension	(J):250kW::SA (1000–1200)
	VOICE OF AMERICA, Via M'rovia, Liberia	(J):250kW::SAF (1600–1800)
	VOICE OF AMERICA, Via Philippines	50kW::SEA (1800–2000)
17815	**BRAZIL** — RADIO CULTURA, São Paulo	7.5kW:DS (continuous)
	EGYPT — RADIO CAIRO, Kafr Silim-Abis	250kW::ANZ
	GERMANY (FR) — DEUTSCHE WELLE, Türkheim-Wertach'l	(J):500kW::ME/SAS
	DEUTSCHE WELLE, Via Cyclops, Malta	(D):250kW::SEA
	ISRAEL — KOL ISRAEL, Tel Aviv	(J):45kW:USB::WEU/ENA
	KOL ISRAEL, Tel Aviv	(J):100/300kW::WUSSR/EEU
	KOL ISRAEL, Tel Aviv	(D)Sa/Su:300kW::WUSSR/EEU
	MOROCCO — RTV MAROCAINE, Tangier	50kW:DS:NAF/ME
	USSR — RADIO MOSCOW/RP&P, Frunze, Kirgiziya	(J):100kW::SEA
17820	**CANADA** — RADIO CANADA INTL, Sackville, NB	250kW::WAF/SAF
		(D):250kW::EU/WUSSR
		(J):250kW::EU/WUSSR
		(J):250kW::WEU/NAF
		M-F:250kW::WAF/SAF
		(J)M-F:250kW::WEU/NAF
		Sa/Su:250kW::WAF/SAF
		(J)Sa/Su:250kW::WEU/NAF
	USA — VOICE OF AMERICA, Via Philippines	(J):100kW::EAS
	USSR — RADIO MOSCOW/RP&P, Kiev	500kW:WS:EAF/SAF
	RADIO MOSCOW/RP&P, Novosibirsk	250kW::EAS (with jamming)
17825	**BULGARIA** — RADIO SOFIA, Rebrovo-Sofia	250kW::ANZ
	GERMANY (FR) — DEUTSCHE WELLE, Jülich	(J):100kW::EAS
	DEUTSCHE WELLE, Türkheim-Wertach'l	500kW::SEA
	DEUTSCHE WELLE, Türkheim-Wertach'l	(J):500kW::EAF/SAF
	DEUTSCHE WELLE, Via Sri Lanka	250/300kW::ME/SAS
	USA — RFE-RL, Via Pals, Spain	(J):250kW::WUSSR
	VOICE OF AMERICA, Via Kaválla, Greece	(J):250kW::ME/SAS
17830	**INDIA** — ALL INDIA RADIO, Delhi	100kW::SEA
	SWITZERLAND — RED CROSS BC SVC, Schwarzenburg	Irr:(J)M/Th:150kW::SEA/ANZ
	RED CROSS BC SVC, Schwarzenburg	Irr:(J)M/Th:150kW::SAS/SEA
	RED CROSS BC SVC, Sottens	Irr:(J)M/Th:500kW::ME/EAF
	RED CROSS BC SVC, Sottens	Irr:(J)M/Th:500kW:FRENCH, ENGLISH:ME/EA
	SWISS RADIO INTL, Schwarzenburg	(J):150kW::SEA
	SWISS RADIO INTL, Schwarzenburg	(J):150kW::SAS/SEA
	SWISS RADIO INTL, Schwarzenburg	(J)M-Sa:150kW::SAS/SEA
(con'd)		

Legend: ENGLISH ▬ GERMAN/DEUTSCH ◊◊◊◊ FRENCH/FRANÇAIS ═ PORTUGUESE/PORTUGUÊS ▬ SPANISH/ESPAÑOL ▬ ARABIC/عربي ▬ RUSSIAN/РУССКИЙ ▬ CHINESE/中文 □□□□ JAPANESE/日本語 ▭▭▭ MULTILINGUAL ▭▭▭ OTHER ▬ SUMMER ONLY (J) WINTER ONLY (D) JAMMING ∧∧∧ or / or \ EARLIEST HEARD ◢ LATEST HEARD ◣ † TENTATIVE

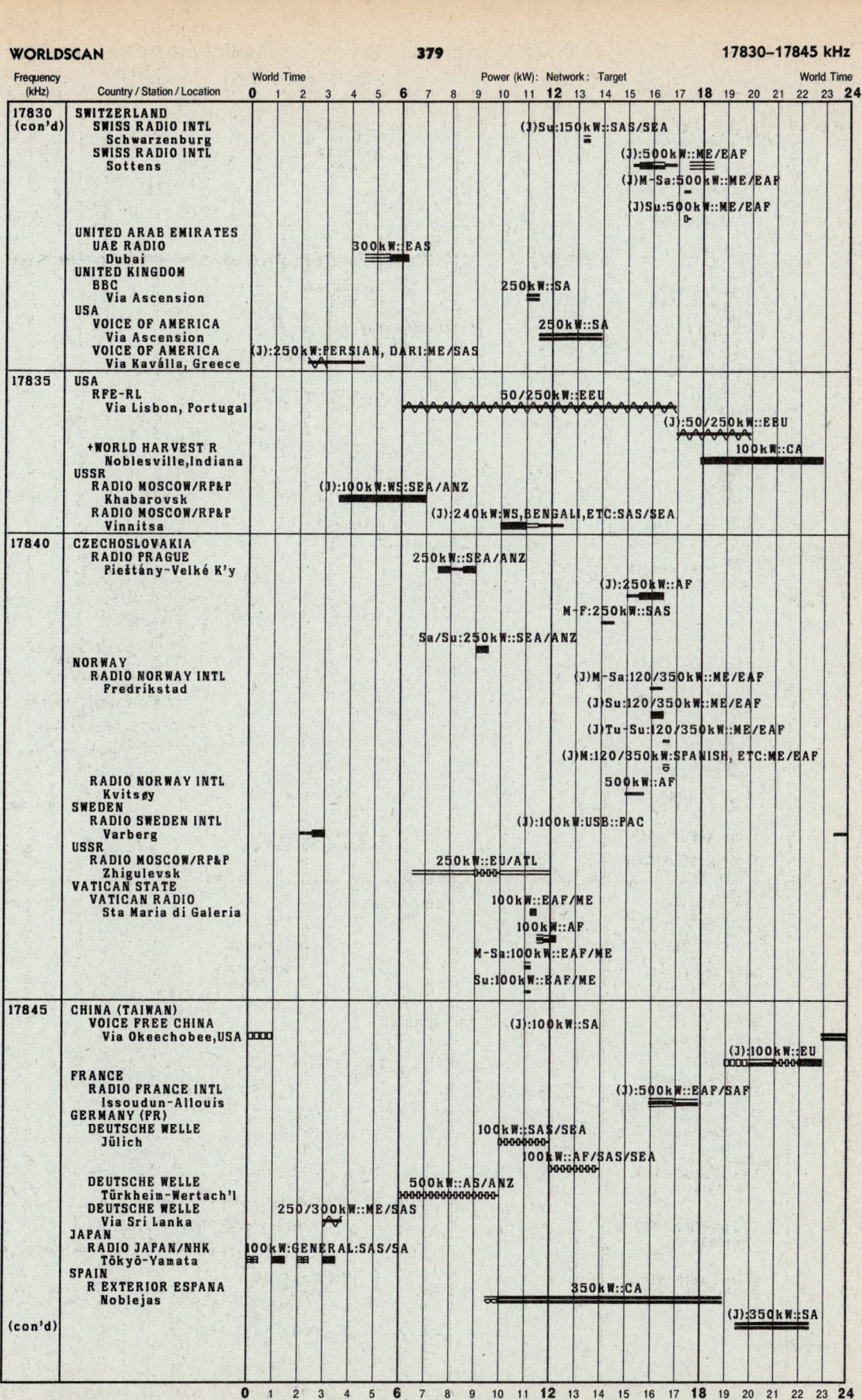

17845–17865 kHz

Frequency (kHz)	Country / Station / Location	Schedule
17845 (con'd)	USA — RFE-RL, Via Pals, Spain	(D):250kW::WUSSR
	WYFR-FAMILY RADIO, Okeechobee, Florida	100kW::EU; (D):100kW::WAF; (J):100kW::SA; (J):100kW::EU
17850	FRANCE — RADIO FRANCE INTL, Issoudun-Allouis	Su:500kW::AF; 500kW:FRENCH, ENGLISH:AF; M-Sa:500kW:MEDIAS FRANCE:AF
	KUWAIT — RADIO KUWAIT, Jadādiyah	250kW:DS-MAIN PROGRAM:CAF
	SOUTH AFRICA — RADIO RSA, Meyerton	(J):500kW::EU
	SRI LANKA — SRI LANKA BC CORP, Colombo-Ekala	35kW::SEA; M:35kW::SEA; M/Tu/Th/F:35kW::SEA; Sa:35kW::SEA; Su/W:35kW::SEA; Tu-Su:35kW::SEA
	USA — VOICE OF AMERICA, Greenville, NC	(J):500kW::EAF
	USSR — RADIO MOSCOW/RP&P, Khabarovsk	(J):100kW:WS:SEA/ANZ
17855	GUAM — +ADVENTIST WORLD R, Agat	100kW::EAS
	INDIA — ALL INDIA RADIO, Aligarh	250kW::SEA
	UNITED KINGDOM — BBC, Various Locations	Su:300/500kW::WUSSR
	USA — VOICE OF AMERICA, Via Kaválla, Greece	(D):250kW::WUSSR
	VOICE OF AMERICA, Via Tangier, Morocco	35/100kW::WUSSR; (J):100kW::WUSSR; (J):35kW::EEU
	VOICE OF AMERICA, Via Woofferton, UK	300kW::WUSSR; (J):300kW::WUSSR
17860	FRANCE — RADIO FRANCE INTL, Via French Guiana	500kW:AMERICAIN:CA
	GERMANY (FR) — DEUTSCHE WELLE, Via Kigali, Rwanda	250kW::WAF/CA
	USA — VOICE OF AMERICA, Via M'rovia, Liberia	(J)M-F:250kW::CAF/EAF
	USSR — RADIO MOSCOW/RP&P, Chita	(J):500kW:WS:ANZ
17865	GUAM — ADVENTIST WORLD R, Agat	100kW::EAS
	USA — RFE-RL, Via Lisbon, Portugal	(J)Sa/Su:50/250kW::EEU
	RFE-RL, Via Pals, Spain	(D):250kW::WUSSR
	VOICE OF AMERICA, Via Kaválla, Greece	250kW::ME/SAS; (D):250kW::WUSSR/ME; (J):250kW::WUSSR/ME
(con'd)		

Languages: ENGLISH, GERMAN / DEUTSCH, FRENCH / FRANÇAIS, PORTUGUESE / PORTUGUÊS, SPANISH / ESPAÑOL, ARABIC, RUSSIAN / РУССКИЙ, CHINESE / 中文, JAPANESE / 日本語, MULTILINGUAL, OTHER

SUMMER ONLY (J)   WINTER ONLY (D)   JAMMING ∧∧ or / or \   EARLIEST HEARD ◢   LATEST HEARD ◣   + TENTATIVE

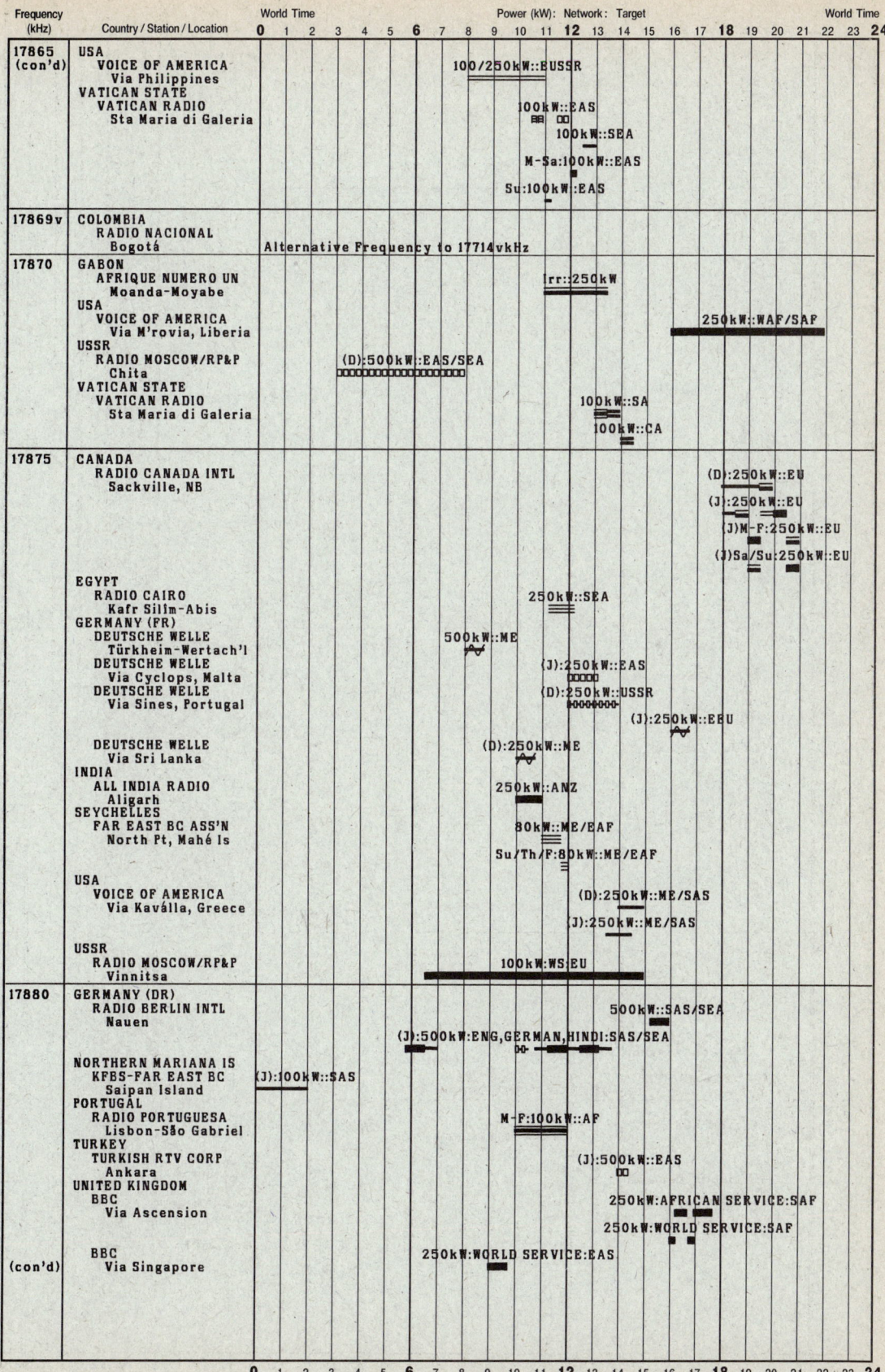

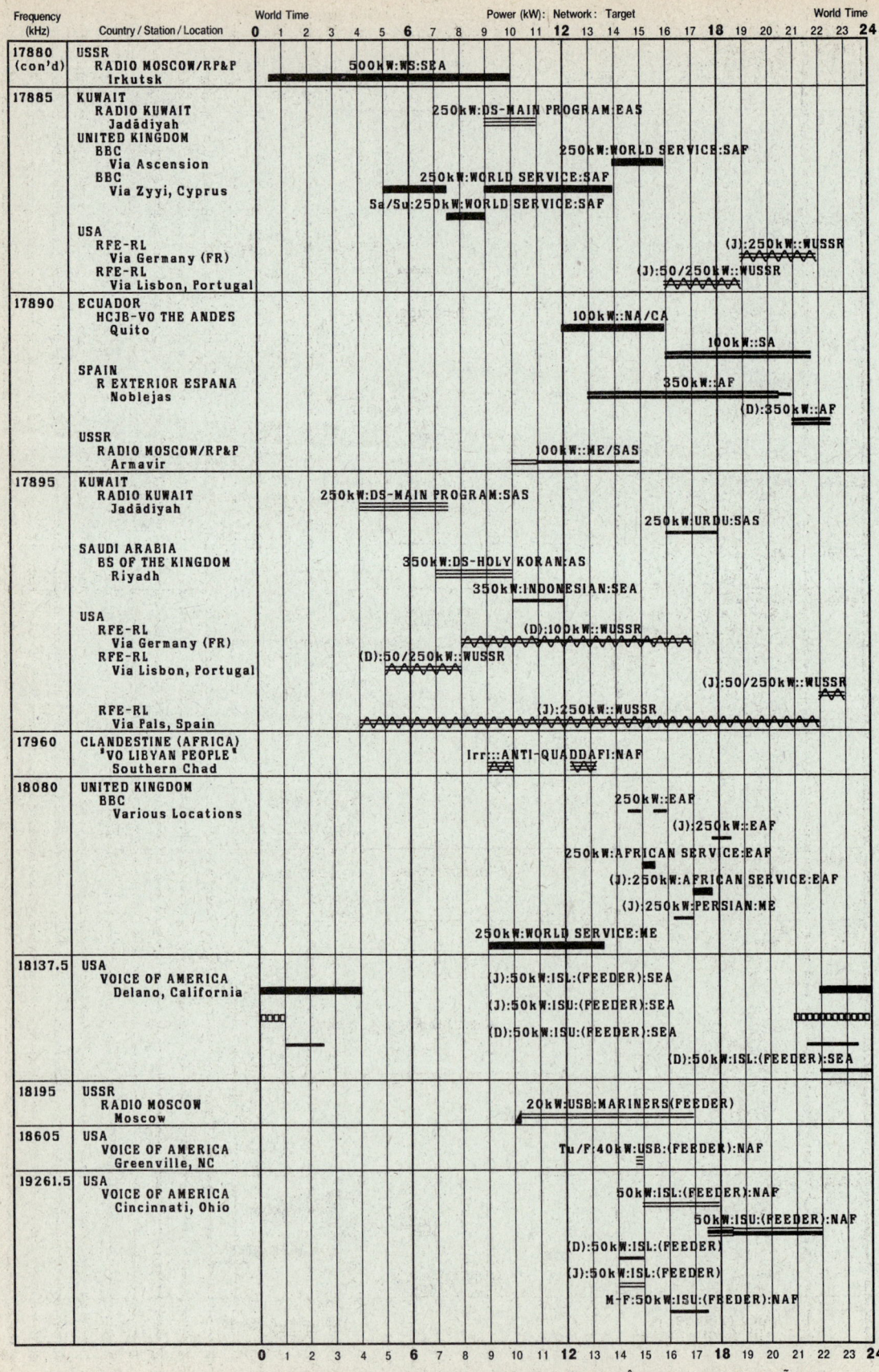

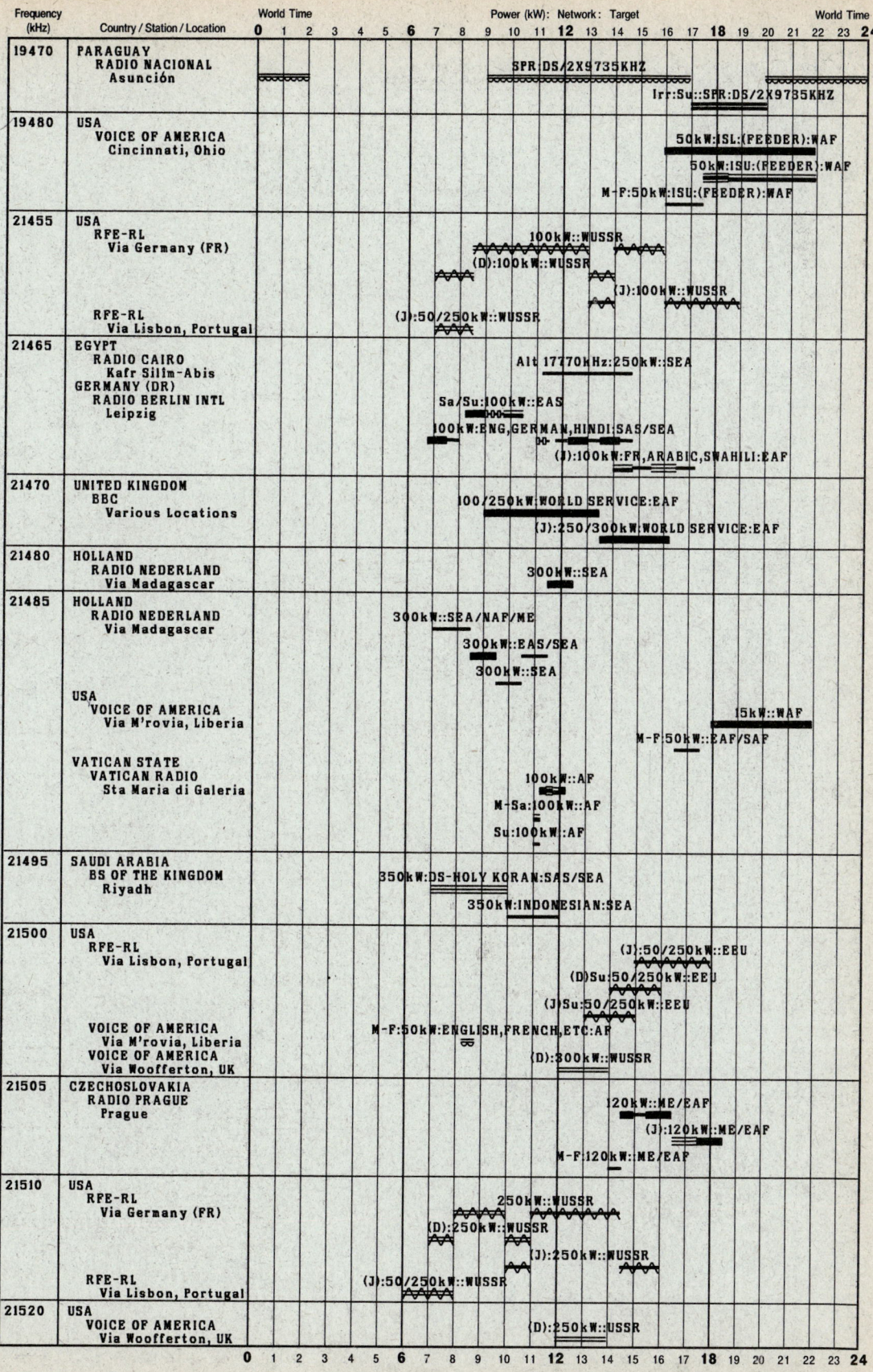

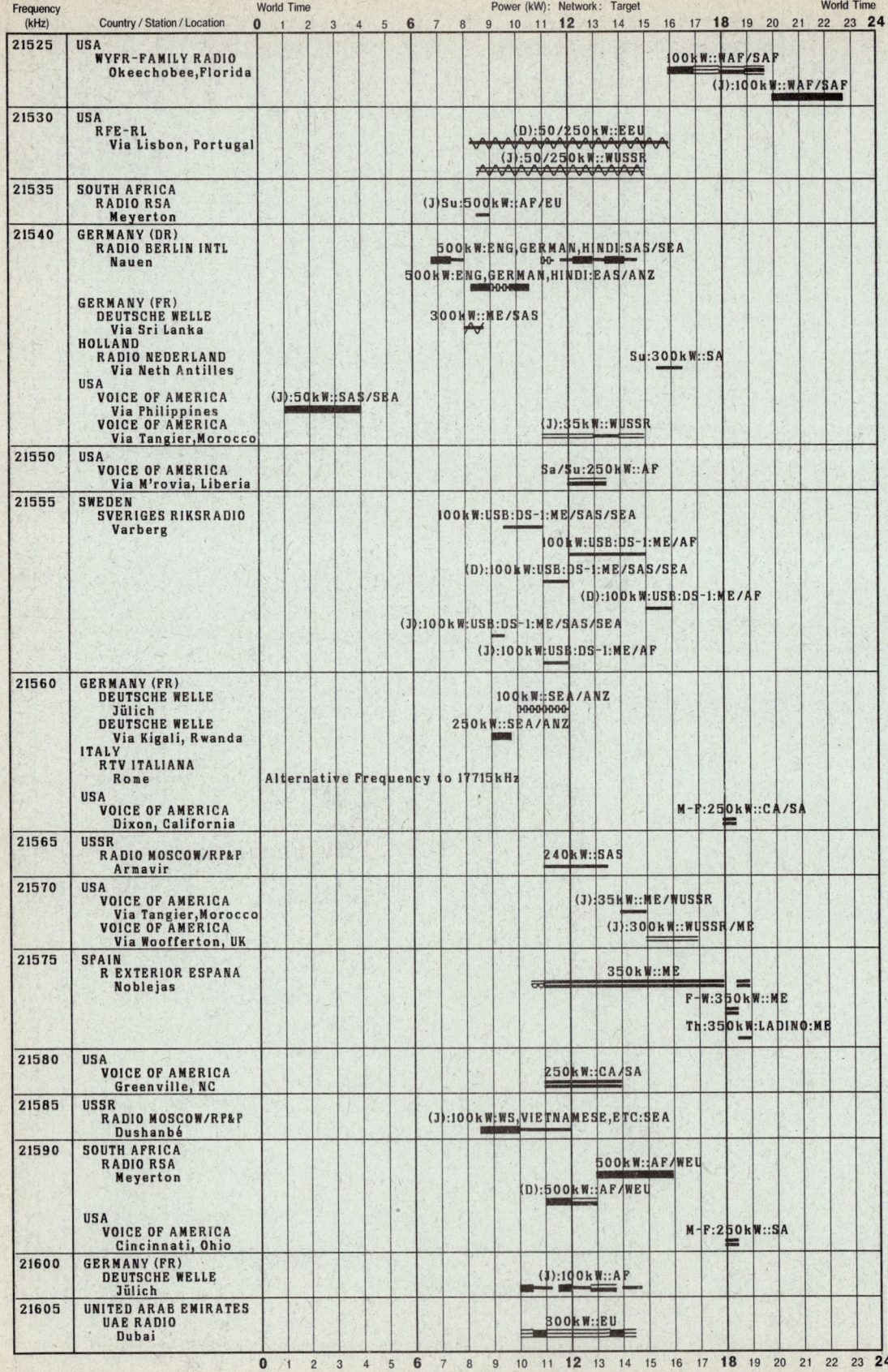

LEXICONS AND GUIDES
- English
- French
- Spanish
- German

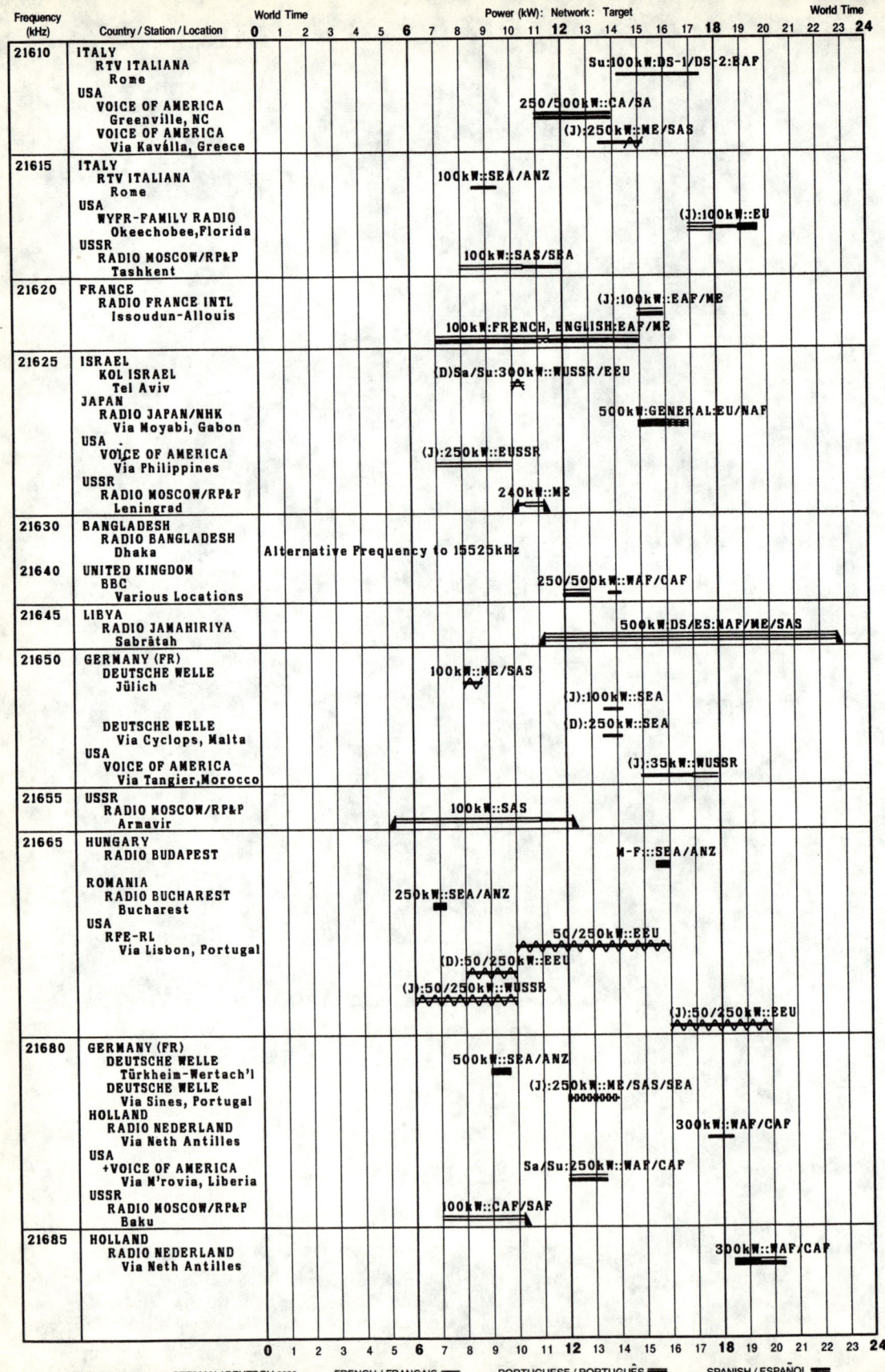

Lexicon of Terms

Alt. Alternative Frequency: Frequency that may be irregularly or unexpectedly brought into use in lieu of the regularly scheduled frequency.

AV. A Voz.

BC. Broadcasting, Broadcasting Company, Broadcasting Corporation.

BS. Broadcasting Service, Broadcasting Station.

Cd. Ciudad.

Channel. The norm for channel spacing within the shortwave broadcasting spectrum is 5.0 kHz. A small proportion of stations does not operate within this norm, so Radio Database International provides frequency resolution to better than one kHz to aid in station location and identification.

Cl. Club, Clube.

Country. Country designations of officially recognized nations are those of the International Telecommunication Union. Unofficial broadcasters and geographical entities are included for reference purposes only and do not imply endorsement or recognition by IBS or the ITU.

Cu, Cult. Cultura, Cultural, Culture.

(D). Winter schedule only.

DS. Domestic Service.

Earliest Heard, Latest Heard. If the Radio Database International monitoring team cannot establish the definite sign-on (or sign-off) of a station, the earliest (or latest) time the station could be traced is entered, instead.

Ed, Educ. Educational, Educação, Educadora.

Em. Emissora, Emisora, Emissor, Emetteur.

EP. Emissor Provincial.

ER. Emissor Regional.

ES. External Service.

F. Friday.

Feeder, shortwave. A Fixed service (point-to-point) transmission, usually in single or independent sideband mode, beamed to a relay site. Although not intended for reception by the public, shortwave feeder transmissions, which can be received on better communications receivers, sometimes provide the best or only means for reception of a particular broadcaster at a given time. For this reason, selected shortwave feeder data is included in Radio Database International, even though any given actual useage may be irregular. Transmission modes for shortwave feeders are:
  LSB—Lower Sideband (single-sideband signal)
  ISL—Lower Sideband (of independent-sideband signal)
  USB—Upper Sideband (single-sideband signal)
  ISU—Upper Sideband (of independent sideband signal)
  RAM—Double Sideband, Reduced Carrier signal

FR Federal Republic.

Frequency. The measurement most widely accepted to indicate the location of a station within the radio spectrum. Frequency is most often expressed in kilohertz (kHz) or Megahertz (MHz), although wavelength is still used in Eastern Europe and certain other parts of the world. The frequency of a station may be determined from the wavelength via the following formula: Frequency (kHz) = 299,792/Wavelength (Meters). Also, cf. "Wavelength" and "Channel".

Irr. Irregular operation or hours of operation; i.e., schedule tends to be unpredictable.

Is. Island(s).

(J). Summer schedule only.

Jamming. Deliberate interference to a transmission with the intent of making reception impossible. The majority of jamming currently emanates from the Soviet Union and its Eastern European allies and is directed against broadcasts in Armenian, Azerbaijain, Belorussian, Bulgarian, Czech, Estonian, Georgian, Hebrew, Latvian, Lithuanian, Polish, Russian, Slovak, Tatar-Bashkir, Turkestani, Ukrainian, and Yiddish. Soviet/Eastern European jamming consists largely of a sound not unlike that of a roomful of truck or lorry engines roaring away. Coded messages appear now and then amidst the roaring, and in some instances distorted programming from the Soviet "Mayak" (DS-2) home service is mixed in, as well. One means to cope with jamming interference is to determine which bands are most likely to provide good reception from the part of the world you wish to hear, while at the same time are not likely to provide good reception from the Soviet Union and Eastern Europe. Directional antennas also can improve the signal-to-jamming ratio under certain circumstances.

kHz. Kilohertz (cf. Frequency).

kW. Kilowatts(s) (cf. Power).

Loc. Local.

Location. The physical location of the transmitter/antenna complex. This location may not correspond to the location of the studios in which the broadcasts are prepared. Data on location names are derived, in most instances, from the Map Section of the Comprehensive Edition of *The Times Atlas of the World*. As transmitter complexes of major international broadcasters are sometimes located in sparsely populated areas, the exact transmitter site may be up to several kilometers from the location indicated.

LV. La Voix, La Voz.

M. Monday.

MHz. Megahertz (cf. Frequency).

Multilingual. A combination of any "graphics language" (Arabic, Chinese, English, French German, Japanese, Portuguese, Russian, Spanish) with another language. Thus, a

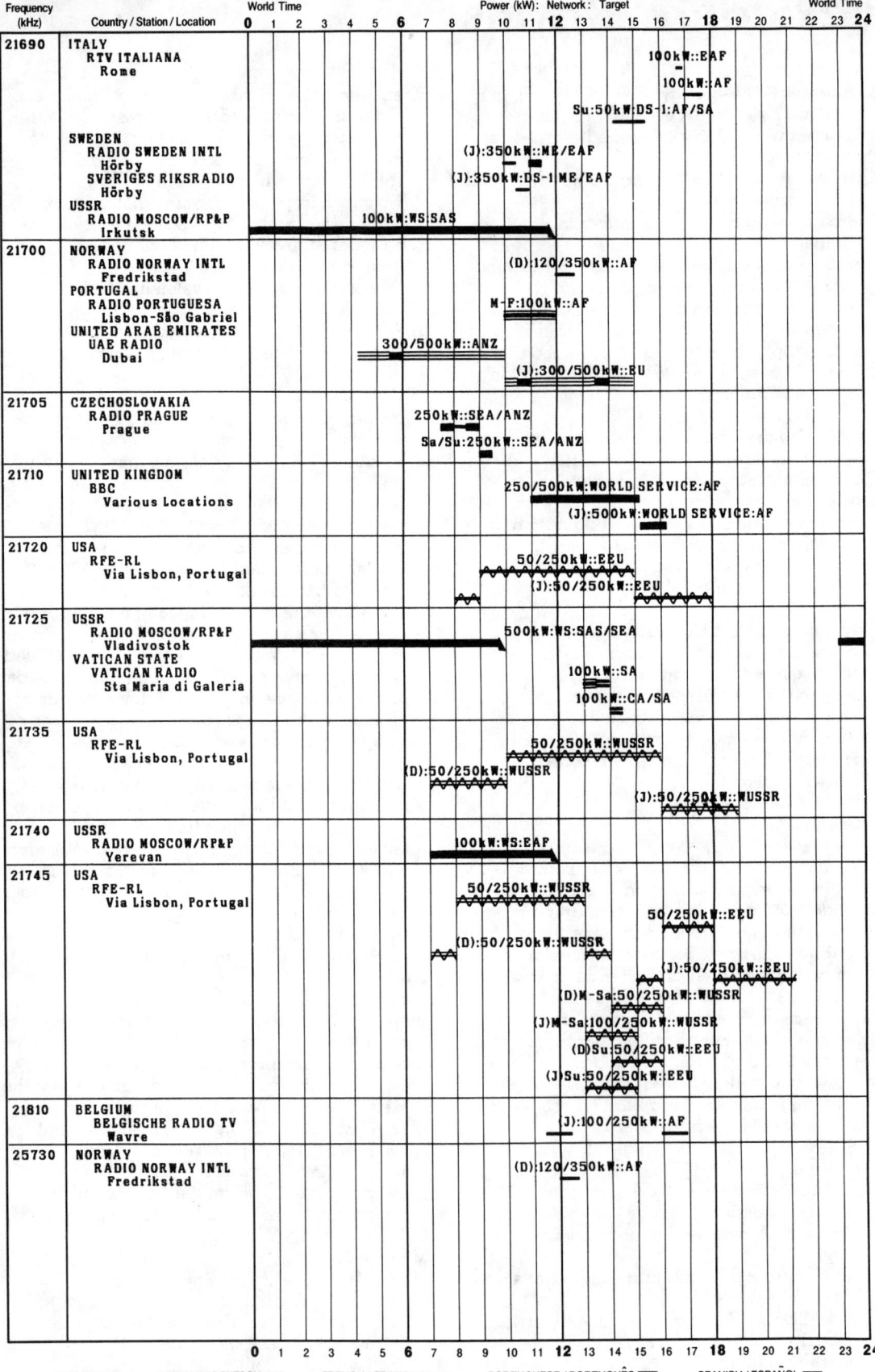

Lexico de Terminos

Alt. Frecuencia alternativa; una frecuencia cuyo uso sea imprevisto o irregular y en lugar de la frecuencia asignada.
AV. A Voz.
BC. Broadcasting, Broadcasting Company, Broadcasting Corporation.
Cd. Ciudad.
Channel. Canal. La norma para la separación de canales dentro de la gama de radiodifusión de onda corta es de 5.0 kHz. Un pequeño porcentaje de estaciones no opera dentro de esta norma, así que el Radio Database International proporciona resolución de frecuencia de menos de un kilohertz, para facilitar la ubicación e identifación de las estaciones.
Cl. Club, Clube.
Country. País. Las desgnaciones, en el caso de las naciones oficialmente reconocidas, son las de la Unión Internacional de Telecomunicaciones. Emisoras y entidades geográficas no oficiales se incluyen para los fines de referencia y no presupone ningún apoyo o reconocimiento por parte de la IBS o UIT.
Cu, Cult. Cultura, Cultural, Culture.
(D). Indica que los detalles refieren solamente al período "D" que corresponde al invierno en el hemisferio del norte y al verano en los paises del sur.
DS. Domestic Service (Servicio Doméstico).
Earliest heard, Latest heard. Si el equipo de monitores del RDI no puede establecer definitivamente el comienzo (o fin) de una transmisión, se entra en cambio la hora más temprana (o tarde) en que pudo rastrearse la estación.

Ed, Educ. Educação, Educadora.
Em. Emissora, Emisora, Emissor, Emetteur.
EP. Emissor Provincial.
ER. Emissor Regional.
ES. External Service (Servicio al Exterior).
F. Friday (Viernes).
Feeder, shortwave. Una transmisión de servicio fijo (punto a punto) usualmente en modo de banda lateral única o independiente, dirigida a la estación de redifusión desde el país de origen de la transmisión. Aunque no destinadas para la recepcion general, las transmisiones de eslabón de onda corta, que pueden ser recibidas por los mejores receptores de comunicaciones, a veces proporcionan la mejor o única oportunidad para la recepción de una emisora determinada a una hora dada. Por esta razón se incluyen en el Radio Database International datos sobre este tipo de transmisión, aunque el uso actual no sea regular. Los modos de transmisión para los eslabones de onda corta son:
   LSB—Banda lateral inferior (señal banda lateral única)
   ISL—Banda lateral inferior (de una señal banda lateral independiente)
   USB—Banda lateral superior (señal banda lateral única)
   ISU—Banda lateral superior (de una señal banda lateral independiente)
   RAM—Banda lateral doble, con la portadora reducida.
FR. Federal Republic (República Federal)
Frequency. Frecuencia. La medida más ampliamente aceptada para indicar la ubi-

COMO UTILIZAR EL WORLDSCAN

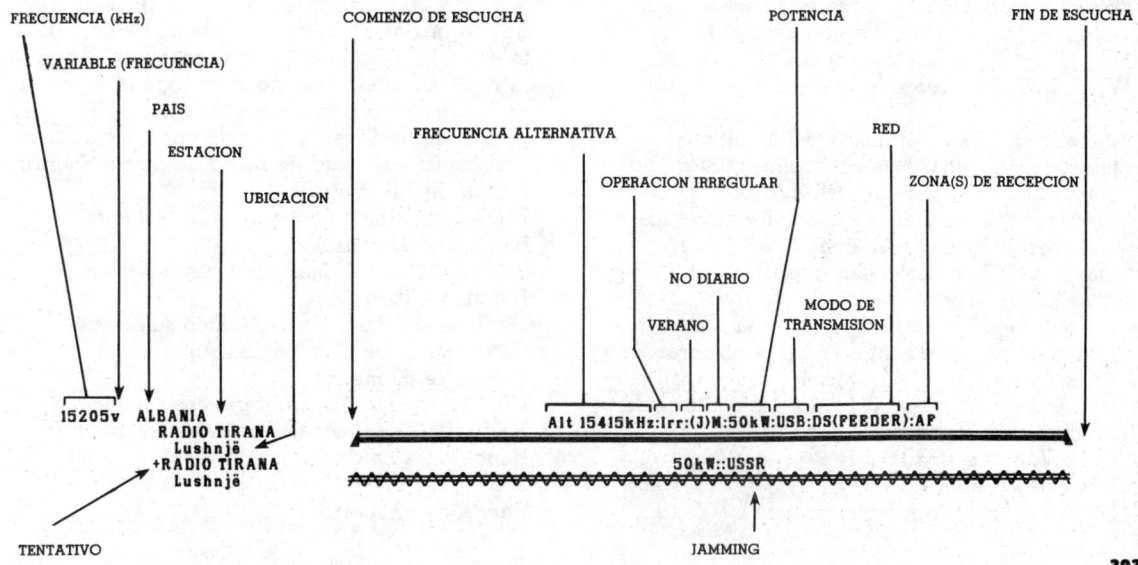

cación de una estación dentro de la gama radial. La frecuencia se expresa más a menudo en kilohertz (kHz) o Megahertz (MHz), aunque la longitud de onda se utiliza todavía en Europa oriental y algunas otras partes del mundo. La frecuencia de una estación puede determinarse de la siguiente fórmula:

Frecuencia (kHz) = 299,792 ÷ Longitud de onda (metros).

Vease también 'Wavelength' y 'Channel'.
Irr. Operación irregular; horas de operación irregulares.
Is. Isla(s).
(J). Indica que los detalles refieren solamente al período "J" que corresponde al verano en el hemisferio del norte y al invierno en los paises del sur.
Jamming. Interferencia intencional a una transmisión con el propósito de hacer imposible la recepción de la misma. La mayoría del jamming proviene de la Unión Soviética y sus aliados de Europa oriental, y es dirigida contra emisiones en armenio, azerbaijano, beloruso, búlgaro, checo, estonio, georgiano, hebreo, letón, lituano, polaco, ruso, slovaco, tatar-bashkir, turquestano, ucraniano e yiddish. El jamming proveniente de la URSS y Europa oriental consiste mayormente de un ruido no muy diferente de lo de un depósito lleno de camiones los cuales tienen sus motores en plena marcha. De vez en cuando se agrega a este ruido algunos mensajes en clave o programación distorsionada del servicio doméstico soviético, Mayak (DS-2). Una forma de contender con la interferencia jamming es determinar cuales de las bandas podrían ser las más apropiadas para proporcionar buena recepción de la parte del mundo que se quiere escuchar, mientras que al mismo tiempo no darían buena recepción de la Unión Soviética y Europa oriental. Antenas direccionales también pueden mejorar la razón de señal a jamming, bajo ciertas circunstancias.
kHz. Kilohertz (vease 'Frequency').
kW. Kilovatio(s) (vease 'Power').
Loc. Local.
Location. La ubicación física del complejo transmisor/antena. Esta ubicación pudiera no corresponder a la de los estudios donde se preparan los programas. Los datos sobre los nombres geográficos se completan en su mayoría de la Edición Comprensiva del *Times Atlas of the World*. Como sucede a veces que los complejos de transmisores de las emisoras internacionales más grandes se encuentran en áreas de escasa población, el sitio exacto del transmisor podría estar a varios kilometros de la localidad indicada.
LV. La Voix, La Voz.
M. Monday (Lunes).
MHz. Megahertz (vease 'Frequency').

Multilingual. Una combinación de cualquier idioma representado por símbolo gráfico (alemán, árabe, chino, español, francés, inglés, japonés, portugués, o ruso) con otro idioma. Así una transmisión en español y quechua sería designada como 'Multilingual', mientras que una emisión en quechua y aymara sería designada como 'Other' (= Otro).
N. New, Nueva, Nuevo, Nouvelle, Novo, Nova, Nacional, National, Nationale.
Nac. Nacional.
Nat. National, Nationale.
Network. La red o el servicio de una emisión. Transmisiones de servicios domésticos son indicadas por "DS".
PBS. People's Broadcasting Station (Estación de Radiodifusión del Pueblo).
Power. La potencia de salida de RF (radiofrecuencia), eso es, antes de la amplificación, de onda corta es 0.04–600 kW.
PR. People's Republic (República Popular)
PS. Provincial Station (Estación Provincial), Pangsong.
Pto. Puerto, Pôrto.
R. Radio, Radiodiffusion, Radiodifusora, Radiodifusión, Radiodifusão, Radioemisora, Radiofonikos, Radiostantsiya, Radyo, Radyosu, etc.
Reg. Regional.
Relay. Instalaciones transmisoras extraterritoriales que retransmiten programas recibidos por satélite, eslabón de onda corta, o en forma de grabación.
Rep. Republic, República, République.
RT, RTV. Radiodiffusion Télévision, Radio Televisión, etc.
RN. Vease R y N.
S. San, Santa, Santo, São, Saint.
Sa. Saturday (Sábado).
Shortwave Spectrum. La gama de onda corta es esa porción de la gama radial que se encuentra entre 3 y 30 MHz (3,000–30,000 kHz); sin embargo, de ordinario se ubica la gama de onda corta entre 2 y 30 MHz aproximadamente. Aunque la radiodifusión tiene lugar en toda la gama de onda corta, los parametros nominales de la UIT para la radiodifusión de onda corta son (mundialmente, salvo ser indicado de otra manera):
Banda de 120 metros:
2300–2498 kHz (Zonas tropicales solamente)
Banda de 90 metros:
3200–3400 kHz (Zonas tropicales solamente)
Banda de 75 metros:
3900–3950 kHz (Asia & Pacífico solamente)
3950–4000 kHz (excepto las Américas)
Banda de 60 metros:
4750–4995 kHz (Zonas tropicales solamente)
5005–5060 kHz (Zonas tropicales solamente)
Banda de 49 metros:
5950–6200 kHz
Banda de 41 metros:

7100–7300 kHz (excepto las Américas)
Banda de 31 metros:
9500–9900 kHz
Banda de 25 metros:
11650–12050 kHz
Banda de 21 metros:
13600–13800 kHz
Banda de 19 metros:
15100–15600 kHz
Banda de 16 metros:
17550–17900 kHz
Banda de 13 metros:
21450–21850 kHz
Banda de 11 metros:
25670–26100 kHz
St. Saint.
Sta. Santa.
Sto. Santo.
Su. Sunday (Domingo).
Target. Las entidades geográficas internacionales a las cuales se dirige una transmisión, con tal que está destinada fuera del país de origen. Las zonas de recepción son las siguientes:
    AF—África
    AM—Américas
    ANZ—Australia & Nueva Zelandia
    AS—Asia
    ATL—Atlántico
    CA—América Central, Caribe & México
    CAF—África Central
    EAF—África Oriental
    EAS—Asia Oriental
    EEU—Europa Oriental
    ENA—América del Norte (Este)
    EU—Europa
    EUSSR—URSS Oriental
    ME—Oriente Medio
    NA—América del Norte
    NAF—África del Norte
    PAC—Pacífico
    SA—América del Sur
    SAF—África Meridional
    SAS—Sud Asia
    SEA—Asia (Sudeste)
    USSR—URSS
    WAF—África Occidental
    WEU—Europa Occidental
    WNA—América del Norte (Oeste)
    WUSSR—URSS Occidental
Th. Thursday (Jueves).
Tu. Tuesday (Martes).
UTC. Hora Universal Coordinada, también conocida como Hora Meridiano de Greenwich (GMT) o Zulu (Z): la referencia internacional de 24 horas para la radiodifusión, basada en el marco de tiempo del Observatorio de Greenwich en el Reino Unido. La diferencia entre UTC y la hora local puede determinarse escuchando los avisos rutinarios de la hora UTC transmitidos por emisoras internacionales.
v. Frecuencia variable, eso es, una que es inestable.
Vo. Voice of (Voz de).
W. Wednesday (Miércoles).
Wavelength. Longitud de onda. Una alternativa antigua a 'Frecuencia' (vease) para precisar la ubicación de la estación dentro de la gama radial. La longitud de onda de una estación puede determinarse atraves de la siguiente fórmula: Longitud de onda (metros) = 299,792 × Frecuencia (kHz). Esta fórmula puede utilizarse también para precisar la banda (en metros) de una estación determinada (vease 'Shortwave Spectrum).
World Time. Vease 'UTC'.
WS. World Service (Servicio Mundial).
(+) Tent. Tentativo (-a); eso es, la exactitud de algunos detalles (p.ej. el nombre exacto de la estación) está incierta.

Glossar

Channel; Kanal. Die Norm für den Abstand verschiedener Sendungen im Kurzwellenspektrum ist 5.0 kHz. Da aber ein kleiner Teil der Stationen nicht gemäss dieser Norm funktioniert, bietet *Passport* Frequenzgenauigkeit von besser als einem kHz für die Lage und Identifizierung von Stationen.

Country; Land. Die Namen für die verschiedenen Länder sind diejenigen der ITU. Inoffizielle Sender und geographische Einheiten werden nur zur Bequemlichkeit erwähnt und bedeuten nicht formale Anerkennung oder Endorsement von IBS oder ITU.

Earliest Heard, Latest Heard; Hineinschwinden, Herausschwinden. Falls die Moniteure der Databasis nicht den Anfang oder das Ende einer Sendung genau bestimmen können, dann wird die früheste oder späteste Zeit angegeben, wann die Station gehört werden kann.

Feeder, shortwave; Kurzwellenspeisesender oder Versorger. Eine Punkt-zu-Punkt bestimmte Sendung zu einer Relaisstation, gewöhnlich durch einzelnes oder unabhängiges Seitenband. Manchmal kommt es vor, dass gewisse Rundfunkprogramme am besten oder nur durch "feeder" gehört werden können, obgleich sie eigentlich nicht für die Öffentlichkeit beabsicht sind und auch bessere Empfänger benötigen. Aus diesem Grunde enthält Radio Database International Information über Kurzwellen "feeder", obwohl es möglich ist, dass tatsächlicher Gebrauch nicht immer regelmässig ist.

Sendearten für Kurzwellen "feeder" sind wie folgt:
- LSB—Lower Sideband (single-sideband signal); Unteres Seitenband
- USB—Upper Sideband (single-sideband signal); Oberes Seitenband
- ISL—Lower Sideband (of independent-sideband signal); Unteres Seitenband (von einem unabhängigen Seitenband Signal)
- ISU—Upper Sideband (of independent sideband signal); Oberes Seitenband (von einem unabhängigen Seitenband, Signal)
- RAM—Double Sideband, reduced carrier signal; Doppeltes Seitenband, reduziertes Träger Signal

Anmerkung: Mittelwellen Sendungen, die hauptsächlich dem Inlandsdienst dienen, bestehen aus einem Träger (carrier) und zwei Seitenbändern, die beide dieselbe Information als Spiegelbilder enthalten. Diese Sendeart wird AM genannt und wird auch auf Kurzwelle gebraucht. Indem ein Rundfunksender nur ein Seitenband oder sogar auch keinen Träger gebraucht, spart er Energie, oder falls er doch noch mehr kW gebraucht, dann reicht die Sendung weiter stärker und klarer.

Frequency; Frequenz. Das weitverbreitetste Mass, um die Lage einer Sendung innerhalb des Radiospektrums anzugeben. In Osteuropa und einigen anderen Teilen der Welt wird jedoch noch Wellenlänge gebraucht. Frequenz wird hauptsächlich in Kilohertz (kHz) oder Megahertz (MHz) ausgedrückt. Die Frequenz einer Sendung kann durch die folgende Formel von der Wellenlänge errechnet werden: Frequenz (kHz) = 299,792/Wellenlänge (Meter). Vgl. "Wavelength" (Wellenlänge) und "Channel" (Kanal).

Jamming; Mischmaschen. Die Störung einer Sendung mit der Absicht ihren Empfang unmöglich zu machen. In der Hauptsache kommt "Jamming" heutzutage von der Sowjetunion und ihren osteuropäischen Verbündeten. Es ist gegen Sendungen in den folgenden Sprachen dirigiert: Armenisch, Aserbeidschanisch, Belorussisch, Bulgarisch, Tschechisch, Estonisch, Georgisch, Hebräisch, Lettisch, Litauisch, Polnisch, Russisch, Slowakisch, Tatarisch-Baschkirisch, Turkestanisch, Ukrainisch und Jiddisch. Sowjetischosteuropäisches Mischmaschen bestetht hauptsächlich aus einem Geräusch wie von vielen, lauten Maschinen. Manchmal kann man Identifikationen in dem Morsealphabet hören oder seltener auch verzerrte Programme des Sowjetischen "Mayak" (DS-2) Inlandsdienstes. Ein Mittel zur Minimierung oder zum Abschwächen des Mischmaschens besteht in der Benutzung des Kurzwelleverbreitung. Man tut dies, indem man herausfindet, welche Bände wahrscheinlich guten Empfang für den Teil der Welt geben, den man hören will; aber zur selben Zeit wahrscheinlich nicht guten Empfang von der Sowjetunion oder Osteuropa. Durch Antennen, die auf verschiedene Richtungen eingestellt werden können, kann man zuweilen auch die Beziehung zwischen Signal und "Jamming" unter gewissen Umständen verbessern.

Location; Standort. Der Platz, wo sich die Sendeanlagen, einschliesslich des Senders und der Antennen, befinden. Dieser mag nicht derselbe sein wie derjenige der Studios, wo die Rundfunkprogramme hergestellt werden. Data für die Namen von Standorten wurden in der Hauptsache mit Hilfe der Karten in *The Times Atlas of the World,* Comprehensive Edition, vorbereitet.

Network: Rundfunknetz. Der Dienst oder Service einer Rundfunkorganisation. Ausser solchen offensichtlichen Anwendungen wie 'Weltdienst' kann dieser Ausdruck sich auf Sprachen beziehen, wenn dies nicht durch die graphischen Symbole für Arabisch, Chinesisch, Englisch, Französisch, Deutsch, Japanisch, Portugiesisch, Russisch, und Spanisch angegeben ist. Inlandsdienste sind durch 'DS' (Domestic Service) bezeichnet.

Power; Kraft. Senderleistung in Kilowatt (kW) vor der Amplifikation durch die Antenne. Zur Zeit gebrauchen Kurzwellensender von 0,04 bis 600 kW, aber manchmal werden zwei Sender mit nur kleinen Unterschieden in der Richtung gebraucht.

Relay; Relais. Ausserterritoriale Sendestation, gewöhnlich durch Tonbänder, Satelliten oder Kurzwellendienst versorgt (vgl. 'Feeder').

Shortwave Spectrum; Kurzwellenspektrum. Gemäss der wissenschaftlichen Definition liegt das Kurzwellen zwischen 3 und 30 MHz oder 3,000 und 30,000 kHz. Gewöhnlicher Gebrauch diktiert jedoch, dass es ungefähr zwischen 2 und 30 MHz liegt. In der Praxis kann man Rundfunksendungen fast überall im Kurzwellenspektrum finden. Aber gemäss vor kurzem angenommenen Prinzipien der ITU sind, mit gewissen Begrenzungen, die in den Klammern angegeben sind, die folgenden die nominalen Parameter für Kurzwellen Rundfunkbände.

120 Meterband:
 2300–2498 kHz (nur tropische Zone)
90 Meterband:
 3200–3400 kHz (nur tropische Zone)
75 Meterband:
 3900–3950 kHz (nur Asien und Pazifik)
 3950–4000 kHz (mit Ausnahme Amerikas)
60 Meterband:
 4750–4995 kHz (nur tropische Zone)
 5000–5060 kHz (nur tropische Zone)
49 Meterband:
 5950–6200 kHz
41 Meterband:
 7100–7300 kHz (mit Ausnahme Amerikas)
31 Meterband:
 9500–9900 kHz
25 Meterband:
 11650–12050 kHz
21 Meterband:
 13600–13800 kHz
19 Meterband:
 15100–15600 kHz
16 Meterband:
 17550–17900 kHz
13 Meterband:
 21450–21850 kHz
11 Meterband:
 25670–26100 kHz

UTC. Coordinated Universal Time; Koordinierte Universalzeit; auch Greenwich Mean Time (GMT) oder Zulu (Z) genannt; die 24-stündige internationale Rundfunkreferenz, die sich auf das Observatorium von Greenwich in Grossbritannien basiert. Man kann den Unterschied zw. UTC und örtlicher Zeit bestimmen, indem man internationalen Rundfunksendungen zuhört. Einige Kurzwellenempfänger und Digitaluhren können auf das vierundzwanzigstündige Format gesetzt werden, sodass man nicht umrechnen braucht.

Wavelength; Wellenlänge. Ein veraltender Ausdruck anstatt Frequenz (vgl. 'Frequency'), um die Lage einer Sendung im Radiospektrum anzugeben. Die Wellenlänge einer Sendung kann durch die folgende Formel errechnet werden: Wellenlänge (Meter) = 299,792/Frequenz (kHz). Diese Formel kann auch gebraucht werden, um das Meterband einer Sendung zu bestimmen (vgl. 'Shortwave Spectrum').

World Time. Siehe UTC.

Abkürzungen

Alt. Alternate; abwechselnd, alternativ.
AV. A Voz.
BC. Broadcasting; Rundfunksendung-sender, oder-gesellschaft.
BCS, BS. Broadcasting Service; Rundfunkdienst
Cd. Ciudad; City; Stadt.
Cl. Club; Clube; Klub.
Cu. Cult-Culture; Cultura; Kultur.
(D). Winter.
DS. Domestic Service, Inlandsdienst
Ed. Educação, Educadora.
Em. Emissora, Emisora; Emetteur Strahlung.
ES. External Service, Auslandsdienst.
FR. Federal Republic; Bundesrepublik.
F. Friday; Freitag.
Hz. hertz—A unit of frequency equal to one cycle per second; Eine Frequenzeinhiet, die eine Schwingung pro Sekunde bezeichnet.
Irr. Irregular operation or hours of operation; unregelmässige Sendung.
Is. Island(s); Insel(n).
(J). Sommer.
kHz. Kilohertz, 1,000 Hertz (cf. Frequency); vgl. Frequenz.
N. New; Nueva; Nuevo; Nouvelle; Novo, Nova, Nacional, Nationale; Neu.
Nac. Nacional; National.
Nat. National, Nationale.
PBS. People's Broadcasting Station Volksrundfunkstation.
PR. People's Republic; Volksrepublik.
Pto. Puerto; Pôrto; Hafen.
R. Radio, Radiodiffusion, Radiodifusora, Radiodifusíon, Radiodifusão, Radioemisora, Radiofonikos, Radiostantsiya, Radyo, Radyosu, etc.

Rep. Republic; République; Republik.
RT, RTV. Radiodiffusion Télévision, Radio Television, etc.
S. San, Santa, Santo, São, Saint; in Heiligennamen verschiedener Sprachen und auf solche zurückgehenden Ortsnamen.
Sa. Saturday; Sonnabend, Samstag.
St. Saint; vgl. S.
Sta. Santa, vgl. S.
Sto. Santo, vgl. S.
Su. Sunday; Sonntag
Target. Zielgebiet, spez. die internationalen geographischen Gebiete für die eine Sendung beabsichtigt ist, falls sie ausserhalb des Landes gehört werden soll:
    AF — Africa; Afrika
    AM — Americas; die amerikanischen Kontinente
    ANZ — Australia & New Zealand; Australien u. Neuseeland
    AS — Asia; Asien
    ATL — Atlantic; Atlantik
    CA — Central America, Carribean & Mexico; Mittelamerika, Karibik u. Mexiko
    CAF — Central Africa: Zentralafrika
    EAF — Eastern Africa; Ostafrika
    EAS — Eastern Asia; Ostasien
    EEU — Eastern Europe; Osteuropa
    ENA — Eastern North America; Ostnordamerika
    EU — Europe; Europa
    EUSSR — Eastern USSR; UdSSR (Ost)
    ME — Middle East; Nahost
    NA — North America; Nordamerika
    NAF — Northern Africa; Nordafrika
    PAC — Pacific; Pazifik
    SA — South America; Südamerika
    SAF — Southern Africa; Südliches Afrika
    SAS — Southern Asia; Südasien
    SEA — South East Asia; Südostasien
    USSR — USSR; UdSSR
    WAF — Western Africa; Westafrika
    WEU — Western Europe; Westeuropa
    WNA — Western North America; Westnordamerika
    WUSSR — Western USSR; UdSSR (West)
Th. Thursday; Donnerstag.
Tu. Tuesday; Dienstag.
v. Variable frequency; variierende Frequenz, d.h. eine, die nicht stabil ist oder sich verändert.
Vo. Voice of... Die Stimme von...
WS. World Service; Weltdienst.

WORLDSCAN: GEBRAUCHSANWEISUNG

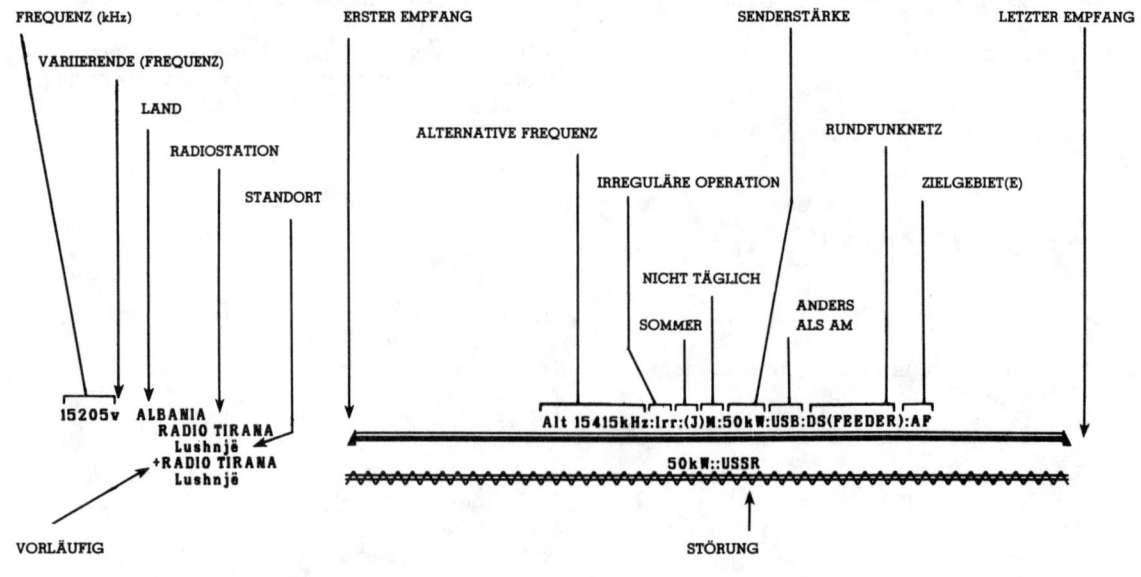

Directory of Advertisers

Advanced Electronic Applications, 21
Alpha Delta Communications, 30, 112
American Radio Relay League, 32, 126
Antenna Supermarket, 108
Atlantic Ham Radio, 42
Barry Electronics Corp., 58
Blaze International Inc., 53
Champs-Elysees, Inc., 40
Com-Rad Industries, 60
Com-West Radio Systems, Ltd., 62
Dressler Active Antennas, 56
Electronic Equipment Bank, 10, 12, 16, 24, 25, 114
Electronics Center, 66
Glenwood Trading Company, 48
Grove Enterprises, Inc., 50
Grundig AG, 2, 3
Hobbytronique, 34
ICOM America, Inc., 7
Imprimé: World Radio Marketplace, 26, 27
Interbooks, 86
Japan Radio Company, Ltd., 65
Kenwood USA Corporation, 400, Inside Back Cover
Klingenfuss Publications, 33
MFJ Enterprises, Ltd., 38, 39
Microlog Corporation, 48
Phase Track Ltd., 43
Philips Consumer Electronics, Inside Front Cover
Radio Canada International, 34
Radio Database International, 94
Radio Infoservices, 118
Radio West, 128
Radio WRNO Worldwide, 116
Radio WWCR, 52
RDI White Papers, 110
Sangean America, Inc., 31
Schau ins Land, 120
Sherwood Engineering Inc., 34
Solar Light Co., 64
Sony Corporation of America, 35
Spectronics, Inc., 14, 15
Stoner Communications, 8
Universal Shortwave Radio, 28, 44, 66
World Press Review, 46
Yaesu USA, 61

Advertising representative: Mary Kroszner, IBS, Ltd., Box 300, Penn's Park, PA 18943 USA. Telephone 215-794-8252

KENWOOD
...pacesetter in Amateur Radio

Scan the World

R-2000
All-mode receiver

Superior engineering, quality, and performance describe Kenwood's multi-mode communications receiver.

The R-2000 receiver has the most often-needed features for the serious or casual shortwave broadcast listener. Listen in on overseas news, music, and commentary. "Listen up" on the VHF public service and Amateur radio frequencies, as well as aircraft and business band communications with the R-2000 and VC-10 option. The R-2000 has a muting circuit so you can monitor your Amateur radio station's signal quality.

- Covers 150 kHz – 30 MHz in 30 bands.
- All mode: USB, LSB, CW, AM, FM.
- Digital VFO's. 50-Hz, 500-Hz or 5-kHz steps. F. LOCK switch.
- Ten memories store frequency, band, and mode data. Each memory may be tuned as a VFO.
- Lithium batt. memory back-up.
- Memory scan.
- Programmable band scan.
- Fluorescent tube digital display of frequency (100 Hz resolution) or time.
- Dual 24-hour quartz clocks, with timer.
- Three built-in IF filters with NARROW/WIDE selector switch. (CW filter optional.)
- Squelch circuit, all mode, built-in.
- Noise blanker built-in.
- Large front mounted speaker.
- RF step attenuator. (0-10-20-30 dB.)
- AGC switch. (Slow-Fast.)
- "S" meter, with SINPO scale.
- High and low impedance antenna terminals.
- 100/120/220/240 VAC operation.
- RECORD output jack.
- Timer REMOTE output (not for AC power).
- Muting terminals.

Optional accessories:

- VC-10 VHF converter for R-2000 covers 118-174 MHz
- YG-455C 500 Hz CW filter for R-2000
- HS-4 Headphones
- HS-5 Deluxe headphones
- HS-6 Lightweight headphones
- HS-7 Micro headphones
- DCK-1 DC cable kit for 13.8 VDC operation

Additional information on Kenwood all-band receivers is available from authorized dealers.

KENWOOD
KENWOOD U.S.A. CORPORATION
2201 E. Dominguez St., Long Beach, CA 90810
P.O. Box 22745, Long Beach, CA 90801-5745

Service manuals are available for all receivers and most accessories.
Specifications and prices subject to change without notice or obligation.